CHALLENGES TO SCIENCE

Life Science
Physical Science
Earth Science
Chemistry: A Humanistic Approach

CHALLENGES TO SCIENCE

LIFE SCIENCE
SECOND EDITION

WILLIAM L. SMALLWOOD
Biology Teacher
Ketchum–Sun Valley Community School
Sun Valley, Idaho

WEBSTER DIVISION, McGRAW-HILL BOOK COMPANY
New York   St. Louis   San Francisco   Auckland   Bogotá
Düsseldorf   Johannesburg   London   Madrid   Mexico
Montreal   New Delhi   Panama   Paris   São Paulo
Singapore   Sydney   Tokyo   Toronto

PRONUNCIATION GUIDE

Throughout this book, pronunciation of words that may be unfamiliar is given in the text. See line 35 on page 5 for an example. The diacritical markings used and the pronunciation guide below are from *The American Heritage School Dictionary*, copyright © 1972, 1977 by Houghton Mifflin Company. They are reprinted by permission of the publisher.

ă	pat	î	dear, deer, fierce, mere	p	pop	zh	garage, pleasure, vision
ā	aid, fey, pay			r	roar		
â	air, care, wear	j	judge	s	miss, sauce, see	ə	about, silent, pencil, lemon, circus
ä	father	k	cat, kick, pique	sh	dish, ship		
b	bib	l	lid, needle	t	tight		
ch	church	m	am, man, mum	th	path, thin	ər	butter
d	deed	n	no, sudden	th	bathe, this		
ĕ	pet, pleasure	ng	thing	ŭ	cut, rough		
ē	be, bee, easy, leisure	ŏ	horrible, pot	û	circle, firm, heard, term, turn, urge, word		
f	fast, fife, off, phase, rough	ō	go, hoarse, row, toe				
		ô	alter, caught, for, paw	v	cave, valve, vine		
g	gag			w	with		
h	hat	oi	boy, noise, oil	y	yes		
hw	which	ou	cow, out	yōō	abuse, use		
ĭ	pit	ŏŏ	took	z	rose, size, xylophone, zebra		
ī	by, guy, pie	ōō	boot, fruit				

STRESS

Primary stress ′
bi·ol′o·gy
|bī ŏl′ə jē|

Secondary stress ′
bi′o·log′i·cal
|bī′ə lŏj′ĭ kəl|

Library of Congress Cataloging in Publication Data

Smallwood, William L
    Life science.

    (Challenges to science)
    Includes index.
    SUMMARY: A life science textbook emphasizing environmental considerations and ecosystems.
    1. Biology—Juvenile literature. 2. Ecology—Juvenile literature. [1. Biology. 2. Ecology] I. Title.
QH308.7.S64   1978   547   77-10965
ISBN 0-07-058420-6

Copyright © 1978, 1973 by McGraw-Hill, Inc. All Rights Reserved. Printed in the United States of America. No part of this publication may be reproduced, stored in a retrieval system, or transmitted, in any form or by any means, electronic, mechanical, photocopying, recording, or otherwise, without the prior written permission of the publisher.

Acknowledgments

The first edition manuscript for *Challenges to Science: Life Science* was reviewed in its entirety by William Harmer, Director of the Learning Disability Center at the University of Texas, Austin. Selected portions of the second edition manuscript were reviewed by the following:

Charlotte A. Szucs
Litchfield Junior High School
Akron, Ohio

Marshall Tsuruda
John Muir Junior High School
San Jose, California

Gerry L. Woods
Hurst Junior High School
Hurst-Euless-Bedford Independent School District
Bedford, Texas

We are grateful to these educators for their insights and interest.

The reading on pages 271–273 is abridged from *Microbe Hunters*, copyright 1926, 1954 by Paul de Kruif. Reprinted by permission of Harcourt Brace Jovanovich, Inc.

Cover and Text Design: Aspen Hollow Artservice
Front Cover: Mountain lion, photo by Tom McHugh from Photo Researchers
Back Cover: Mountain lion drinking, photo by John S. Crawford from Photo Researchers
Frontispiece: Banding snow geese along the Missouri River, photo by David Hiser from Photo Researchers

Editorial Development: Irwin Siegelman
Editing and Styling: Linda Richmond
Design: James Darby
Production: Ellen Leventhal

Photo Research: Pat Vestal

Preface

Most students study life science concepts in their elementary science courses. Then, in the secondary schools, they often take one course in life science prior to, or in place of, the standard high school biology course. This book is designed to serve the needs of students who are taking this introductory course.

The emphasis in this book is primarily upon those concepts that will have the greatest personal value for the students and the society that they will help create during the years of their adult lives. The laboratory investigations are integrated with the text, and some of them are "you are on your own" investigations designed to give the student a feel for what it is like when a scientist devises his or her own strategy for conducting an investigation. All investigations are designed so that they can be carried out with a minimum of equipment and laboratory facilities.

Countless people have helped in some way in the preparation of this second edition. I am especially grateful to Melissa Sousley Flaherty, who developed many of the laboratory investigations. Chapter 16 has been reviewed by many drug workers and doctors. Of these, Dr. Larry Halpern at the University of Washington School of Medicine was most helpful.

Numerous science teachers and supervisors from schools using the first edition have supplied important feedback. Special thanks are due Charles E. "Kelly" Carpenter, former Science Supervisor of the Spokane, Washington, Public Schools; John Brennan, Supervisor of Science of the Denver, Colorado, Public Schools; and Otho Perkins, Supervisor of Science of the Columbus, Ohio, Public Schools. These persons and their teaching colleagues spent many hours collecting feedback and offering suggestions for this second edition.

Also deserving special thanks are James Schmidt and Richard Del Grosso of the University-Liggett School, Grosse Pointe, Michigan. Their comments on the laboratory investigations were especially helpful, and their game called "food-chain tag" inspired the laboratory activity in Chapter 3, *Play the Surviving Pelican Game*. The latter was devised by John C. Smallwood.

Finally, I acknowledge the cooperation and assistance of Dr. Kenneth Russell, Director of the Cooperative Wildlife Unit at Colorado State University. Dr. Russell arranged the interview with his student, Mary Jean Currier, which makes up most of the introductory chapter of this book. He also provided consultation regarding the content of the interview, and I am grateful for his insights.

wls

Contents

1 LIFE SCIENCE— AN INTRODUCTION 1

CHAPTER 1 WHAT IS LIFE SCIENCE? 3
What do life scientists do? • Life scientists produce data • What happens to data? • Life science in textbooks • Selecting facts for textbooks • Lesson Review

2 POPULATIONS AND ECOSYSTEMS 17

CHAPTER 2 POPULATIONS 19
Populations, Habitats, and Niches 19
What is a population? • Every population has a habitat • Every population has a niche • Available niches are filled in nature • People create niches • Lesson Review
Populations and Pesticides 25
Pesticides create more problems • Pesticides help create "tougher" pests • Vacant niches can also cause problems • Lesson Review
How Can Pests Be Controlled? 28
"Fly swatter" and "shotgun" methods are least desirable • How are populations controlled in nature? • Diversity in nature helps control population size • Biological control may be used on pests • Insect attractants can be used to control pests • Radiation can be used to control pests • Are pesticides necessary for controlling pests? • Lesson Review
Laboratory Activity
Can You Attract Insects with Food? 32
Applying What You Have Learned 36

CHAPTER 3 ECOSYSTEMS 39
The Earth Is an Ecosystem 39
The community is a part of an ecosystem • The ecosystem has two parts • The biosphere is the largest community • The ecosphere is the largest ecosystem • Lesson Review
The Three Basic Niches 44
Producers make food and release oxygen • Consumers are the animal populations • Decomposers occupy the third basic niche • Decomposers rid ecosystems of dead producers and consumers • Lesson Review
Interdependency in the Ecosystem 48
How are populations dependent upon each other? • What are food chains and food webs? • The food pyramid is a useful idea • Persistent chemicals are in the ecosystem • Food chains cause biological magnification • Two problems remain • Lesson Review

Laboratory Activity
Play the Surviving Pelican Game 54
Applying What You Have Learned 56

CHAPTER 4 FRESHWATER ECOSYSTEMS 59
The Water 59
What is fresh water? • Fresh water varies in temperature • Fresh water differs in transparency • Current is important in freshwater ecosystems • Microhabitats protect organisms in strong currents • There is a difference in oxygen in fresh water • There is a difference in minerals in fresh water • Lesson Review
Laboratory Activity
What Happens When Warm Water Meets Cold Water 61
The Organisms 65
What are the organisms in fresh water? • The neuston live on the surface • The periphyton cling to plants • The benthos live on the bottom • Plankton and nekton live free in the water • The nekton includes fish and other animals • There are two kinds of plankton • Lesson Review
Applying What You Have Learned 72

CHAPTER 5 MARINE ECOSYSTEMS 75
The Marine Environment 75
The marine environment is continuous • There are barriers in the marine environment • The sea has currents • Upwelling supports populations • When is the sea not a sea? • Lesson Review
Laboratory Activity
Why Is the Ocean Restless? 78
Marine Populations 82
The neritic zone houses the most life • Diatoms and dinoflagellates are the key marine producers • Sometimes neritic phytoplankton "bloom" • There is great diversity in marine ecosystems • Lesson Review
Applying What You Have Learned 87

CHAPTER 6 TERRESTRIAL ECOSYSTEMS 89
The Terrestrial Environment 89
Water is a key factor in terrestrial ecosystems • Temperature is another key factor • Oxygen is usually not a problem • Most terrestrial organisms need strong support systems • Terrestrial life is linked to the soil • How do terrestrial ecosystems differ from each other? • Lesson Review

Laboratory Activity
What's in Soil? 92
Major Terrestrial Ecosystems 94
What is a major terrestrial ecosystem? • The tundra is cold and barren • The taiga is a vast northern coniferous forest • Giants live in the moist coniferous forest • Many people live in the temperate deciduous forest • The temperate grassland is easily abused • The tropical savanna supports large animals • The tropical rain forest has the greatest variety of life • The desert covers much of the Earth • The mountains are complex ecosystems • Lesson Review

Applying What You Have Learned 103

3 LIFE INSIDE ORGANISMS 105

CHAPTER 7   CELLS AND CELL PRODUCTS 107

Cells: Through the Light Microscope 107
What is a cell? • Two types of microscopes are used to study cells • Most cells have two main regions • Plant cells can be recognized • Animal cells have many shapes • Some cells are organisms • Lesson Review

Laboratory Activity
What Are Organisms Made Of? 112

Cells: Through the Electron Microscope 117
Organelles do the work in the cell • The cell membrane encloses the cell • The mitochondrion is the powerhouse • The endoplasmic reticulum partitions the cell • The nucleus and the ribosomes work together • The Golgi complex exports proteins • The lysosome is a digestive organelle • The centrioles help animal cells divide • Cells produce many products • Lesson Review

Applying What You Have Learned 124

CHAPTER 8   TISSUES, ORGANS, AND SYSTEMS 127

Tissues 127
What is a tissue? • Animal tissues do a variety of jobs • Cancer is a tissue disease • Lesson Review

Organs 131
What is an organ? • An organ is defined by the function it performs • An organ may have one or more functions • What are some other animal organs? • Some organs can be transplanted • Lesson Review

Organ Systems 135
What is an organ system? • Most complex animals have ten systems • Life is organized in levels • Life and death occur at different levels of organization • Lesson Review

Laboratory Activity
How Many Tissues and Organs Can You Find in a Chicken Wing? 138

Applying What You Have Learned 141

4 HUMAN BODY FUNCTIONS 143

CHAPTER 9   DIGESTION, BREATHING, AND TRANSPORT 145

Digestion 145
What is digestion? • Digestion begins in the mouth • The esophagus moves food after it is swallowed • The stomach stores and digests food • The small intestine is the main digestive organ • Water is absorbed in the large intestine • Lesson Review

Laboratory Activity
How Can Food Particles Pass through All Those Walls? 150

Breathing 153
Gas exchange is the basic problem • What happens during breathing? • Gas exchange occurs in the alveolus • Muscles and elastic tissue aid breathing • Lesson Review

Transport 157
What is blood? • What does the blood do? • The heart is the pump • Arteries, veins, and capillaries carry the blood • Your capillaries leak • Plasma gets back to the blood in two ways • What is the lymphatic system? • Blood takes two circulatory routes • How are wastes removed from the blood? • Lesson Review

Applying What You Have Learned 166

CHAPTER 10   CONTROL IN THE BODY 169

Hormone Control 169
Hormones are chemical messengers • The adrenal glands produce other hormones • The thyroid and parathyroid are close together • Some hormones aid digestion • The pancreas produces two hormones • Your body has a "master" endocrine gland • Lesson Review

Nerve Control 174
The nervous system works faster • The nerve cell is the key unit in the nervous system • Nerve cells are organized • Nerves are like telephone cables • Nerve control is both voluntary and involuntary • Lesson Review

Laboratory Activity
How Do Your Eyes Adjust to Bright Light? 178

Brain Control 180
The brain is the main control center • The brain has three main regions • The spinal cord is a switching center • Lesson Review

Applying What You Have Learned 185

CHAPTER 11   SKIN, MUSCLES, AND BONES 187

The Integumentary System 187
The skin has two main layers • The integumentary system has organs besides the skin • There are two types of glands in the skin • The skin has three main functions • Lesson Review

The Muscular System 191
There are three kinds of muscles in the human body • Cardiac muscle is found in the heart • Skeletal muscle moves bones • Skeletal muscles can be conditioned • What is muscle fatigue? • Oxygen debt is the price of exercise • Lesson Review

The Skeletal System 196
The skeletal system is more than bones • There are joints in the skeletal system • What are the functions of the skeletal system? • Lesson Review

Laboratory Activity
Do You Know Your Own Body? 200

Applying What You Have Learned 203

CHAPTER 12 REPRODUCTION 205

The Nature of the Process 205
Cells can reproduce themselves • Cells can grow • Cells can specialize • Lesson Review

Asexual and Sexual Reproduction 209
Asexual reproduction is simple and fast • Sexual reproduction is more complicated • Sexual reproduction is the more common method • Genes enter the picture • Sexual reproduction is a safety mechanism • Lesson Review

Laboratory Activity
How Fast Does Yeast Reproduce Asexually? 212

The Human Reproductive Process 215
Human beings reproduce only sexually • The male system produces sperm • The female system has three functions • The ovaries produce eggs and estrogen • Other organs aid the reproductive process • The female has a menstrual cycle • Lesson Review

Growth, Birth, and Early Development 220
The embryo stage lasts two months • The fetus develops into a baby • Labor causes birth • What are the needs of a newborn baby? • Lesson Review

Applying What You Have Learned 225

CHAPTER 13 GENES AND HEREDITY 227

Genes 227
What are genes? • What do your genes do? • What are proteins? • Lesson Review

The Gene Pool 232
What happens to genes? • Each population has a gene pool • Genes can mutate • What happens when mutated genes do survive? • Lesson Review

Predicting Inheritance 235
Human traits are difficult to predict accurately • Human ABO blood type can be predicted • The big question: genes or environment? • Lesson Review

Laboratory Activity
What ABO Blood Type Genes Could You Have? 238

Applying What You Have Learned 240

CHAPTER 14 GROWTH AND DEVELOPMENT 243

How Growth Occurs 243
What is growth? • Your growth occurred in three ways • Development is growth and change of form • How do you grow taller? • How do your muscles grow? • How does your brain grow? • What causes growth? • Lesson Review

Stages of Growth and Development 248
Everyone has an internal clock • Childhood development is slow and steady • Adolescence brings rapid growth • Puberty also arrives during adolescence • Adolescence brings changes in body form • Adolescents usually have a muscle problem • Adolescents do not grow proportionately • The slow developer has a special problem • Lesson Review

Laboratory Activity
How Do They Grow? 252

Applying What You Have Learned 255

5 CHALLENGES TO HUMAN SURVIVAL 257

CHAPTER 15 DISEASE 259

What Causes Disease? 259
Foreign invaders cause disease • Fungi cause ringworm and athlete's foot • Viruses cause many common diseases • Worm parasite diseases are declining • Some diseases are inherited • Improper diet causes deficiency diseases • Protein deficiency disease is tragic • Lesson Review

How Are Diseases Treated and Prevented? 267
The body combats invaders in two ways • Antibiotics kill bacteria • Surgery cures a variety of diseases • A disease may create immunity • We can create immunity • Another organism's antibodies can be used • Killer number one is yet to be conquered • What causes atherosclerosis? • Lesson Review

Laboratory Activity
Discovery: How Life Scientists Can Create Immunity to Disease 271

Applying What You Have Learned 277

CHAPTER 16 PEOPLE AGAINST THEMSELVES 279

The Real Enemy 279
Placebos tell us something about ourselves • The mind can protect or destroy • The mind can be a disease agent • Heart surgery can sometimes cure the disease • Lesson Review

The Drug Problem 282
What is the drug problem? • We have a human problem • Drugs are not good or bad • Drugs are different and so are individuals • What is drug abuse? • People do become dependent upon drugs • Can drug dependency be broken? • Lesson Review

Laboratory Activity
Do You Live in a "Drug Society"? 286
Problem Drugs 291
What are problem drugs? • Alcohol is a social drug • Tobacco is a new problem • The opioid user faces serious problems • Barbiturates are depressants • Amphetamines are stimulants • The psychedelics change perception • A last word about drugs • Lesson Review
Applying What You Have Learned 298

CHAPTER 17 POLLUTION OF THE ENVIRONMENT 301
What Is Pollution? 301
Pollution is a relative term • What is pollution? • Pollution can affect or destroy a person's health • Pollution can upset the balance in ecosystems • Pollution can destory scenic beauty • Lesson Review
Water Pollution 306
The nature of the problem is people • Decomposers use oxygen • Disease organisms lurk in polluted water • Rapid decomposition speeds eutrophication • Warm water can pollute • Human carelessness can cause water pollution • Radioactive wastes are a special problem • Lesson Review
Laboratory Activity
Why Isn't the Water Blue Anymore? 308
Air Pollution 317
Air pollution is a difficult problem • How does air become polluted? • A temperature inversion can increase pollution • Fog can also increase air pollution • There are two types of smog • There is a variety of pollutants in the air • What do pollutants do to the human body? • Air pollution probably causes emphysema • Other diseases may be caused by air pollution • How much is clean air worth? • There are hard decisions ahead • Lesson Review
Applying What You Have Learned 325

6 ORGANISMS 327
CHAPTER 18 SIMPLE ORGANISMS 329
Viruses and Monerans 329
Viruses are not true organisms • Bacteria are true organisms • The blue-green algae are also monerans • Lesson Review
The Producer Protists 334
Most algae are protists • Green algae are common in fresh water • Diatoms are the main producers in aquatic ecosystems • Most brown and red algae are marine • Lesson Review
The Consumer Protists 339
Consumer protists are like small animals • The paramecium is a common protozoan • The ameba is a very simple protozoan • The cyst enables protozoans to survive drought • Lesson Review

Laboratory Activity
Can You Find These Organisms? 340
Applying What You Have Learned 345

CHAPTER 19 FUNGI, MOSSES, AND FERNS 347
Fungi 347
What are fungi? • Molds are common fungi • Molds are important fungi • There are many other kinds of fungi • Lesson Review
Laboratory Activity
How Many Mold Spores Can You Collect in Five Minutes? 350
Mosses and Ferns 353
What are mosses? • Why do moss plants grow close together? • Mosses have an important niche in terrestrial ecosystems • Ferns are vascular plants • Ferns once dominated the landscape • Lesson Review
Applying What You Have Learned 357

CHAPTER 20 SEED PLANTS 359
Conifers 359
Conifers are cone-bearing plants • Most conifers are trees • The oldest and the largest organisms are conifers • The conifers are important plants • Lesson Review
Flowering Plants 364
What is a flower? • The ripened ovary is a fruit • The seed contains a dormant plant embryo • Lesson Review
The Kinds of Flowering Plants 368
There are two groups of flowering plants • The grass family is an important group of monocots • The lily family has attractive flowers • The pea family is a large family of dicots • The rose family has many important fruit producers • The composite family is very large • Lesson Review
Laboratory Activity
How Are Flowers Alike? How Are They Different? 372
Applying What You Have Learned 376

CHAPTER 21 SPONGES AND COELENTERATES 379
Sponges 380
Sponges are the simplest animals • Most sponges live in marine ecosystems • Sponges are full of holes • What is the sponge's niche in aquatic ecosystems? • Sponges still have commercial value • Lesson Review
Coelenterates 384
Coelenterates are aquatic animals • Coelenterates have a gastrovascular cavity • Coelenterates have two body forms • The mesoglea serves as an internal skeleton • Several coelenterates have an external skeleton • Coelenterates have varied niches • Lesson Review
Laboratory Activity
Observe a Freshwater Coelenterate 388
Applying What You Have Learned 391

CHAPTER 22 WORMS AND MOLLUSKS 393

Flatworms 393
Most flatworms have a flat body • Turbellarians are mostly free-living • The planarian is a typical turbellarian • Flukes are parasitic • Some flukes lead complex lives • Some flukes cause human diseases • Tapeworms are internal parasites • Humans are the host for some tapeworms • Lesson Review

Roundworms 399
Roundworms are almost everywhere • Nematodes are the most abundant group of roundworms • Nematodes have an important niche • Parasitic nematodes are a problem • Rotifers are the "wheel animals" • Gastrotrichs are also interesting creatures • Lesson Review

Mollusks 404
Mollusks are soft-bodied animals • Chitons have shells of eight plates • Bivalves have no head • Gastropods are the largest group of mollusks • Cephalopods are the largest mollusks • Lesson Review

Annelids 409
Annelids are segmented worms • Polychaete worms are mostly marine • Oligochaetes are freshwater and terrestrial • Leeches are parasites • Lesson Review

Laboratory Activity
Find the Body Systems in an Earthworm 412

Applying What You Have Learned 415

CHAPTER 23 ARTHROPODS AND ECHINODERMS 417

Arthropods 417
What are arthropods? • Millipedes have many legs • Centipedes are predators • The insects make up the largest group of animals • Insects develop in stages • Some insects live together in social groups • Crustaceans are used for food • The arachnids have four pairs of legs • Lesson Review

Laboratory Activity
Can You Catch an Insect? 423

Echinoderms 426
The echinoderms are marine animals • The starfish is a common echinoderm • The sea urchin looks like a ball of spines • There are three other groups of echinoderms • Lesson Review

Applying What You Have Learned 431

CHAPTER 24 VERTEBRATES 433

Fish 433
Fish are the oldest vertebrates • Hagfish and lampreys are the most primitive fish • Sharks and rays would win another contest • The bony fish are the most plentiful • Lesson Review

Amphibians and Reptiles 439
Most amphibians are extinct • The wormlike amphibians are terrestrial • Newts and salamanders are amphibians with tails • Frogs and toads are amphibians without tails • Most reptiles are extinct • The reptiles are adapted for terrestrial life • Tortoises and turtles have shells • Crocodiles and alligators are aquatic animals • Lizards and snakes are the most successful reptiles • Lesson Review

Laboratory Activity
How Far Will a Cold Frog Jump? 442

Birds and Mammals 449
Birds have feathers, wings, and beaks • Many adaptations enable birds to fly • Birds have many other adaptations • Birds and mammals are warm-blooded animals • Mammals are the newest animals • Mammals have a variety of habitats • Mammals can be classified by what they eat • Care by the mother is essential for mammal survival • Mammals are the most intelligent animals • Lesson Review

Applying What You Have Learned 456
Glossary 457
Index 463
Illustration Sources 468

Unit 1

Life Science— An Introduction

Life.
You have it.
Elephants and bees and cherry trees have it.
But rusty car bodies
Will never have it.

Life is the Earth's most precious treasure.

Figure 1.1: Mary Jean Currier is a life scientist working in Colorado. She is studying mountain lions, which she captures and later releases. Sometimes she captures mountain lion kittens, such as this one, which she named "Gerhart."

Chapter 1

WHAT IS LIFE SCIENCE?

This year you will study life science. You have a textbook with that name. But what are you going to study? What is life science?

If you asked some friends they might tell you that life science is about worms and frogs—or bugs and leaves. Are they right?

There is a story about a little boy who went to the circus. He was only able to look under the edge of the big tent. When he came home his mother asked him what the circus was about. He told her that the circus was about people who sat on benches and ate popcorn. Was he right?

Life science is about worms and frogs—and bugs and leaves. But life science is also about many other things. Life science is about people. Life science is about your **environment** and its problems. And life science is about mysteries that nobody knows anything about.

environment:
all things outside the bodies of living things that affect their lives

WHAT DO LIFE SCIENTISTS DO?

Life science is also about what life scientists do when they work. How can you find out what life scientists do when they work? You could go watch a life scientist while he or she is working. Or you could find life scientists and ask them questions about what they do. Both methods might be interesting for you. But they might also be hard for you to use.

Instead, the author of your textbook will take your place and use one of the methods for you. He is a life science teacher who asked a life scientist questions about her work. The answers to these questions will help you understand what life scientists do when they work. They will also help you understand what you will be studying this year.

The life scientist who answered the author's questions is named Mary Jean Currier. She was in Fort Collins, Colorado, when the author spoke with her.

TEACHER: What can you tell us about your work?

LIFE SCIENTIST: I'm a field biologist. This means that most of my work is outdoors. Right now, I'm studying the mountain lion here in Colorado.

TEACHER: Isn't the mountain lion also called by other names?

LIFE SCIENTIST: Yes, it has several common names. It is also called cougar, puma, and catamount.

TEACHER: Before telling us more about your work, will you explain how you got started as a biologist?

LIFE SCIENTIST: I was born here in Colorado and spent a lot of my childhood outdoors. I had a horse, and I was active in 4-H Club. My mother was a high school biology teacher.

Figure 1.2: The colored areas on the map are places where mountain lions have been seen. The lines indicate approximate regions where different common names are in use. There are many exceptions in the use of common names.

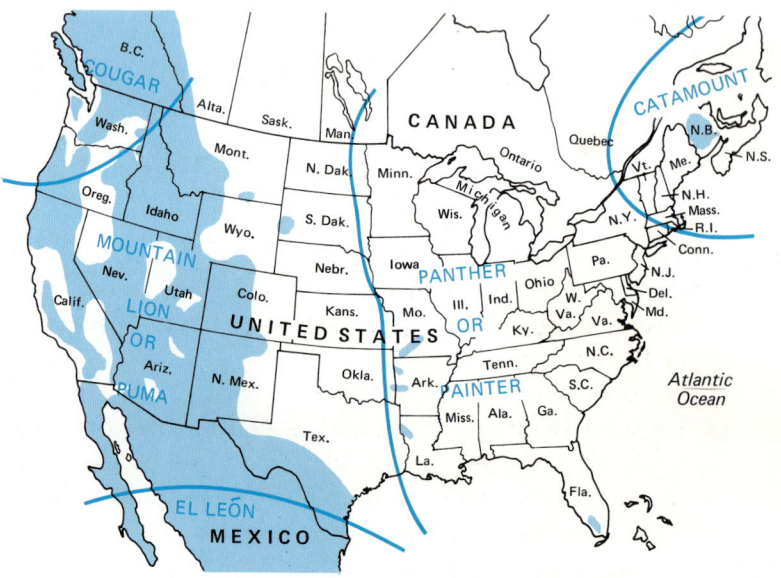

She taught my sister and me to be curious about plants and animals that we observed.

TEACHER: Do you think your mother influenced you to become a biologist?

LIFE SCIENTIST: Perhaps a little. But she died when I was 10 years old. She mainly helped develop my curiosity. After she died, my father, who is a doctor, took us backpacking as often as he could. The three of us hoped to climb all of the 14,000-foot mountains in Colorado.

TEACHER: How many have you climbed?

LIFE SCIENTIST: Well, there are fifty-two of them in Colorado. I climbed my first one at age 12. I have since climbed seven different ones. And I've climbed one—Long's Peak—three times.

TEACHER: Were you totally an outdoor girl? Did outdoor activities help you prepare for the work you are now doing?

LIFE SCIENTIST: Not really. I was not entirely an outdoor girl. I played violin in the school orchestra and enjoyed that. I also enjoyed reading books and doing other things that weren't related to the outdoor life.

TEACHER: Did you plan on being a field biologist when you went to college?

LIFE SCIENTIST: No, I didn't know for sure what I wanted to do. I just knew that I liked science, especially biology. I decided to go to a college in Wisconsin where students could take time out to work at some job.

TEACHER: What work did you do?

LIFE SCIENTIST: I got a job in Ireland. I helped test soaps to see if they could kill **bacteria** on dentists' hands. It was scientific research and I enjoyed it. I also enjoyed living with an Irish family and learning about another culture.

bacteria: microscopic living things. Some cause disease.

TEACHER: Then what did you do?

LIFE SCIENTIST: I finished college and worked in a research laboratory in Germany for a year. I worked with a team that was trying to find a cure for *leukemia* (lo͞o kē′mē ə). This is a cancer of the white blood cells.

TEACHER: You didn't enjoy that work?

LIFE SCIENTIST: Yes, I did. But I soon learned that I needed lots more education. However, when I came home I was confused. I still didn't know exactly what kind of scientific work I wanted to do.

TEACHER: What did you do then?

LIFE SCIENTIST: Well, my sister was in medical school in Denver. I lived with her and found a job in the **biochemistry** department of the medical school. I thought that I should work a while and think about my future.

biochemistry: the study of chemicals that make up living things

TEACHER: What did you decide?

Life Science—An Introduction

LIFE SCIENTIST: While working in the biochemistry laboratory, I regained an interest in a problem that I had thought about in college. For some years I had been interested in the problem of how certain chemicals like *morphine* (**môr′fēn′**) could relieve pain in a person's body. Now I had a renewed interest in the problem. I wanted to study the brain and how pain-relieving chemicals affected it.

TEACHER: What did you do then?

LIFE SCIENTIST: I began checking around and found that the University of California in Berkeley had one of the best brain research programs in the country. I applied there and was accepted. I spent the next year studying in Berkeley.

TEACHER: But still you weren't satisfied?

LIFE SCIENTIST: I liked my studies and research ok. But most experiments were with some kind of animal. Usually dogs, rats, and mice were used. In the experiments, we had to kill many of them. There was no choice. Scientists can't do many experiments with humans. Anyhow, I didn't want to kill any more animals. Also, I missed the out of doors that I had known all through my years of growing up. I wanted to be outside again, and to work with animals that were alive and healthy.

TEACHER: So you came back to Colorado?

LIFE SCIENTIST: Yes, and I worked hard to get the job that I have now. I went to the director of the Cooperative Wildlife Research Unit here at Colorado State University. I had to convince him that I could be a good field biologist. Also, at this time, I married a man who was just completing his education to become an engineer.

TEACHER: Will you explain the first research problem that you were assigned?

LIFE SCIENTIST: Yes, but first I'll have to give you some background information. There used to be a bounty on mountain lions in Colorado. People were given a prize in money for each one they killed. Then, in the mid-sixties, people began to worry about the animal. They were afraid that too many mountain lions might be killed by bounty hunters. So the Colorado State Game Department was given control of the state's mountain lion population. It set hunting seasons in the hope of regulating the number killed. However, there were many problems in trying to control the size of the population. First, the Game Department had no idea how many mountain lions there were in the state. So it didn't know how long a hunting season to allow. Second, it didn't know much about the **life cycle** of the lions. For example, no one knows at what age the lions start reproducing in the wild. We still don't know.

life cycle: the series of changes in a living thing from birth to death. A key change occurs when the living thing can reproduce its own kind.

6

Figure 1.3: Mary Jean Currier discusses her study area with her supervisor, Dr. Kenneth Russell. The study area was in the rugged mountains southwest of Pike's Peak, one of the 52 mountains in Colorado that is over 4200 meters (14,000 feet) high.

TEACHER: So what was your research problem?

LIFE SCIENTIST: Actually, I had two problems and worked on both of them at the same time. The first problem was to find out how many mountain lions lived in a specific area of Colorado. The Game Department believed it had a high concentration of lions. The second problem was to develop a way to tell the age of a mountain lion.

TEACHER: You mean that no person had yet learned how to tell the age of a mountain lion?

LIFE SCIENTIST: That's right. With animals like deer, bear, and bobcats one can pull a tooth and make a cross section of the root. During each year of the animal's life a layer forms around the root. By counting the annual rings you can tell the age of those animals.

TEACHER: That's like counting tree rings, isn't it?

LIFE SCIENTIST: It's just about the same. But it doesn't work for all animals, including the mountain lion.

TEACHER: Well, how did you proceed with your problems?

LIFE SCIENTIST: We first selected my specific study area. It was in the mountains near Cripple Creek, Colorado. Then I spent lots of time reading all of the research information that had been published on mountain lions. I also spent lots of time planning what I would do and need. And I gathered equipment.

TEACHER: Did you plan to work alone?

LIFE SCIENTIST: No, I couldn't have caught the lions by myself. A hunting guide was hired to help me. He was an expert hunter whose dogs could follow the scent [smell] of a mountain lion. They could follow the scent until the lion climbed a tree.

Figure 1.4: When possible, horses were used to trail the mountain lions. The hunting guide had the main responsibility for tracking the animals and always rode ahead with the hounds.

7

Life Science—An Introduction

hibernate:
to spend the winter in a sleeplike state

kilo:
short for kilogram, a metric weight equal to 2.2 pounds

TEACHER: What time of year did you start your field work?

LIFE SCIENTIST: We started in December and worked all winter. That was the best time of the year because the dogs had been trained to trail bears as well as mountain lions. In the winter, the bears **hibernate** and we didn't have to worry about the dogs trailing them instead of the lions.

TEACHER: How did you follow the dogs once you had them trailing a lion?

LIFE SCIENTIST: The guide and I rode horses some of the time. But many times we had to walk, sometimes with snow up to our waists.

TEACHER: You carried a backpack, too?

LIFE SCIENTIST: Yes, with up to 12 **kilos** (kē′lō) of equipment.

TEACHER: What did you do after the mountain lion was treed by the dogs?

LIFE SCIENTIST: I used a special gun called a "Cap-Chur" gun. It shoots a device that injects a drug into the lion. After a short time the lion can't move its legs. Then it is safe to handle.

TEACHER: Then what did you do?

LIFE SCIENTIST: Some researchers tie a rope around the lion and lower it out of the tree. We found an easier way. Before the drug took effect, we threw snowballs and things at the lion in the tree. Usually, if we kept the dogs back, the lion would leap out of the tree and start to run away. We would track it until the drug caused it to stop running.

Figure 1.5: This is the 12 kilograms of equipment that sometimes had to be carried through waist-deep snow. It includes a fishing vest filled with test tubes, syringes, a stopwatch, tranquilizing drugs, and antibiotics. Other equipment includes the Cap-chur gun, darts, a walkie-talkie, a pistol, a tattooing instrument, a slingshot, collars, a spring scale, and footprinting equipment.

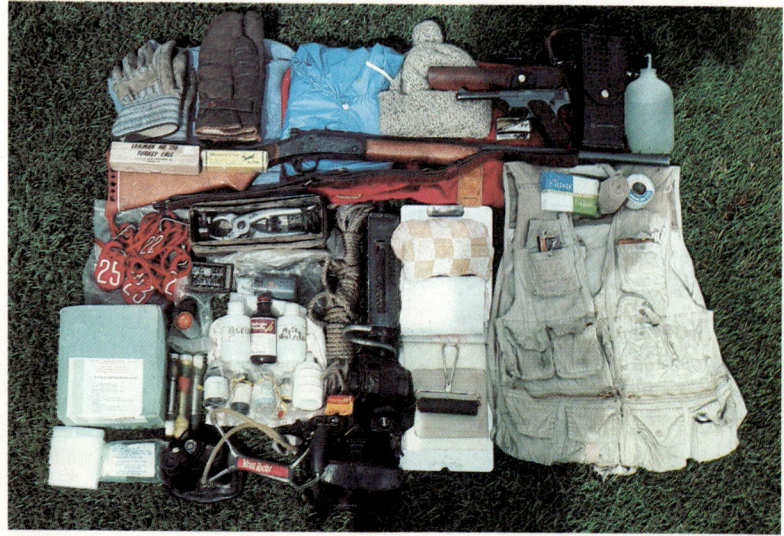

Figure 1.6: One of the hounds has treed a mountain lion (left). A treed animal may react with violence (middle) or with boredom (right).

TEACHER: What did you do to the lions after you captured them?

LIFE SCIENTIST: Many things. I weighed them and took measurements of different parts of their body. I tagged them with a special collar and tattooed each ear. I took prints of their paws and samples of their hair and blood. I also examined their teeth and gums.

TEACHER: Why did you take all this information?

LIFE SCIENTIST: I wanted to use this information to find a way to tell the age of the lions.

TEACHER: What were the results of your field work?

LIFE SCIENTIST: The first year was a real disappointment. That whole winter we only captured two lions. The main problem was a heavy snow early in the winter. That snow drove the deer to lower places in the mountains. Mountain lions prey (prā) on deer so they also moved down to the lower places. Unfortunately, those places were outside my study area. For a while I was afraid that my supervisors might think I was a failure.

TEACHER: What happened them?

LIFE SCIENTIST: I was very determined. I knew that I could catch the lions and solve the problems. Really, I didn't need to worry. My supervisors were very understanding.

TEACHER: So you continued.

LIFE SCIENTIST: Yes. That following summer I visited several zoos around the United States. I collected hair samples, blood samples, and other information from their mountain lions.

Figure 1.7: An ear of each mountain lion was tattooed with an identification number. To do that, black ink was first rubbed on the inside of the ear. Then the tattooing instrument was clamped down. Little needles in it, arranged in the form of numerals, penetrated the skin. A second coat of black ink was applied. Now, if any of the marked animals are ever captured or killed, they can be identified by the number tattooed in their ear.

9

Figure 1.8: Each animal was weighed and tagged with a number. The tags were a temporary way to identify different lions in case they were treed a second time. The heaviest lion weighed 70.5 kilograms (155 pounds).

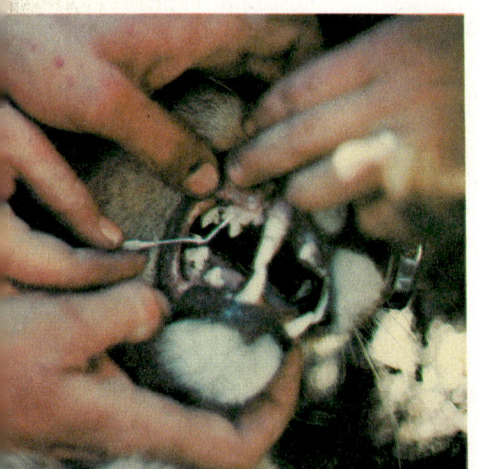

Figure 1.10 A special dental probe, marked off in millimeters, was used to measure how far the gum line had drawn back from the lions' teeth. In humans, the gum line draws back toward the jawbone as aging occurs. Mrs. Currier is trying to find out if the same thing happens in mountain lions.

TEACHER: Why did you do that?

LIFE SCIENTIST: In many cases zoo managers know the exact age of their mountain lions. By collecting information from their "zoo lions," I could compare that with information I would get from wild mountain lions. This would help me to find a way to tell the age of the wild animals.

TEACHER: But you had to capture more wild mountain lions.

LIFE SCIENTIST: Right, and I started that project again the next winter. This time I had good success. We captured twenty lions—about one for every 4 days we hunted. That's a very good record.

TEACHER: What did you do differently?

Figure 1.9: Red paint (washable and nonpoisonous) was applied to the hind feet (top). Footprints were then made on paper (bottom). Later, the exact measurements of each foot were taken. The hunting guides helped with most phases of the field work.

Figure 1.11: These are mule deer in Mrs. Currrier's study area in Colorado before heavy snowfall came. Often, it is the old or sick deer that are killed by the mountain lions.

LIFE SCIENTIST: We hired a new guide with dogs trained to track mainly mountain lions. Also, I picked a slightly different study area, where the deer and the mountain lions were more plentiful.

TEACHER: Have you now solved the two problems?

LIFE SCIENTIST: My first problem was to determine how many mountain lions live in a specific area of Colorado. I've done that. There is one mountain lion for every 30 to 56 square **kilometers** in a specified area. Also, I used my and other people's information to estimate the total number of mountain lions in Colorado. My estimate is that there are between 1100 and 1500 of them in the state.

TEACHER: How about your second problem?

LIFE SCIENTIST: I have not solved that one yet. I have spent hundreds of hours working on ways to tell the age of a mountain lion. Sometimes I am encouraged and think I'm making progress. Then I find something wrong in what I've been doing and know it won't work. So I begin again. Other people are working on the same problem, so I'm not disappointed.

TEACHER: So your work isn't finished?

LIFE SCIENTIST: A scientist's work is never finished. That is one way a scientist's work is different from other peoples'. The more we learn, the more we see there is to learn.

TEACHER: That would make some people unhappy.

LIFE SCIENTIST: Probably, but that's what I like most about my work. There is always something new to be discovered. And it's a thrill when I can be the first person to know something.

TEACHER: Thank you for the interview. Maybe the students who read this will remember you and your work if they ever see a mountain lion.

You have just read what one life scientist does when she works. This life scientist studies a **population** of mountain lions. Right now other life scientists are studying populations of elephants, coral, and chestnut trees.

kilometer:
a metric distance equal to 0.6 mile

Figure 1.12: This mountain lion is ready to be released. Mrs. Currier and the hunting guides have taken all the measurements needed. (The red color on Mrs. Currier's hands is from the footprinting ink.) In a short while the effects of the tranquilizing drug will wear off and the lion will be on its way.

population:
in life science, a group of living things of the same kind that are found in a particular place

11

Do all life scientists study populations? No. Many life scientists study individual plants and animals. Many others study the things that go on inside plants and animals—things that go on in cells, tissues, and organs.

Right now you probably know very little about populations, cells, tissues, and organs. These are topics that you will study in the next few chapters. What is important now is this: populations, cells, tissues, and organs are all a part of the life on this planet. Anything that is alive, or has ever been alive, is either being studied now, has been studied, or may be studied in the future by some life scientist.

Life scientists produce data. Dressmakers work and produce dresses. Photographers work and produce pictures. What do life scientists produce when they work?

Life scientists produce **data** when they work. Data are facts. Facts can be called "true statements." It is a fact that the Earth is round. That is a true statement. It is a fact that your body temperature is about 37 **degrees Celsius.** That is a true statement.

The data of any scientist must meet one important test. Another scientist must be able to prove the data. How are data proved? Generally by one of two methods. One method is by experiment. For example, a life scientist thinks he or she has found a chemical that will stop cells from dividing. The scientist does an experiment by trying the chemical on a number of cells. The scientist's findings—the data—show that the chemical does stop cells from dividing. Next comes the important step. Another

data: observed facts, including measurements. More generally, "data" has come to mean information of any kind.

degree Celsius: a metric temperature equal to $9/5$ of a degree Fahrenheit. To convert from degrees Celsius (°C) to degrees Fahrenheit (°F), use the formula °F = $9/5$ °C + 32°. To convert from °F to °C, use the formula °C = $5/9$ (°F − 32°).

Figure 1.13: This life scientist (below) is studying plant and animal life he has collected from the sea. This life scientist (left) is exploring what happens inside the cells of living things.

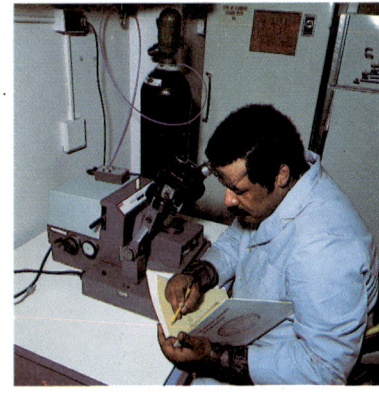

Figure 1.14: Taking notes (above), or recording data, is an important part of a scientist's work. Scientists publish what they observe or measure (left) so that other scientists may read about the work, discuss it, and often try to plan ways to support it or disprove it.

life scientist must be able to do the same experiment, in the same way, and get the same result. By experiments, other scientists prove the data and help the data become accepted.

The second method of proving data is by direct observation. The life scientist studying the mountain lion population does not always do experiments. She gets some of her data by direct observation—she looks, she sees, and she describes. Such data are better accepted when other life scientists can make the same or similar observations.

What happens to data? Data gathered by life scientists are normally collected in notebooks. Then the data are organized and written into a scientific paper. (Scientists use the term "paper" only to describe such writings.) The paper is usually published in a magazinelike journal. Most journals are specialized. For example, a journal might include only papers about research work done on birds. Life scientists all over the world read the journals that relate to their own studies.

Scientists also "get together" to discuss their work. Many of these gatherings are formal meetings called "congresses" or "conventions." At these formal meetings some scientists read papers describing their latest work. Others listen or take part in discussions of various kinds. In such discussions, scientists often challenge each others' conclusions. They describe new ideas and experimental methods. Some discussions end up as debates, some of them heated. Newspaper reporters and other journalists usually "cover" the larger scientific meetings. Scientists often use press releases to announce results that are of wide interest to the public.

Life Science—An Introduction

Life science in textbooks. Now you may ask: What is it that I will study within the covers of this textbook? At times you will be given instructions for doing some of your own experiments, and for conducting your own observations. This way you can collect and organize data the way a scientist does it.

Seldom will you study data as you read the chapters of this book. Data from science are much like reports from a battlefield. The reports of a battle arrive in bits and pieces. Rarely are they meaningful until the battle is over. Then the reports can be put together and one can describe what happened. In a like manner, data are meaningful only when they are put together and organized. Mainly you will study facts about life science that have been organized for you.

Selecting facts for textbooks. What facts were selected for this textbook? This question can almost be answered by another. If you were selecting facts of life science to study, which ones would you pick? Probably you would pick those that are important or meaningful to you. That same approach was used in selecting materials for your textbook. Right now, at your age, and at your time of living, there are facts of life science that are important to you. Some will help you understand yourself. Some will help you understand your environment. Both kinds of understanding are important.

You may ask: How do I know that the things I will study in this textbook are important to me? This question cannot be answered now. You will have to ask it over and over—as your study proceeds this year. Just do not forget the question.

LESSON REVIEW *(Think. There may be more than one answer.)*

1. Mary Jean Currier's first problem was to
   a. find out why mountain lions kill deer.
   b. find out what mountain lions eat in a certain area of Colorado.
   c. find out how many mountain lions live in a certain area of Colorado.
   d. find out how to capture mountain lions without killing them.
2. Mary Jean Currier's second problem was to find a way
   a. to count the annual rings on a mountain lion's teeth.
   b. to tell the age of a mountain lion.
   c. to train wild mountain lions to be as tame as those found in zoos.
   d. to take blood samples from mountain lions to find out if they are diseased.

3. A life scientist
    a. can only obtain data after doing an experiment.
    b. can only obtain data after making a direct observation.
    c. can obtain data from an experiment or direct observation.
    d. cannot obtain data from experiments or direct observations.
4. Mary Jean Currier's data could be
    a. organized and written into a scientific paper.
    b. published in a scientific journal.
    c. read at a scientific meeting.
    d. released to newspaper and television reporters.
5. In life science textbooks
    a. data are seldom found.
    b. data will be found everywhere.
    c. data have been organized in meaningful ways.
    d. data can never be found.

KEY WORDS

environment (p. 3)
bacteria (p. 5)
biochemistry (p. 5)
life cycle (p. 6)
hibernate (p. 8)

kilo (p. 8)
kilometer (p. 11)
population (p. 11)
data (p. 12)
degree Celsius (p. 12)

Unit 2

Populations and Ecosystems

Earth is one large sphere. But a glance at any world globe reminds you that the surface of this sphere is anything but uniform. Almost three-fourths of the surface is covered with salty waters. Land areas are speckled with lakes and cut by rivers. Polar regions are capped with huge sheets of ice. Mountains and canyons interrupt vast prairies and deserts. The surface is burned by the sun, sprinkled with rain, battered by wind, and buried in snow. Yet everywhere, filling every possible role in life, there is a community of living things. Each community is made up of populations living within an ecosystem.

Figure 2.1: A small population of Japanese beetles is destroying the leaves of a sassafras tree.

Chapter 2
POPULATIONS

In Chapter 1, you read about a life scientist who was studying a population of mountain lions. We might just as easily have interviewed a life scientist studying a population of goatweeds or grasshoppers. The study of populations—of all kinds—is important. In this chapter you will learn why.

POPULATIONS, HABITATS, AND NICHES

What is a population? The population of Indianapolis, Indiana, is growing. So is the population of Dallas, Texas. In both cases, we are talking about populations of people—human beings. But what does the term "population" mean when we speak of mountain lions or gorillas?

The word "population" comes from the Latin word *populus*, which means "people." For many years, the word "population"

Populations and Ecosystems

has been used to refer to the people living in an area, such as a city or state. About 60 years ago, life scientists borrowed the term. Now life scientists use the term "population" to refer to a group of plants or animals in a specific area.

Keep one thing in mind when you think of a plant or animal population. Such a population includes plants or animals of only *one* kind. For example, it would not be correct to speak of the eagle population of an area. The reason: There may be two different kinds of eagles in the area. The area may have bald eagles. It may also have golden eagles. To be correct, you would have to speak of the bald eagle population or the golden eagle population.

Every population has a habitat. Every plant or animal population has a home, or a place where you could go and find any of its members. This is the population's **habitat.** The habitat of a population of cypress trees could be a swamp in Florida. The same swamp could also be the habitat of a population of alligators.

habitat: the area where a plant or an animal population has its home

There are many different kinds of habitats. Besides swamps, there are prairies, forests, ponds, rivers, oceans, deserts, and many more.

Sometimes it is useful to have a general term to describe the habitat of a population. Thus, we use the term **marine** (mə rēn′) **habitat** for a population that lives in the ocean. Some populations live in bodies of water other than the ocean. We say that they live in **freshwater habitats.** A third general kind of habitat is the **terrestrial** (tə rĕs′trē əl) **habitat.** *Terrestrial* means "of the land." Your own population, and all others that live on land, are said to live in a terrestrial habitat.

marine habitat: an ocean area where a plant or an animal population has its home

freshwater habitat: an area in a body of water other than an ocean where a plant or an animal population has its home

terrestrial habitat: an area on land where a plant or an animal population has its home

Figure 2.2: The black grouper (top left) lives in a marine habitat. The dry grassland (lower left) is one of several kinds of terrestrial habitats. A group of students is studying a freshwater habitat (right).

Figure 2.3: These pigeons have a habitat and a niche. Their niche is their total way of life in the city environment.

Every population has a niche. Anyone who has ever lived in a city probably remembers the pigeons that live there. Their habitat usually includes the buildings, streets, and parks of the downtown area. Pigeons are well suited for this habitat. They are descendants of the rock dove, which still lives on rocky cliffs in Europe. The pigeon's feet, like those of the rock dove, are not suited for grasping the limbs of a tree—an ordinary feat for most birds. But pigeons are at home on the flat ledges of city buildings.

Suppose you wanted to tell people who had never been to a city about the pigeons that live there. If you just told them about the pigeons' habitat, would they know much about these birds? Not really. They would never know how pigeons strut and swagger up and down busy streets and sidewalks. They would never know that pigeons gather in flocks and always seem to be present wherever food is available. They would never know that there is a dominant pigeon in each flock that will always be allowed to eat first. They would never know about the messes pigeons make with their body wastes on statues and building walls. They would never know about people who get pleasure from feeding and watching pigeons in the city parks.

A pigeon's way of life in the city is called its **niche** (nĭch). The niche includes all the things that a pigeon does. The niche also includes all the effects that a pigeon has on its environment. As you can see, the term "niche" has a much wider meaning than the term "habitat." The habitat describes only where a plant or animal lives. The niche describes the plant's or animal's total way of life. You will learn more about the niche in your study this year. For now, just remember this: Every population has a habitat, *and* it has a niche.

niche:
the total way of life for a living thing

21

Populations and Ecosystems

Available niches are filled in nature. Use your imagination. Picture a large cardboard box. We are going to fill that box with watermelons. Let us say that we fit five of them into the box. Could we say that the box is full? Not really. Suppose we had some baseballs that we wanted to put in the same box. Could we do it? Probably. By moving the watermelons, we could fit the baseballs into the spaces between the watermelons. Now could we add some golf balls? There should be some space left. Could we also add some marbles? Probably. Now suppose the box was full of watermelons, baseballs, golf balls, and marbles. How much sand could we pour into the box? We cannot say for sure. But we could pour in quite a bit. Why? Because there would still be some space left after the box is "filled" with watermelons, baseballs, golf balls, and marbles.

What does a box filled with watermelons and other things have to do with populations and niches? Quite a lot—in a general way. Imagine that we have a bare piece of land, several acres in size. Assume that the soil is good, that the land is in a warm climate, and that it receives enough rainfall. Next, we'll plant some trees in the area and let them grow to their natural height. Is the population of trees the only population we could have in the area? No, for one good reason. By planting the trees, we would create niches for many other populations of plants and animals. Moss plants would now be able to maintain their way of life on the bark of the trees. Flowering plants that grow well in shaded places could then live under the trees. Populations of different kinds of insects could carry on their way of life on the trees and under them. And wouldn't there be a population of squirrels and several small populations of birds?

By planting the trees, we really have helped create a lot of different niches that can be filled by other populations. And when they are filled, we would have a forest instead of a grove of trees.

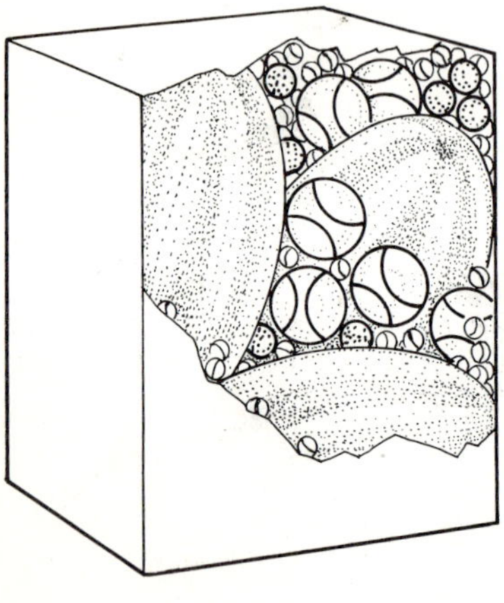

Let us look more closely at our forest. Keep in mind the imaginary box filled with watermelons, baseballs, and other things. Think of that box as we look at one of the birds that lives in our forest. If we looked under its feathers, we would probably find a population of fleas or mites. If we could examine the bird's intestine, we would find several different populations of microscopic life. Probably we would find more populations of microscopic life in the bird's blood. One bird might easily provide niches for ten or fifteen different populations. And we would find the same thing if we looked closely at any large animal or plant.

All of nature is much like the box that we started filling with watermelons. The spaces in the box compare to the niches that are available in nature. The only difference is: **In nature, all of**

Figure 2.4: We can put baseballs, golf balls, and marbles in the spaces between watermelons in a box. In somewhat the same way, all available niches are filled in nature.

Populations

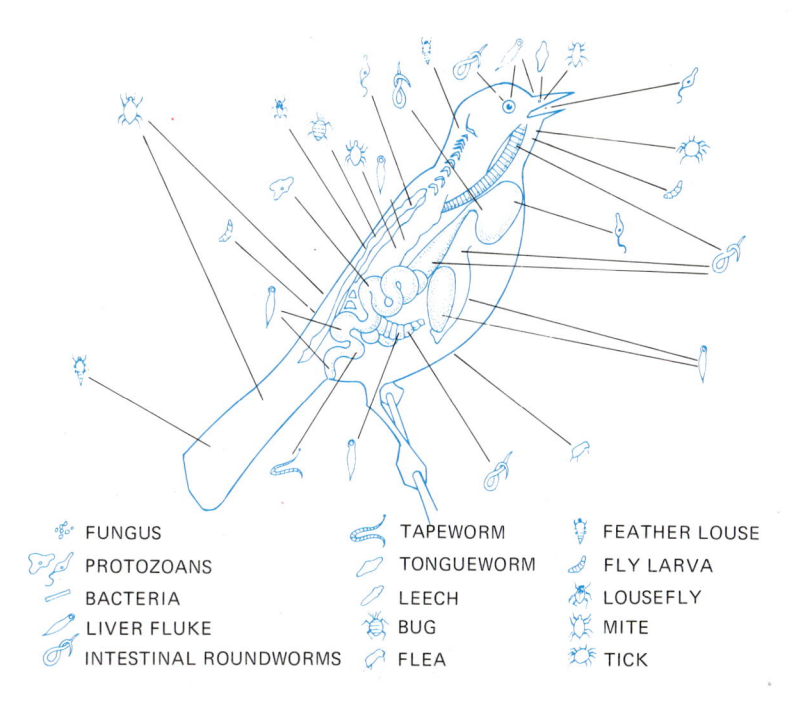

Figure 2.5: In nature, you seldom find one animal or plant by itself. Most plants and animals have millions of individual living creatures on and in their bodies. So do you!

FUNGUS
PROTOZOANS
BACTERIA
LIVER FLUKE
INTESTINAL ROUNDWORMS
TAPEWORM
TONGUEWORM
LEECH
BUG
FLEA
FEATHER LOUSE
FLY LARVA
LOUSEFLY
MITE
TICK

the available niches are usually filled. Nature has few vacant niches.

People create niches. For centuries, humans have been destroying the plants in their natural environment and have been replacing them with crops. There is a good reason why people do this. We all need the food and other raw materials that are produced by **agriculture.** But in replacing the natural environment with crops, humans created a number of new niches. These new niches are in fields of crops such as corn, wheat, cotton, rice, and potatoes.

By creating these new niches, people have created problems for themselves. Why? Because it is very likely that these new niches will be filled. When a new niche is filled, the new populations begin to compete with us for our crops. Such populations, when they are large enough to be troublesome, are called **pests.** Many pests are insects. However, there are a number of other animals, plants, and microscopic forms of life that are also pests.

agriculture: most simply, farming, including the raising of livestock

pest: a plant or an animal whose population is so large that it is a nuisance to people

LESSON REVIEW *(Think. There may be more than one answer.)*

1. Which one or more of the following best describes one population?
    a. all the insects within 1.6 kilometers (1 mile) of your school

23

Populations and Ecosystems

      b. all the ladybird beetles ("ladybugs") within 1.6 kilometers (1 mile) of your school
      c. all the ladybird beetles and green-winged grasshoppers within 1.6 kilometers (1 mile) of your school
      d. all the beetles and grasshoppers within 1.6 kilometers (1 mile) of your school

2. Which one or more of the following could be a habitat for some population?
      a. the oxygen in your classroom
      b. the leaves of a maple tree
      c. the courtship dance of the sharp-tailed grouse
      d. the skin of a dog

3. A bird
      a. can never be a habitat for a population.
      b. can be a habitat for many populations.
      c. provides several niches for other populations.
      d. can never provide a niche for a population.

4. Which one or more of the following could be a part of a population's niche?
      a. sharp-tailed grouse doing a courtship dance
      b. salmon swimming up a river to breed
      c. robins on the ground listening and searching for earthworms
      d. mother lions lying down to nurse their kittens

5. Which of the following terms best matches each phrase?
      a. population   e. terrestrial habitat
      b. habitat   f. niche
      c. marine habitat   g. agriculture
      d. freshwater habitat h. pest

_____ a population's total way of life

_____ an ocean area where populations of living things have their home

_____ a group of plants of one kind growing in a certain area

_____ the planned growing of food crops

_____ a nonmarine body of water that is a home to populations of living things

_____ the area where a plant or an animal population lives

_____ a plant or an animal that is a nuisance to humans

_____ a land area where populations of living things make their home

KEY WORDS

habitat (p. 20)   terrestrial habitat (p. 20)   pest (p. 23)
marine habitat (p. 20)   niche (p. 21)
freshwater habitat (p. 20) agriculture (p. 23)

Populations

POPULATIONS AND PESTICIDES

What do people do about pests? You already know the answer to that question. We are constantly at war with pests of all kinds. What are the weapons in this war? There are several. One of the most important weapons is the **pesticide.** Pesticide means "pest killer."

pesticide: a chemical used to kill pests

Pesticides create more problems. Pesticides are chemical poisons. These poisons can cause illness or death in animal populations other than those against which they are used. Even people are in danger from the use of some pesticides. The problem of possible illness or death from a pesticide is an obvious one. The use of pesticides creates other problems that are not so easy to recognize.

Pesticides help create "tougher" pests. In the early 1970s, there was a popular song called "A Boy Named Sue." Perhaps you have heard it. It is about a father who named his son "Sue." The father then left and never returned. "Sue" grew to manhood defending himself against boys and men who made fun of his name. Later, when "Sue" was a grown man, he found his father. He wanted to kill his father for giving him a name that had caused him so much trouble. He changed his mind after hearing his father's explanation.

Figure 2.6: The individuals in a population may look alike, but *they are not alike*. In a harsh environment, some will survive and reproduce. The survivors have the niche to themselves.

A varied population at the time of spraying

Two days later — those that resisted pesticide are still alive

Next year — most of the insects inherited resistance from those that survived

25

His father explained that he had known that his son would have to grow up without a father. He named him "Sue" because he knew that the boy would be forced to live a much harder life with such a name. He also knew that the hard life would help develop the boy into a much "tougher" man. That way, the boy could better overcome the handicap of not having a father.

Now let us apply the idea from "A Boy Named Sue" to pests and pesticides. Suppose we have an apple orchard. We want to kill a population of insects that is eating our apples. We will use the normal procedure to destroy them. We will spray the orchard with a pesticide. What happens?

The pesticide will probably kill most of the pest insects. But note that we used the word "most." Even though insects in a population are all of the same kind, the individuals are not exactly alike. **It is rare for any two individuals to be exactly alike.** This means that there are probably going to be some individuals in the population that will not be killed by the pesticide. Those individual insects that can resist the pesticide will remain alive. They will also continue to produce offspring. Then what happens?

The offspring of the insects that survive will probably inherit their parents' resistance to the poison. More important, the offspring have the whole niche to themselves. All of the competitors that were once there have been destroyed. So what kind of population do we have in our orchard next year?

If we use the words from the song, we would have a "tougher" population. With the pesticides, we produce a "tougher" population in about the same way as the father who named his boy "Sue." We created a harsh environment for the pests. By doing so, we guaranteed that the population, if it survived, would be more difficult to destroy!

Vacant niches can also cause problems. Suppose we spray an orchard and manage to kill all the pest insects. Would that solve our problems? The answer is no. All that would do is invite more problems.

Remember this: **Vacant niches do not last long in nature.** If we completely destroy a population of pests in our orchard, we can usually be sure of two things. First, some other population will eventually fill the vacant niche. Second, the population that does fill the vacant niche will probably be able to survive whatever killed the original population.

There are many cases where people created problems for themselves by creating vacant niches. The case of the codling (**kŏd′**ling) moth is a good one. The codling moth used to be a serious pest in some apple orchards. Pesticide spraying greatly reduced these populations. But this created a vacant niche that was filled by the orchard mite. Now apple growers are fighting even harder to control the orchard mite.

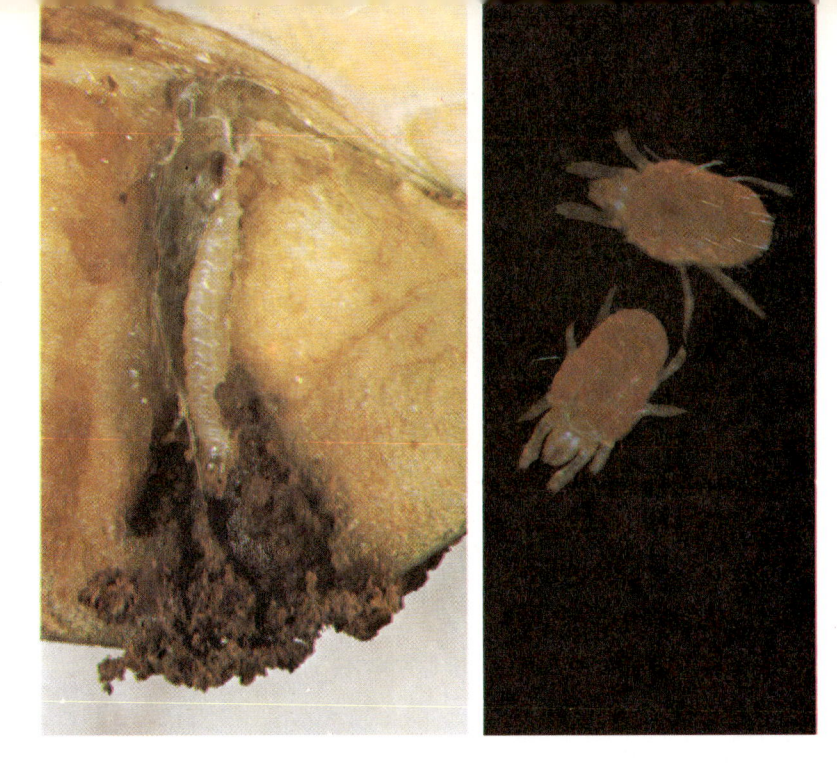

Figure 2.7: The larva of the codling moth (left) is a serious pest in apple orchards. If it is removed from its niche, the niche may be filled with orchard mites (right). The mites are harder to control.

Another example involves **antibiotics** (ăn'tē bī ŏt'iks) and you. Antibiotics are somewhat like pesticides. They are chemicals used to kill some bacteria that cause disease. During the years that antibiotics have been used, some of the bacteria treated with them stayed alive. The offspring of these bacteria are now living and causing diseases. Like the boy named "Sue," the bacteria populations that survived are "tougher." They are more difficult to kill with antibiotics. Even worse, some bacteria that used to be fairly harmless are now serious agents of disease.

Now you may be wondering: Are antibiotics bad? Most life scientists would probably say no. Antibiotics have saved millions of lives. Perhaps they have even saved your own. Doctors must continue to use them. What you must remember is this: We are paying a price for using antibiotics. As they are used over and over through the years, "tougher" bacteria populations will continue to develop. These populations will continue to cause trouble for us.

antibiotic:
a natural or laboratory-made chemical used to kill disease-causing bacteria

LESSON REVIEW *(Think. There may be more than one answer.)*

1. Within any population of living things
    a. the individuals are exactly alike.
    b. most of the individuals are exactly alike.
    c. some of the individuals are exactly alike.
    d. rarely any of the individuals are exactly alike.

2. When a pesticide is used against a pest population
   a. most of the pests will probably be killed.
   b. some of the pests will probably survive.
   c. the niche occupied by the pests could become vacant.
   d. the habitat of the population will be destroyed.

3. Some pests in a population survive after being sprayed with a pesticide.
   a. They cannot live in their niche because the habitat has been destroyed.
   b. They will have the niche of the whole population all to themselves.
   c. Their offspring may survive if they, too, are sprayed with the same pesticide.
   d. The niche will eventually be filled with their offspring.

4. All the pests in a population are killed by a pesticide.
   a. As a result, a vacant niche will be created.
   b. There will never be any more pests in that population's niche.
   c. The vacant niche will probably be filled with another population.
   d. Any new population in the niche will probably be more resistant to the pesticide that killed the original population.

5. In your own words, describe what is meant by:
   a. a pesticide.
   b. an antibiotic.

KEY WORDS

pesticide (p. 25)   antibiotic (p. 27)

HOW CAN PESTS BE CONTROLLED?

If pesticides are bad, what can we do to control pests? Life scientists do not know enough about pest control to answer this question completely. But there are other methods that are effective.

"Flyswatter" and "shotgun" methods are least desirable. Two common names are often applied to certain pest-control methods. By the "flyswatter" approach, pests are killed by the quickest and easiest method. This usually means that chemical sprays are used. The problem with the "flyswatter" approach is that we always need more and bigger flyswatters. For example, some apple growers must now spray their orchards twelve times a year, and with as many as five chemicals! Then they still worry about pest outbreaks.

The "shotgun" approach to pest control applies when broad-spectrum poisons are used. *Broad-spectrum* means that the poison "kills a number of different kinds of individuals." DDT is a broad-spectrum pesticide that was once widely used in the United States to kill insects. One main problem with DDT is that it kills harmless or helpful insects as well as pests. In this way, it creates new vacant niches. For this and other reasons, the use of DDT is now banned in the United States. (It can be used in an emergency, however.) Its use is not banned in other countries, where tons of it are still sprayed each year.

How are populations controlled in nature? Pests cause problems for people mainly because they are present in large numbers. A pest outbreak is often called a **population explosion.** Pest-control methods can be best understood after we consider natural population explosions.

Population explosions are very rare in tropical rain forests. They are fairly common in arctic regions. Why are they rare in one habitat and fairly common in another? There are a great number of different populations in a tropical rain forest. We say that there is great **diversity** in this kind of habitat. There are few different populations in an arctic habitat. There is little diversity in the Arctic.

Picture yourself as a small animal. If you live in a tropical rain forest, there are many different kinds of animals waiting to eat you. If the population of one of your enemies declines, your chances of survival are not much improved. You are still likely to be eaten before you can reproduce and leave more offspring in your habitat. Now picture yourself as a small animal in a very

population explosion:
a rapid and great increase in the population of a certain living thing

diversity:
most simply, variety. In life science, diversity is typical of a habitat with many different kinds of populations.

Figure 2.8: Lemmings live in an arctic habitat, where population explosions are fairly common. For unknown reasons, the individuals of a population may migrate—in all directions. Their paths may take them across rivers or to the sea, where they drown.

Figure 2.9: On most farms, such as this one where lettuce is growing, natural diversity is carefully destroyed. This is a good way to bring about population explosions of pests.

cold habitat. There may be only one animal that is likely to eat you. What if the population of that animal gets smaller for some reason? Then you and others like you will multiply rapidly. This may cause a population explosion within your population.

Diversity in nature helps control population size. Let us apply what we have learned about diversity to pests. For example, when farmers plant crops, they carefully rid their fields of all other plants. They usually spray fencerows and hedgerows to get rid of weeds. By doing this, farmers destroy habitats and niches for a variety of animals that might eat the pests in their fields.

When a farmer plants crops, usually only one kind of seed is planted in a single field. This practice of planting a single crop is called **monoculture.** With monoculture, a pest has an unlimited supply of food. It is also unlikely that there will be other animals in the habitat that will eat the pest.

Pest problems are created because farmers carefully *destroy the diversity* in their agricultural environment. We can say that another way. Farmers use tractors and machinery and sprays to create an environment more like the Arctic and less like a tropical rain forest.

Biological control may be used on pests. Life scientists use the term **biological control** when pests are eaten or destroyed by natural enemies. Biological control works best against pests that have crossed a border into new territory. A good

monoculture:
the practice of planting only one kind of crop in a field

biological control:
a pest-control method in which pests are killed by natural enemies

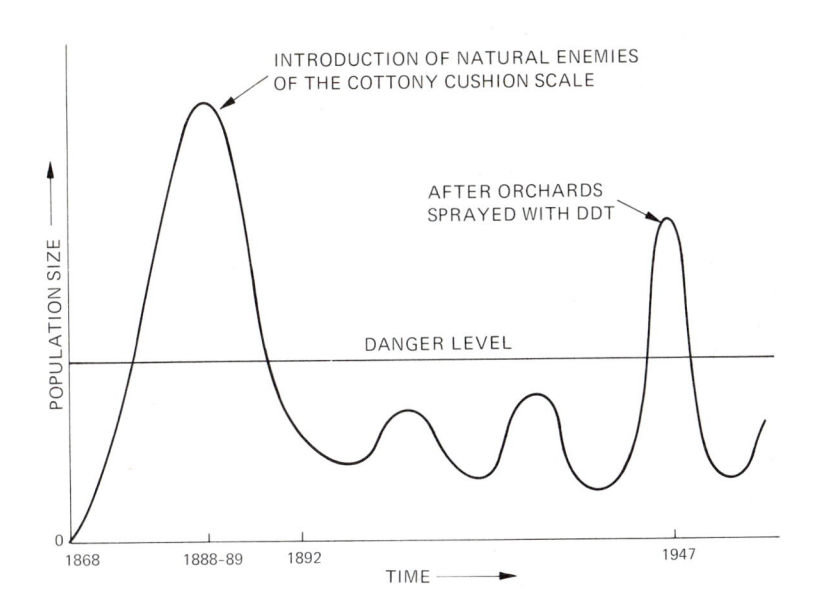

Figure 2.10: This graph shows how a natural enemy can be used to control a pest population. Note what happened in 1947, when the orange groves were sprayed with DDT. The chemical killed the ladybird beetles, natural enemies of the cottony cushion scale. For nearly 60 years, the beetles had kept the scale population in check.

Figure 2.11: The ladybird beetle is a useful insect in home gardens. In fact, many gardeners buy a supply of the beetles to keep aphids (ā'fĭd) under control in flower gardens. Here you see a ladybird beetle on a soon-to-blossom flower. If you look closely, you can see a pale-green aphid on the upper, right side of the unopened flower. Gardeners who use ladybird beetles to control aphids do not need to spray their flower gardens to be rid of that pest.

example is that of the cottony cushion scale. This is an insect pest that was accidentally brought to California in 1868. Its native habitat is in Australia. When it arrived in California, it left behind all of its natural enemies and diseases. The pest found a new home in California's citrus crops. Within 20 years, a population explosion of the scale occurred. And it almost destroyed the California citrus industry.

In 1888, two of the scale's natural enemies were imported from Australia and released in its new environment. Both were insects. One, an Australian ladybird beetle, proved to be extremely effective. Within months, the scale population in the citrus groves was reduced to a point where it was no longer a pest. The orange shipment from one county—Los Angeles County—jumped from 700 to 2000 carloads in 1 year. As you can see from the graph in Figure 2.10, there was one more population explosion of the scale—in 1947. This was caused when DDT was sprayed on the citrus trees, killing the ladybird beetles.

Insect attractants can be used to control pests. Some female insects give off an odor that attracts males during breeding season. The odor is caused by chemicals produced by the female insect. The chemicals are called **attractants.** The attractants of some insect pests have been produced in chemical laboratories. Some of these have been used successfully as bait in traps. Male insects lured to the traps die when they get inside.

attractant:
in some female insects, a chemical with an odor that attracts males during the breeding season

Populations and Ecosystems

Laboratory Activity

Can You Attract Insects with Food?

PURPOSE

Insects have a great ability to detect odors and to follow them to their source. In this investigation, you will observe insects in action. You will experiment to see how great their ability to detect odors is.

MATERIALS

short jars with lids
plastic sandwich bags
fingernail polish of different colors
straight pins
knife

cardboard square
watch with a second hand
canned dog or cat food
can opener
fresh kidney or liver

PROCEDURE A

Can you attract insects with food?

1. Place some meat in short jars. Use canned dog or cat food in some. Use fresh kidney or liver in others. Cut the meat into small pieces. (**Caution:** Be careful with the knife.) Place the lids on the jars.
2. Place the jars outdoors. Note and record the time. Remove the lids.
3. Watch the jars to see how soon an insect finds the food. What kind of insect appears first? How much time passes before each insect appears? Which food seems to be most attractive? Record your observations.

PROCEDURE B

Will an insect come back after being captured once? If so, from how far?

1. Capture some of the insects. To do that, place the lids on the jars when the insects are inside. Transfer the insects to individual plastic sandwich bags. Roll the top of each bag. Secure each one with a paper clip, as shown.
2. Lay a bag with an insect in it on a square of cardboard. Trap the insect in a small section of the bag. Use straight pins to do that, as shown.

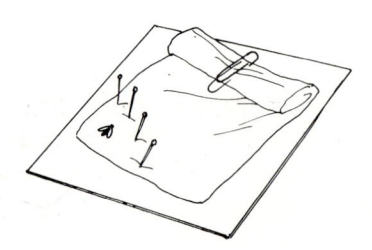

Populations

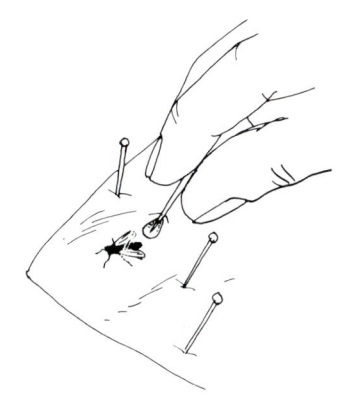

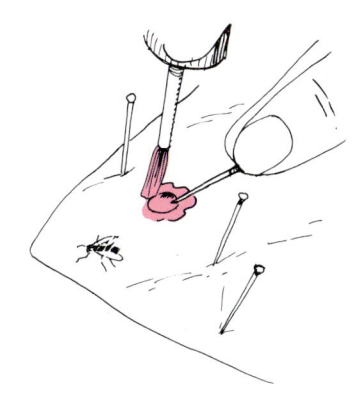

3. Make a hole in the bag right behind one of the insect's wings. Use a pin to do that, as shown. (**Caution:** Do not stick the insect.)
4. Place a dab of fingernail polish over the hole. Spread it back and forth with a pin, as shown.

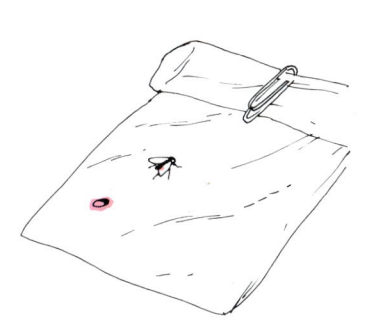

5. Move the plastic so the hole will be over one of the insect's wings. The nail polish should mark the wing. One or both wings should be marked. For insects of the same kind, use different colored nail polish. Check to see that the wing is marked. To do that, remove the pins to let the insect move around inside the bag, as shown.
6. Release the marked insects at measured distances from the food. Note the time when each one is released. Record your data.
7. Which insects, if any, return? How long does it take them? From how far do they return? Record your observations.

QUESTIONS

1. Why do you think the insects were attracted to the food? Describe an experiment to test your answer.
2. How might the air temperature affect how insects are attracted to food? Describe an experiment to test your answer.

33

Radiation can be used to control pests.

Another promising pest-control method uses atomic radiation. Male pest insects are exposed to radiation (are irradiated). Then they are released into their population. The radiation destroys or damages their reproductive cells. But it does not affect their mating ability. Thus, when they are among the population, they compete for females along with the normal males. If an irradiated male mates with a female, there are two likely results: (1) the female will produce no offspring, or (2) the female's offspring will be defective in some way.

The release of irradiated males into a pest population is a fairly new method of pest control. But some of the results so far have been spectacular. The screwworm fly used to be a serious pest of cattle in Florida. Male screwworm flies were irradiated, then released. The females that bred with them failed to produce offspring. This caused a sudden drop in the population and solved a serious pest problem. The use of irradiated males shows promise of controlling the cabbage looper, tobacco budworm, pink bollworm, and corn earworm. All four insects are now serious pests that have not yet been effectively controlled by other methods.

Are pesticides necessary for controlling pests?

We should conclude our discussion of pest control with one question: Must we continue using pesticides? Most life scientists would probably answer yes to this question. But they would also insist that they be allowed to make further comments. Most would argue against using broad-spectrum pesticides. Such pesticides kill too many plants or animals. (We will have more to say about harmful pesticides in the next chapter.)

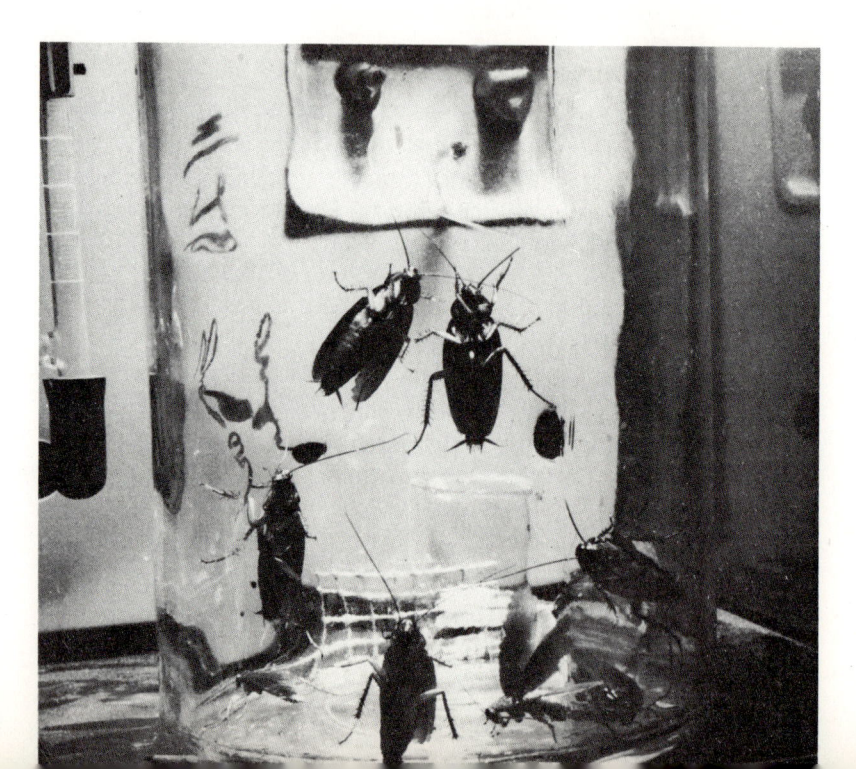

Figure 2.12: In a laboratory experiment, these male cockroaches are reacting to an extremely small amount of female cockroach sex attractant. The raising and spreading of wings by the males is a characteristic sexual response. For this experiment, life scientists needed 10,000 female cockroaches and 9 months to isolate 12.2 milligrams (0.0004 ounce) of pure attractant.

Figure 2.13: This life scientist with the United States Department of Agriculture is opening a box of 2000 sterile male screwworm flies. In practice, the box would be dropped from an airplane and open automatically. In 1972, a serious screwworm epidemic broke out in Texas. At that time, the USDA produced 200 million sterile screwworm flies a week, using radiation, in its sucessful effort to end the epidemic.

Life scientists also stress that pesticides should be used only as a temporary solution to our problems. Other methods should be used when possible. Most important, more research must be done to learn how populations are controlled in nature. With such knowledge, people should be able to use methods that nature has used successfully for millions of years.

LESSON REVIEW *(Think. There may be more than one answer.)*

1. A farmer sprayed a field with a pesticide. Later, it was found that bees and other helpful insects had been killed along with the pest insects. It is likely that the farmer used
    a. biological control.
    b. monoculture.
    c. a broad-spectrum pesticide.
    d. a pesticide like DDT.
2. Where there is diversity in nature,
    a. there are few populations.
    b. there are many different populations.
    c. population sizes usually will not increase or decrease greatly.
    d. population explosions and declines are more likely to occur.
3. A farmer's cornfield
    a. has little diversity of populations.
    b. has great diversity of populations.
    c. can be compared to the Arctic because of its small number of populations.
    d. can be compared to the tropics because of its large number of populations.

Populations and Ecosystems

4. Ladybird beetles, attractants, and irradiated males
   a. are used to control pests without pesticides.
   b. are examples of monoculture.
   c. are examples of broad-spectrum pesticides.
   d. are other ways that pesticides may be used.

5. Which of the following terms best matches each phrase?
   a. population explosion   d. biological control
   b. diversity   e. attractant
   c. monoculture

   ____ the growing of one crop over a large area

   ____ insect pest control by natural methods

   ____ a sudden, great increase in population size

   ____ a chemical produced by a female insect that attracts a male

   ____ many different kinds of populations

KEY WORDS

population explosion (p. 29)   biological control (p. 30)
diversity (p. 29)   attractant (p. 31)
monoculture (p. 30)

Applying What You Have Learned

Questions 1 and 2 are based on this paragraph:

Isle Royale is an island in Lake Superior, between Michigan and Canada. The moose population on this island was not being controlled by nature. The moose population became so large that it almost ate itself "out of house and home."

1. From what you have read about the moose population on Isle Royale, what niche do you think was vacant on that island?

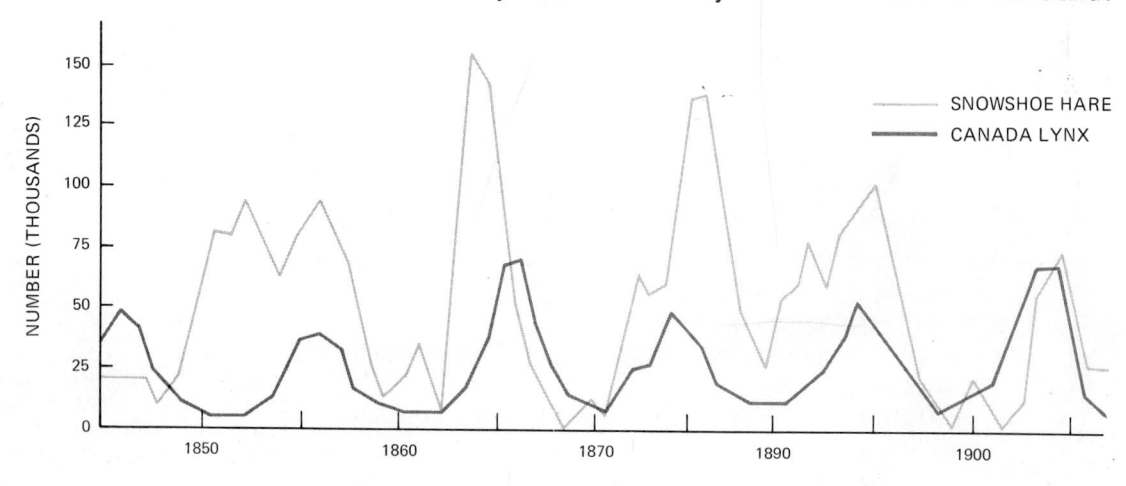

2. Can you think of a way that people could have filled this niche?
3. Name five pests (nonhuman, of course) that you know of in your own environment. Explain briefly how each is a problem.
4. For many years, a farmer who kept "clean fencerows" was thought to be a good farmer. Why might it be a better practice to allow weeds to grow along fencerows?
5. The following is a true statement: Individuals do not develop resistance to pesticides, but populations do. Why don't individuals develop resistance to pesticides?

Questions 6–11 are based on this paragraph:

The graph below was made from the records of the Hudson's Bay Company. This company bought furs from trappers. One line of the graph shows the number of horseshoe hare furs that they bought. The other shows the number of Canada lynx (lĭngks) that were bought.

6. In what year did the Hudson's Bay Company buy the largest number of snowshoe hare furs?
7. What was probably happening to the lynx population between the years 1862 and 1866?
8. What kind of general statement could you make about the purchase of snowshoe hare furs as compared to lynx furs?
9. The lynx is a natural enemy of the snowshoe hare. Therefore, life scientists think this graph of fur purchases shows the actual population changes of these two animals. If that is so, about how often does the snowshoe hare population reach maximum size?
10. When there is a population rise of hare and lynx at about the same time, which population increases first?
11. When there is a population decline of both the hare and the lynx, which population usually declines first?
12. The use of chemical attractants and radiation are really forms of biological pest control. Explain why that is so.

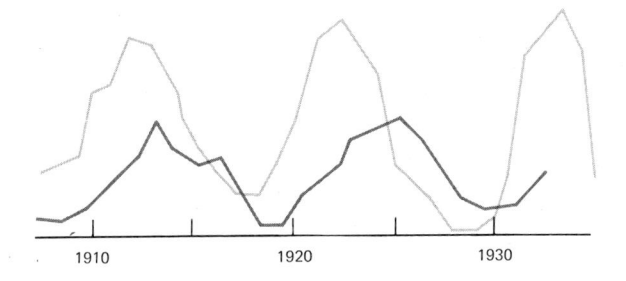

This graph shows the number of snowshoe hare furs and Canada lynx furs bought by the Hudson's Bay Company over a period of about 90 years.

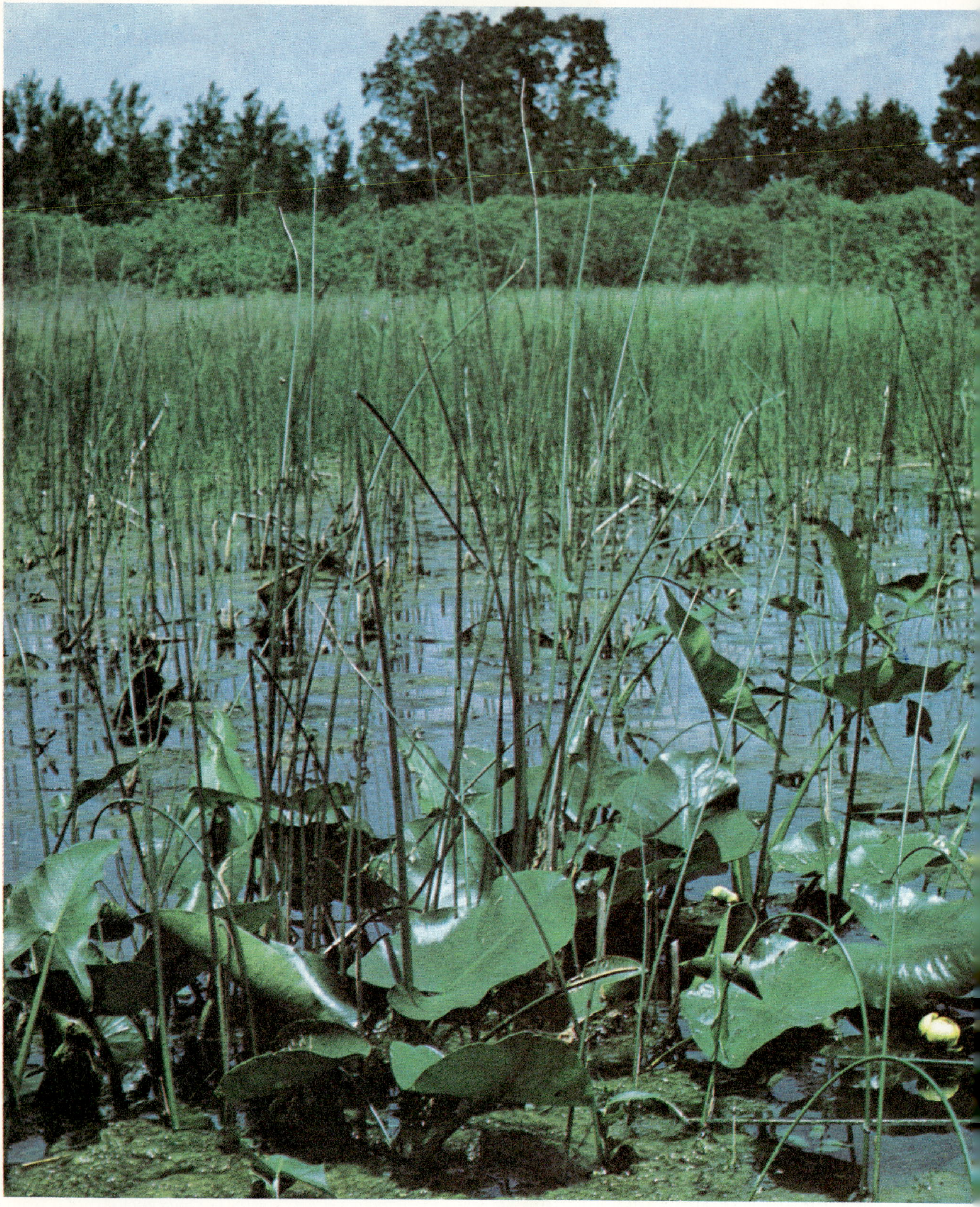

Figure 3.1: All the populations living in or near this pond belong to the pond community.

Chapter 3

ECOSYSTEMS

Some life scientists are called ecologists. They work in a specialized field of life science called **ecology** (ĭ kŏl′ə jē). What is ecology? One definition is this: Ecology is the study of *ecosystems* (ĕk′ō sys′təmz). In this chapter you will learn about ecosystems. You will learn what they are. And you will learn why the ecologist is a very important life scientist.

ecology:
the study of ecosystems

THE EARTH IS AN ECOSYSTEM

The community is a part of an ecosystem. Before you can understand the ecosystem, you must first understand the term **community**. A community is a group of populations that lives in a particular area. We normally name a community by the type of habitat in which the populations are living. For example, we would call the populations living in or near a pond a *pond community*. A *forest community* would include all of the populations living in or near a forest.

community:
a group of populations that lives in a particular area

Populations and Ecosystems

Communities can exist within communities. For example, think of all the populations in a forest as one single community—the forest community. Within the forest there are many different habitats. The soil, for example, is one habitat. It has many populations living in it. Thus, we could refer to these populations as the *soil community*. We could also refer to communities within the forest that live in caves, mud puddles, or rotten logs.

The ecosystem has two parts. The community is the *living* part of an ecosystem. The other part of the ecosystem is the **physical environment.** What is the physical environment? One way to answer such a question is to think of all the things in an area such as a forest that are *not* alive. This includes the gases in the air. It includes all of the water and minerals in the soil. It also includes all of the dead and decaying materials, such as old leaves and animal bodies.

physical environment:
any part of an ecosystem that is not alive

Now we can answer the question: What is an ecosystem? An **ecosystem** is a community interacting with its physical environment. *Interacting* means that "the community affects the physical environment and the physical environment affects the community."

ecosystem:
a community together with its physical environment

For example, think of a forest. Only now think of the forest as an ecosystem. To do this, think of what the forest community does to the physical environment. Think also of what the physical environment does to the forest community.

First, what does the forest community do to the physical environment? It does many things. For example, think what populations of earthworms, ground squirrels, chipmunks, and mice do to the soil. All of these animals dig or bore through the soil. They create holes and tunnels in it.

Now think what these and other animals do to the air. Animals take oxygen from the air when they breathe. They also give off carbon dioxide as a waste product. Plants affect the physical environment, too. Like animals, they breathe by taking in oxy-

Figure 3.2: The varying widths of tree rings show how the physical environment affects a forest community from year to year.

Figure 3.3: This tree growing out of rock shows how a living thing can affect the physical environment.

gen and releasing carbon dioxide. (When plants make food, though, they use carbon dioxide from the air and release oxygen.)

Plants also change the soil. They bore into it as they grow. They take minerals and water from it. Some plants even break rocks when they grow. (See Figure 3.3.)

What does the physical environment do to the forest community? We can answer that with another question: What happens to a forest community when there is plenty of moisture and when the air is warm? The plants grow and turn green. The animals are active. They feed and reproduce. Under these conditions the forest is alive with activity. Now think what happens when there is little moisture, or when the air is cold. Low moisture can slow down or stop growth. Extremely low moisture can cause death. Low temperature does the same thing. Just think of the changes that occur in a forest community during the autumn months.

The biosphere is the largest community. Think again about what a community is. It is defined as those populations living within a particular area. There is always one problem with this definition. Few populations live within a single area

Figure 3.4: Part of the west coast of North America can be seen in this *Apollo 10* view of nearly half of the largest community—the biosphere. This community includes all life on Earth. The biosphere and the physical environment interacting with it is the ecosphere.

biosphere:
the community of *all* living things on Earth

ecosphere:
the biosphere interacting with the Earth's total physical environment

that can be accurately described. Many animal populations move from one area to another.

Is there a community for which the area can be accurately described? Yes, there is one. This is the largest community. It is called the **biosphere.** The area of this community is the entire surface of the Earth. The biosphere is the community that includes all forms of life on Earth.

The ecosphere is the largest ecosystem. It is difficult to define the area of any single ecosystem. For example, air does not stay within the boundaries of a single forest. It circulates and spreads over the entire surface of the Earth. Thus, it is useful to think of the biosphere—the largest community—interacting with the Earth's total physical environment. This is the largest ecosystem. It is called the **ecosphere.**

LESSON REVIEW *(Think. There may be more than one answer.)*

1. Which describes a community?
    a. all the mice in a farmer's barn
    b. all the trees and bushes in a forest
    c. all the living things in Wahoo, Nebraska
    d. all the animals in the San Diego Zoo

2. A mouse digs a hole, winter comes, and the mouse hibernates in the hole. This is an example of
    a. how community members affect their environment.
    b. how the physical environment affects members of a community.
    c. interaction in an ecosystem.
    d. how both parts of an ecosystem can affect one another.

3. You are one member of
    a. a community.
    b. an ecosystem.
    c. the biosphere.
    d. the ecosphere.

4. All living things on Earth belong to
    a. a community.
    b. an ecosystem.
    c. the biosphere.
    d. the ecosphere.

5. Which of the following words best matches each phrase?
    a. ecology   d. ecosystem
    b. community   e. biosphere
    c. physical environment f. ecosphere

    ____ the largest ecosystem

    ____ air, soil, water

    ____ the living things in a cornfield

    ____ all the living things on Earth

    ____ a pond community and its physical environment

    ____ the study of the interaction of living things and their environment

KEY WORDS

ecology (p. 39)   ecosystem (p. 40)
community (p. 39)   biosphere (p. 42)
physical environment (p. 40)   ecosphere (p. 42)

Populations and Ecosystems

PRODUCERS

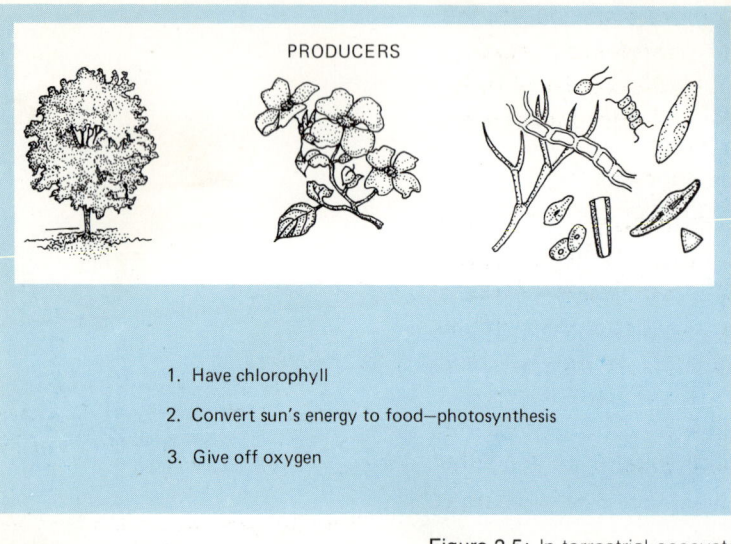

1. Have chlorophyll
2. Convert sun's energy to food—photosynthesis
3. Give off oxygen

Figure 3.5: In terrestrial ecosystems, the producers are visible. They are the green plants. In aquatic ecosystems, most of the producers are microscopic.

THE THREE BASIC NICHES

There are three basic niches in every ecosystem. Every population fits into one of the three niches.

Producers make food and release oxygen. Suppose someone were to ask you this question: What is a tree? You might answer: A tree is a large plant. Put the same question to an ecologist. He or she might answer: A tree is a large producer. Anyone who studies ecosystems is likely to think first of an individual's or a population's niche within an ecosystem. A tree is a plant, but its niche is that of a producer.

What is a producer? A **producer** is any living thing that makes food and releases oxygen. The process by which food is made and oxygen is released is called **photosynthesis.** In an individual like a tree, only certain cells carry on the process of photosynthesis. These cells are in the leaves of the tree. In smaller plants, photosynthesis may take place in cells of the stem as well as in the leaves.

During photosynthesis, carbon dioxide and water are converted to food. Sunlight is the source of energy. Oxygen is a byproduct. How important are the producers in an ecosystem? Think what it would be like without them. It is easy to see that all the food would be used up. But also realize that the oxygen in our atmosphere would be used up and not replaced.

Who are the producers in an ecosystem? In a terrestrial ecosystem, such as a forest or a grassland, the producers are

producer: a living thing that makes food and releases oxygen. Producers make up one of the three basic niches in an ecosystem.

photosynthesis: the process by which certain individuals make food and produce oxygen

green plants. These producers are easy to recognize. Most of the producers in freshwater and marine ecosystems are small or microscopic in size. Many have bodies which consist of only a single cell. You will learn more about the different kinds of producers in later chapters.

Consumers are the animal populations. In every ecosystem there are animals that eat producers. Others eat products made by the producers. Many different kinds of insects occupy such niches. Grazing animals, such as the cow, buffalo, and deer, also occupy these niches. These animals are **consumers.** They are specifically called *first-order consumers.*

Some consumers eat other consumers. An example would be a robin that eats a grasshopper. The grasshopper is a first-order consumer. The robin would be called a *second-order consumer.*

Are there *third-* and *fourth-order consumers* in ecosystems? Usually, but they are always fewer in number than first- and second-order consumers. The robin that eats the grasshopper is not a good example. The robin is not likely to be eaten by another animal. But consider a weasel that has eaten a grasshopper. Within minutes, this weasel may find itself in the claws of an eagle or large hawk. In this case, the eagle or hawk would be called a third-order consumer.

The bald eagle is often a fourth-order consumer. It may dive to the water and catch and eat a large fish, such as a bass. That bass may have eaten a smaller fish. The smaller fish may have eaten an insect. The insect may have eaten plant material.

consumer: an animal that eats a producer or the product of a producer. Consumers make up one of the three basic niches in an ecosystem.

Figure 3.6: Most consumers in terrestrial ecosystems are animals. Many consumers in aquatic ecosystems are microscopic. How can you tell that the pronghorn antelope is a first-order consumer?

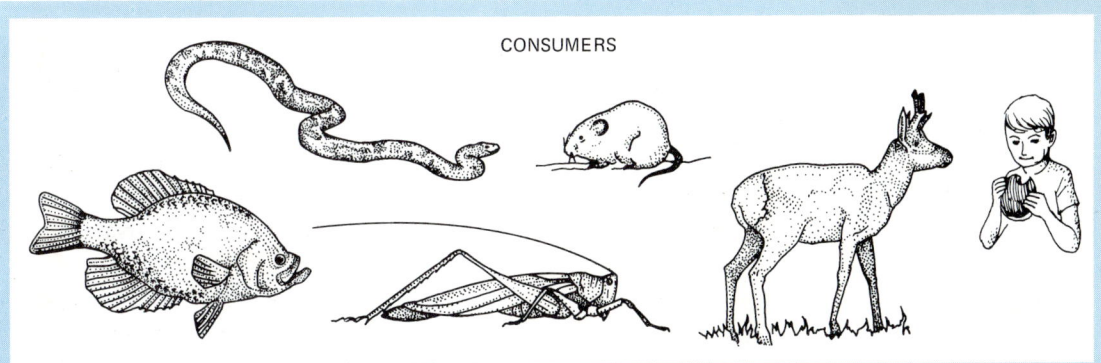

CONSUMERS

1. Are animals
2. They feed on producers (first-order consumers), or
3. They feed on animals that feed on producers (second-order consumers), or
4. They feed on animals that feed on animals that feed on producers (third-order consumers), or
5. They feed on animals that feed on animals that feed on animals that feed on producers (fourth-order consumers)

45

Populations and Ecosystems

Decomposers occupy the third basic niche. What happens to all the producers and consumers in any ecosystem? Are all of them eaten? Certainly not. Then what happens to them? In time they die. Then what happens to them? Their bodies do not remain in the ecosystem. If they did, by now there would be stacks of bodies hundreds of meters deep.

Decomposers rid ecosystems of dead producers and consumers. What are the **decomposers?** They are mostly microscopic bacteria and fungi (fŭn′jī). You will study about them in later chapters.

decomposer: a living thing that digests dead producers and consumers. Decomposers make up one of the three basic niches in an ecosystem.

How do decomposers do their job? Their bodies produce chemicals that *digest* the dead material. The digested materials end up as simple chemicals. Carbon dioxide is one such chemical. It is a gas released into the air. Other chemicals may remain in the soil. Decomposers use some of the chemicals to grow.

Decomposers have two important roles in any ecosystem. One is to rid the ecosystem of dead bodies. The second is to release basic chemicals. Carbon dioxide has already been mentioned. Let us use it as an example. All producers must have a supply of carbon dioxide. They take it from the air. During photosynthesis they use it to make food. Really, there is very little carbon dioxide in the air at any time. It would be used in a short time if it were not for the decomposers. Carbon dioxide taken from the air by producers is replaced by the actions of decomposers.

The decomposers release other important chemicals besides carbon dioxide. Nitrates and other nitrogen-containing chem-

Figure 3.7: The growths on this log are fungi. They will help to decompose the log, thus aiding its removal from the ecosystem. But the basic chemicals in the log will be recycled.

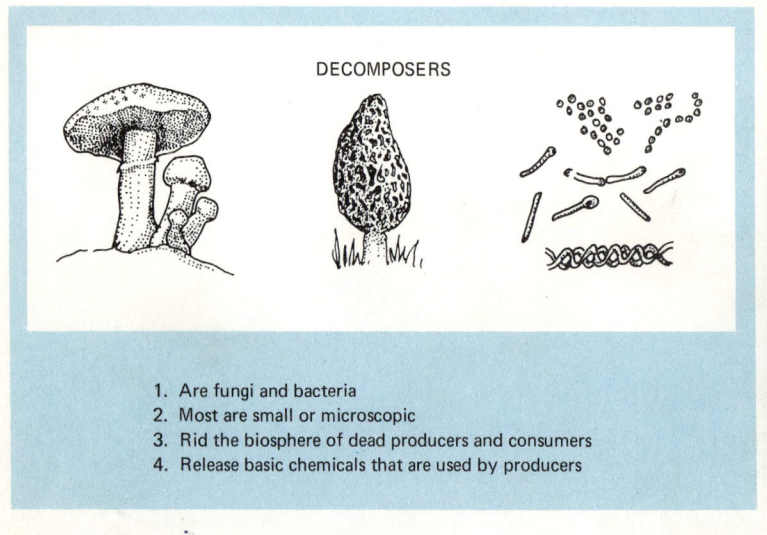

DECOMPOSERS

1. Are fungi and bacteria
2. Most are small or microscopic
3. Rid the biosphere of dead producers and consumers
4. Release basic chemicals that are used by producers

Ecosystems

icals are released and remain in the soil. Such chemicals are called **nutrients**. Nutrients are taken up by the roots of plants and used in making proteins and other plant food.

nutrient: most simply, food. Anything that is needed by a plant or an animal for growth.

LESSON REVIEW *(Think. There may be more than one answer.)*

1. In Figure 3.6,
    a. the producer is feeding upon the consumer.
    b. the decomposers are clearly visible.
    c. producers are present in greatest numbers.
    d. a consumer is feeding upon producers.

2. A producer
    a. is a green plant.
    b. eats consumers.
    c. carries on photosynthesis.
    d. may eat green plants.

3. A consumer
    a. may eat producers.
    b. may eat other consumers.
    c. may carry on photosynthesis.
    d. may eat living things that carry on photosynthesis.

4. A living thing that gives off oxygen and uses carbon dioxide may be
    a. a producer.
    b. a consumer.
    c. a decomposer.
    d. a green plant.

5. Decomposers
    a. are mostly small or microscopic.
    b. rid the biosphere of dead producers and consumers.
    c. cause basic chemicals to be released that are used by producers.
    d. are fungi and bacteria.

6. Which of the following terms best matches each phrase?
    a. producer   d. decomposer
    b. photosynthesis e. nutrient
    c. consumer

    __c__ a horse is an example
    __a__ an oak tree is an example
    __d__ organism that breaks down dead plants and animals
    __e__ chemical necessary for life
    __b__ process by which producers make food

47

Populations and Ecosystems

KEY WORDS

producer (p. 44)
photosynthesis (p. 44)
consumer (p. 45)

decomposer (p. 46)
nutrient (p. 47)

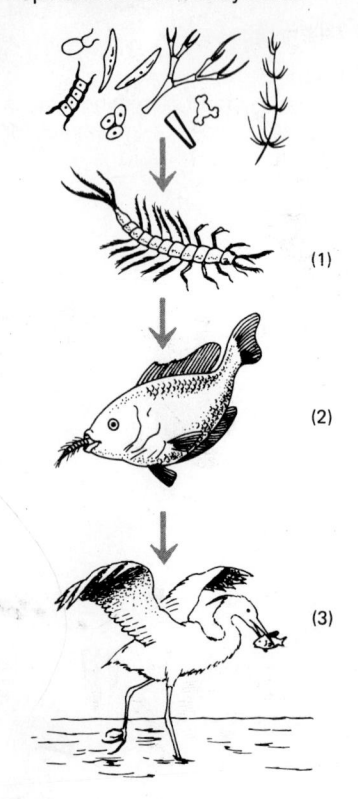

Figure 3.8: This is a food chain in a pond ecosystem. The insect, a mayfly nymph (1), is a first-order consumer. It feeds mostly on microscopic producers. The sunfish (2) is a second-order consumer. The great blue heron (3) is a third-order consumer.

INTERDEPENDENCY IN THE ECOSYSTEM

In your study of life science this year, you are going to learn many important facts about life. Of all these, the one you are about to learn may be the most important: **Every population in an ecosystem depends on other populations for its survival.**

How are populations dependent on each other?

Let us use the squirrel as an example to show how animal populations are dependent. First we can ask: What does the squirrel eat? Answer: Nuts. Next question: Where do the nuts come from? Answer: From trees. Final question: Could the squirrel live without food produced by trees? Answer: No. Now you can understand one way that a squirrel population is dependent on populations of trees.

But let us ask two more questions. First: What is the squirrel's habitat? Answer: A tree. Second question: Where does the squirrel population reproduce itself? Answer: In trees. Now you know two more ways that the squirrel population is dependent on trees. All together, the squirrel population depends upon tree populations for food, habitat, and reproduction. **All other animal populations depend upon plant populations for at least one of these three things: food, habitat, or reproduction.**

What about plants? How are they dependent upon other populations? Stop and think. Can a plant grow without nutrients in the soil? The answer is no. All plants must take in certain nutrients through their roots. You already know how the nutrients get into the soil. They get there through the actions of bacteria and fungi—the decomposers. Thus, plants depend on the different populations of decomposers.

What would happen to populations of decomposers if there were no plants or animals? There would be nothing for them to decompose. They would have no way of getting the nutrients that they need. Decomposers depend on both plant and animal populations for their food.

48

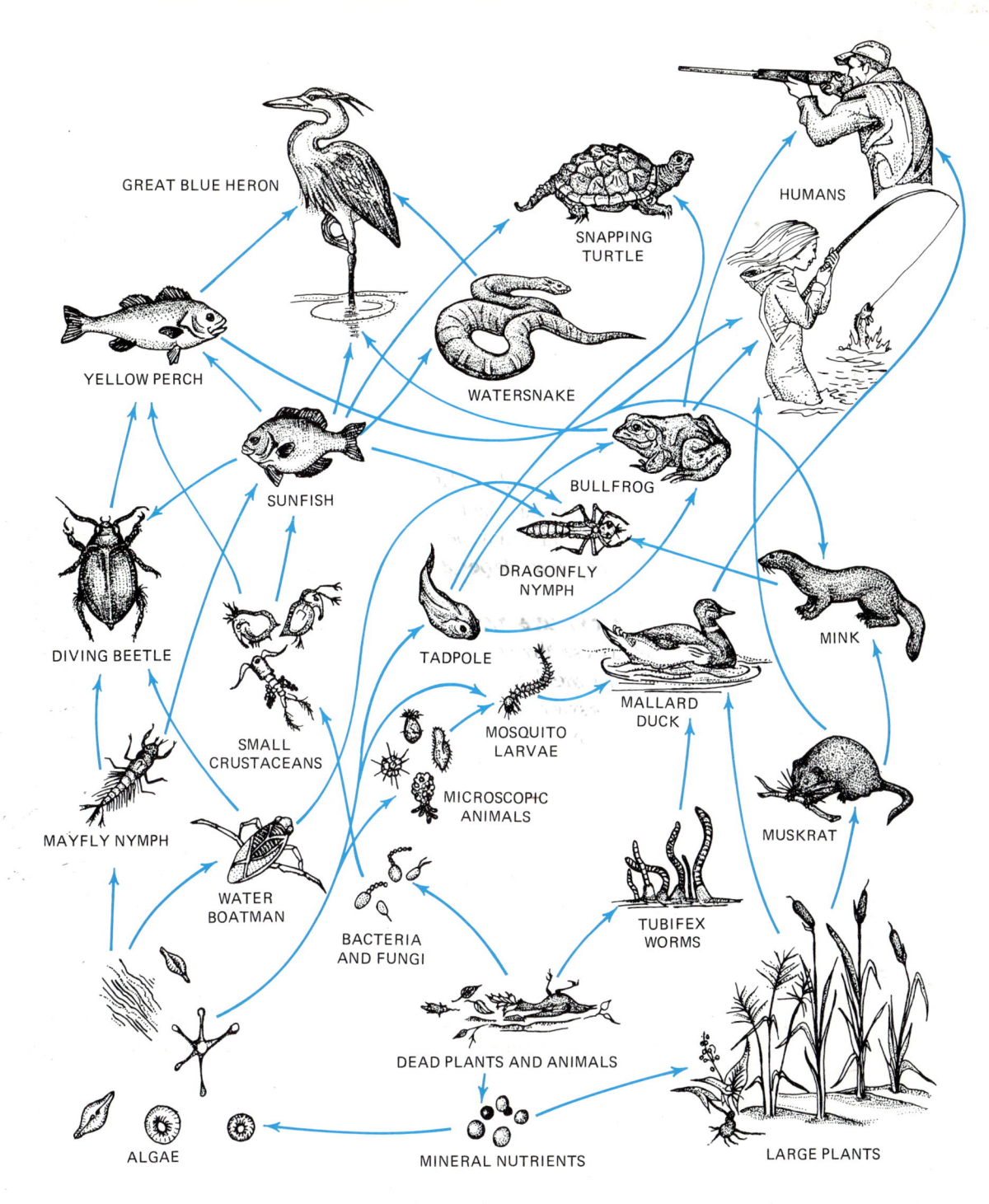

Figure 3.9: This food web shows who eats whom in a pond ecosystem.

What are food chains and food webs?

A food chain is a way of describing who eats whom in an ecosystem. Earlier in this chapter we used an example of a food chain. The grasshopper that ate the plant was then eaten by a weasel. The weasel

food chain:
a description of who eats whom in an ecosystem

49

Populations and Ecosystems

was eaten by an eagle. The plant, grasshopper, weasel, and eagle —listed in that order—tell us who is eaten by whom. This is a food chain.

The food chain is a useful idea. However, it may tell only part of the story. For example, the grasshopper may eat several different kinds of plants. Likewise, many animals besides weasels eat grasshoppers, and eagles eat animals other than weasels. Such a pattern is typical in most ecosystems. Usually, the food chain is used to describe who eats whom *most often*. If we can describe *all* the populations that feed on each other, we construct a **food web.** Think of a spider's web and you can understand why the term "web" is used. Like each part of a spider's web, each population is linked to others in several ways. The food web is a more complex way to describe interdependency in an ecosystem than a food chain. It is also more accurate.

food web:
a description of how all the populations in an ecosystem feed on each other

The food pyramid is a useful idea. There is another way of showing dependence relationships in an ecosystem. This is with a **food pyramid,** which may also be called the *pyramid of numbers*. The diagram in Figure 3.10 shows this idea better than it can be described in words. Here are the important facts that are expressed by the food pyramid: (1) In most ecosystems, the number of decomposers is much greater than the number of

food pyramid:
a way to show the relationship among decomposers, producers, and consumers in an ecosystem. Decomposers are at the base of the pyramid. High-order consumers are at its top.

Figure 3.10: Relationships in a grassland ecosystem are shown by this food pyramid.

50

producers. (2) The number of producers is greater than the number of consumers. (3) As consumers move farther along the food chain, their numbers decrease.

If we think in human terms, being an animal at the top of a food pyramid has rewards and it has problems. The reward, if one can think this way, is that such an animal is more likely to die of old age. The "top dog," as such an animal is often called, is not likely to be eaten by another animal. But the top dog also has less food available. Hunger and starvation can be problems. Now, because of long-lasting chemical poisons in their ecosystems, the top dogs have an even greater problem.

Persistent chemicals are in the ecosystem. A **persistent chemical** is one that is not broken down or changed for a long period of time. Poisonous chemicals of this kind are in our ecosystems. They are creating serious problems. Now that you know about food chains, food webs, and food pyramids, you can understand why.

DDT is a pesticide that has been used since 1945. At that time, it was a life saver. Back then, malaria was a disease that was the number one killer of people. Malaria was spread by mosquitoes. DDT was used to kill those mosquitoes. And DDT helped to wipe out malaria. Since that time, DDT has been widely used to kill other insect pests, such as those which destroy crops. Besides saving human lives it has also been a great help to agriculture.

DDT still is used in most countries of the world. But it is now against the law to use it in the United States without permission from the federal government. Why should DDT be illegal in the United States?

The problem with DDT—or any persistent chemical—is that it does not stay where it is put. If it is applied to a field of cotton, it does not stay in the field. It washes into rivers and streams. Then it moves into the oceans and throughout the world in ocean currents. DDT applied on a farm in Alabama or Brazil can be found later in the waters near England or Antarctica.

DDT and other persistent chemicals have now been spread throughout the world. However, they are present in extremely small amounts. In fact, it is sometimes difficult to measure their presence. Then why should they be a problem?

Food chains cause biological magnification. Very small amounts of DDT have not proven to be harmful to most living things. However, that only tells part of the story. When an animal eats food with DDT in it, the DDT may remain in the animal's body fat. That can create problems.

persistent chemical: a chemical that persists, or lasts unchanged, for a long period of time

Figure 3.11: Future generations may have to see our national bird, the bald eagle, in museums. It may be on its way to extinction because it is at the end of food chains in which the concentration of DDT has been magnified.

For example, suppose a person drank a glass of milk with a very small amount of DDT in it. That DDT could be stored in the person's body fat. The DDT in one glass of milk would not be a problem. But who drinks just one glass of milk? Some young people drink as many as 500 glasses of milk per year. What if each glass had the same amount of DDT in it? The DDT from 500 glasses could be stored in that person's body fat. Can that cause a health problem? No. There is no firm evidence now to prove that DDT in human food can create any health problem. But other organisms are not so lucky. Those who are at the end of long food chains—the top dogs—have problems. We'll use a freshwater food chain to explain.

Microscopic organisms are at the beginning of freshwater food chains. They may have an extremely small amount of DDT in their bodies. These microscopic organisms are eaten by insects. But an insect does not eat just one of them. It eats thousands. And in the process, the DDT from all the microscopic organisms becomes stored in the insect's body.

Next in the food chain is the small fish that eats the insect. But the small fish does not eat just one insect. It eats hundreds. In the process, all the DDT from those insects becomes stored in the fish's body. Then a big fish comes along and eats not one, but hundreds, of the smaller fish in its lifetime. DDT from all of them is stored in the big fish. Finally, at the end of the food chain, are one or more birds, such as the blue heron, pelican, or bald eagle. During their lifetimes, such birds will eat many large fish. In the process, they may build up amounts of DDT in their bodies that are 1000 times greater than the amount found in their environments.

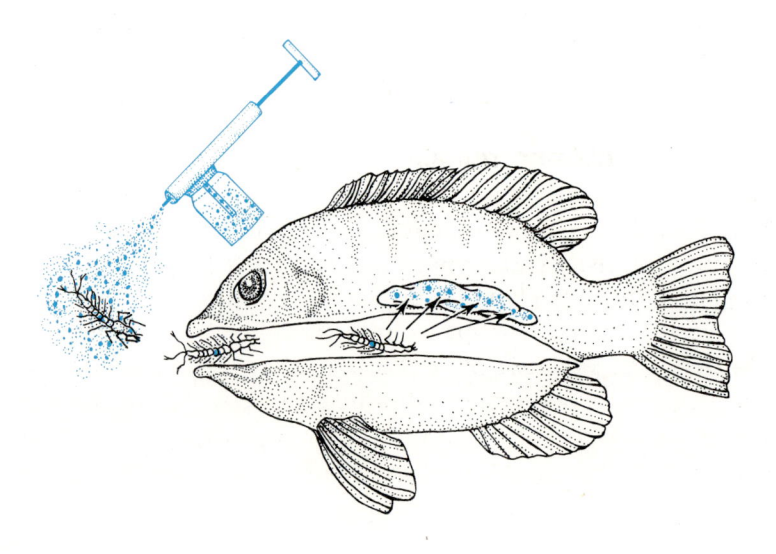

Figure 3.12: A sunfish concentrates the tiny amounts of DDT in the bodies of insects. DDT is stored in the fish's body fat.

52

Figure 3.13: The diagram shows how DDT can be concentrated, or magnified, through a food chain. The great blue heron is a "top dog."

The buildup of a persistent chemical through a food chain is called **biological magnification.** Now you can understand why life scientists worry about very small amounts of a persistent chemical in the environment. Through food chains, these very small amounts can build up in animals along the chains. This can mean disaster for the animals at or near the end of food chains. For example, large amounts of DDT in birds can cause them to lay defective eggs. The shells of such eggs are much thinner than normal. They crush when the parent bird sits on them. This problem probably has caused the severe decline observed in populations of birds such as hawks, falcons, eagles, herons, and pelicans. One bird, the peregrine (pĕr′ə grēn) falcon, almost has died out in the eastern United States. Life scientists blame the biological magnification of DDT and other chemicals as the major cause of its decline.

Two problems remain. There is no easy answer to the problem of harmful persistent chemicals. For example, suppose Canada, the United States, and Mexico agreed to ban their use in North America. Would that solve the problem? No. For example, that would not stop such countries as Argentina or Nicaragua, or India or Spain from using them. **And persistent chemicals do not recognize the borders of any country.** They travel almost everywhere—through the air, in the oceans—and need no passports. So one major problem is to get worldwide control of harmful persistent chemicals. So far this has not been possible.

biological magnification: the buildup of a persistent chemical as it passes through a food chain

Populations and Ecosystems

Laboratory Activity

Play the Surviving Pelican Game.

PURPOSE
This game will help show you how biological magnification of a persistent chemical occurs along a food chain in a freshwater ecosystem.

MATERIALS
decks of cards   paper and pencil or pen

PROCEDURE
1. Each student draws a pelican to use in the game.
2. Discard the jokers from a deck of cards. With pencil or pen label the 52 cards. Label 30 as algae, 15 as little fish, and 7 as big fish. In marking a card, disregard the numbers or letters on the individual card, as well as its suit.
3. Shuffle the deck and place it face down in the middle of the players. Any number can play, but four is recommended.
4. The game is played according to "who eats whom." In this game: (a) Algae eat nothing. (They are producers.) (b) Little fish eat algae. (c) Big fish eat only little fish. (d) Pelicans eat only big fish.
5. To start the game, the first player draws a card and turns it over. The card remains in front of the player if it is algae or a little fish. If it happens to be a big fish, the next player will not draw a card, but will use the next turn by having his or her pelican eat the big fish (take it), and *remove it from play*. (Put the big fish card aside.)
6. The next player turns over a card (if her or his pelican didn't eat). If the next player gets a little fish, and if there are algae in front of any players, the little fish must eat all of them (take them). This stack of cards is left in front of the player with the little fish on top.
7. If a player turns over a big fish, it must eat (and can only eat) all of the little fish that are in front of the other players—*along with all the algae that the little fish ate.*
8. After a big fish has eaten, the next person's pelican must eat whatever big fish are showing (and all the little fish and algae under them). These cards are taken from in front of the players and kept aside, out of the game, by the student whose pelican ate them.
9. After the cards are all played and taken, each person counts how many doses of pesticide (how many cards) her or his pelican has eaten. The pelican who has eaten the most has to drop out of the game. Then the surviving pelicans shuffle the cards and play as many other rounds as needed to see which one will be the final survivor.
10. Each class plays to see which one pelican in the class is the final survivor. That pelican can be displayed on the bulletin board, or a playoff can be held with the final survivors in other classes.

QUESTIONS
1. How does chance play a part in the survival of pelicans in the pelican game?
2. How may chance play a part in the survival of pelicans in the real world?
3. Assume that each algae absorbs one unit of chemical. Then, during the game, how much chemical does each little fish take in when it feeds? How much chemical does each big fish take in when it feeds? At the end of the game, how much chemical does each pelican contain?

54

A second problem is to make chemicals that are not persistent. For example, suppose a new pesticide breaks down shortly after it is used. If the breakdown products were harmless, their biological magnification would not matter. The production and use of nonpersistent chemicals is an ideal solution. Scientists are working right now on that type of solution. Yet there is no promise that such chemicals can be developed. Even if they can be, who is going to force all countries in the world to use them?

LESSON REVIEW *(Think. There may be more than one answer.)*

1. In Figure 3.9, the sunfish could be in which of the following food chains?
    a. algae—water boatman—diving beetle—yellow perch—sunfish
    b. dead plants and animals—small crustaceans—sunfish—yellow perch—great blue heron
    c. algae—mayfly nymph—sunfish—snapping turtle
    d. algae—water boatman—dragonfly nymph—sunfish—water snake

2. In *b* of question 1, the great blue heron would
    a. be a producer.
    b. be a "top dog."
    c. be the animal most likely to starve.
    d. probably have the highest amount of DDT in its body.

3. In Figure 3.9, which would most likely have the greatest amounts of DDT stored in their bodies?
    a. diving beetle, sunfish, dragonfly nymph, mink, human
    b. great blue heron, human, water snake, bullfrog, mink
    c. water snake, yellow perch, great blue heron, snapping turtle, human
    d. mayfly nymph, water boatman, bacteria and fungi, tubifex worms, muskrat

4. Most ecosystems have a food pyramid. It will show that there are
    a. more animals than plants.
    b. more plants than animals.
    c. more plants than decomposers.
    d. more decomposers than plants.

5. Which of the following terms best matches each phrase?
   a. food chain   d. persistent chemical
   b. food web   e. biological magnification
   c. food pyramid

   _____ a chemical that remains unchanged in an ecosystem for a long time

   _____ simplest way to show who eats whom in ecosystems

   _____ a way to show number relationships among decomposers, producers and consumers

   _____ fairly complete way to show who eats whom in an ecosystem

   _____ process that builds up a persistent chemical in higher members of food chains

KEY WORDS

food chain (p. 49)
food web (p. 50)
food pyramid (p. 50)

persistent chemical (p. 51)
biological magnification (p. 53)

Applying What You Have Learned

1. How could each of the following help to solve the problem caused by the biological magnification of harmful, persistent chemicals?
   a. congresspersons   c. farmers   e. international groups
   b. scientists   d. teachers   f. yourself

2. Refer to the food web on page 49. Use a separate sheet of paper. List each member of this food web in the niches listed below. Some members may fill more than one niche.
   Decomposers   Fourth-Order Consumers
   Producers   Fifth-Order Consumers
   First-Order Consumers   Sixth-Order Consumers
   Second-Order Consumers   Seventh-Order Consumers
   Third-Order Consumers

3. Place the following niches in their proper order on a food pyramid:
   a. producer   c. second-order consumer
   b. first-order consumer d. third-order consumer

4. For each level of the pyramid in question 3, select the best population size from the following choices: 10,000 individuals, 50 individuals, 1 individual, and 10 individuals.

5. Producers are able to trap the sun's energy and store it as food energy. Referring to the pyramid in question 3, do the second-order consumers have more or less food energy available to them than was available to the first-order consumers? Why?

6. A circular energy chain is shown below in Figure A. It shows how different populations depend upon each other for necessary energy and nutrients. Match the following individuals with the numbers in Figure B: hawk, snake, bacteria, cricket

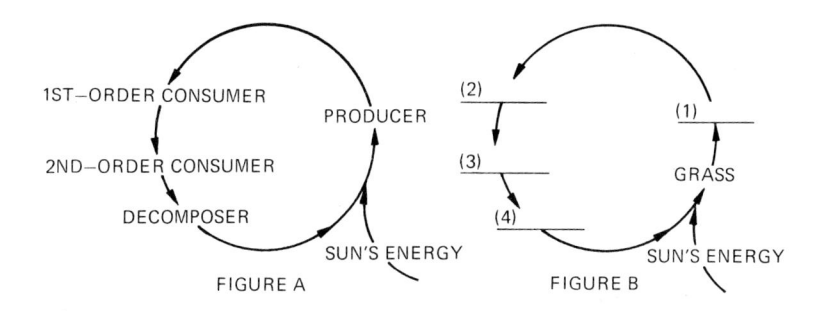

7. What do you think might happen if the first-order consumer niche in Figure A of question 6 became vacant?

8. Many harmful persistent chemicals are broad-spectrum pesticides. What other type(s) of pest-control measures might be used in place of them?

9. Why isn't our use of pesticides to improve our comfort always the best thing? Why then, don't we limit or do away with pesticides for this purpose?

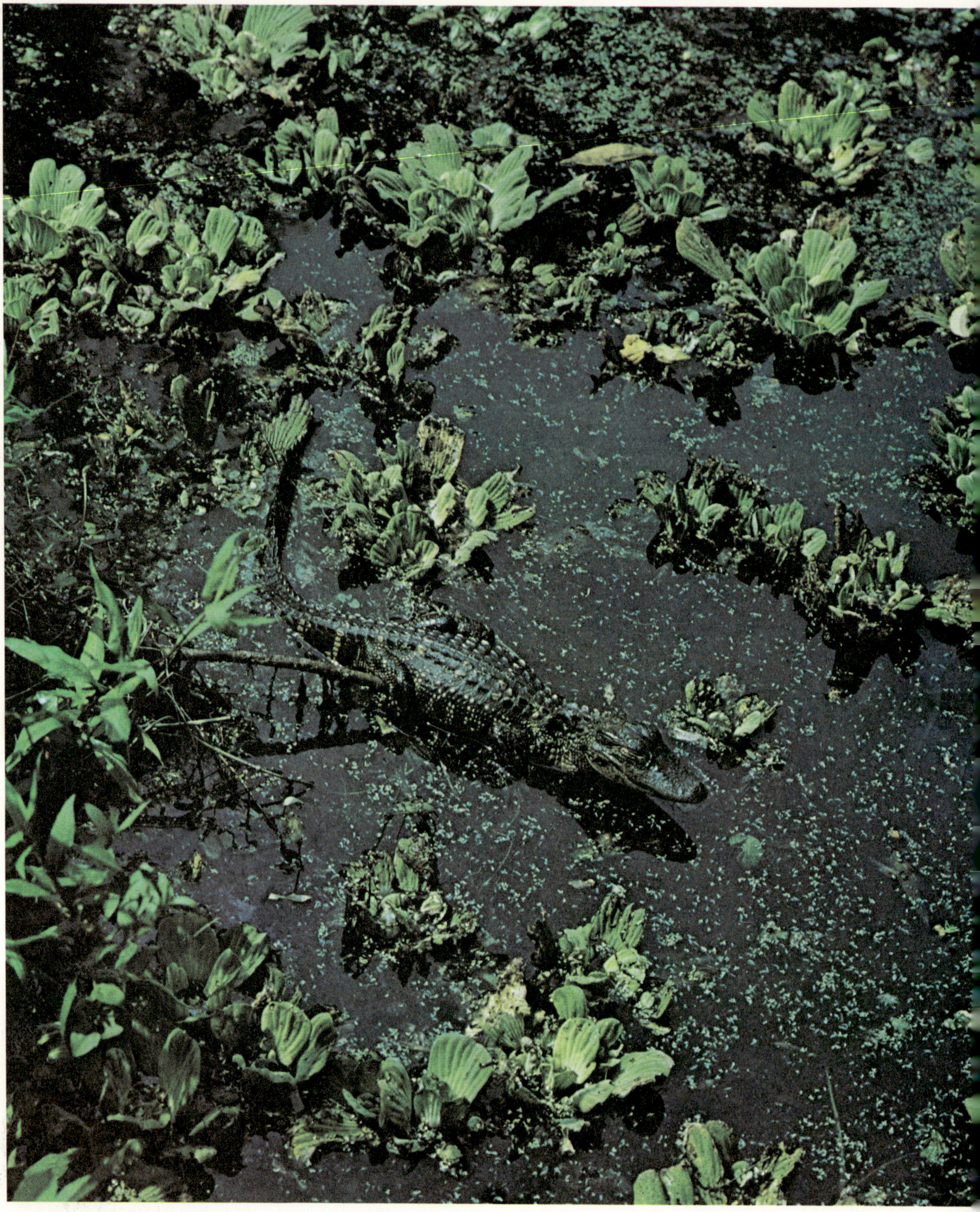

Figure 4.1: The habitat of this alligator is Florida's Corkscrew Swamp Sanctuary. Many swamps are fragile ecosystems that dry up if less water is available.

Chapter 4

FRESHWATER ECOSYSTEMS

In this chapter you will study freshwater ecosystems. This will include a study of their physical environments and of the population of things that live in them.

THE WATER

What is fresh water? The ecologist—the life scientist who studies ecosystems—would give you one description of fresh water. A mining prospector walking across a hot desert might give you another. Much of the water called *fresh water* by the ecologist would be refused by the thirsty prospector. To an ecologist, any water that is not in an ocean or a sea is fresh water. The water in a river is called fresh water, even if it is polluted. The water in a lake is fresh water, even if the lake is a salt lake. The water in "mudholes," "potholes," and swamps is also fresh water.

Populations and Ecosystems

organism:
an individual living thing of any kind

Not all fresh water is the same. If it were, we would find the same kinds of ecosystems everywhere there is fresh water. And we might expect all of these ecosystems to have the same kinds of **organisms** living in them. What are the differences in fresh water?

Fresh water varies in temperature. You would expect different bodies of fresh water to have different temperatures. But that is not the whole story. Suppose someone gave you the job of finding the temperature of the water in a pond, a river, and a lake. How would you do a job like this?

Suppose you start out with the pond. You take a thermometer and walk up to the edge of the pond. You put the thermometer in the water for a few minutes. Then you read the temperature. Suppose it is 20° C (68° F). Would you return to whomever gave you the job and report the temperature of the pond as 20° C? What do you think the person might say about your result—and the way you got it? What would be a better way?

Suppose you used the same procedure to find out the temperatures of a river and a lake. What do you think the person who sent you to do the job might say about your results—and your procedure? Can you think of a better procedure?

Natural bodies of water are not like swimming pools. Usually the water temperature is about the same anywhere in a pool. You dive into the water and do not feel much difference between the water temperature at the top and at the bottom. This is not true of the water in ponds, rivers, and lakes. You could measure and record the temperature of the water near the edge of a pond, where the water is shallow. Then you could take a boat to the center of the pond. There you could measure and record the temperature near the surface of the water. Then, with a string tied to your thermometer, you could drop it to the bottom. You could measure and record the temperature of that water. (You would need a special thermometer that would not change temperature as you brought it back up.) Then you could compare the results of your measurements.

What could you expect to find? The water near the edge of the pond, where it is shallow, would be the warmest. The water near the surface but in the center of the pond would be cooler. And the water at the bottom of the pond would be the coldest water in the pond.

You could use the same procedure to measure the temperatures in a lake. However, you would find greater temperature differences in the lake, especially if it were a deep lake. There are very great temperature differences from the top to the bottom of a deep lake. Careful studies have also shown that there

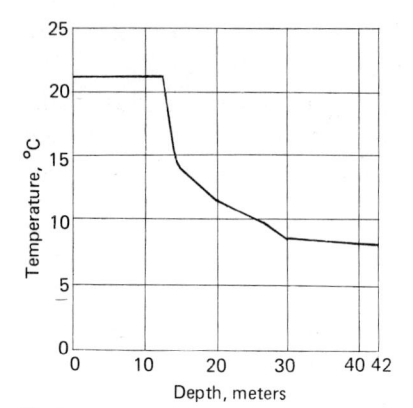

Figure 4.2: You can see here the variation of temperature with depth in Lake Geneva, Wisconsin.

Freshwater Ecosystems

Laboratory Activity

What Happens When Warm Water and Cold Water Meet?

PURPOSE In a deep lake, the water at different temperatures forms different layers. In this investigation, you will observe the layering that occurs where warm water and cold water meet.

MATERIALS
2 wide-mouth jars
2 small bottles with caps
food coloring
warm water

PROCEDURE
1. Pour cold water into one wide-mouth jar until it is half filled.
2. Pour warm water into another wide-mouth jar until it is half filled. (**Caution:** Handle the warm water with care. Do not burn yourself.)
3. Place three drops of food coloring into each of the small bottles. Fill one with warm water. (**Caution!**) Fill the other with cold water. Cap each bottle securely.
4. Lower the small bottle containing the warm, colored water into the wide-mouth jar containing the cold water. When the bottle is resting on the bottom, remove the cap as gently as possible. Slowly remove your hand from the wide-mouth jar. *Do not create any currents.* Observe and record what you see happening in the jar. Draw a diagram of what you see.
5. Lower the small bottle containing the cold, colored water into the wide-mouth jar containing the warm water. Continue as you did in step 4. *Do not create any currents.* Observe and record what you see happening in the jar. Draw a diagram of what you see.

QUESTIONS
1. In steps 4 and 5, you were cautioned not to create any currents when you removed your hand from the wide-mouth jar. Why?
2. Explain your observations in steps 4 and 5.
3. Other than differences in temperature, what differences would you expect to find at different depths? Explain your answer.
4. Draw a diagram to show what happens when the water from a cold mountain stream flows into a large shallow lake.

are *temperature layers* in deep lakes. These layers are somewhat like the layers of a cake. But they act more like the walls of a hotel. The walls in a hotel keep people separated from each other. In a similar way, the temperature layers in a lake help keep different populations of living things separated from each other.

Fresh water differs in transparency. When something is *transparent*, light rays can pass right through it. Air is transparent (unless it contains something such as smoke or smog). Most window glass is transparent. Water can be transparent.

Most fresh water is not completely transparent. In most fresh water some of the light rays are absorbed as they pass through it. A clean mountain lake or stream is one of the most transparent bodies of fresh water. A muddy pond or river is one of the least transparent.

The transparency of fresh water has an important effect upon the number and kinds of organisms that can live in it. Here is why. The consumers in a freshwater ecosystem depend on its producers. The producers need light to carry on the food-making process, photosynthesis. How much light are the producers going to get? That depends on the transparency of the water. If the water is fairly transparent, the producers can live well below the surface. Unless the water is very deep, some producers may even live at the bottom of a freshwater ecosystem.

If the water is not transparent, the producers live in a limited area near the surface. Such a body of water will not contain as many producers as cleaner or more transparent water. And with fewer producers for consumers to feed on, there will be fewer consumers.

As you can see, the transparency of the physical environment—water—affects the organisms in freshwater ecosystems. In turn, the organisms can have a real effect on this part of the physical environment. Water that contains many living things will not be completely transparent.

Current is important in freshwater ecosystems. Picture yourself in the middle of a fast river. You are sitting in a small rubber boat. How easy would it be for you to stay in the same place in that river—without an anchor? Some animals are equipped to keep their places in a strong current. Many fish are good examples. But many other animals would be washed away.

Differences in current may separate whole groups of organisms in a river. If it is a winding river, there will be quiet pools of water in some places. One group of plants and animals will use that area for a habitat. In other places, the river will be moving

Figure 4.3: A clean mountain lake will have more producers than will a muddy pond or lake. The sunlight cannot penetrate very far into muddy water.

faster. There may be shallow rapids. There may be waterfalls with deep rumbling pools below them. Each area—all in the same river—may have its own populations of plants and animals living there.

Microhabitats protect organisms in strong currents. Currents are usually strongest in the middle of a stream or river. However, there may be some spots that are exceptions. To understand them, it will help to "think small." Think of yourself as a small insect. You are down near the bottom of a river, behind a rock. The rock causes the strong current flowing by it to pass on each side of you. As long as you stay behind the rock, you will be protected from the current. If you could find food down there behind that rock, this might be a good place for you to make your home. An ecologist studying your home would call it a **microhabitat**—a *small home*. There are many microhabitats in freshwater ecosystems. Like the insect behind the rock, many microhabitats enable organisms to live in places that do not seem suitable for them.

microhabitat: a small habitat of one kind that is within a larger habitat of another kind

There is a difference in oxygen in fresh water. You probably know that salt and sugar dissolve in water. So does oxygen, but not as easily. Much of the oxygen found in water comes from the air. It touches the water everywhere on its surface. Some of the oxygen comes from producers that live in the water. Some of the oxygen they produce during photosynthesis dissolves.

Think of a mountain stream that tumbles down over rocks. It has much more dissolved oxygen than most other bodies of fresh water. The tumbling exposes more of its water to the oxygen in the air. Quiet bodies of water, such as ponds and swamps, have much less oxygen than fast-flowing streams.

Pollution can affect the quantity of oxygen in fresh water. Heat pollution affects it in one way. Warm water cannot hold as much dissolved oxygen as cold water can. When warm water is dumped into lakes or streams, the amount of dissolved oxygen is lowered.

Decomposers of waste materials also use up oxygen. When there are many of them, decomposing large amounts of waste, the amount of dissolved oxygen may be very low. Only a few animals can live in such water.

There is a difference in minerals in fresh water. You will not find any fish living in the Great Salt Lake in Utah. You can probably guess the reason. The water contains too much dissolved salt. The salt acts as a poison. If you put a fish in this water, it would die.

Figure 4.4: The organisms that live in a fast-flowing stream are different from those that inhabit slow or still water.

Populations and Ecosystems

Most freshwater ecosystems have lesser amounts of dissolved minerals than the Great Salt Lake. Dissolved minerals are usually not harmful to freshwater plants and animals. In fact, all of them need some dissolved minerals. Completely pure water would not be a good habitat for most freshwater populations. Bony animals, such as fish, need calcium and phosphorus for their bones. Snails and clams need calcium for their shells. Many other minerals are needed in small quantities. Some fresh water is rich in all the minerals needed by freshwater plants and animals. If other conditions are right, such water is likely to have large populations of living things.

LESSON REVIEW *(Think. There may be more than one answer.)*

1. To the ecologist, fresh water is water that
    a. does not have salt in it.
    b. is not an ocean or a sea.
    c. humans can drink.
    d. has fallen as rain or snow.

2. In a typical lake,
    a. the water near the surface is warmer than water below the surface.
    b. the water near the middle is warmer than the water near the shore.
    c. the temperature at the deepest part would be the lowest.
    d. the temperature throughout all parts is about the same.

3. In Lake Geneva, Wisconsin, as shown in the graph in Figure 4.2,
    a. the temperature at the surface is about 24° C.
    b. the temperature drops evenly from the surface to the bottom.
    c. there is a layer of water near the bottom that is about 8° C.
    d. the temperature of the water is about the same at any depth.

4. In a body of fresh water,
    a. the transparency will affect the numbers of producers and consumers.
    b. all organisms can live in strong currents.
    c. water that moves fast will usually have more oxygen than slow-moving water.
    d. dissolved minerals usually cause harm to the plants and animals that live in it.

5. Define what is meant by a microhabitat. Describe one or more microhabitats in which you live.

64

Freshwater Ecosystems

KEY WORDS

organism (p. 60)   microhabitat (p. 63)

THE ORGANISMS

What are the organisms in fresh water? There are thousands of different freshwater ecosystems. And there are thousands of different organisms that live in them. It would be impossible to name and describe all of these in one book. But we do not have to know about all of them to understand how they fit into their ecosystems.

Ecologists have found five basic niches in freshwater ecosystems. So ecologists have placed all the freshwater organisms into five basic groups. All the living things in any one of the groups have a similar way of life. And all will be pretty much alike. There may be different kinds of organisms in different ecosystems. But if you know some of those in each group, you can predict what you might find in any freshwater ecosystem.

The neuston live on the surface. Find any body of fresh water and you can probably find some organisms that live on its surface. Ecologists call such organisms the **neuston** (noo′stən).

Walk up to any pond or other quiet body of water. You can usually observe the neuston. If you are at a pond, and it is summer, you are likely to see duckweed. These tiny plants float on the surface. They can reproduce rapidly. In late summer, they may completely cover a pond.

neuston: all together, the things that live on the surface of a body of water

Figure 4.5: Duckweed (left) and water hyacinth (right) clog the surfaces of the bodies of water where they grow.

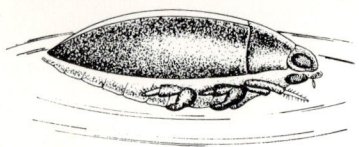

Figure 4.6: Water striders (top) are supported by the surface tension of the water. The whirligig beetle (bottom) swims partly above the surface and partly submerged. It has two sets of eyes, for seeing above and below water.

periphyton:
all together, the things that live clinging to shallow-water plants

The water hyacinth is a beautiful floating plant. It is also a pest. In many southern states, the water hyacinth has completely covered thousands of kilometers of rivers and canals. The plants form thick mats that choke waterways and stop boats.

The neuston include animals as well as plants. You can almost always find water striders and whirligig beetles on a pond's surface. If you look closely at the water striders, they look like they are skating across the top of the water. The whirligig beetles race round and round in circles as if they were trying to get away from something chasing them.

The periphyton cling to plants. Almost every body of fresh water has some plants growing in it. These are plants that have their roots in the bottom. These are land plants, in a way. They are attached to the earth. But such plants provide a variety of habitats for organisms that cling to them. Such clinging individuals are called the **periphyton** (pĕr′ĭ fī′tən).

Many ponds have water lilies. These rooted plants have huge leaves that float on the surface. If you have a chance to look at water lilies, lift one of the leaves and look closely at it. Chances are that you will see a variety of animals on the underside of the leaf. Hydra (hī′drə), small, freshwater animals, commonly live on such a plant. But you may also find various kinds of snails, worms, and immature insects. You might scrape the stem or the

Freshwater Ecosystems

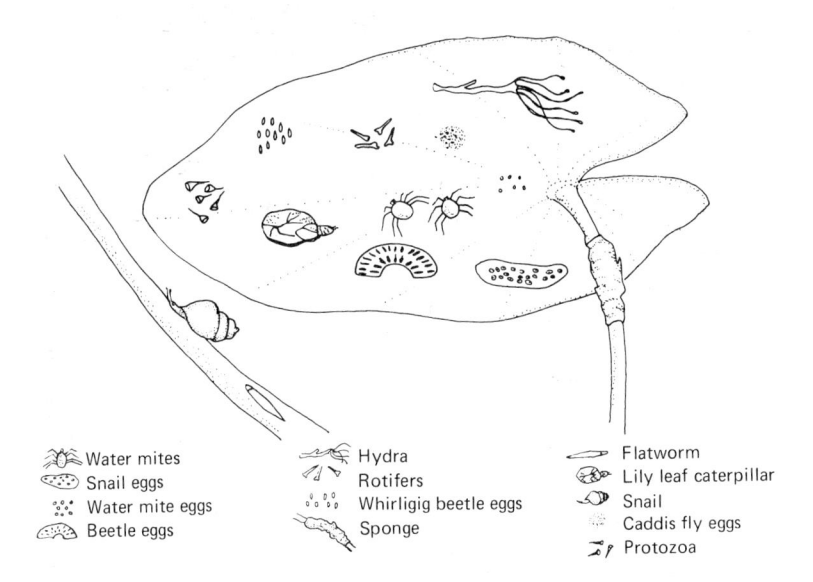

Water mites
Snail eggs
Water mite eggs
Beetle eggs
Hydra
Rotifers
Whirligig beetle eggs
Sponge
Flatworm
Lily leaf caterpillar
Snail
Caddis fly eggs
Protozoa

Figure 4.7: The underside of this lily leaf is a highly populated microhabitat.

underside of the leaf of the lily. If you look at the scrapings through a microscope, you will see a whole new world of creatures.

The benthos live on the bottom. Bodies of fresh water have many different kinds of bottoms. They range all the way from soft mud, clay, and sand to various kinds of rocky bottoms. Each kind of bottom has characteristic organisms that live there. They are called the **benthos** (bĕn′thōs).

Many ponds have soft muddy bottoms. There you could find a number of insects and worms that burrow in the mud. Many kinds of **algae** (ăl′jē) can also often be found there. In some ponds large crayfish may live in holes they have made in the bottom.

In rivers, bottom-living animals often make their own microhabitats. This lets them live in strong river currents. A good example is the caddis fly larva (lär′və), which is a wormlike stage in the development of an insect. This larva usually builds a case around itself (See Figure 4.8). Each kind of caddis fly larva makes its own specific kind of case. And it always uses the materials that are on the bottom of the water in which it lives. These materials may include such things as leaf fragments, pine needles, twigs, and small rocks.

If you ever visit a mountain stream, one with waterfalls and a strong current, pick up a rock and look at it. You may find several *rock cases* stuck to it. If you look at these closely, you will find that there is a creature in each of them—a caddis fly larva. (Or maybe a *pupa* [pyoo′pə], the stage where the larva is changing into an adult insect.) The caddis fly larva can live in

benthos:
all together, the things that live on the bottom of a body of water

algae:
simple plantlike organisms. All algae carry on photosynthesis.

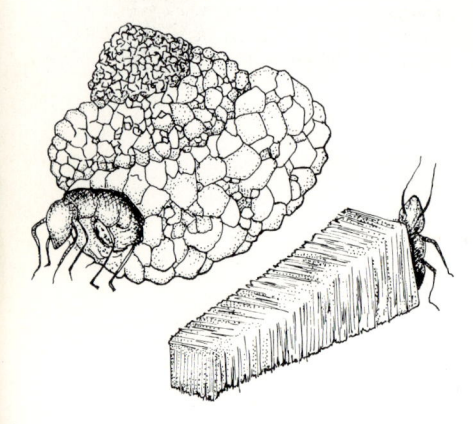

Figure 4.8: This caddis fly larva in its case (above) is camouflaged against the streambed. These larvae build their cases (left) of almost any available material.

plankton: very small plantlike or animallike organisms that live free in a body of water

nekton: as a group, all animals larger than plankton that live free in a body of water

waterfalls and strong currents because it can cement its case to the rocks on the bottom of the stream.

Plankton and nekton live free in the water. Suppose your home is a pond. You do not live on the surface. You do not live on a rooted plant. Where else could you live? You could live in the water, free and not attached to anything. If you did live like this, an ecologist would say that you are part of either the plankton or the nekton.

What would you be if you were part of the **plankton** (plăngk′tən)? You would be small. You would probably be floating in the water. You might be able to swim just enough to keep yourself from settling to the bottom. But you would not be able to swim well enough to keep a current from carrying you away. If an ecologist pulled a fine silk net through the water, you would be caught in it. You could not avoid such a trap.

If you belonged to the **nekton** (něk′tən), a human being would have a more difficult time catching you. However, one might try. A person might put some attractive bait on a hook and dangle it over the side of a boat. You might be fooled and think the bait-covered hook was a prize bit of food just waiting to be eaten. If so, you might find yourself being reeled into the boat. And if you are large, before you die from lack of oxygen, you might hear the person bragging about catching you.

The nekton include fish and other animals. By now, you have probably figured out that fish belong to the

68

Figure 4.9: This life scientist is collecting specimens with a plankton net. The small-mesh net allows the scientist to catch the suspended plankton.

nekton. Actually, the nekton include all the animals that move freely through the water—and so escape the little silk nets that ecologists pull through it.

The kinds of animals found in the nekton vary greatly with the current and depth of the water. In a pond or lake, the deeper water will be occupied mostly by fish, and perhaps a few turtles. Large populations of fish make up most of the nekton in rivers and streams. But the picture changes when you examine the shallow areas of a lake or pond or a quiet spot in a river. The nekton in such places may include quite a variety of animals. There will be several kinds of insects, often including diving beetles, water boatmen, backswimmers, and giant water bugs. You are also likely to find frogs and their tadpoles, water snakes, and turtles, as well as one or more kinds of fish.

Figure 4.10: This photo shows part of a food web. The giant water bug has captured and is eating a mayfly nymph. The water bug in turn may become food for another organism.

Populations and Ecosystems

There are two kinds of plankton. The plankton have a very large niche in many freshwater ecosystems. What is large about this niche? You can guess from this hint: The ecologist often divides the plankton into two groups. One group is called the **phytoplankton** (fī'tō plăngk tən). The name means "plantlike plankton." The other group is called the **zooplankton** (zō'ə plăngk tən). The name means "animallike plankton."

phytoplankton:
plantlike plankton

zooplankton:
very small animals and animallike plankton

The phytoplankton are made up mostly of different kinds of algae and diatoms. (Diatoms are mostly single-celled organisms. You will read more about them in the next chapter and in Chapter 14.) These are usually the major producers in an *aquatic* ecosystem, an ecosystem that is in or involves a body of water.

Using light energy from the sun, algae and diatoms manufacture food by photosynthesis. In this niche, phytoplankton are at the beginning of many different food chains. And the oxygen that they produce during photosynthesis helps keep all other aquatic organisms alive. Now you can understand why ecologists worry about phytoplankton in aquatic ecosystems. Any change that affects the phytoplankton will eventually affect all the organisms in a freshwater ecosystem.

The zooplankton are made up of small animals and animallike creatures. They swim around in the water and feed on the phytoplankton—or on each other. Usually, they feed on the phytoplankton.

In this niche, the zooplankton are somewhat like the grazing animals in a terrestrial ecosystem. Grazing animals on the land feed on grass and other plants. So do other animals in terrestrial ecosystems. Many of the grazing plant-eaters, in turn, are eaten by second-order and third-order consumers.

In a freshwater ecosystem, the zooplankton "graze" on phytoplankton. In turn, the zooplankton are eaten by larger animal consumers. These may be many different kinds of animals, such as insects, fish, and turtles.

Figure 4.11: Phytoplankton are producers. They are the first step in the food web of a freshwater ecosystem. These phytoplankton are (a) *Hydrodictyon,* (b) *Anabaena,* (c) *Closterium,* (d) *Pediastrum,* (e) *Oscillatoria,* (f) *Fragilaria,* (g) *Microcystis,* and (h) *Pleurosigma.*

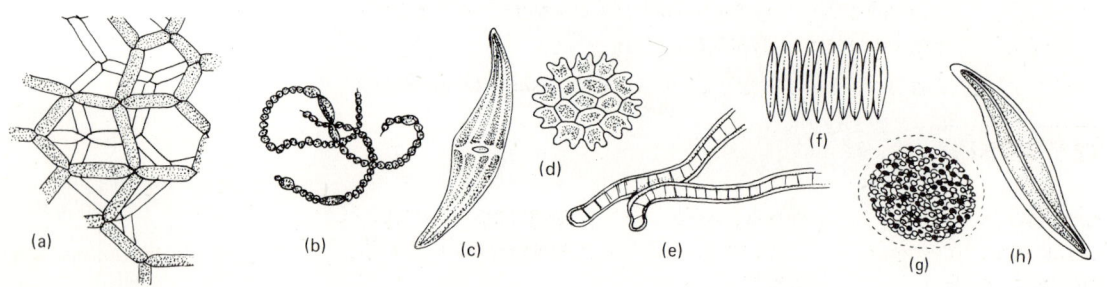

Freshwater Ecosystems

LESSON REVIEW *(Think. There may be more than one answer.)*

1. Which of the following would be part of the neuston of a pond?
    a. water strider
    b. turtle
    c. whirligig beetle
    d. duckweed

2. To be part of the periphyton in a pond, organisms
    a. could be clinging to the stem of a water plant.
    b. could be clinging to the shell of a turtle.
    c. could be swimming free in the water.
    d. could be crawling on the bottom of the pond.

3. The plankton of a pond
    a. include fish, turtles, and swimming insects.
    b. are made up entirely of producers.
    c. include both producers and consumers.
    d. live only on the bottom.

4. Zooplankton could be called
    a. the main producers in freshwater ecosystems.
    b. the largest consumers in freshwater ecosystems.
    c. grazers on phytoplankton.
    d. the smallest consumers in freshwater food chains.

5. Which of the following terms best matches each phrase?
    a. neuston   e. plankton
    b. periphyton f. nekton
    c. benthos   g. phytoplankton
    d. algae   h. zooplankton

    _____ organisms that live at the bottom of a freshwater ecosystem

    _____ organisms that move freely through a freshwater ecosystem

    _____ the smallest inhabitants of a freshwater ecosystem

    _____ organisms that live clinging to shallow-water plants

    _____ the major producers of a freshwater ecosystem

    _____ organisms that live on the surface of a freshwater ecosystem

    _____ plantlike organisms that carry on photosynthesis in a freshwater ecosystem

    _____ the smallest "grazers" in a freshwater ecosystem

KEY WORDS

neuston (p. 65)
periphyton (p. 66)
benthos (p. 67)
algae (p. 67)
plankton (p. 68)
nekton (p. 68)
phytoplankton (p. 70)
zooplankton (p. 70)

71

Populations and Ecosystems

Applying What You Have Learned

1. What nonliving factors cause the aquatic environment to vary from place to place?
2. Where in a lake would you expect to find the warmest water and the coldest water? Why does water temperature vary from place to place in freshwater lakes?
3. Review the five groups of freshwater organisms discussed in this chapter. Which of those groups will have the largest territory and will show the greatest amount of movement? Explain your answer.
4. Explain how ice and snow on a frozen lake would affect the population size of the phytoplankton, neuston, and benthos in a freshwater ecosystem.

Questions 5–7 are based on this paragraph:

During the summer, temperature layers form between plankton and available nutrients and minerals on the bottom of a lake. Winter weather stirs up some of these minerals and nutrients. This seasonal mixing may also help break up the different temperature layers. The graph shows the changes that occur in plankton population size in one year's time.

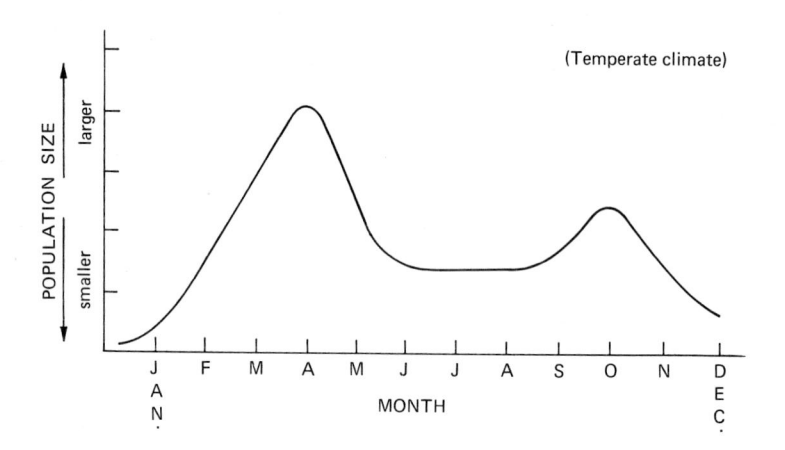

(Temperate climate)

POPULATION SIZE (smaller → larger) vs MONTH (JAN–DEC)

5. Describe the change in plankton population size during November and December when nutrients and minerals may be available. How can this be explained?
6. Describe what happens to the size of plankton populations from April through June. How can this be explained?
7. What seasons of the year show the greatest increases in the number of plankton? Why?

Freshwater Ecosystems

8. How do pollutants, such as industrial wastes and eroded soil, change the transparency and amounts of dissolved oxygen in a freshwater ecosystem?

9. How do pollutants, such as industrial wastes and eroded soil, affect the size of plankton populations? Explain your answer.

10. How would you place each of the following populations in a food pyramid: (a) zooplankton, (b) trout, (c) phytoplankton, (d) minnows (small fish)?

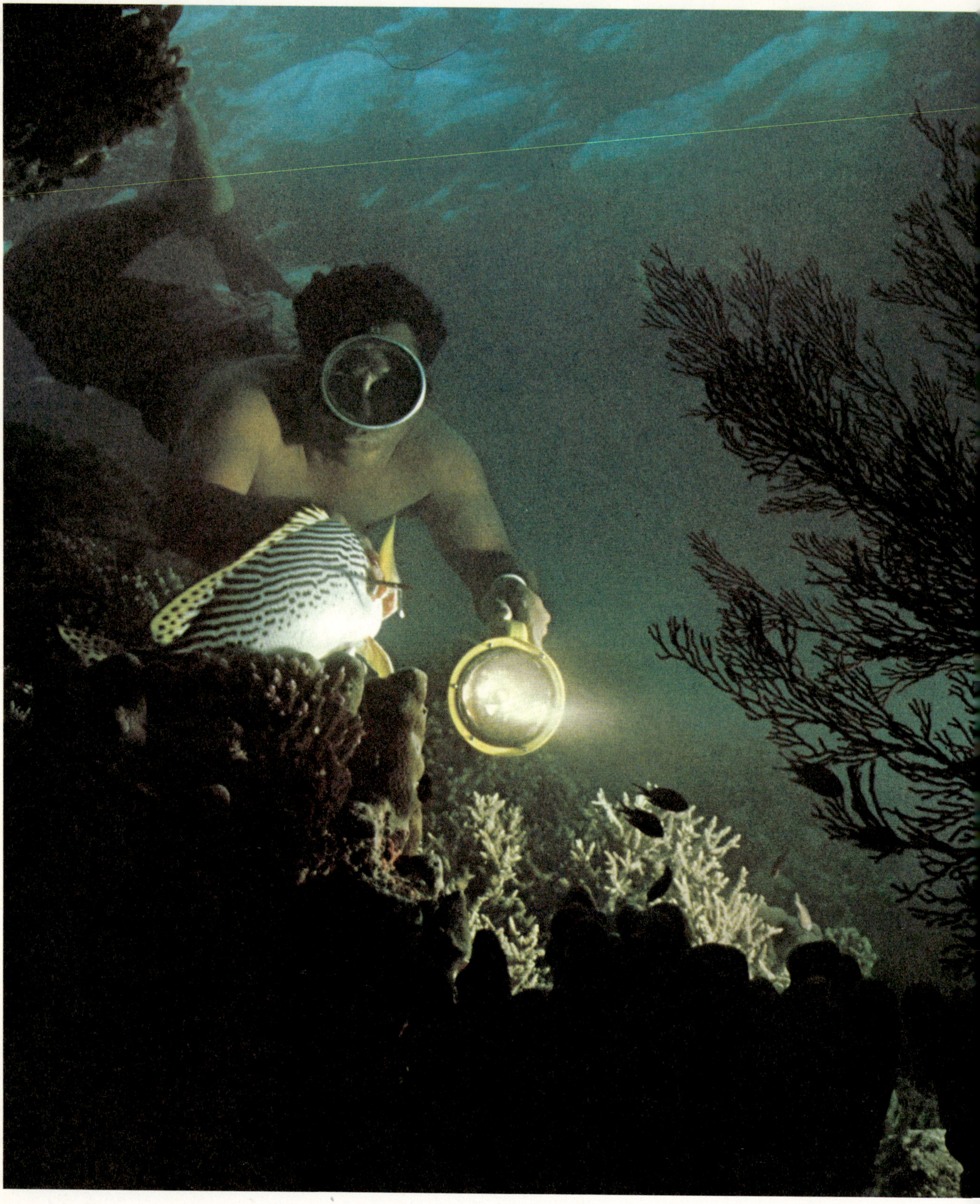

Figure 5.1: There are many marine ecosystems in the marine habitat. Of all the kinds of habitats on Earth, the marine habitat is the largest.

Chapter 5

MARINE ECOSYSTEMS

Seventy percent of the Earth is covered with seawater. This water, along with the land near it and beneath it, makes up the marine habitat. The marine habitat is larger than all the terrestrial habitats and freshwater habitats put together. It also contains more individual organisms than all the other habitats put together.

There are many ecosystems within the marine habitat. Like all other ecosystems, they differ in the nature of their physical environment. They also differ in the populations that live in them. You will explore both the environment and the populations of marine habitats in this chapter.

THE MARINE ENVIRONMENT

The marine environment is continuous. When you think of freshwater ecosystems, you usually think of separate bodies of water—lakes, rivers, or perhaps ponds. The marine environment is different. The Earth's entire marine environment is like the air in this way: It is one single continuous body. It is true that geographers divide the marine waters into oceans, seas, and bays. But they are all of one piece. They are all parts of one huge body of water.

There are barriers in the marine environment. The marine environment is of one piece. But that does not mean that there are not barriers that keep populations confined to limited areas. There are barriers. And they are effective.

One barrier is temperature. Temperature differences in the marine environment are very great. Polar temperatures are cold enough to put a cap of ice over the seawater. Tropical bays are a warm delight even for humans in bathing suits.

The *salinity* (sə lĭn′ĭ tē), or salt content, of seawater is another barrier. The salinity varies from place to place. In the open seas, there is little variation. But near the mouths of rivers, where fresh water is added, there is a wide range in the salinity. Such an area is called an *estuary*. The water is said to be *brackish water*. The salt content of brackish water is between that of normal seawater and typical fresh water.

A third barrier is depth. The marine environment is by far the "thickest" life-supporting environment on the Earth. In some places, water is 11 kilometers (about 7 miles) deep, or thick, depending upon how you want to think of it. Producers cannot live at extreme depths because there is no light. This severely limits the animals that live there, since they must rely upon another source for their food.

The sea has currents. You can fill a bathtub with water. Then you can let the water stand for awhile. The water

Figure 5.2: This satellite picture shows the Sacramento River flowing into San Pablo Bay, an inlet of the Pacific Ocean. San Francisco Bay is at the lower left of the photo and is bordered by the city of San Francisco on the west and Oakland, California, on the east. The salinity of the river's mouth and of the two bays is less than that of the Pacific. Organisms in the less saline waters are different from those in the more saline regions.

Marine Ecosystems

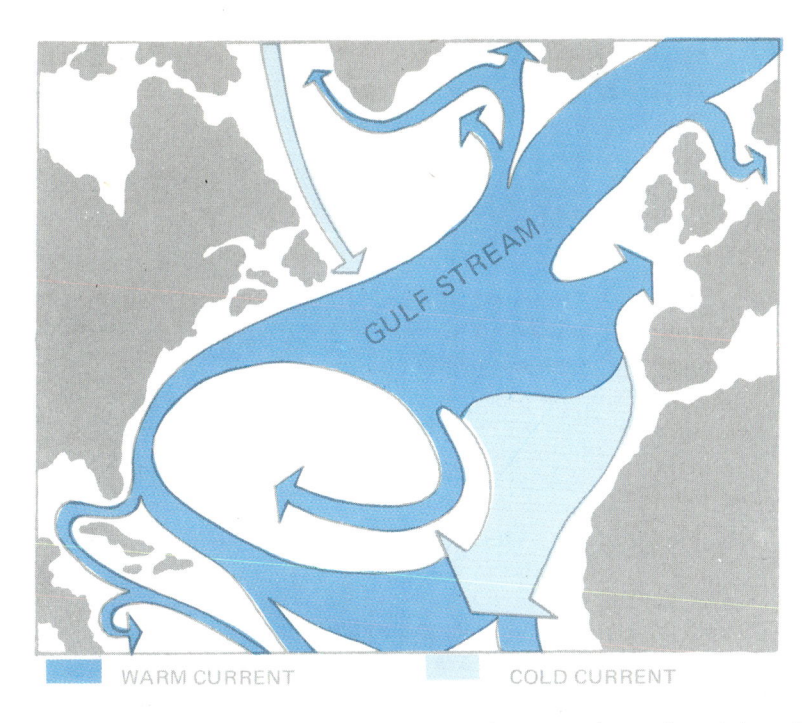

Figure 5.3: The Gulf Stream is a huge current of water moving northward along the eastern shore of the United States. This current extends across the Atlantic Ocean to Europe.

will be quiet. It will have no visible movement. Would this water be a good model of what the water in the sea is like? Is the sea like a huge bathtub full of quiet water? What do you think?

The sea is not quiet. Many writers have used another word to describe it. They have said that it is "restless." That means the water in the sea is always moving. Most of these movements are called *currents*.

There are two general kinds of currents. One kind of current is like a river in the sea. An example of this kind is the Gulf Stream. This is a huge current that flows from the Gulf of Mexico across the Atlantic toward northwestern Europe. The Gulf Stream brings warm water to the lands of northwestern Europe. The result is that a country like England has a much warmer climate than it would have without this warm current.

Upwelling supports populations. A second kind of current is called *upwelling*. This is a movement of water in an upward direction, from below the surface to the surface. Most upwelling occurs where strong winds blow on the surface waters. The surface water moves away, in the direction of the wind. Water from below comes up and replaces the water which moved away.

Populations and Ecosystems

Laboratory Activity

Why Is the Ocean "Restless"?

PURPOSE

The water in the ocean seems to move constantly. In this investigation, you will study the effect of temperature and wind on the movement of water. You will then apply what you observe to the movement of the oceans.

MATERIALS

rectangular pan
cardboard
water (warm and cold)
ice cubes
medicine dropper
food coloring
hairdryer

PROCEDURE

1. Cut a cardboard strip so that its length fits the width of the pan. The height of the strip should be about the same as the sides of the pan.
2. Place the cardboard strip in the pan so that the strip is 8 cm from one end. The pan should now be divided into one large section and one small section, as shown in Figure A.
3. Pour warm water into the pan to form an 8-cm layer on both sides of the divider.
4. Fill the small section of the pan with ice cubes. Let the water settle.
5. Add one drop of food coloring to the side farthest from the ice, as shown in Figure B.
6. Observe the water. What happens to the food coloring? Diagram the results. Explain your observations. Empty the contents of the pan, including the cardboard divider.
7. Pour water into the pan to a depth of about 8 cm. Let the water settle.
8. Add one drop of food coloring to the water at bottom-center of the pan. Try not to disturb the water as you do this.
9. Turn on the hairdryer to "high." Direct a stream of air across the surface of the water, as shown in Figure C. You should see ripples in the water.
10. Observe the water. What happens to the food coloring? Diagram the results. Explain your observations.

QUESTIONS

1. How does temperature affect the movement of ocean water?
2. How does the wind affect the movement of ocean water?

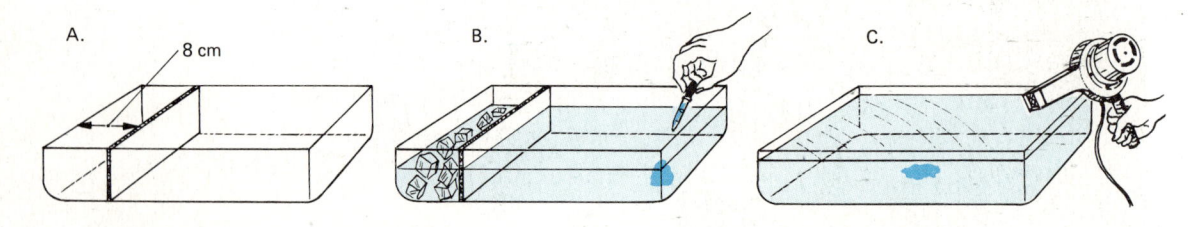

A. 8 cm

B.

C.

Career Spotlight

Oceans cover more than two-thirds of the Earth's surface. It is no surprise that many people want to study the secret world under the seas. This is what a *marine biologist* does. She or he studies the plants and animals that live in our oceans—the marine ecosystem.

Marine biologists do much of their work at sea, but they also work in laboratories on land. They use diving gear and special equipment so that they can explore underwater. In the picture, you see three marine biologists as they prepare to launch a diving ship, or submersible.

In recent years, marine biologists have become greatly concerned about the damage being done to marine ecosystems by people polluting them heavily. Oil spills are one example of this. Also, chemical and radioactive wastes are too often dumped carelessly in our oceans. These pollutants kill plant and animal life and upset the ecological balance in the marine environment. Marine biologists try to discover the problems caused by pollution and try to work with others to find ways to solve these problems.

A marine biologist is a scientist. To be one, therefore, you must complete at least 4 years of college. Many marine biologists continue with more advanced study. In advanced study, they work on a specific problem under the guidance of an experienced scientist. Often they earn a higher university degree as a result.

Populations and Ecosystems

Figure 5.4: This green water off the Aleutian Islands of Alaska is rich in phytoplankton because of upwelling. Nutrients needed by the marine producers are brought up from dead matter on the bottom.

Upwelling is a very important process in marine ecosystems. There is much more life in the sea because of it. Here is why. If it were not for upwelling, all dead matter would settle to the bottom of the sea. Decomposers would release the nutrients in this dead matter. But these nutrients would remain on the bottom. Producers, which must live in the upper waters because they need sunlight, would not be able to use these nutrients. You know, of course, what would happen to any ecosystem without producers. There could be no consumers.

If you had a satellite picture of the whole sea, you could tell where the greatest amount of upwelling is. All you would have to do is count and mark the location of the fishing boats in the picture. Generally, the greatest number of fishing boats would be in areas of strong upwelling—for one simple reason. That is where one is likely to find the greatest numbers of fish. Fish are there because that is where food is most plentiful. But remember, the fish are consumers that are near the end of most food chains. Producers that use nutrients brought up by the upwelling make it possible for these consumers to exist.

When is the sea not a sea? The answer to this question is: When the seawater disappears. In some places, the seawater does disappear. You would realize this better if you were a small animal living on the bottom of the sea near a coastal beach. Once or twice a day, depending upon your specific location, the water over you would disappear. You would be left "high and dry." Why? Because of the *tides*.

What are tides? Tides are movements of water that are caused by the pull of the moon and the sun. The water of the

Marine Ecosystems

sea is pulled up and down as the Earth, moon, and sun change positions. The raising or lowering of the water level in the open sea has little effect upon the life there. But the raising or lowering of the water level along a coastline has important results. It means that some of the organisms will be in and out of the water.

How often do the tides occur? There is not a simple answer to this question. The time of the tides varies—as does the distance that the water moves. Look at the map in Figure 5.5. The coastal areas marked in blue have regular twice-daily tides. These occur regularly about every 12½ hours. But you can also see that there are coastal areas where there is only one regular tide per day—and others where the tides are irregular.

Figure 5.5: Solid red indicates areas with once-daily tides; barred red, irregular once-daily tides. Solid blue has twice-daily tides; barred blue, irregular twice-daily tides.

ONCE DAILY
IRREGULARLY ONCE DAILY
IRREGULARLY TWICE DAILY
TWICE DAILY

Populations and Ecosystems

No matter how often the tides occur, the coastal regions have their own unique populations. These and other marine populations will be discussed later.

LESSON REVIEW *(Think. There may be more than one answer.)*

1. The marine habitat
    a. is larger than all terrestrial and freshwater habitats put together.
    b. is continuous.
    c. has the same populations everywhere.
    d. has about the same temperature everywhere.

2. Three main barriers that separate marine populations are
    a. salinity, depth, and upwelling.
    b. temperature, currents, and salinity.
    c. salinity, temperature, and depth.
    d. currents, upwelling, and tides.

3. In the deepest parts of the ocean,
    a. living things that die sink to the bottom.
    b. decomposers release nutrients from dead plants and animals.
    c. released nutrients stay at the bottom unless strong winds blow over the surface.
    d. released nutrients never reach the surface.

4. Tides
    a. occur twice daily along the eastern coastline of the United States.
    b. are always regular.
    c. occur twice daily along the western coastline of the United States.
    d. occur irregularly once daily along the Texas coast.

MARINE POPULATIONS

The neritic zone houses the most life. Look at the diagram in Figure 5.6. Note the shallow-water region above the *continental shelf*. This is the **neritic zone.** The neritic zone of the sea has a very large number of organisms. It is truly the richest part of the sea. There are at least two good reasons why this is true. First, light penetrates all the way through the neritic zone. Therefore, the entire area can support populations of producers. Second, nutrients, which are released by decomposers, are more easily obtained in shallow water. Minor upwelling currents are all that is necessary to bring nutrients up from the bottom of the shallow water.

neritic zone: the shallow-water region above the continental shelf in a marine ecosystem

Marine Ecosystems

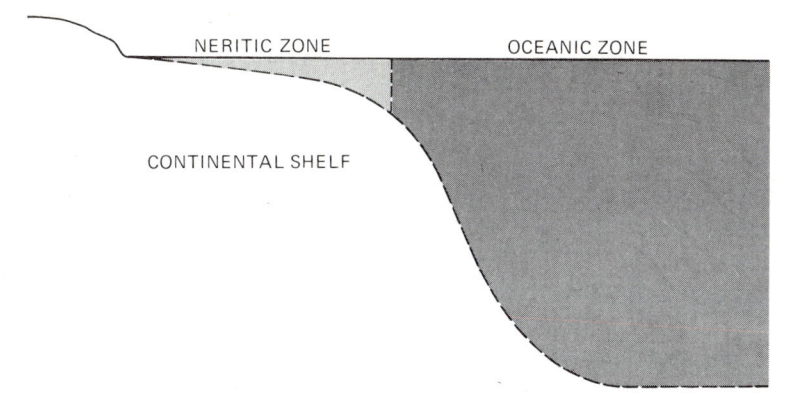

Figure 5.6: The neritic zone is the region of ocean over the continental shelf, where the sunlight penetrates to the bottom. The oceanic zone is the region of deeper water.

The second major zone in the sea is the **oceanic zone.** As you can see in Figure 5.6., this is the deep-water part of the sea. Ecologists know much less about the kinds of life in the oceanic zone—for one simple reason. It is much more difficult to study populations that live at the greater depths of this zone.

oceanic zone:
the deep-water region in a marine ecosystem

Diatoms and dinoflagellates are the key marine producers.

Diatoms (dī′ə tŏm′) and **dinoflagellates** (dī nō flă′ jə lāt) are single-celled organisms. They make up the major part of the marine phytoplankton. It is almost impossible to believe how important they are. They are the key producers in both zones of the marine environment—the neritic and the oceanic zones. As such, they are at the beginning of most food chains in marine waters. They also produce oxygen during their photosynthetic activities. Not all of this oxygen stays in the water.

diatom:
one of a large number of golden-brown algae

dinoflagellate:
a kind of single-celled marine phytoplankton

Figure 5.7: The greatest number of producers in the seas are the diatoms and the dinoflagellates. The picture on the left shows a number of diatoms. The picture on the right shows some magnified *Ceratium*, marine dinoflagellates.

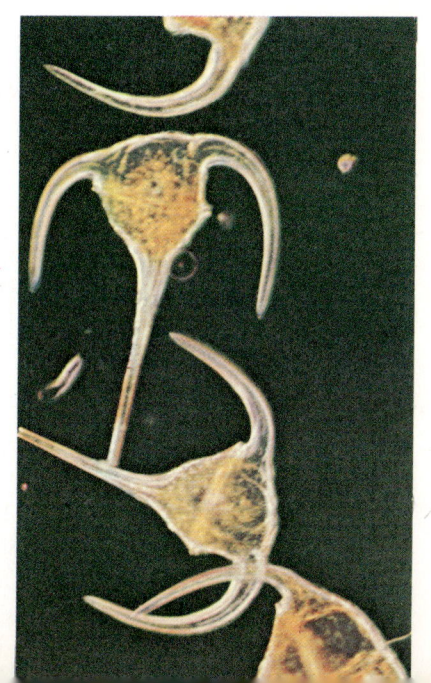

Much of it is released into the air. This replaces the oxygen that is constantly being taken from the air. Chemical activities on the Earth—like the rusting of iron—take oxygen from the air. So do the breathing activities of all terrestrial consumers. (This includes you!) Ecologists have a great fear that we might accidentally kill off these phytoplankton. This could happen by any one of several kinds of pollution.

What would we do if our oxygen supply began to disappear? Could we do without it? Could we make enough to keep ourselves alive? Could we keep other terrestrial consumers alive? Such questions used to be topics for science fiction stories. (You might still write a good story or two if you have a good imagination.) But now they are questions that we cannot dismiss as being "impossible."

Sometimes neritic phytoplankton "bloom." You know what happens when a flower blooms. This same term is also used to describe a population explosion of phytoplankton. Usually a phytoplankton "bloom" is made up of one specific population. Very often this is a population of dinoflagellates. Blooms occur in both freshwater and marine environments. The blooms in marine waters cause greater problems.

Some dinoflagellates produce a poison that is released into the water. In normal populations, the poison is not enough to kill other organisms. But when such a population explodes, the amount of poison increases greatly. Then it is enough to kill other organisms, such as fish.

The Gulf Coast of Florida sometimes experiences blooms of poisonous dinoflagellates. During such a bloom, called a **red tide** because of the color of the dinoflagellates, thousands of fish are killed. These fish wash up on the beach. Rows and rows of dead and decaying fish completely destroy the recreational value of such a beach.

red tide: a "bloom" of a poisonous, red dinoflagellate

What causes a phytoplankton bloom? Ecologists would like to know all the answers to that question. Probably one major cause is a sudden increase in nutrients in coastal waters. For

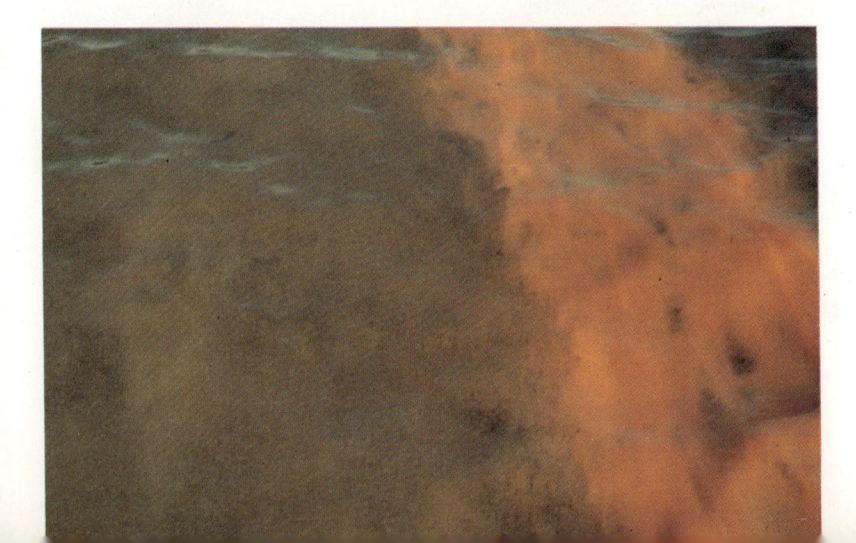

Figure 5.8: This is a red tide. A bloom of red-colored dinoflagellates is responsible for this condition. Poisons from these phytoplankton kill many fish.

Figure 5.9: The organisms in this tidal pool can live either in water or out of it.

example, there is evidence to show that heavy rains in a coastal area can trigger a seasonal bloom. The rains wash nutrients from the land into rivers. The rivers carry the nutrients into the shallow waters of the neritic zone.

There is great diversity in marine ecosystems. We have said that the sea is not a uniform environment. You do not have to be a marine scientist to understand this fact. All you have to do is look at a picture book of marine life. As you turn the pages, you can see fantastic examples of diversity in marine ecosystems. You can see the ways marine organisms are suited to the many different habitats in the sea.

The organisms of the *intertidal community* offer one good example for study. These **intertidal organisms** are those that live near the shores, where the tide waters come and go. They have to be able to live in the water. They also have to be able to survive when the water is absent.

Deep-water fish are probably the most fantastic in all the marine environment. These fish live in a cold, dark environment where food is scarce. How do they manage to find food? Some have huge mouths and can swallow a fish larger than themselves. Such a fish does not have to find food as often. Some fish have greatly enlarged eyes, which help them locate food. Others have no eyes and rely on other senses, such as taste and smell. Some have organs that can produce light. Some fish produce light and use it to help them see and locate food. Others produce light but use it like a bait to attract fish, which they then capture and eat.

intertidal organism: an organism that lives near the ocean's shore, where the water comes and goes with the tides

Populations and Ecosystems

Figure 5.10: Deep-sea fish have some of the most bizarre features of all marine organisms. At the top is a picture of a hatchetfish. At the bottom is an anglerfish. Both fish live in dark, deep-sea waters and have organs that glow sticking out between their eyes. Of what value to each fish's survival is each of the unique features you see?

LESSON REVIEW *(Think. There may be more than one answer.)*

1. Of the two marine zones in Figure 5.6,
    a. the neritic zone is richer in living things.
    b. the oceanic zone is richer in nutrients.
    c. the neritic zone is filled with sunlight.
    d. the oceanic zone is less understood in terms of its populations.

2. Marine plankton
    a. are the biosphere's main producers.
    b. are the biosphere's main consumers.
    c. produce most of the oxygen in the air.
    d. could be killed by pollutants.

3. A bloom of poisonous phytoplankton could
    a. kill fish.
    b. destroy the recreational value of a beach.
    c. lower our oxygen supply.
    d. prevent upwelling.

4. Intertidal organisms
    a. are caused by upwelling.
    b. live mainly in the neritic zone.
    c. live mainly in the oceanic zone.
    d. must be able to live on land as well as in the water.

5. The neritic zone is rich in living things because
    a. it is filled with light during the day.
    b. it is closest to the shore.
    c. nutrients released by decomposers are easier to get.
    d. it is most affected by the tides.

6. Which of the following terms best matches each phrase?
    a. neritic zone   d. dinoflagellate
    b. oceanic zone   e. red tide
    c. diatom   f. intertidal organism

    _____ population explosion of a poisonous dinoflagellate

    _____ offshore shallow zone of an ocean

    _____ deep zone of an ocean

    _____ organism that lives on the seashore, where the water comes and goes with the tides

    _____ most important producers; part of the oceans' phytoplankton

KEY WORDS

neritic zone (p. 82)
oceanic zone (p. 83)
diatom (p. 83)

dinoflagellate (p. 83)
red tide (p. 84)
intertidal organism (p. 84)

Applying What You Have Learned

1. Which of the two main zones of the ocean offers the best fishing? Why?
2. What important producers are found in both marine and freshwater ecosystems?

Questions 3 and 4 are based on this paragraph:

The graphs in Figures A and B show the changes in population size of both phytoplankton and zooplankton. Figure A represents the change in these populations from month to month in a tropical climate. Figure B shows the changes in population size from month to month in an arctic climate.

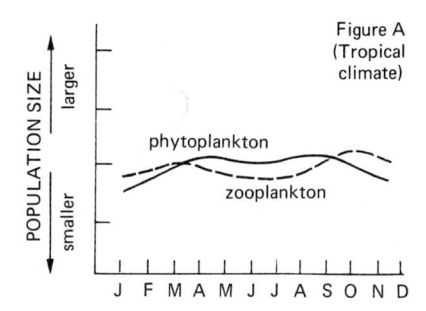

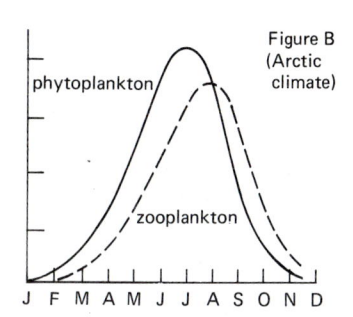

3. What environmental factor do you think is responsible for the sharp changes in population size of arctic plankton? Explain your answer.
4. Why do you think there is very little change in the size of plankton populations found in tropical climates?

Questions 5 and 6 are based on this paragraph:

The diagram in Figure C indicates where most of the phytoplankton are likely to be found during the day. Figure D indicates how these same phytoplankton will be distributed during the night.

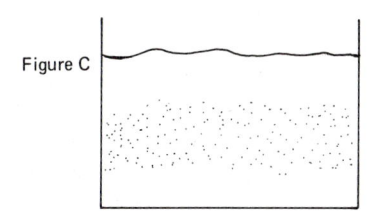

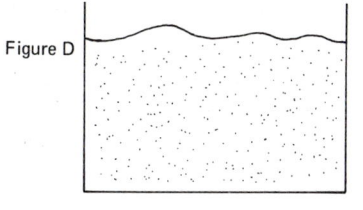

5. What may cause the phytoplankton to remain a few meters below the surface during the day?
6. Draw two diagrams, one to show where you think zooplankton would be found during the day and a second to show where you think the same zooplankton would be found during the night. Explain why you think the zooplankton are located where you have positioned them.

87

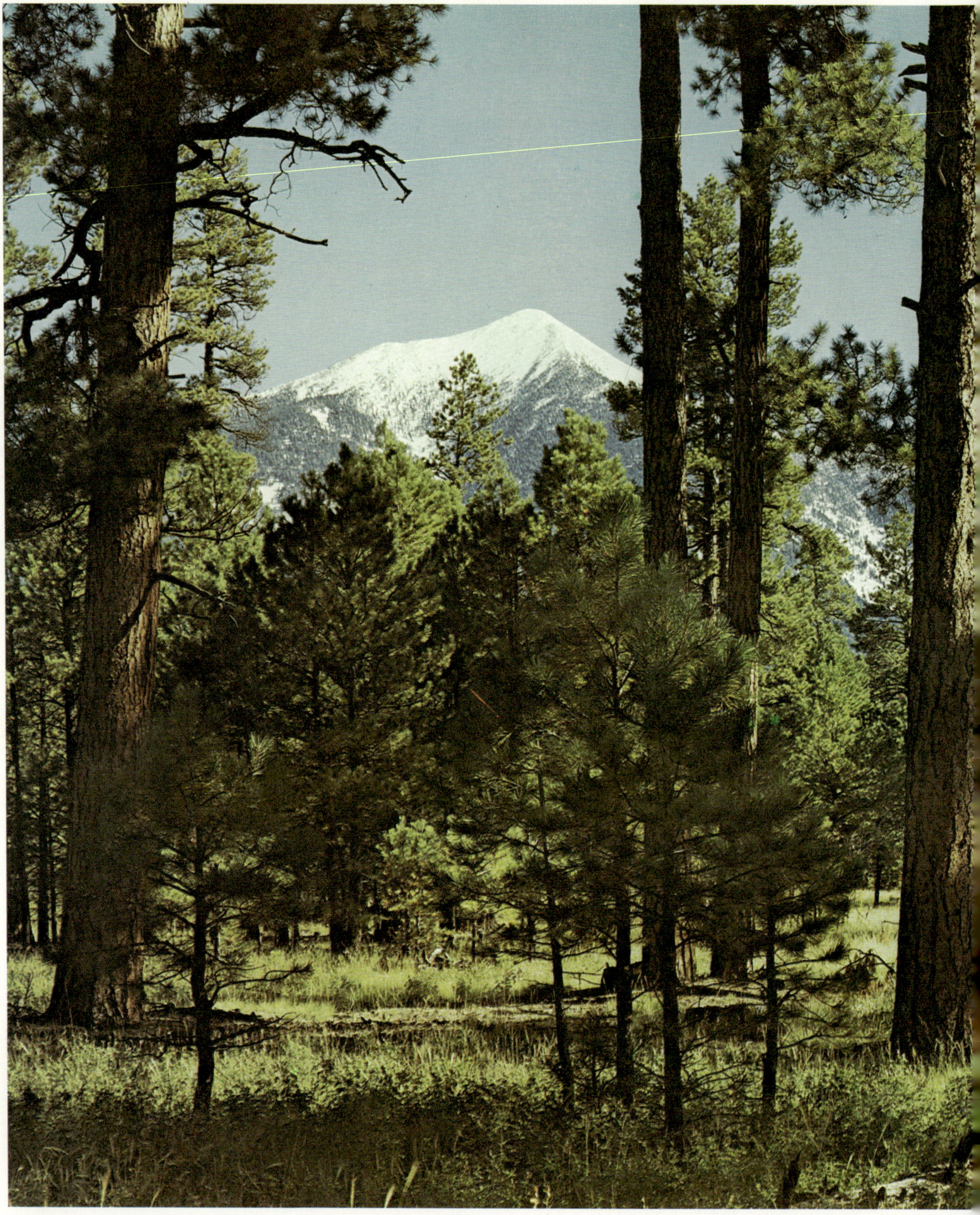

Figure 6.1: This photo shows one of the ecosystems of the terrestrial habitat. These trees are ponderosa pines. The snowcapped San Francisco mountains are in the background.

Chapter 6

TERRESTRIAL ECOSYSTEMS

The third major habitat on the Earth is the terrestrial habitat. It is the "land" habitat. It is your habitat. It is also the habitat for the greatest variety of organisms on the Earth.

THE TERRESTRIAL ENVIRONMENT

Water is a key factor in terrestrial ecosystems. Water is the main part of the habitat in both freshwater and marine ecosystems. Water is also an important part of terrestrial ecosystems. Organisms living on the land are, by definition, living out of water. But they are not free of their dependence upon it. Without it, they would die. In any study of a terrestrial ecosystem, we must always ask one question: What water is available to this ecosystem? The amount of water available is a factor that will determine to a great extent the kind of ecosystem it will be.

Figure 6.2: This polar bear is catching its dinner. Even in the frozen areas of the north there is a well-established ecosystem. The bear can exist in temperatures as low as −45°C (−49°F); the temperature of the water is about −1°C (30°F).

Temperature is another key factor. You have already learned that the temperature in freshwater and marine ecosystems is never uniform. However, the temperature does not vary as much in those two ecosystems as it does in terrestrial ecosystems. In the aquatic ecosystems, the organisms are surrounded by water. In terrestrial ecosystems, organisms are surrounded by air. (Some burrowing organisms would be an exception.) The temperature of air varies greatly. In some places, it is very warm, warmer than the water in most aquatic ecosystems. But in other places, air is very cold, much colder than water could ever be and still be a liquid.

Oxygen is usually not a problem. Organisms living in aquatic environments depend upon the oxygen that is dissolved in the water. As you learned, many things can cause that oxygen supply to decrease. This can make life impossible for many consumers in aquatic habitats. Usually the problem of a lack of oxygen does not exist for terrestrial organisms. (An exception is in habitats near the tops of very high mountains.) They are surrounded by the air, which is in constant motion. The oxygen content of air is about 20 percent. This is more than enough for any terrestrial organism.

Most terrestrial organisms need strong support systems. Life is different on land from what it is in water. The land is like a platform upon which most terrestrial organisms must live. This creates problems. Terrestrial organisms need better support systems than aquatic organisms.

What is a support system? It is a system that supports the body of an organism. In animals, it is also the system that enables the animal to move. Your muscles and bones make up your support system. Without this system, you would be a blob. With your support system, you can stand erect. And you can move.

Many terrestrial animals have skeletons and muscles. Some animals, like you, have skeletons inside the body. Others, like the insects, have their skeletons on the outside. The support systems in plants are not as complex as in animals. They do not need to be. Rigid tubes in the stem and branches of a plant are enough to keep the plant erect.

Terrestrial life is linked to the soil. In most aquatic ecosystems, the key producers are the phytoplankton. These organisms, in turn, get their nutrients from a variety of sources. Thus, you are apt to find phytoplankton almost anywhere in water, except in the great depths of the ocean.

In terrestrial ecosystems, the major producers are the seed plants. But we cannot say that you are likely to find these producers "almost anywhere." They will usually be found in one place—attached to the top layer of the soil. The soil provides a place for the attachment of the plant. And through the plant's root system, the soil feeds the plant and allows it to develop.

Figure 6.3: Terrestrial producers need good soil in order to grow well. Poor, eroded soil like this will not support a good crop of seed plants.

Populations and Ecosystems

Laboratory Activity

What Is in Soil?

PURPOSE

Soil is a mixture of substances. Some of the substances in soil are more important than others for feeding plants. In this investigation, you will find out which parts of the soil are the most important for providing plant nutrients.

MATERIALS

liter (quart) jars
small flowerpots (half-pint milk
   cartons or soup cans may
   be used instead)
dark, rich soil
a long spoon
sand
seeds of plants which sprout quickly
   (example: lima beans)
a pan suitable for heating

PROCEDURE

1. You are on your own with this investigation. Here are some facts and hints that may help you:
    a. A sample of rich, dark soil and water will separate into several layers if they are shaken together in a quart jar. These layers will contain different parts of the soil.
    b. You may use some or all of the materials listed above.
    c. After water and soil are shaken together, some parts of the soil will probably remain dissolved in the water. You will not be able to see these parts of the soil so long as they stay dissolved in the water.
    d. Soils in different areas may vary in their structure and in their ability to support plant life.
    e. Add the water and dissolved minerals to sand. Try to grow seeds in this mixture.
    f. Try to grow seeds in sand that has "plain" water added to it.
2. Write out the plan you expect to follow. Discuss it with your teacher. Then carry out the plan.
3. Record all that you do and all that you observe.

QUESTIONS

1. What are the most important parts of the soil for plant growth?
2. Explain why plants grow in some desert areas but not others.

Since the key producers—the seed plants—are attached to the soil, the terrestrial consumers must also be linked to the soil. This is an important fact to keep in mind. Ecologists worry about the soil in terrestrial ecosystems. What if it were accidentally poisoned? What if it were washed away by erosion? People who do not understand food chains are likely to say: "Who cares if the soil is washed away? It's only dirt." It is important for you to remember that soil is not "just dirt." Soil nourishes and supports all forms of terrestrial life.

How do terrestrial ecosystems differ from each other? In general, there are three main factors in the physical environment that cause ecosystems to differ: (1) the available moisture, (2) the temperature of the air, and (3) the nature of the soil. There are many combinations of these three factors on the Earth. Each one forms a unique habitat.

LESSON REVIEW *(Think. There may be more than one answer.)*

1. Terrestrial ecosystems
   a. usually have more uniform temperatures than marine or freshwater ecosystems.
   b. usually have an adequate supply of oxygen.
   c. differ in the amount of water available for organisms.
   d. are about the same size as marine ecosystems.

2. Terrestrial organisms
   a. need stronger support systems than freshwater organisms.
   b. need stronger support systems than marine organisms.
   c. if animals, cannot exist without an internal skeleton.
   d. if plants, cannot exist without an external skeleton.

3. The soil in a terrestrial ecosystem
   a. is a producer.
   b. is at the start of every food chain.
   c. provides nutrients for producers.
   d. indirectly supports all terrestrial life.

4. A factor that causes terrestrial ecosystems to differ from each other is
   a. the amount of available light for producers.
   b. the amount of available moisture for producers.
   c. the temperature of the air.
   d. the nature of the soil.

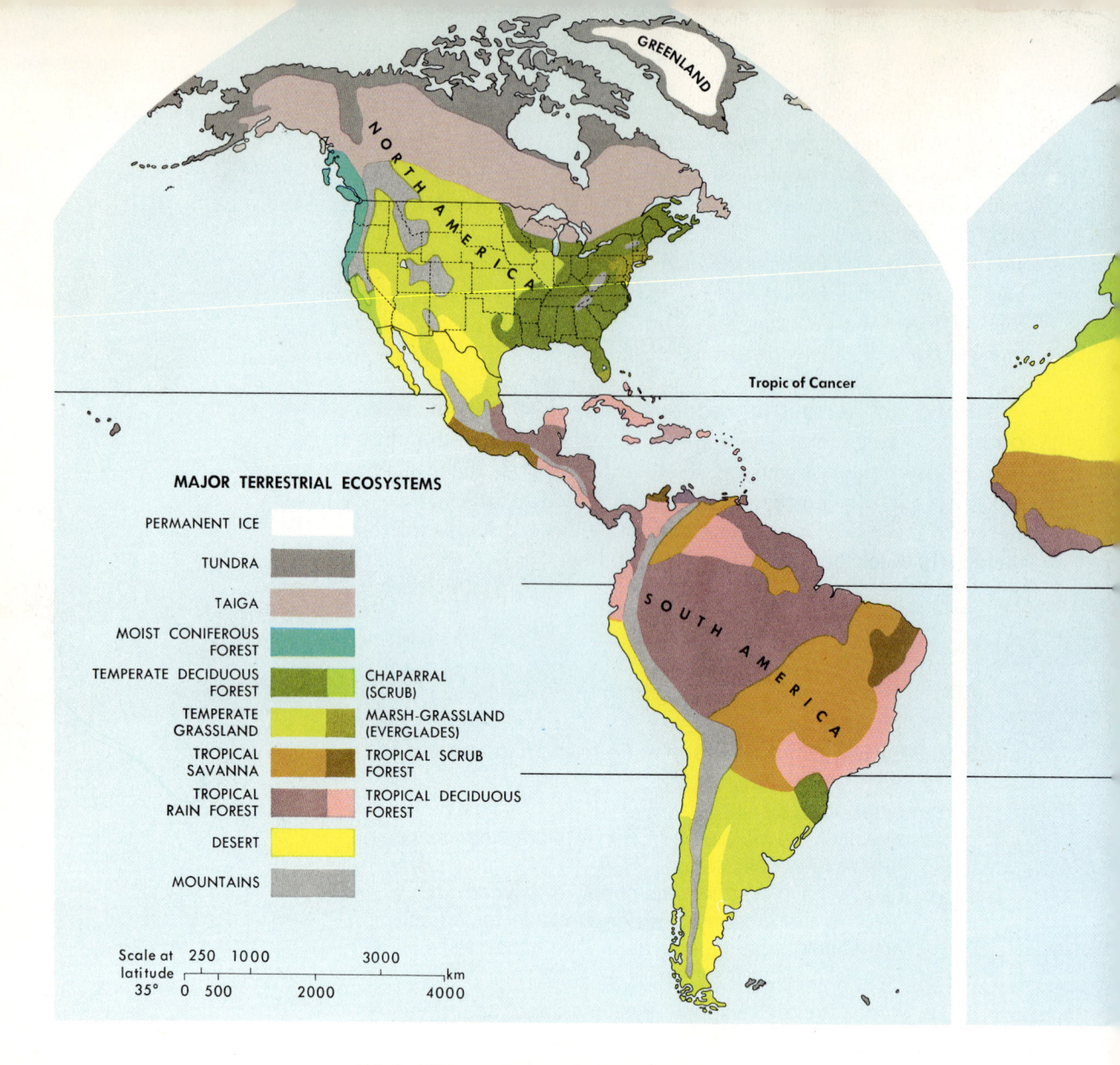

MAJOR TERRESTRIAL ECOSYSTEMS

What is a major terrestrial ecosystem? There are certain large areas of the Earth that have similar physical environments. The organisms living in these areas are also similar. Such areas are called major terrestrial ecosystems.

An example of a major terrestrial ecosystem is a desert. The physical environment of all deserts is about the same. The organisms that live in all deserts are also similar. What if you learn some of the important facts about a desert ecosystem in Arizona? Then you can later use that same knowledge to understand a desert ecosystem in Egypt or China. You can apply the

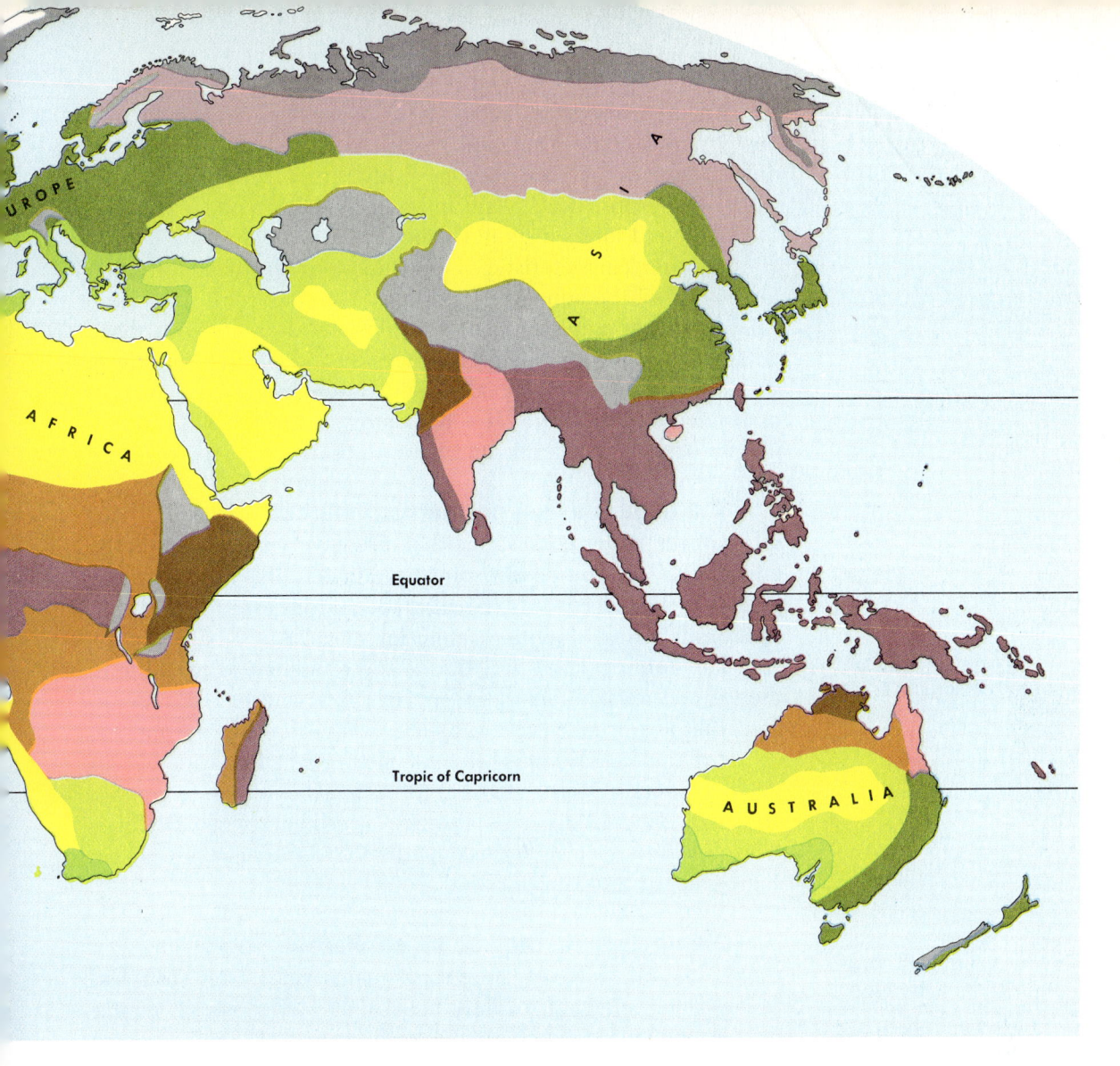

Figure 6.4: The major ecosystems of the world are shown on this map. What ecosystem do you live in? How has it been modified by people?

same idea to any of the different types of terrestrial ecosystems you will study next.

The tundra is cold and barren. Examine Figure 6.4, which shows the location of the different types of major terrestrial ecosystems of the world. The **tundra** is one of the most widespread. It is also the coldest. In fact, its low temperature is the factor that most affects the life of the region.

The tundra is a vast, nearly treeless area. (There are a few stunted trees growing here and there.) Because it is so cold, the tundra has a short growing season.

tundra:
the coldest major terrestrial ecosystem

95

Populations and Ecosystems

An interesting fact about the tundra concerns the decomposer populations. They are there. But because of the cold during most of the year, they have little opportunity to do their work. Therefore, dead plant and animal matter builds up over the years and forms a spongy type of mat over the soil.

Another characteristic of the tundra is the wide variation in the size of the animal populations. You may remember from Chapter 2 that arctic population explosions alternate with sharp declines in population. (Reread pages 29 and 30.) Despite these variations, many warm-blooded animals and insects still manage to survive in the tundra. It is not empty of animal life, as some people believe that arctic and near-arctic regions are.

The taiga is a vast northern coniferous forest. Another of the major ecosystems that also covers large areas is the **taiga** (tī gä′). If you do not want to memorize the word "taiga," just learn the three words "northern coniferous forest." Northern coniferous forest is another name for taiga.

taiga:
the northernmost forest ecosystem of Earth, also called the northern coniferous forest

conifer:
a plant (or tree) that produces its seeds in cones

The **conifer** (kŏn′ə fər) tree is the dominant form of life in the taiga. Such conifers as the spruce, pine, and fir are so dominant that few other plants can survive. Because of this, and because of the long winters, there is little plant life available for grazing animals. Hence, most of the animals that live in the taiga are small. They live on such things as seeds and twigs. The taiga is useful to us mainly because of the vast lumber supplies that can be harvested from it.

Giants live in the moist coniferous forest. You probably quit believing in giants many years ago. You had better start believing in them again. They do exist. Of course they do not go around chopping down forests, or climbing down beanstalks. Instead, they stand quietly. And they grow and grow. Consider, for example, the "General Sherman" tree. This is a

Figure 6.5: The tundra (left) is a vast flat area that is almost treeless. Small plants and a short growing season are characteristic of the tundra. The taiga (right), or northern coniferous forest, in the western hemisphere stretches from northern United States up through Canada. Its dominant trees are spruce, fir, and pine.

Terrestrial Ecosystems

giant sequoia (sĭ **kwoi′ə**) living in the **moist coniferous forest** in California. Right now its bark is about 1.2 meters (4 feet) thick. Its largest limb, when measured 20 years ago, was 2 meters (7 feet) thick. Its height is estimated to be 83 meters (277 feet). Its weight is estimated to be more than 5.5 million kilograms. That is more than 12 million pounds! And the General Sherman is believed to be between 3500 and 4000 years old.

The moist coniferous forest is somewhat like the taiga. Conifer trees dominate both ecosystems. Populations of large grazing animals are few and far between. However, the moist coniferous forest differs from the taiga in two main ways. First of all, this forest ecosystem gets much more moisture. Its yearly rainfall is up to 375 centimeters (150 inches). The second difference is that its average yearly temperatures are higher. More rainfall and warmer temperatures combine to provide excellent growing conditions for trees—hence the giants.

The moist coniferous forest has been a favorite of people for many years. After all, where else can we find so much wood that is easy to harvest? But now, after many years of lumbering, there is a problem. Our appetite for this wood has been so great that some moist conifer forests could be wiped out in a short time. Many groups are at work trying to preserve some of these forests. This is especially so for the beautiful redwood areas of California.

Many people live in the temperate deciduous forest. What is a **temperate deciduous forest?** "Temperate" refers to a middle latitude climate. A middle latitude, or temperate, climate is one that has four seasons of about equal length. The winters are cold and the summers are warm or hot. Spring and fall are in between. There is a good chance that you know this kind of climate very well. A great many people on Earth live in a temperate climate.

The term "deciduous" refers to the way the dominant vegetation loses its leaves in the winter. An "evergreen" plant keeps its leaves the whole year. A deciduous plant loses its leaves in the winter. The temperate deciduous forest is dominated by trees that lose their leaves in the winter. Look at the map in Figure 6.4. You can see that almost all of the United States east of the Mississippi River is a temperate deciduous forest ecosystem.

People have had a great impact on the temperate deciduous forest. They have cleared much of the forest area. In place of the trees, they have built cities, factories, farms, and highways. A few natural areas still exist in regions of low population.

The temperate grassland is easily abused. Except for the tundra, all the major ecosystems so far discussed have been dominated by trees. Now we are discussing a grassland

moist coniferous forest: a major terrestrial ecosystem marked by a moist climate with moderate temperatures. Large conifers dominate this forest ecosystem.

Figure 6.6: This picture of the fall season in the temperate deciduous forest in Vermont shows the leaves just before they begin to fall from the trees.

temperate deciduous forest: a major terrestrial ecosystem marked by seasonal changes. Deciduous trees dominate this ecosystem.

Figure 6.7: The prairie of grasslands (left) is the habitat of large grazing animals, such as the buffalo. People have introduced their own particular types of grasses, such as wheat, to these regions. These animals (right) are grazing in a tropical savanna area. In what ways is a tropical savanna like a temperate grassland? How do they differ?

grassland: a major terrestrial ecosystem of low moisture. Grasses and very low plants dominate this ecosystem, in which large populations of grazing animals used to live.

ecosystem. Why don't trees grow in a **grassland?** A few do grow there. But not many. The reason is that grasslands do not receive enough moisture to support the growth of forests.

Animal populations in grasslands are very different from those in forested areas. If you had been alive 150 years ago and visited the grasslands of western America, you would understand this fact better. The *prairie* (one of the names for a grassland) supported large populations of buffalo, deer, elk, and antelope. These are large animals that graze. They need large quantities of grass or other grassland plants. Such animals cannot exist, at least not in any large numbers, in a forested region.

Burrowing animals form another major group of animals living in grasslands. These include many rodents, such as the "prairie dog," as well as snakes, badgers, gophers, and owls.

In the last few hundred years, people have completely taken over just about all of the Earth's natural grasslands. Through experimental breeding, we have taken such natural grasses as corn and wheat and changed them to suit our needs. With plows and other machinery, we have destroyed the natural grasses and have planted our new grasses—*our* corn and wheat. In the process, we were a little too eager. We have created serious problems in some of the dry grasslands.

Before people began to change them, the dry grasslands had plant populations that could withstand one or more years of very dry weather. Our new grasses—mostly wheat—cannot. The result is that in very dry years wheat crops do not grow. And the native grasses are not there to grow. So the soil is bare. When the winds blow, the soil goes with the wind. Now many such regions are barren and are called "dust bowls."

Terrestrial Ecosystems

The tropical savanna supports large animals. If you want to see lions, giraffes, zebras, and elephants in nature, you will have to go to Africa. You will have to visit the **tropical savanna** (sə văn′ə) ecosystem.

What is a tropical savanna ecosystem? A tropical area is both warm and moist. A savanna is a grassland that has scattered trees, or clumps of trees. Why, you may wonder, is a tropical area with lots of moisture not completely covered with trees? That is a good question. The answer helps explain the difference between a tropical savanna and a tropical rain forest, which will be discussed next.

The tropical savanna gets large amounts of rainfall, but the rainfall comes mostly during one season. The tropical savanna has many months when there is little if any rain. Such a climate cannot support heavy growth of trees. So it is largely a grassland area. Since it is a grassland, it will support populations of large grazing animals and the second-order consumers that live on them. This ecosystem is also in danger from the actions of people.

The tropical rain forest has the greatest variety of life. The **tropical rain forest** exists in very warm areas that receive heavy rainfall during much of the year.

You already know some things about the tropical rain forest. This is the "jungle" that you have seen so many times on your television screen and in movie theaters. Still, it is very hard to understand what a jungle is really like until you have seen one. It is a teeming mass of life—of every kind that you can imagine. There is a greater variety of life in this major ecosystem than in any other.

tropical savanna: a major terrestrial ecosystem that is a warm, moist grassland. It has one rainy season a year, but is dry the rest.

tropical rain forest: a major terrestrial ecosystem that is very warm and wet year round. It is the ecosystem with the greatest variety of living things.

Figure 6.8: This tropical rain forest, or jungle, in Ecuador, South America, is teeming with animal life from under the ground to the tree tops. Life in the jungle is stratified, with certain types of organisms occupying certain vertical levels.

Populations and Ecosystems

The jungle is also the darkest of all the ecosystems. This is because it has a "roof" over it—almost a solid roof like on a building. This roof consists of the tops of trees, and the thousands of vines and other plants that grow in and among the tree branches.

The desert covers much of the Earth. Most people think of a **desert** as a place that is hot and dry. They may be only partly right. Some deserts are hot. But some deserts are cold. All deserts are dry. It is the low moisture that counts, not the average temperature. It is dryness that gives the desert its main characteristics.

desert:
a major terrestrial ecosystem marked by its very low moisture

Surprisingly, many different kinds of plant and animal life manage to survive in desert ecosystems. All of them have special structures or abilities that enable them to survive on low amounts of water. For example, some plants survive by being able to grow rapidly and reproduce during the few days that the desert receives moisture. Others drop their leaves during dry periods. Some, like the cactus, store water in their trunks and stems. Many desert plants have root systems that grow deep and spread wide under the ground.

Desert animals also have special structures and abilities. Most animals in hot deserts are *nocturnal*. This means they move about only during the night, when it is cool. Most animals in the desert have bodies that lose little water. Many survive by obtaining water from the vegetation that they eat.

mountain ecosystem:
a major terrestrial ecosystem made up of different ecosystems that change with altitude

The mountains are complex ecosystems. Only one major type of ecosystem remains to be discussed. This is the **mountain ecosystem**. What can be said about mountains? One fact is most important: Mountains do not make up one general type of ecosystem. Mountains include many different ecosystems.

Figure 6.9: This is Death Valley, California, one of the hottest places on Earth. Parts of this desert are covered by sand dunes, but the major portion has desert plants growing on it.

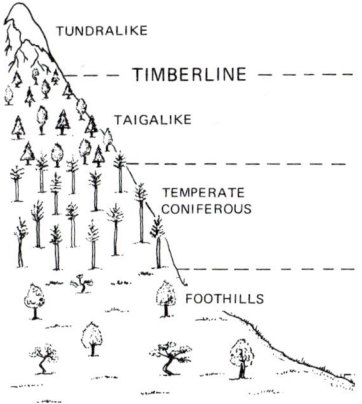

Figure 6.10: This picture (left) was taken from above the timberline in the Colorado Rocky Mountains. Note the tree line on the distant mountain. A mountain ecosystem diagram (right) includes many different ecosystems.

Picture a range of mountains. One side of this range is near the sea. The base of the mountains on this side is covered with a lush temperate coniferous forest. Now let us look on the other side of the mountain range. This other side is very dry. The base of the mountains on this other side is a very dry grassland.

Now, mentally, start to climb to the top of the mountain range. If it is a tall range, you would need a coat. The temperature decreases as you go up. If it is a very tall range, you would also need oxygen tanks. The amount of oxygen also decreases with altitude. The decrease in oxygen probably has little effect on the plant and animal life of the mountains. But the decrease in temperature does have an effect. On high mountains, we could observe a distinct line where the trees stop growing. This is called the **timberline.** Farther up you can see grasslands blend into areas almost like the tundra. If the mountains are tall enough, you would finally reach a very cold region almost barren of life.

There are several mountain ranges throughout the world. In these mountains, it is possible to find ecosystems of every type that has been described in this chapter. In fact, in some mountain ranges near the equator it is possible to find everything from a tropical rain forest to a tundralike ecosystem. Differences in altitude mean differences in temperature. Temperature differences mean differences in ecosystems.

timberline:
the height above which trees do not grow on a mountain

Populations and Ecosystems

LESSON REVIEW *(Think. There may be more than one answer.)*

1. The tundra
   a. is almost without animal life.
   b. has no decomposers.
   c. is cold during most of the year.
   d. is not one of the ecosystems of the United States.

2. Conifer trees
   a. dominate the taiga.
   b. dominate the moist coniferous forest.
   c. lose their leaves in winter.
   d. are important sources of lumber.

3. Many other types of ecosystems are included in
   a. the taiga.
   b. the temperate deciduous forest.
   c. the tropical savanna.
   d. the mountain ecosystem.

4. The temperate deciduous forest is found in
   a. Florida and Texas.   c. Africa.
   b. South America.   d. China and England.

5. Tropical savannas are unique because
   a. they grow more trees than grass.
   b. they grow more grass than trees.
   c. they get low amounts of rainfall.
   d. they get heavy rainfall for only a short period each year.

6. Which of the following terms best matches each phrase?
   a. tundra   g. tropical savanna
   b. taiga   h. tropical rain forest
   c. conifer   i. desert
   d. moist coniferous forest   j. mountain ecosystem
   e. temperate deciduous forest   k. timberline
   f. grassland

   ____ area in Africa where you would find lions and zebras

   ____ major ecosystem on coastal northwestern United States

   ____ prairie

   ____ northern coniferous forest

   ____ coldest major terrestrial ecosystem

   ____ highest point on a mountain where you would find trees

   ____ an evergreen in most cases

   ____ the dryest major ecosystem

   ____ most complex ecosystem on land

   ____ jungle

   ____ most people live in this ecosystem

Terrestrial Ecosystems

KEY WORDS

tundra (p. 95)
taiga (p. 96)
conifer (p. 96)
moist coniferous forest (p. 97)
temperate deciduous forest (p. 97)
grassland (p. 98)

tropical savanna (p. 99)
tropical rain forest (p. 99)
desert (p. 100)
mountain ecosystem (p. 100)
timberline (p. 101)

Applying What You Have Learned

Questions 1, 2, and 3 are based on this paragraph.

The following diagram shows how the shape of the land changes as you travel east from the Pacific Ocean. The high Sierra Nevada mountain range prevents moist air from reaching the intermountain area.

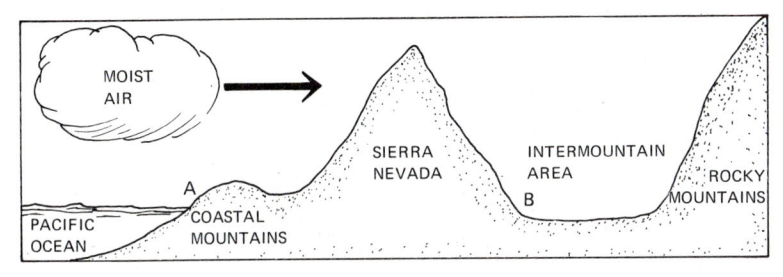

1. Compare the amount of annual rainfall expected at point A with that you would expect at point B.
2. What is the main reason for the different amounts of rainfall at points A and B?
3. Describe the type of terrestrial ecosystem you might find at point A and at point B.
4. Why are strong support systems unnecessary for organisms living in marine or freshwater ecosystems?
5. Complete the following table:

Major Terrestrial Ecosystem	Main Producers	Main Consumers
Savanna		
		Mixed wildlife and main center of population (human)
Taiga		
		Grazers and burrowers

103

Unit 3

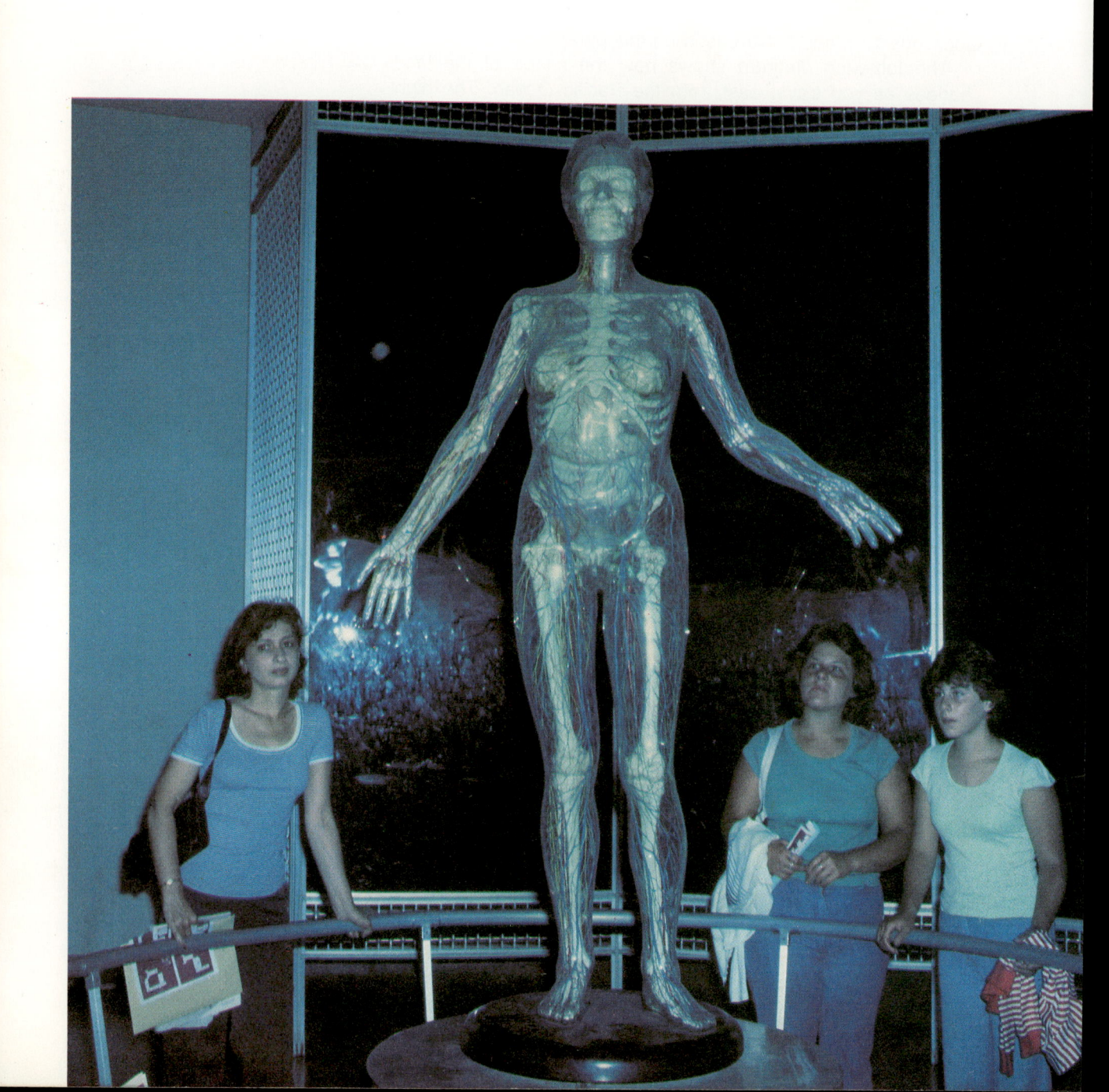

Life Inside Organisms

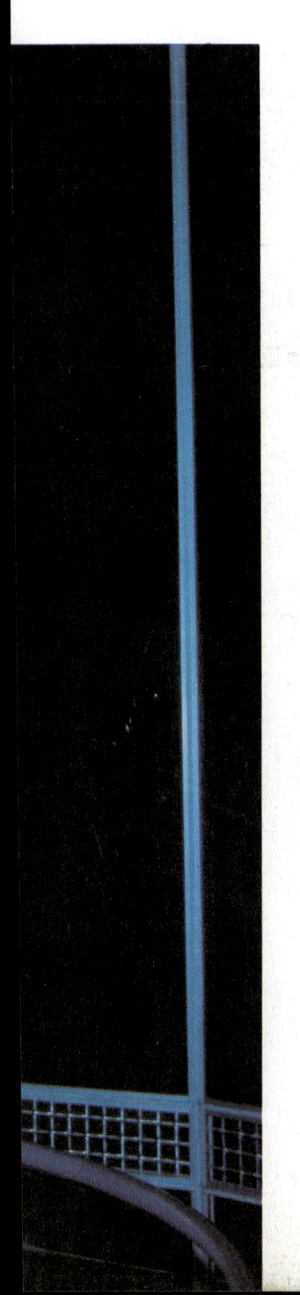

Life has a unique organization. What does that mean? When something is organized it has order: Its parts are arranged in a particular way. Life has a special kind of order. That is why it is unique. In this unit, you will learn something about the inner organization of living things—from the simple to the complex. The two chapters of this unit will lead you to the next unit. There you will learn how life inside organisms is ordered so that the organisms can function in special ways.

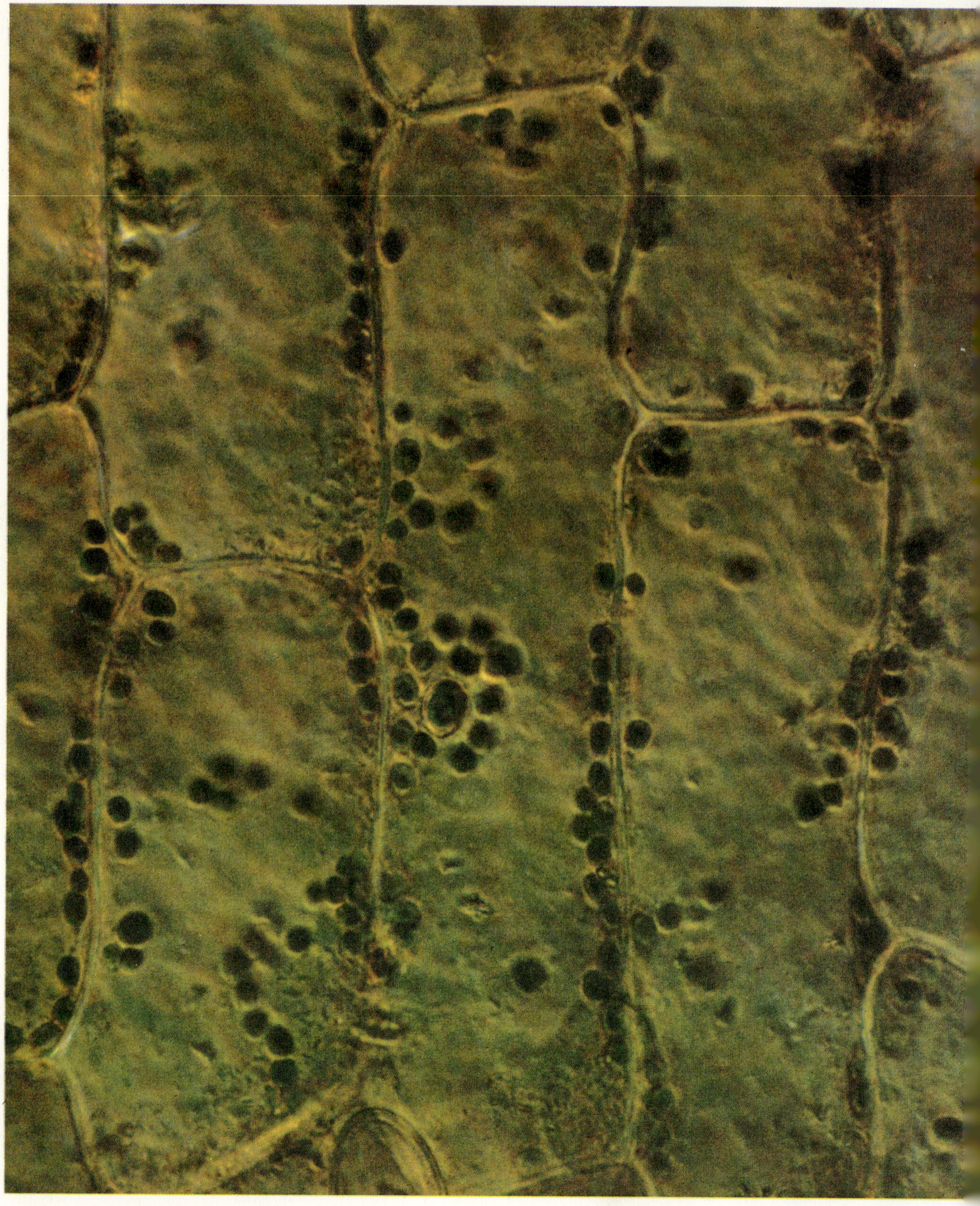

Figure 7.1: Cells are the "building blocks" of organisms. These are plant cells. The green bodies are chloroplasts, which make food.

Chapter **7**

CELLS AND CELL PRODUCTS

After studying Unit 2, you should be able to answer this question: What are ecosystems made up of? Answer: Ecosystems are made up of populations and their physical environments. You should also be able to answer this question: What are populations made up of? Answer: Populations are made up of **organisms.** (You might have said "individuals" instead of "organisms." The life scientist prefers to use the term "organism" to describe an individual living creature.) Here is another question that a student of life science should be able to answer: What are organisms made up of? The title of this chapter gives you the answer: Organisms are made up of cells and cell products. But what are cells and cell products? That is what you will learn in this chapter.

organism:
an individual living thing of any kind

CELLS: THROUGH THE LIGHT MICROSCOPE

What is a cell? About 300 years ago in England, Robert Hooke sliced a piece of cork with his pocket knife. He looked at the cork with his microscope. He saw that it was full of holes. Looking more carefully, he saw that the cells were like little

107

Life Inside Organisms

rooms. "Cell" is another name for a little room. So he used the name cell for the holes in the cork.

Cells are not holes in cork. In fact, they are not holes in anything. Life scientists since Hooke's time found that the holes he had named once contained small living bodies. They also found these kinds of small living bodies in every plant and every animal. What should we call such a small living body? It does not seem like a good idea now to name them after little rooms. But that is what happened. The small living bodies found in all plants and animals were called **cells.** They still are.

cell:
the basic structural unit of organisms

Two types of microscopes are used to study cells.

A microscope is needed to see most cells. The first microscopes seem to have been invented in the late sixteenth century. There is evidence that they were used for the first time in Holland. By 1610, the famous Italian scientist Galileo had made a microscope and had used it to observe small animals.

The usual microscope you see in a school or hospital is called a *light microscope*. It has a lens system that focuses light rays in such a way that a magnified image is created. How much does such a microscope magnify? The best light microscope can magnify a cell 1500 times. This is about like magnifying a postage stamp to the size of a football field.

The second type of microscope used for cell study was invented in the early 1950s. It is the *electron microscope*. In it, magnets are used to focus electrons. (In a TV set, magnets are used to focus electrons to get an image.) Great magnifications are possible with the electron microscope. When special techniques are used, cells can be magnified nearly 500,000 times with the electron microscope. This is like magnifying the dot over this letter "i" to about the size of a baseball field.

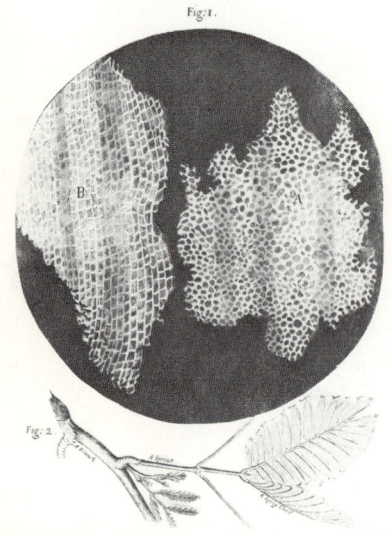

Figure 7.2: This is the microscopic view of cork cells as seen by Robert Hooke, himself. The drawing comes from a book that Hooke published in London in 1667.

Figure 7.3: This student is using a light microscope (left) to study cells. The microscope shown is a research model, with a tube on top for mounting a camera. An electron microscope (right) is large and expensive.

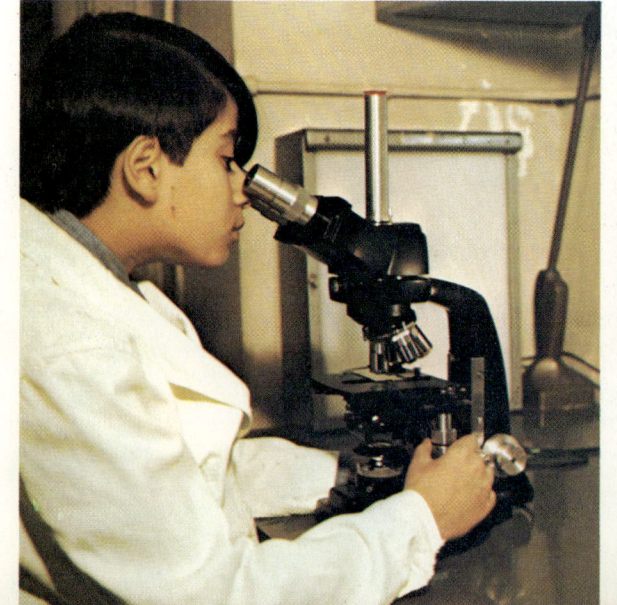

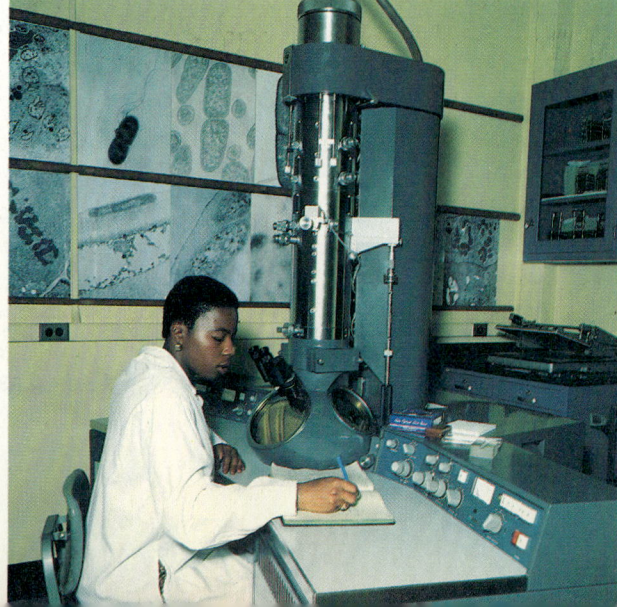

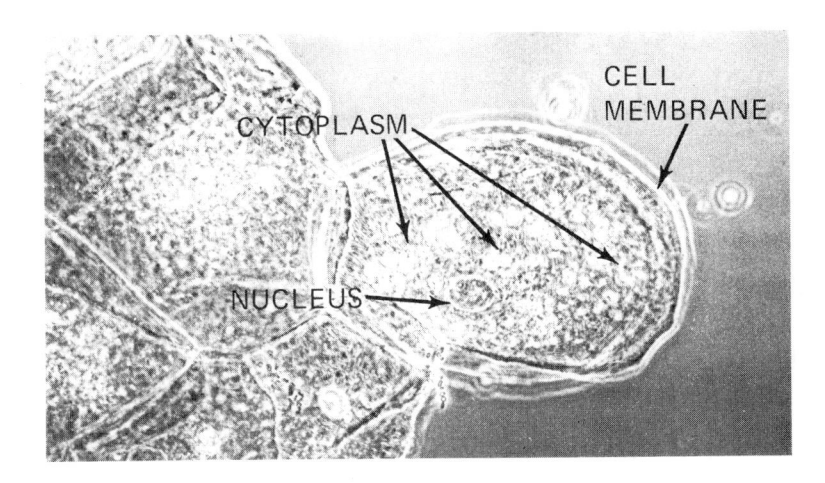

Figure 7.4: Thin flat cells like these line the inside of your cheek. A light microscope was used to take this photo.

The electron microscope has two main disadvantages. First, it is very expensive. Second, almost everything must be killed and treated chemically before it can be viewed under the electron microscope. Therefore, living cells cannot be studied with the microscope.

Most cells have two main regions. Cells are found in many sizes and shapes. But most of them have two regions that can be seen with the light microscope. One such region is the **nucleus** (noo′klē əs). (See Figure 7.4.) In some cells, the nucleus is visible even when viewed through a simple light microscope. In other cells, the nucleus is not visible unless it is stained with a dye. You will learn more about the nucleus later in this chapter.

The second region of the cell includes everything that is outside the nucleus. This is called the **cytoplasm** (sī′tə plaz′əm). Have you ever watched soup boiling in a pot? That is about how living cytoplasm looks when it is viewed with a good light microscope. The cytoplasm is truly a busy place.

Plant cells can be recognized. With a little practice, you can easily tell the difference between most plant and animal cells. Most plant cells have a thick wall around them. It is called the **cell wall.** The plant cell wall is fairly stiff or rigid.

Many plant cells are regular in shape, like little boxes. Also, in the cytoplasm of many plant cells, there are tiny, green bodies called **chloroplasts.** The process of photosynthesis—the food manufacturing process—takes place in the chloroplasts. (See Figure 7.5.) Chloroplasts can be observed with the light microscope.

nucleus:
usually the central portion of a cell. It controls the growth, reproduction, and heredity of the cell.

cytoplasm:
the material outside the nucleus in a living cell

cell wall:
the stiff, outer covering of most plant cells

chloroplast:
a tiny, green body in the cytoplasm of a plant cell. Photosynthesis takes place in the chloroplast.

Life Inside Organisms

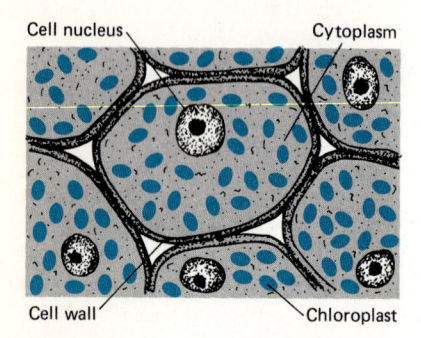

Figure 7.5: Plant cells often have a rigid cell wall and chloroplasts, where food manufacturing (photosynthesis) takes place.

In the stems of plants, there are few living cells. Only the thick cell walls of cells that were once alive remain. These cell walls form the tubes through which water and food materials move up and down the stem. These cell walls also strengthen the stem of a large plant such as a tree. A piece of wood is really a big mass of cell walls packed together. The mass of cell walls in cork is also part of a tree. Cork is the thick bark of the cork oak.

Animal cells have many shapes. Think of a shape—any shape. A life scientist could probably find you an animal cell that has a similar shape. The walls of animal cells are not like the walls of plant cells. They are not as stiff or rigid. Animal cells are also more likely to be irregular in shape.

Certain kinds of nerve cells have the most unusual shape of all the animal cells. Some have their nucleus and the main part of their cytoplasm in your spinal cord. (The spinal cord is inside your backbone.) These same nerve cells have a branch of cytoplasm that is long and very thin. Such branches extend out as far as your fingertips and the tips of your toes. So really you have cells in your own body that are nearly 1 meter (about 39 inches) long.

Some cells are organisms. If you looked at a drop of pond water with a microscope you might see a hundred or more organisms. A few of these may be *multicellular organisms*. *Multi* means "many." A multicellular organism has a body made of many cells. Most of the organisms in the pond water will probably be *single-celled*. A single-celled organism has a body

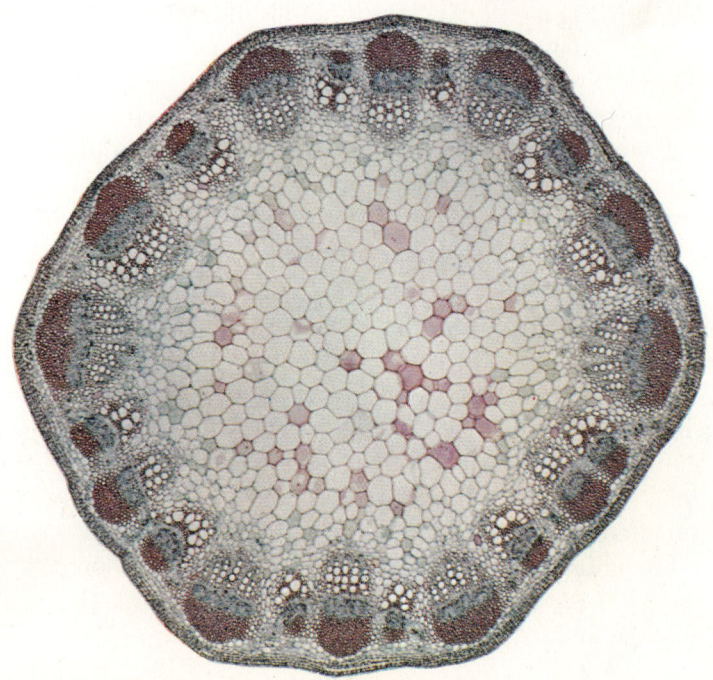

Figure 7.6: Under the microscope, a slice through a plant stem shows its structure. Cell walls make up most of the plant stem.

110

Cells and Cell Products

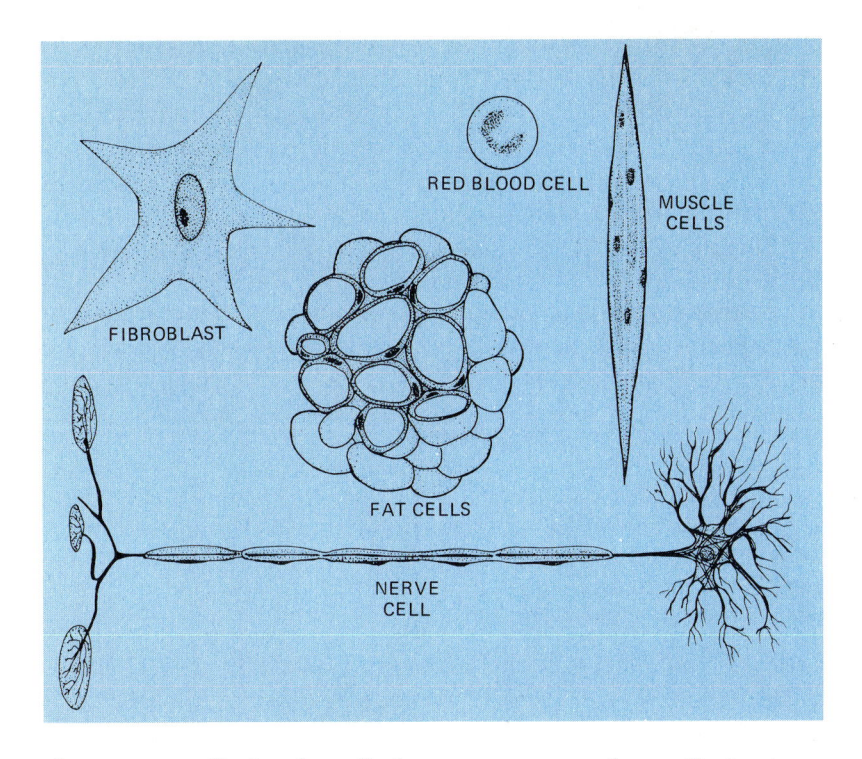

Figure 7.7: Animal cells have a variety of shapes. The star-shaped fibroblast is a cell that helps to form scar tissue.

that is one cell. Single-celled organisms are often called **microorganisms.**

Some single-celled organisms have cell walls and chloroplasts in their cytoplasm. Thus, they are like plant cells. Some single-celled organisms have no cell wall, and they can move quickly. They are somewhat like animals. Many single-celled organisms have bodies that are not like plant or animal cells. (See Figure 7.8).

microorganism:
an organism too small to be seen with the naked eye. Usually one-celled, it can be seen through a microscope.

Figure 7.8: Most single-celled organisms are microscopic. Like an animal, a paramecium (right) moves and has no cell wall. Diatoms (center) have chloroplasts and cell walls—like plants. *Euglena* (left) is different: It moves and also has chloroplasts.

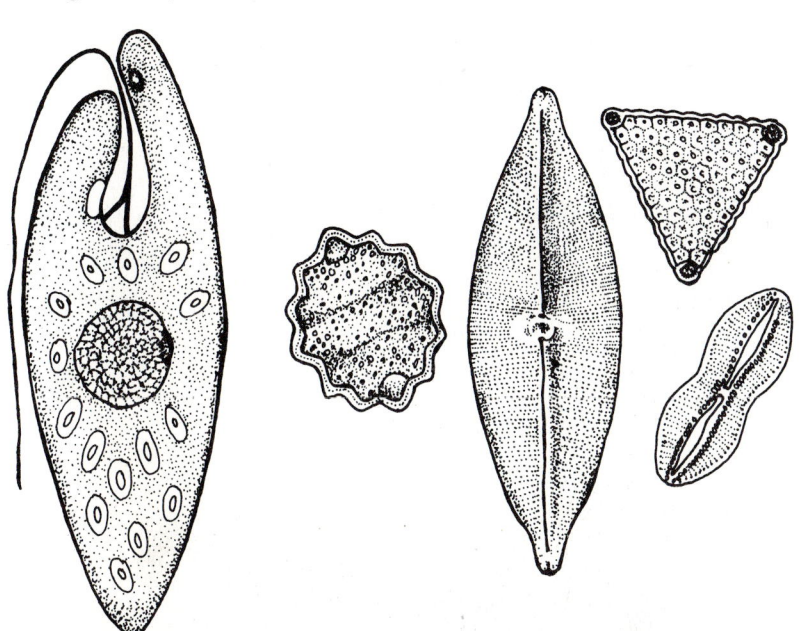

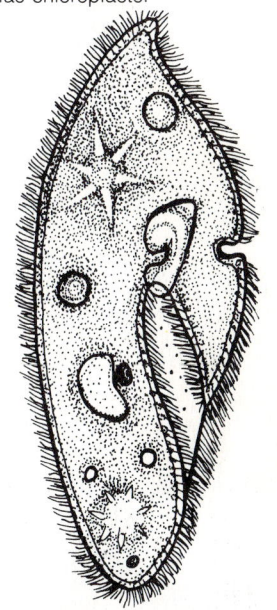

111

Life Inside Organisms

Laboratory Activity

What Are Organisms Made Of?

PURPOSE You know that organisms are made up of cells and cell products. In this investigation, you will explore some organisms with a microscope. In this way, you will get to know different cells and cell products.

MATERIALS
celery
drinking glass
water
red food coloring
medicine dropper
glass slides
onion
sharp knife
tweezers
tincture of iodine
cover slips
microscope
raw meat (including fat)
toothpicks

PROCEDURE

1. Celery Cells. Pour water into a glass to a depth of about 3.0 cm. Add a few drops of red food coloring. Cut the rough edge from the base of a celery stalk. Place the celery stalk in the colored water. Set the stalk in water aside for use later in the investigation.

2. Onion Cells. Place a drop of water on a glass slide. Cut an onion into quarters. Separate several layers from one quarter. Select an inner layer. With tweezers, peel off a small piece of the inner skin, as shown in A. Place the skin in the drop of water, as shown in B. Add a drop of tincture of iodine to the top of the skin, as shown in C. Place a cover slip over the skin, as shown in D. Flatten the skin against the slide. To do that, gently press down on the cover slip, as shown in E.

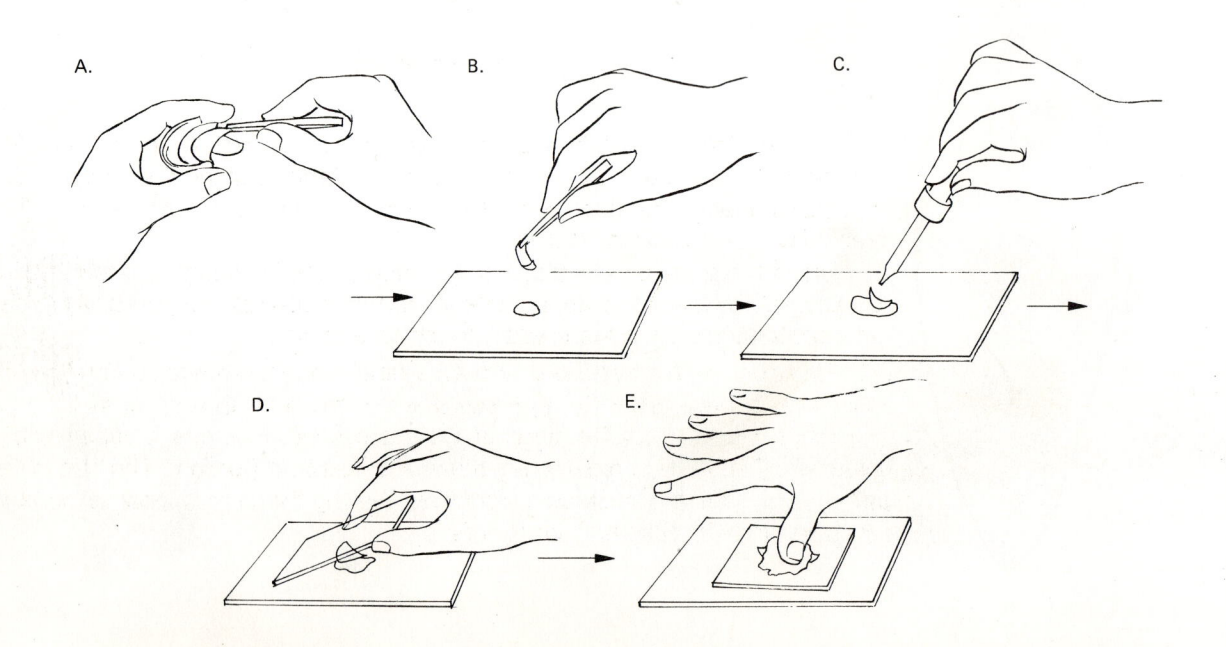

A.

B.

C.

D.

E.

Place the slide on the stage of the microscope. Examine the slide under low power. Observe the individual cells. These cells store food for the developing onion plant. Look at the rigid wall around each cell. The cell wall was formed from materials produced by the cells. Examine the slide under high power. The tiny round yellow mass in each cell is the nucleus.

3. Connective Tissue. Place a small piece of meat on a slide. Gently tear the meat apart with tweezers. Examine the meat to find a layer of cell material that is thin and white. Hold the meat with a toothpick. Grasp the thin, white material with tweezers. Separate it from the meat, as shown in A. Put the meat aside for use in the next investigation.

Place the white material in a drop of water on the slide, as shown in B. Hold down an edge of it with a toothpick. With a second toothpick, flatten the white material on the slide, as shown in C. Add a drop of tincture of iodine to it. Place a cover slip over it. Gently press down on the cover slip.

Place your slide on the stage of the microscope. Examine it under high and low power. You are looking at cell products called *connective tissues*. The long wavy threads are cell products called *collagen* (**kŏl'** ə jən) *fibers*. Connective tissue surrounds muscle groups and holds the muscle fibers together.

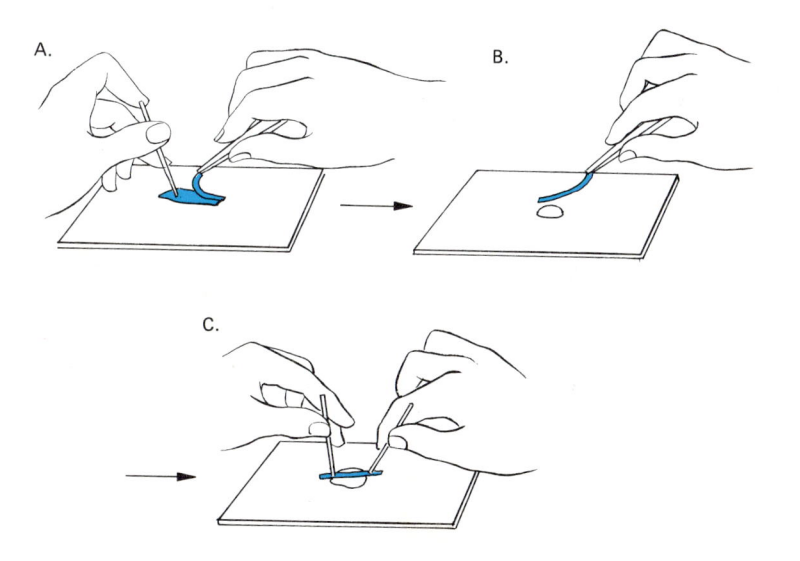

4. Muscle Cells. With tweezers, remove one threadlike section of meat. Place the meat section in a drop of water on a slide. Gently pull the fiber apart with two toothpicks. Flatten the fiber on the slide. Place a cover slip over the fiber. Gently press down on the cover slip.

Place the slide on the stage of the microscope. Examine it under high and low power. The long, thin striped cells are called *striated muscle cells*. When these cells contract, the muscles shorten to produce motion.

5. Fat Cells. With tweezers, remove a small amount of white fat from the meat. Place the fat on a slide. Gently press down on the fat with a cover slip.

Place the slide on the stage of the microscope. Examine it under high and low power. The cells contain a fatty material called *lipid* (**lĭp'** əd). The lipid is transparent. It pushes the nucleus to one side, making the cells appear to be empty. Lipid stored in fat cells is used for energy.

Life Inside Organisms

6. Cheek Cells. Wash and dry your hands. Place a drop of tincture of iodine on a slide. Gently scrape the inside of your mouth with a fingernail. With a toothpick, remove the soft white material from under your fingernail. Smear this material in the tincture of iodine on the slide. Place a cover slip over the smear.

Place the slide on the stage of the microscope. Examine it under high and low power. These flat irregularly shaped cells line the inside of your cheeks.

7. Tomato Cells. Cut into a tomato with a knife. Peel off a thin layer of skin. Place a small piece of tomato skin in a drop of water on a slide. Place a cover slip over the skin.

Place the slide on the stage of the microscope. Examine it under low and high power. Observe especially the cells at the edge of the torn skin. The green cell walls are cell products. They are similar to the cell walls of the onion. The yellow-red bodies in the cytoplasm are called *chromoplasts.* They give the cells their red color.

Remove a small piece of pulp from the tomato. Place the pulp in a drop of water on a slide. Gently "tease" the pulp apart with two toothpicks. Gently press down on the pulp with a cover slip. Examine the slide under the microscope under low and high power. Locate cell walls, nuclei, and chromoplasts.

8. Celery Cells. Remove the celery stalk from the red-colored water. Cut the stalk above the water line. Cut a 1.5-cm slice from the bottom of a long stalk, as shown in A. Look at the cut edge of the 1.5-cm section. The little red circles are called *xylem* (**zī′**ləm) tubes. These tubes carry water up the celery stalk.

With care, slice along the length of one of the red tubes, as shown in B. Make a second slice, close to the first. Now you should have a thin section containing part of the red tube. Place this section in a drop of water on a slide. Place a cover slip over the celery section.

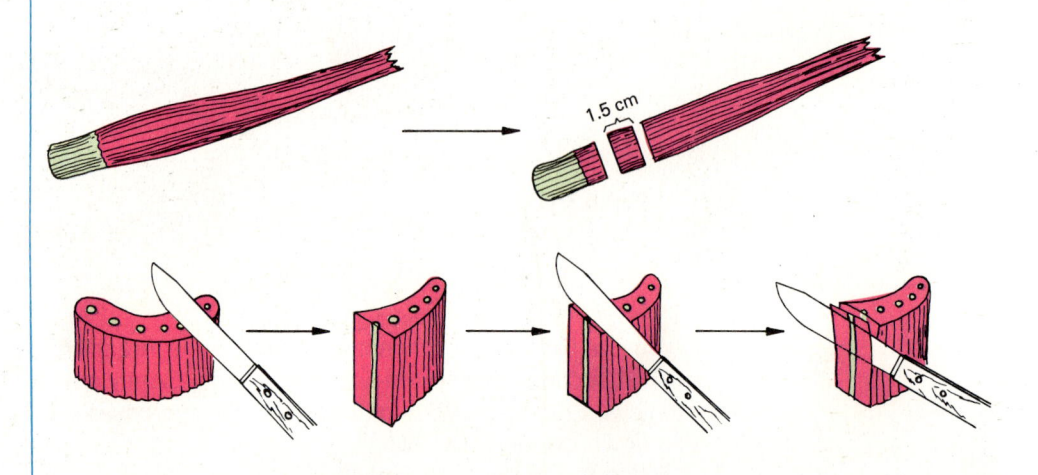

Place the slide on the stage of the microscope. Examine it under high and low power. The long spiraled cells you see make up the xylem tubes. The thick cell walls help to support the weight of the stalk.

Career Spotlight

Laboratory tests play an important part in detecting and treating many diseases. These tests are done by medical laboratory workers, under the direction of doctors. Special equipment is often used for the tests. The findings from medical laboratory tests help doctors treat their patients.

Medical technologists have 4 years of college training and in some states, must be licensed to do their work. They are helped by *medical technicians,* laboratory workers who have finished a 2-year program. There are also *medical laboratory assistants,* trainees who have at least 1 year of college work.

Medical laboratory workers do many different kinds of tests. They draw blood, study it under a microscope, and test it chemically. They check tissue cells to see if there are any harmful bacteria or parasites in a patient's body. They test urine samples that patients give almost every time physical examinations are done. They also perform more complicated tests that were once done by doctors. In the picture, a technologist is testing blood using radioactive chemicals.

Medical care is becoming more widely available, and laboratory tests are becoming a greater part of that care. Thus, there should be many opportunities for you as a medical laboratory worker—at all levels—in the years ahead.

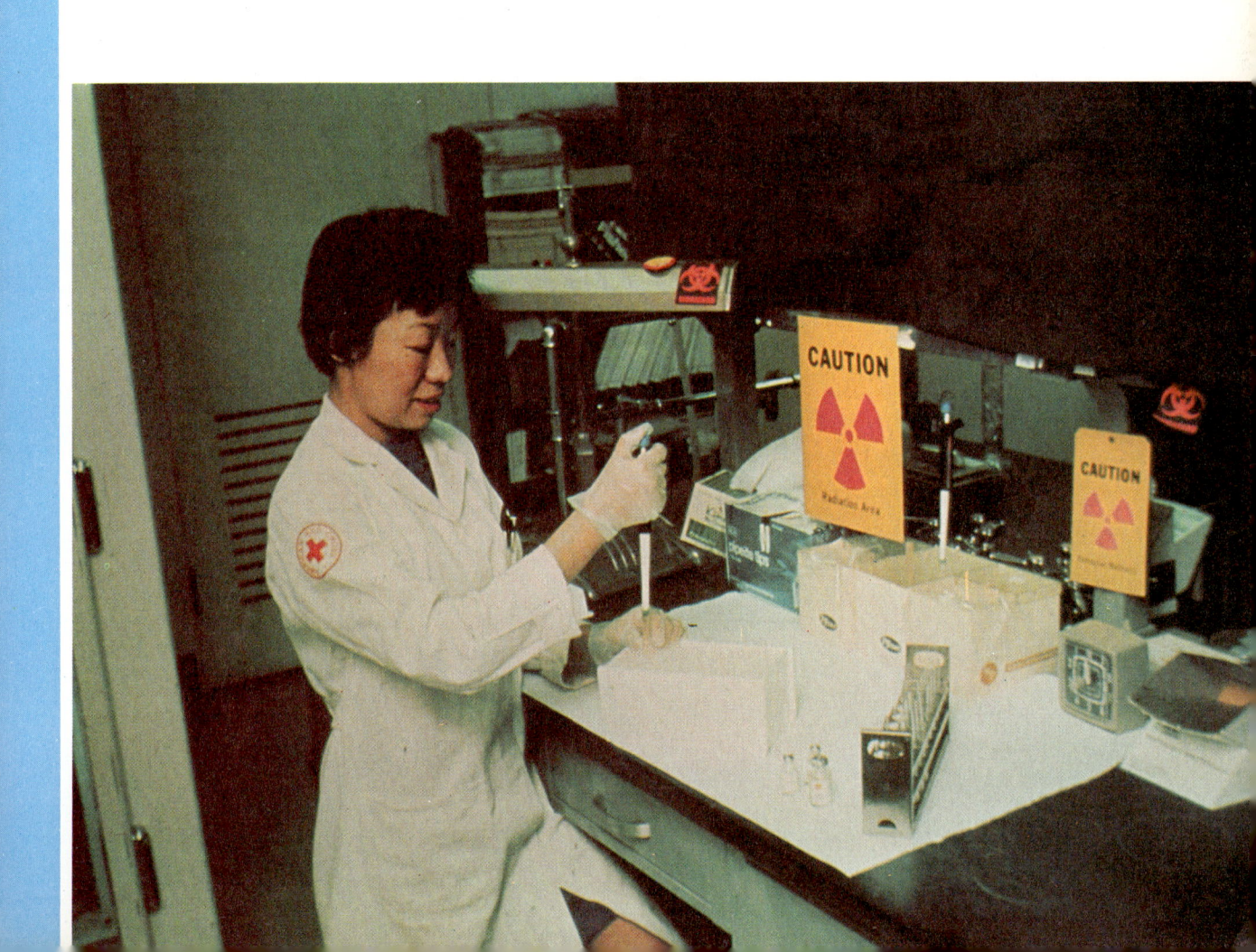

Life Inside Organisms

LESSON REVIEW *(Think. There may be more than one answer.)*

1. Cells
   a. are small living bodies found in organisms.
   b. were named by Robert Hooke.
   c. were first described as holes in cork.
   d. are found in plants but usually not in animals.

2. The electron microscope
   a. can be moved around like a light microscope.
   b. magnifies more than the light microscope.
   c. magnifies objects a maximum of 1500 times.
   d. cannot be used to study living cells.

3. The cytoplasm of a living cell
   a. is the region inside the nucleus.
   b. is the region inside the cell.
   c. is the region inside the cell that is outside the nucleus.
   d. is in constant motion.

4. Look at the drawing of two cells. Of the two,
   a. cell A is more likely to be an animal cell.
   b. cell A is more likely to contain chloroplasts.
   c. cell B is more likely to be surrounded by a cell wall.
   d. cell B is more likely to belong to a producer.

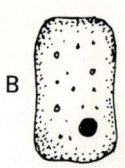

5. Which of the following words best matches each phrase?
   a. organism   e. cell wall
   b. cell   f. chloroplast
   c. nucleus   g. microorganism
   d. cytoplasm

   _____ controls cell growth, reproduction, and heredity

   _____ individual living thing

   _____ region of cell outside the nucleus

   _____ rigid boundary of plant cell

   _____ basic unit of living things

   _____ organism that can only be seen with a microscope

   _____ cell body that carries on photosynthesis

KEY WORDS

organism (p. 107)   cytoplasm (p. 109)   chloroplast (p. 109)
cell (p. 108)   cell wall (p. 109)   microorganism (p. 111)
nucleus (p. 109)

CELLS: THROUGH THE ELECTRON MICROSCOPE

Organelles do the work in the cell. An organelle is a part of the cell that does a specific job. We have already mentioned two organelles, the chloroplast and the nucleus. Life scientists learned much more about chloroplasts, nuclei, and other cell organelles after they started to use the electron microscope.

organelle: any one of several parts of a cell that does a specific job

The cell membrane encloses the cell. One of the least understood organelles is the **cell membrane**. (See Figure 7.9.) It regulates whatever enters and leaves the cell. It is this "gatekeeper" function that is poorly understood.

cell membrane: the outer part of a cell. It encloses and regulates whatever enters or leaves the cell. In a plant cell, the cell membrane is inside the cell wall.

The mitochondrion is the powerhouse. Through a good light microscope, you can see tiny specks dancing in the cytoplasm. Through the electron microscope, these tiny specks are seen as the complex organelle shown in Figure 7.10. This organelle is called the **mitochondrion** (mīt ə kon′drē ən). It is called the powerhouse of the cell. The mitochondrion has this name because work done in it produces most of the energy used by the cell.

mitochondrion: an organelle in which most energy in a cell is produced

Figure 7.9: In this electron micrograph (a photo taken with an electron microscope), you see the cell membrane of one of the cells lining the intestines of an animal body.

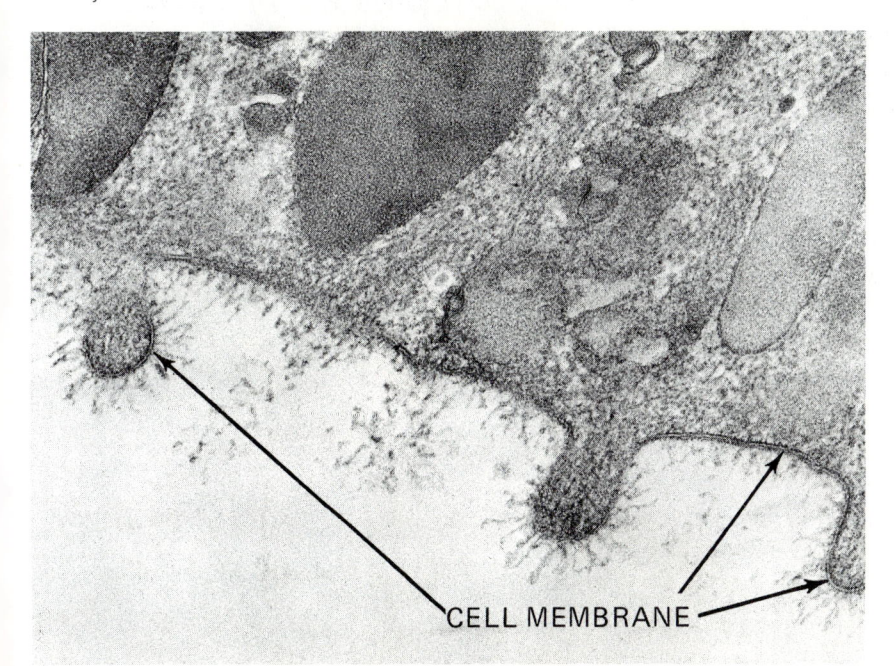

Life Inside Organisms

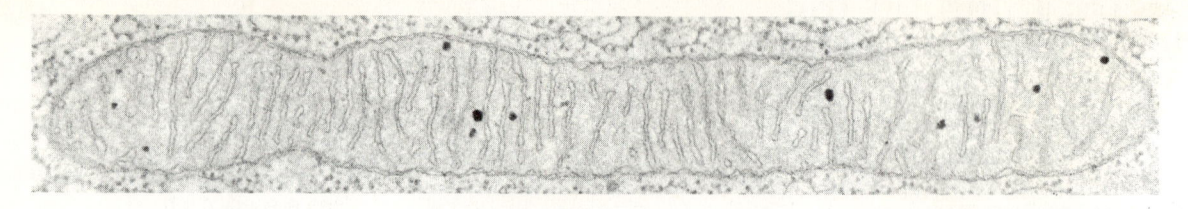

Figure 7.10: The mitochondrion is usually "sausage-shaped." The structures inside it are like partitions or cross-walls.

The endoplasmic reticulum partitions the cell. It was once thought that the cytoplasm was one jellylike mass. That was when the light microscope was the major tool of the life scientist. The electron microscope has since shown that the cytoplasm is more complex. For example, there is a network of thin membranes that criss-crosses the cytoplasm. This network is called the **endoplasmic reticulum** (rə tĭk′yōō ləm). (See Figure 7.11.) No one is exactly sure yet what the endoplasmic reticulum does for the cell. The membranes seem to act like

endoplasmic reticulum: the network of thin membranes across the cytoplasm of cells

Figure 7.11: This is an electron micrograph of the cytoplasm of a cell. The endoplasmic reticulum covers most of the picture.

Figure 7.12: The ribosomes are small round bodies. Most ribosomes are close to membranes of the endoplasmic reticulum.

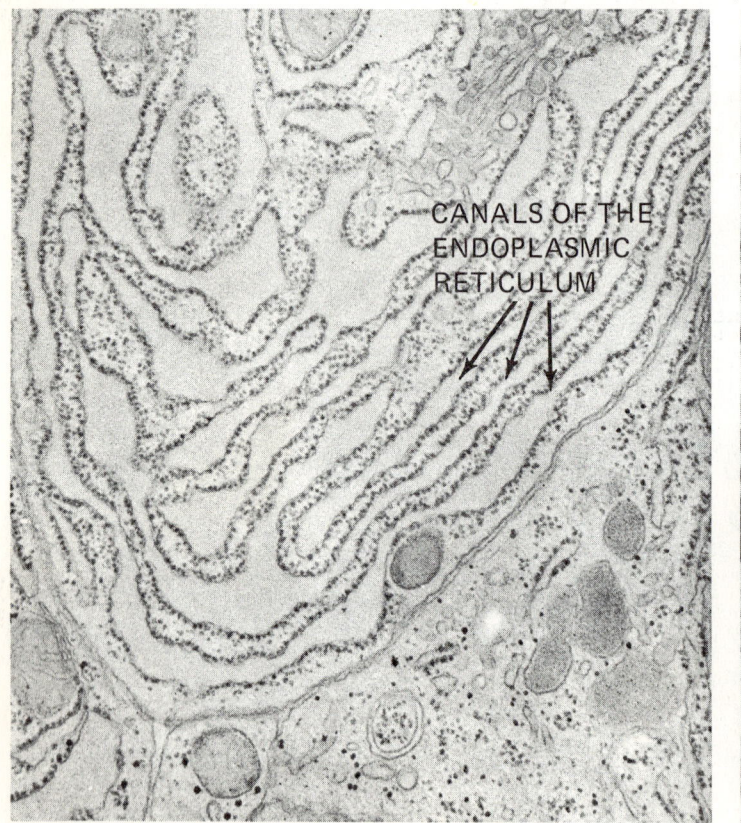

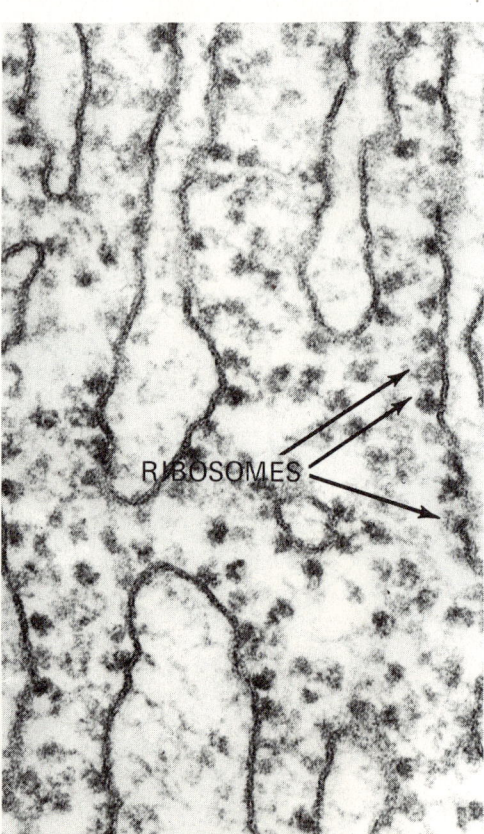

CANALS OF THE ENDOPLASMIC RETICULUM

RIBOSOMES

partitions, such as the walls that divide your house into rooms. The membranes may also act like canals in the cytoplasm. Perhaps chemicals move back and forth from the nucleus to the outside of the cell through these canals.

The nucleus and the ribosomes work together. The **ribosomes** (rī′bə sōm′) are very small bodies that are scattered throughout the cytoplasm. (See Figure 7.12.) Along with the nucleus, they help make **proteins** (prō′tēn). What are proteins? You would have to know a lot more about atoms and molecules before we could tell you exactly what they are. For now, let us say that proteins are the main building blocks of cells and organisms. They also do most of the work inside the cell.

The nucleus and the ribosomes work together to make proteins. The nucleus contains the instructions for each of the different proteins that can be made by a cell. The instructions are coded in one of a group of chemicals that you may have heard of. It is called **DNA** for short. The instructions in the DNA can be copied by chemical "messengers." The messenger chemicals carry the instructions to the cytoplasm. There the ribosomes team up with the messenger chemicals. The ribosomes use the instructions transferred from the DNA to make the proteins. Of course, this is a very simple explanation. You might like to check in your library for more information about DNA and the way proteins are made in cells.

The Golgi complex exports proteins. There is an organelle in the cell that has been called a "stack of sacs." That

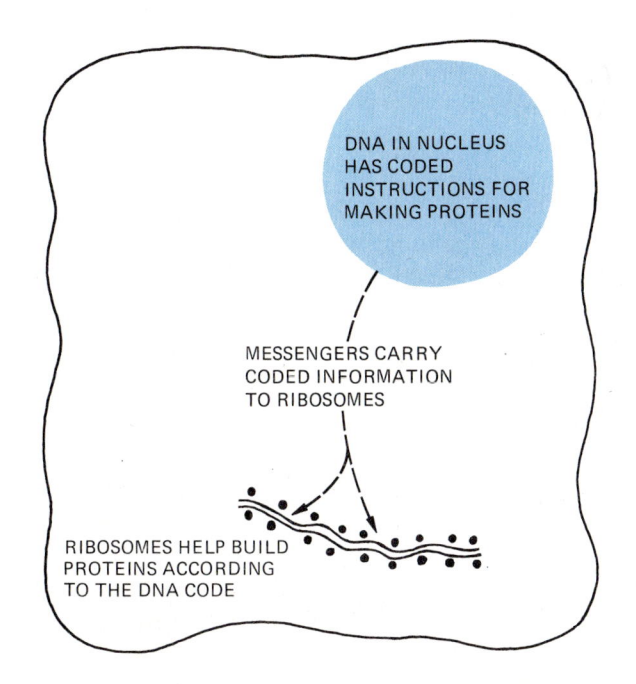

ribosome:
an organelle that helps to make protein

protein:
one of a group of complex chemicals that are the building blocks of cells. They do most cellular work.

DNA:
an abbreviation for the complex chemical in all nuclei that carries "instructions" for heredity. All the body's proteins are "built" according to these instructions.

Figure 7.13: The DNA contains the instructions for making all the organism's proteins. The ribosomes do the actual building.

Life Inside Organisms

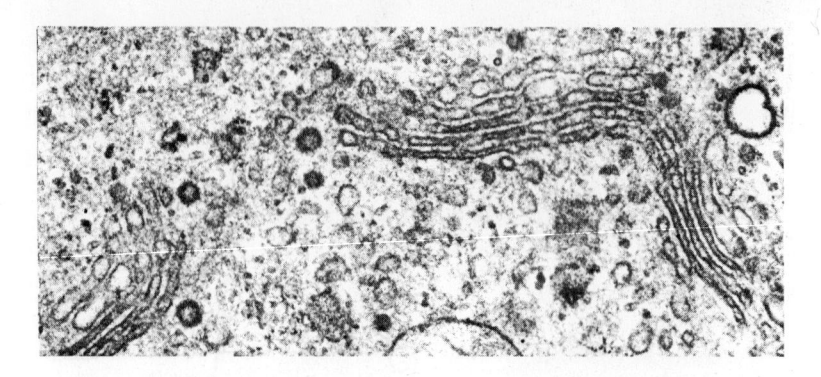

Figure 7.14: The series of hump-shaped structures visible in this electron micrograph is the "stack of sacs" of the Golgi complex.

Golgi complex:
an organelle that prepares proteins for "export" from cells

organelle is the **Golgi** (gōl'jē) **complex,** named after the Italian life scientist who first identified it. (See Figure 7.14.)

After proteins are assembled by the ribosomes, some of them move into the saclike parts of the Golgi complex. Once inside, other chemicals are added to the proteins. (These other chemicals are made within the Golgi complex.) Then an interesting thing happens. The protein and the chemical linked to it are enclosed in a bubblelike container. That breaks off from one of the sacs and moves toward the outside of the cell. It is eventually released from the cell. One could say the Golgi complex wraps and exports proteins from the cell.

The lysosome is a digestive organelle. The **lysosome** (lī'sō sōm) is an organelle that functions like your stomach and intestine. (See Figure 7.15.) The stomach and intestine digest food. They break it down into smaller parts that can be used by your body. The lysosomes do the same thing inside cells. When small particles of food enter cells, they are digested by the lysosomes.

lysosome:
an organelle that digests food in an individual cell

The centrioles help animal cells divide. In animal cells, there are two round, long organelles that look like cylinders. They are called **centrioles.** Centrioles are made up of threadlike fibers. When cells divide, the DNA in the nucleus is duplicated. Then it must be equally divided between the two daughter cells. In animal cells, the centrioles help make this possible.

centriole:
an organelle in an animal cell that helps divide the DNA into two equal parts when the cell divides

Cells produce many products. At the beginning of this chapter, we said that organisms were made up of cells and cell products. What are cell products? Your hair and fingernails are two examples. If you looked at a piece of hair or fingernail with a microscope, you would not see a cell. Hair and fingernails

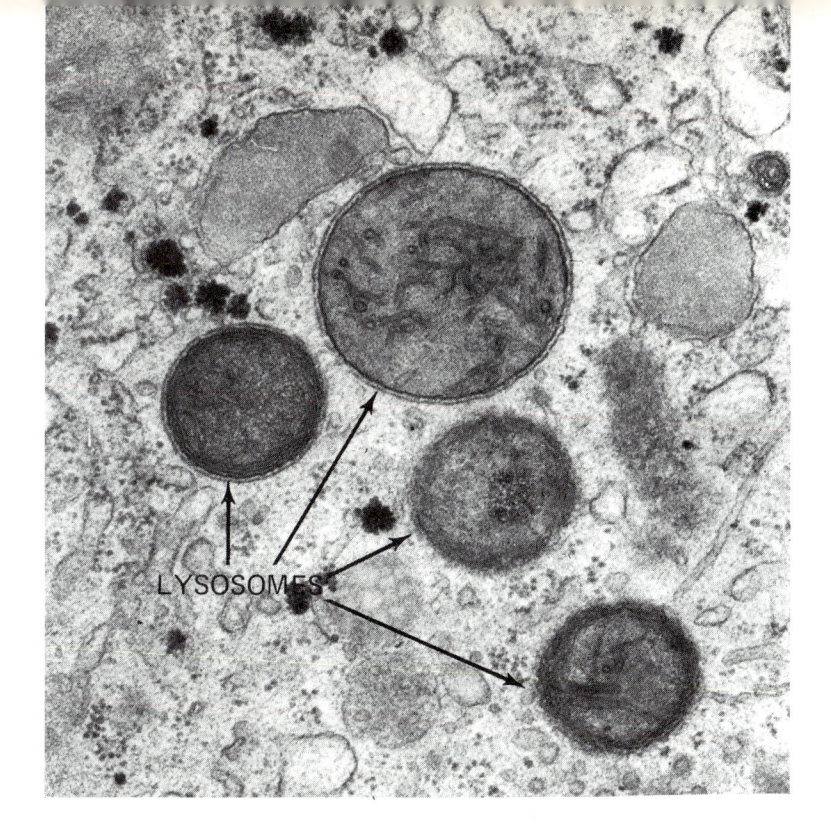

Figure 7.15: The dark circular bodies in this electron micrograph are lysosomes, the digestive organelles of the cell.

are made up of proteins that were made inside cells. (Remember: The nucleus and the ribosomes team up to make proteins.)

The thick cell wall that surrounds a plant cell is also a cell product. Most of the time, the cell wall remains after the cell that made it is dead. The cork that Robert Hooke examined was produced by cells that once lived in the holes that he saw.

Bone is a material that is made up mostly of cell products. If you could examine a thin slice of living bone, you would see

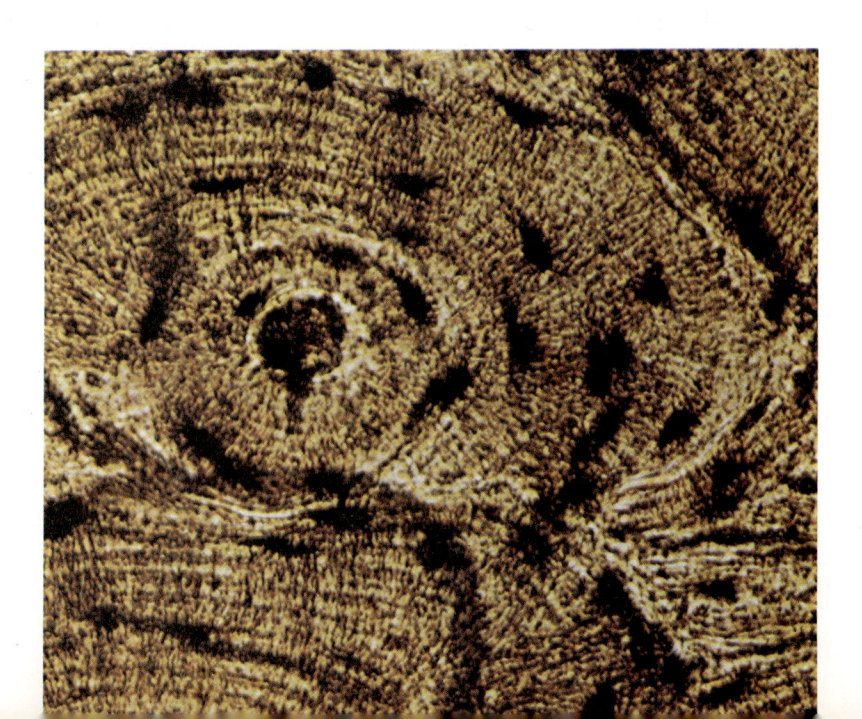

Figure 7.16: This photo of bone was taken with a light microscope. The dark regions are cells. Cell products make up the remainder of the bone. (Courtesy Carolina Biological Supply Co.)

Life Inside Organisms

only a few cells. They would be surrounded by a calcium mineral that makes up most of the bone. The calcium mineral is produced by the bone cells.

Throughout your body, there are special cells that produce tough fibers. These fibers hold your bones together, and they hold your muscles to your bones. These same fibers hold your skin and all your internal organs in place.

Many students are disappointed when they first examine plant or animal materials with a microscope. "Why don't I see cells?" is a question often asked. Just remember this. Rarely is an organism made up entirely of cells. If you are observing plant or animal material, you are about as likely to see a cell product as you are to see a cell.

LESSON REVIEW *(Think. There may be more than one answer.)*

1. Refer to Figure 7.17. Which of the organelles act together to make proteins?
    a. cell wall and cell membrane
    b. cell membrane and mitochondrion
    c. mitochondrion and endoplasmic reticulum with ribosomes
    d. endoplasmic reticulum with ribosomes and nucleus

2. The cell in Figure 7.17
    a. would have to be from a consumer.
    b. would have to be from a producer.
    c. could be from a consumer or producer.
    d. could be a phytoplankton.

3. If the cell in Figure 7.17 was from the flight muscles of a honeybee, it should
    a. have many more mitochondria.
    b. not have a centriole or endoplasmic reticulum.
    c. have many more chloroplasts.
    d. not have a cell wall or chloroplasts.

4. If the cell in Figure 7.17 were that of a single-celled zooplankton, it would have
    a. more nuclei.
    b. no mitochondria.
    c. more lysosomes.
    d. no chloroplasts.

Cells and Cell Products

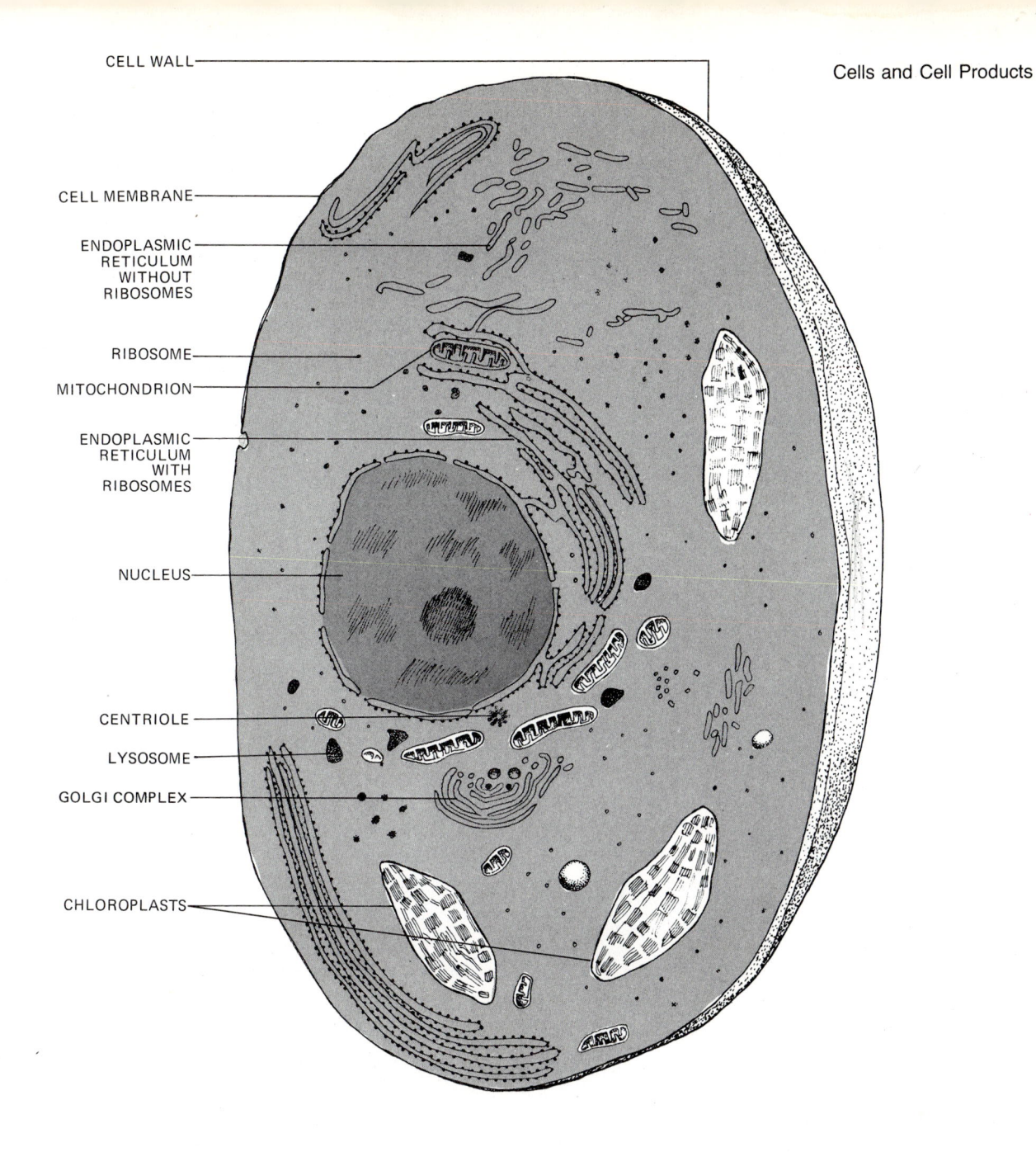

Figure 7.17: This diagram shows the main parts of the cell.

5. Which of the following words best matches each phrase?
   - a. organelle
   - b. cell membrane
   - c. mitochondrion
   - d. endoplasmic reticulum
   - e. ribosome
   - f. protein
   - g. DNA
   - h. Golgi complex
   - i. lysosome
   - j. centriole

Life Inside Organisms

_____ a cell body that helps to make proteins
_____ prepares proteins to leave a cell
_____ cell body that releases energy
_____ any cell body that performs a function
_____ chemical that controls heredity
_____ aids during animal cell division
_____ main chemical building block of organisms
_____ membrane system in the cytoplasm
_____ contains chemicals to digest "food"
_____ regulates passage of materials into and out of a cell

KEY WORDS

organelle (p. 117)   protein (p. 119)
cell membrane (p. 117)   DNA (p. 119)
mitochondrion (p. 117)   Golgi complex (p. 120)
endoplasmic reticulum (p. 118)   lysosome (p. 120)
ribosome (p. 119)   centriole (p. 120)

Applying What You Have Learned

1. What are some differences between plant cells and animal cells?

2. Match the organelle in Column A with the function listed in Column B.

   Column A   Column B
   a. mitochondrion   _____ helps with cell division
   b. Golgi complex   _____ acts like canals in the cytoplasm
   c. centriole
   d. lysosome   _____ wraps and exports proteins
   e. cell membrane   _____ releases energy used by cell
   f. endoplasmic   _____ digests materials in cell
      reticulum   _____ works with nucleus to make
   g. ribosome   proteins

124

Cells and Cell Products

3. Name some products made by cells.
4. What are two uses of the plant cell wall after the cell dies?
5. Look around you at your environment. Name at least five cell products in that environment.
6. The red blood cells that circulate in your bloodstream are the only cells in your body that do not have a nucleus. What normal cellular activity cannot be carried on by these cells? Why?
7. Life scientists would like to watch what goes on inside mitochondria and chloroplasts, but they cannot. Why?

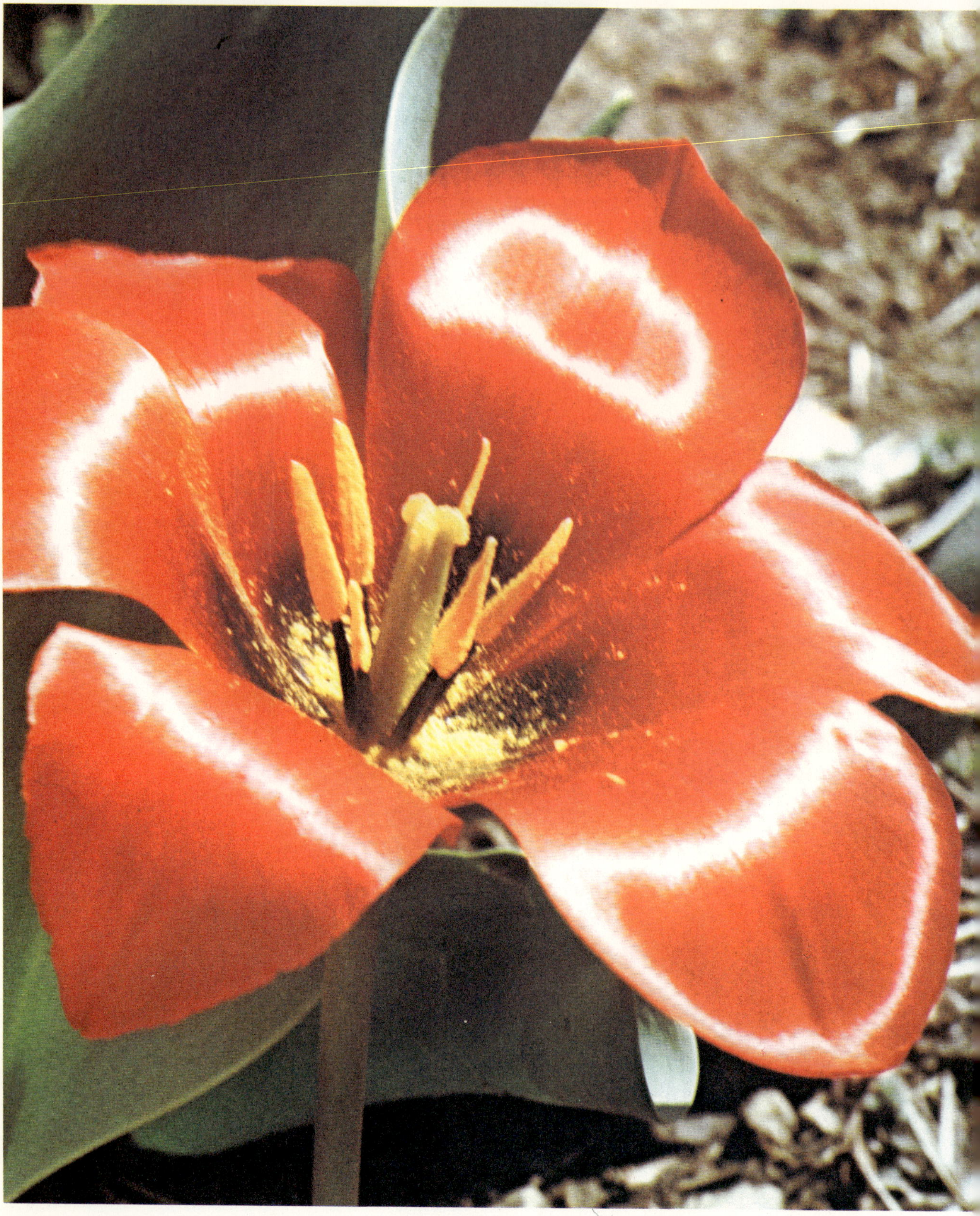

Figure 8.1: The flower is the reproductive system of many plants. It produces more plants.

Chapter 8

TISSUES, ORGANS, AND SYSTEMS

What are the walls of a brick building made of? Answer: Bricks and mortar. What are organisms made up of? Answer: Cells and cell products. Can we compare the two? Can we say that organisms are made up of cells and cell products the way walls are made up of bricks and mortar? The answer is no. When you look at a brick wall you see that the bricks are all the same. The cells in organisms are not like that. Usually there are several different kinds of cells in organisms. And the cells in organisms are organized. They are arranged in groups and they work in teams. How are the cells organized? That is the subject of this chapter.

TISSUES

What is a tissue? A **tissue** is a group of similar cells that work together as a team in doing a specific job. Almost all multicellular organisms have different kinds of tissues in their bodies.

tissue: a group of similar cells that work together to do a specific job

Figure 8.3: The large flat cells in the leaf's outer covering form a tissue that has a protective function. Note their lack of chloroplasts. The two guard cells (the mouthlike structure) have chloroplasts. Besides making food, their most important job is to control water loss.

Figure 8.2: This is how the thin outer covering of a leaf can be removed for study (above). Under the light microscope (photo at left), you can see two types of cells in the leaf's outer layer.

In Figure 8.2, you see the thin outer covering of a leaf being removed. In Figure 8.3, you see how this outer covering looks through a microscope. Most of the cells are thin and flat. Together they form a kind of skin on the leaf. On some leaves, these cells manufacture a waxy material. This coats the leaf and gives it more protection. These leaf cells form one specific kind of plant tissue. There can be many different kinds of tissue in a plant.

Look closely in Figure 8.3 for a structure that looks somewhat like a mouth. This is another type of tissue. Each "lip" of the mouth is a specialized cell called a *guard cell*. The dark space between them is an opening in the leaf. The guard cells, by working together, help to control the loss of water from the leaf. They work in pairs. The two cells may shrink, or grow smaller. If so, they create a small opening in the surface of the leaf. Water will evaporate through this opening. Though the guard cells in a leaf are not all in contact with each other, they do work together. Thus, they form a second kind of specific plant tissue.

Animal tissues do a variety of jobs. Gently scrape the inside of your cheek with your fingernail. You will see that you have collected a whitish material. Figure 8.4 shows you what that material looks like through a microscope. The cells are flat and thin. They form a tissue that covers and protects the inside of your cheek.

In Figure 8.5, you can see what a piece of beefsteak looks

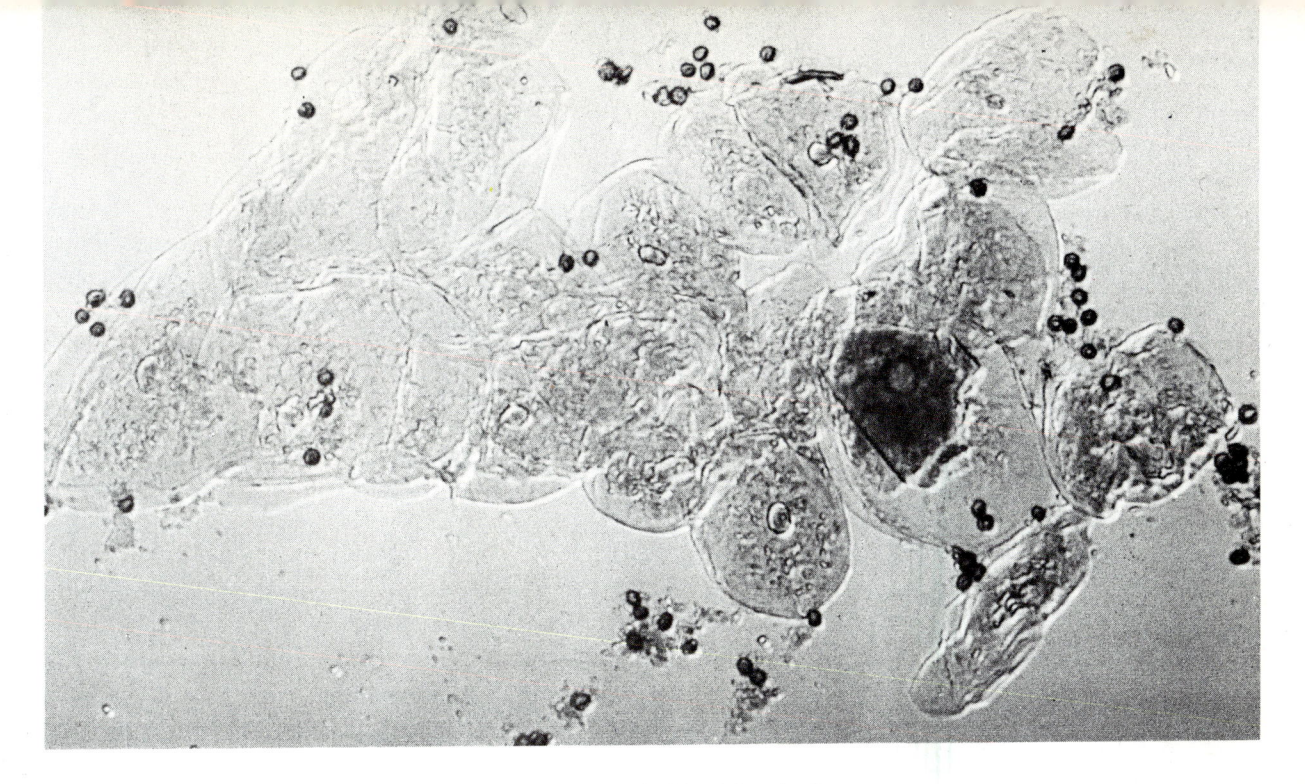

Figure 8.4: These large flat cells form a tissue that covers the inside of the cheek. The small dark circles are red blood cells.

like through the microscope. Beefsteak is mainly the muscle tissue of cattle. The cells are arranged in long, thin, threadlike fibers. The fibers can get shorter or longer. Thus, a muscle can get shorter or longer. Muscle tissue produces the movements of an animal's body.

Figure 8.6 shows you another kind of animal tissue. A sharp knife was used to slice through this tissue. The cells in this tissue lie side by side. They form several hollow ball-like structures. This tissue could be called a "manufacturing tissue." The cells produce saliva, the watery fluid that is in your mouth. The saliva collects in the hollow part of the "ball." Each ball is connected by tiny tubes that are connected to main tubes. The main tubes empty into your mouth. Tissues that produce different substances are found several places in the animal body. Such manufacturing tissues are called **glands.**

gland:
a tissue that produces one or more substances. These serve specific purposes in an animal.

Figure 8.5: When beefsteak is properly cut and stained, it looks like this under the microscope. The muscle fibers have bands across them. That's why this is sometimes called striated muscle.

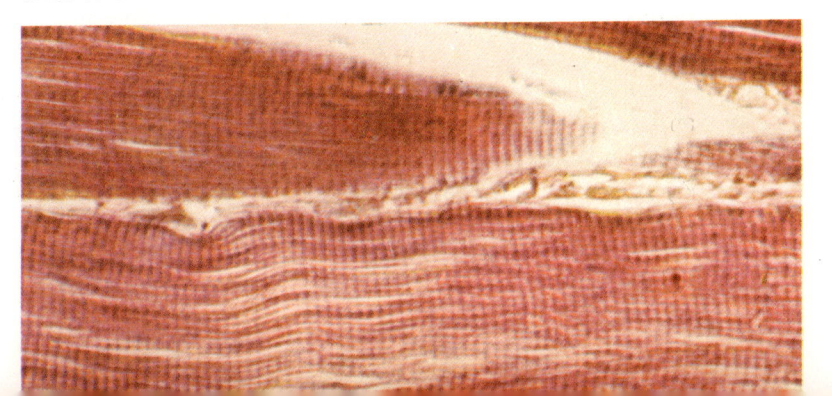

Life Inside Organisms

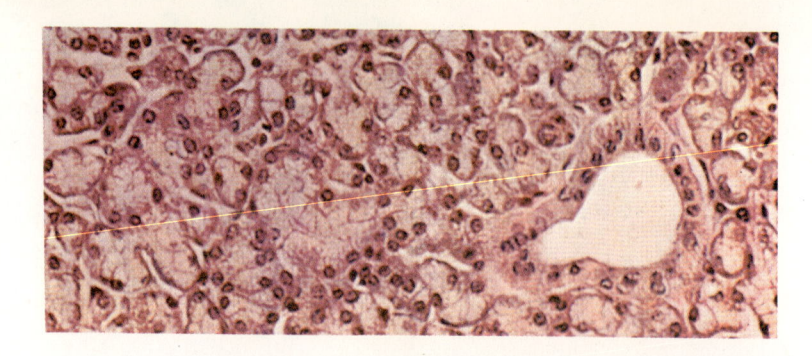

Figure 8.6: This is a slice of normal tissue from a salivary gland. You can see the individual cells by looking carefully.

cancer: a tissue disease in which cells grow and divide out of control. Cancerous tissue loses its normal functions.

Cancer is a tissue disease. Cancer is a disease that kills many people. It is a tissue disease. In Figure 8.7, you see a cancer of saliva-producing tissue. The cells no longer lie neatly side by side. They have grown and divided in an irregular way. The cells no longer work as they did. The tissue is no longer in control of itself. A cancer is sometimes described as a tissue "out of control." Without treatment, a cancer will usually destory all of the tissue in which it is developing.

What causes cancer? Some cancers are caused by viruses. Viruses are extremely small bodies that reproduce inside cells and destroy them. What else causes cancer? Certain chemicals produce cancers in experimental animals. Tobacco tar from cigarette smoke is one such chemical.

How is cancer treated? One way is for surgeons to remove the cancerous tissue from the body. There is a problem with this method. A single tissue is almost impossible to remove without damaging or removing others.

Various kinds of high-energy radiation, such as X rays, can sometimes be used to destroy cancerous tissue. For some reason not yet known, cancer cells are much more likely to be killed by radiation than are normal cells.

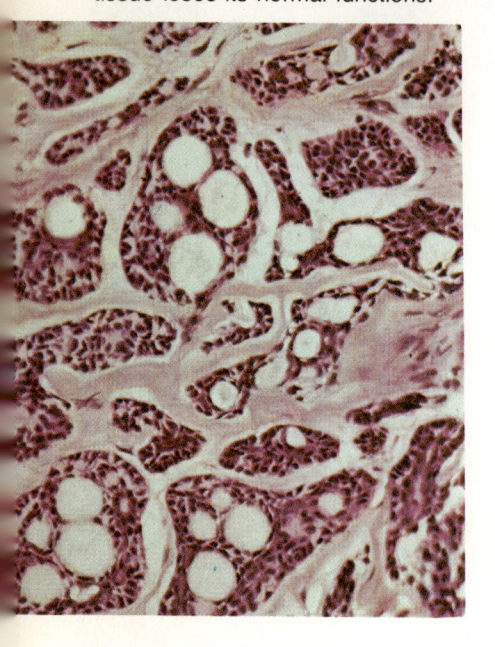

Figure 8.7: This is a section cut through a cancerous salivary gland. The cancer cells, stained dark purple, are smaller than normal and more irregular. The tissue has lost its organization.

LESSON REVIEW *(Think. There may be more than one answer.)*

1. All cells of a tissue
    a. are similar in size and shape.
    b. perform a specific function.
    c. must be in physical contact with each other.
    d. work in two-cell teams.

2. A tissue can
    a. protect other tissues.
    b. regulate some function.
    c. manufacture some product.
    d. create movement.

3. The easiest tissue to observe from your own body would be
    a. from one of your muscles.
    b. from your blood.
    c. from the lining of your cheek.
    d. a fingernail or hair.

4. The large cells in Figures 8.3 and 8.4 function as
    a. manufacturing tissue.
    b. protective tissue.
    c. regulatory tissue.
    d. manufacturing, regulatory, and protective tissue.

5. A gland is a
    a. manufacturing tissue.
    b. protective tissue.
    c. tissue made up of specialized cells.
    d. tissue like muscle tissue.

6. Cancer
    a. is a tissue disease.
    b. may be caused by environmental factors.
    c. may be cured by surgery.
    d. may be cured by radiation.

KEY WORDS

tissue (p. 127)
gland (p. 129)
cancer (p. 130)

ORGANS

What is an organ? An **organ** is a group of tissues that work together in performing a specific job. The leaf is an example of an organ. In Figure 8.3, you saw two tissues that work together in a leaf. There are more. Study Figure 8.8. Underneath the outer tissue lining the leaf, you can see other layers of cells. They perform most of the photosynthesis—the food-making process—in the leaf. Also note the tubes or *veins* in the leaf. Tissues that make up the veins carry water to the leaves, where it is used in photosynthesis. Other tissues in the veins carry food from the leaves to other parts of the plant.

organ:
a group of different tissues that form a specialized structure in a plant or an animal. The structure, or organ, does one or more specific jobs.

Life Inside Organisms

Figure 8.8: This cutaway view shows the different tissues that function together in a leaf.

MOST PHOTOSYNTHESIS OCCURS

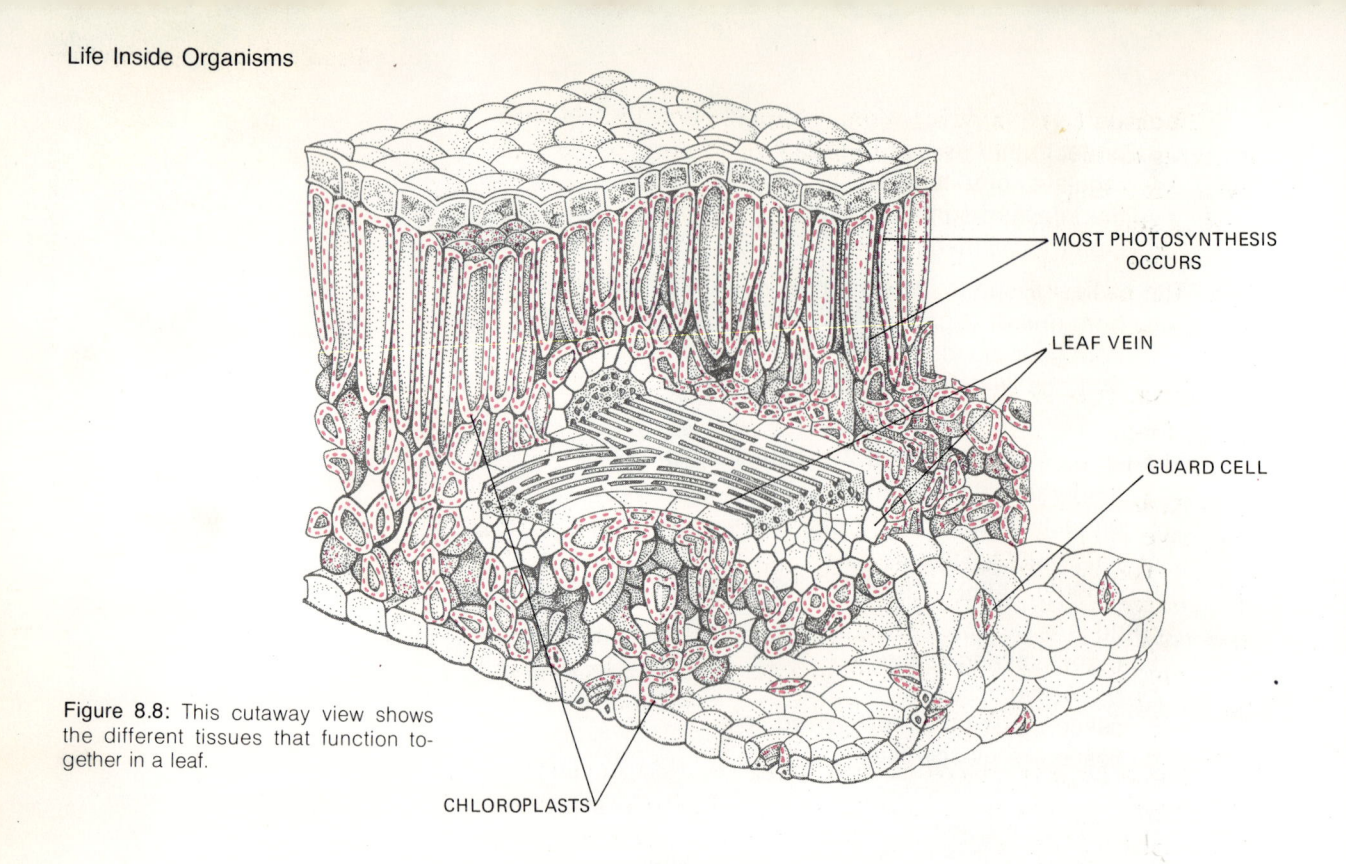

LEAF VEIN

GUARD CELL

CHLOROPLASTS

An organ is defined by the function it performs. An organ is defined by what it does, by the job or process it performs. The life scientist calls the job or process a *function*. We say the arm is an organ because there is a group of tissues in it that perform a function. One function would be to throw a ball. Also, we can say that the hand is an organ, or that a finger is an organ. In each case, we can define a specific function performed by a hand or finger.

An organ may have one or more functions. Some organs are specific in function. The heart is such an organ. It is a specialist. The heart's only job is to pump blood.

The largest organ inside your body is the *liver*. Unlike your heart, it performs a variety of functions. One of its functions is manufacturing. In fact, if the liver were larger, we might call it a "chemical plant." It makes many of the chemicals used in your body.

A second function of the liver could be called a "disassembly" function. Disassembly is the opposite of manufacturing. It is a tear-down function. Every hour, for example, millions of red blood cells are broken down inside the liver. (Somehow the liver is able to select the oldest of the red blood cells!) Chemicals are also disassembled in the liver. These include some chemicals produced in the liver and foreign chemicals such as alcohol.

The liver has a third function. It is a "chemical warehouse." It stores chemicals. Much of the sugar that your body can use for quick energy is stored in the liver. It is stored as a special kind of starch, which can be converted to sugar.

A fourth function of the liver is to aid in the digestion of fats. This is done by a substance called *bile*. The liver produces and releases bile into the digestive system.

What are some other animal organs? The arm, hand, finger, skin, and liver are animal organs. What are some others? In your head, you have eyes, ears, a nose, a throat, a mouth, a brain, and a skull. These are organs. In your chest, you have lungs, a windpipe, a heart, and ribs. These are also organs. You have many organs in the region of your body called the *abdomen*. This is the region between your chest and legs. Some of these organs are your stomach, pancreas, kidneys, small intestine, large intestine, and gall bladder. You will learn about the functions of many of these organs in later chapters.

Some organs can be transplanted. When you transplant a tree, you dig it up from one place and plant it in another. A transplanted organ is one taken from one organism and placed in another. In this sense, we usually think of transplanting an organ from one human being to another.

Skin, kidneys, and the heart are examples of organs that have been transplanted in humans. But each transplant is difficult. And many such efforts are doomed to failure. Why? The main reason is that the body will reject anything foreign that enters it. This rejection process will throw off a skin transplant or destroy a transplanted kidney.

Then how can transplants be successful? There are two main ways. One way is to transplant organs between people who are closely related—brother to brother, for example. Skin from one brother is not so likely to be recognized by the other brother's body as "foreign." The ideal transplant is between identical twins.

Foreign tissue—bacteria and viruses

Rejection process destroys foreign invaders

Foreign tissue—organ transplants

Rejection process rejects or destroys organ transplant

Transplants accepted

Bacteria and viruses accepted

Rejection process destroyed

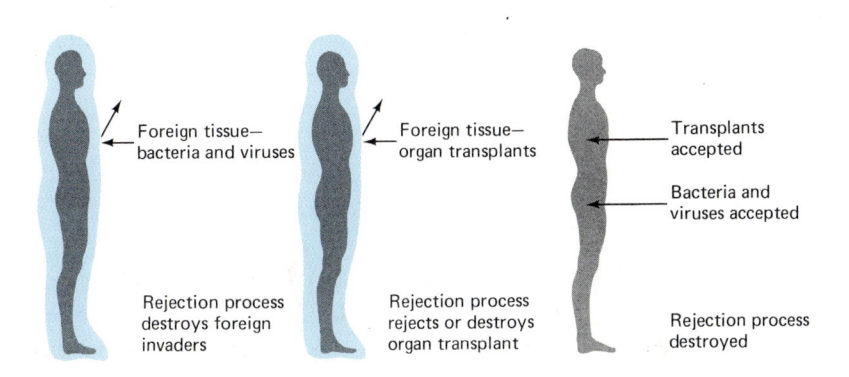

Figure 8.9: The body's rejection process works *for* you when your body is invaded by bacteria or viruses. The rejection process works *against* you when foreign tissues or organs are transplanted.

Life Inside Organisms

A second way to make a transplant successful is to destroy the body's rejection machinery. Certain white cells in the blood start the rejection process. These can be killed by special chemicals. But then the body cannot reject disease organisms. The rejection process usually protects us from them. This is one of several reasons why the second method is a "last resort" method. Life scientists are busy studying the rejection process. They are hopeful that some day it can be controlled so that organs from other animals can be used in the human body.

LESSON REVIEW *(Think. There may be more than one answer.)*

1. An organ is made up of
    a. more than one cell.
    b. more than one tissue.
    c. more than one system.
    d. neither cells, nor tissues, nor systems.
2. Which of the following are examples of an organ?
    a. skin
    b. finger
    c. hand
    d. bone
3. An organ
    a. is any group of cells in contact with each other.
    b. can perform only one specific function.
    c. can be contained within another organ.
    d. can only be found in the abdomen.
4. Your rejection machinery
    a. helps protect you from diseases.
    b. makes it difficult for you to use transplanted organs.
    c. can be destroyed by certain chemicals.
    d. should be permanently destroyed when chemicals are found that will do the job.
5. If the human body
    a. rejects bacteria, it will not reject a transplanted heart.
    b. rejects a transplanted heart, it will also reject bacteria.
    c. is normal, it will reject bacteria and a transplanted heart.
    d. did not have a rejection system, people would probably live longer.
6. A function of the liver is to
    a. assemble chemicals.
    b. disassemble chemicals.
    c. aid in the digestion of fats.
    d. store fats.

KEY WORD

organ (p. 131)

ORGAN SYSTEMS

What is an organ system? An **organ system** is a group of organs working together to perform a specific function. In everyday conversation, the life scientist leaves out the word "organ" and talks only of "systems" of various kinds.

What is an example of a system? The flower is a plant system. It is a reproductive system. Its function is to make more plants. Most flowers have male and female reproductive organs, as well as bright showy organs that attract insects.

Organization into systems is more common among animals than among plants. Even then, it is only the more complex animals that have systems.

Most complex animals have ten systems. You and most other complex multicellular animals have ten organ systems. We will describe each of these briefly now. They will be covered more fully in later chapters.

1. The *integumentary* (ĭn tĕg′yōō mĕn′tə rē) *system* encloses or covers an animal somewhat like the peel on an orange. Your integumentary system includes hair, skin, and the nails on your fingers and toes. Scales, feathers, and hooves are parts of the integumentary systems of some other animals.
2. The *skeletal system* of most animals is made up of bones. This system supports and protects the animal. It also functions in animal movement.
3. The *muscular system* consists of muscles that produce movements of the animal. Muscles also move the internal organs, such as the lungs, heart, and intestines.
4. The *breathing system* moves gases back and forth between the outside and the inside of the body. The lungs are the major organs of your breathing system.
5. The *excretory system* functions to eliminate waste products from the body. The kidney is an important organ in any excretory system.
6. The *nervous system* has several functions. It gathers and makes sense of information from the environment. It sends information to muscles and organs in response to the information it receives. It also functions somewhat like an orchestra leader. It helps keep all the organs in all the systems working in harmony. The eyes, ears, brain, and spinal cord are major organs in the nervous system.
7. The *endocrine system* also performs an "orchestra leader" function in the animal body. This system is made up mostly of specialized glands. These glands produce chemicals called

organ system:
a group of organs that work together to do a specific job

135

hormones. Hormones travel in the bloodstream and regulate organs throughout the body.

8. The *digestive system* changes food into a form that can be used inside the cells of the animal body. Your stomach and intestines are key organs of your digestive system.
9. The *transport system* moves materials throughout the animal body. Blood is the main carrier. The heart and blood vessels are the main organs of the animal transport system.
10. The *reproductive system* of animals produces more animals. Some animals have both male and female organs in the same body. The earthworm is such an animal. Other animals have either male or female reproductive organs.

Life is organized in levels. Have you ever unwrapped a present only to find that there was a smaller box inside? If so, you might expect to open the smaller box and find an even smaller box. This is a favorite trick of someone giving a small gift, such as a ring or pin. You may have to open a whole series of boxes within boxes to find out what you have.

The "boxes within boxes" trick will help you understand something very important about life. Life is organized like boxes within boxes. To show you what we mean, let us start with the biosphere and compare it to a large box. What is in the biosphere? Perhaps you will remember the answer from Chapter 3. The living portion of the biosphere is made up of communities. Now, what is in the community? The answer: Populations. What is in the populations? The answer: Organisms.

Let us continue with this kind of questioning. What is in organisms? Answer: Many organisms are made up of organ systems. What is in an organ system? Answer: Organs. What is in an organ? Answer: Tissues. What is in a tissue? Answer: Cells and cell products.

We could continue this questioning to show that (1) cells are made up of organelles, (2) organelles are made up of molecules, and (3) molecules are made up of atoms. We use the phrase **levels of organization** to describe the way life is organized. In speaking of levels of organizations, the life scientist might refer to the *cellular level*. He or she might also speak of any other level, such as the *population level* or the *tissue level*.

levels of organization: a way to describe the increasing complexity in the way life is organized

Life and death occur at different levels of organization. What do we mean when we say, "Death has occurred"? What is it that is dead? To someone untrained in the life sciences, death is something that occurs only at the organism level. The life scientist knows that life and death can occur at many levels. For example, in your own body, cells are dying by the thousands every hour. Tissues may die during the course of your lifetime.

Tissues, Organs, and Systems

- BIOSPHERE
- COMMUNITY
- POPULATION
- ORGANISM
- SYSTEM
- ORGAN
- TISSUE
- CELL
- ORGANELLE
- MOLECULE
- ATOM

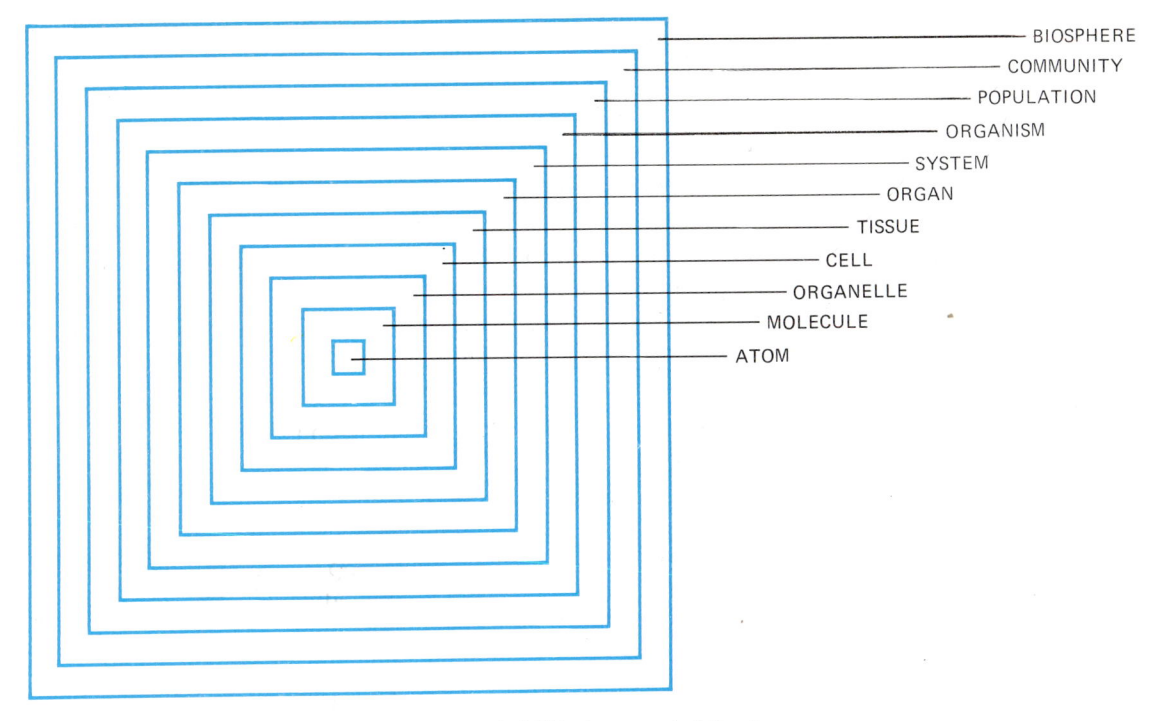

Figure 8.10: Life is organized into a series of "levels." This is somewhat like the organization of a set of boxes within boxes.

Even organs may be removed and thus killed. So within an individual such as yourself, we can have life or death at several levels.

We know that organisms die. But we also know that populations die. We use the term **extinction** as a name for population death. We can also have the death of a community. This has occurred in some lakes as the result of pollution. It can also occur naturally as the result of events such as earthquakes and volcanic eruption.

More than 25 years ago, a woman died of cancer. Before her death, some of the cancer cells were removed from her body. They were kept alive in bottles in special solutions. The cells continued to multiply. They were transferred from bottle to bottle. The cells were ideally suited for experiments where live cells were needed. This woman's cells are still alive, and in most of the major life science laboratories throughout the world!

Death at the organism level does not mean that death must also occur at all other levels below it. Think of the heart transplant as an example. An organism—say a human being—dies. The heart is removed but kept alive. Then it is transplanted into another human being. The organs, tissues, and cells of an organism can go on living after the individual dies.

So keep this in mind: Life and death may occur at any one of several levels of organization.

extinction:
the death of an entire population of plants or animals

Figure 8.11: These cells, called HeLa cells, are cancer cells from a woman who died more than 30 years ago. Cultures of HeLa cells are used in many life science laboratories—for example, to grow viruses, which can multiply only in living cells.

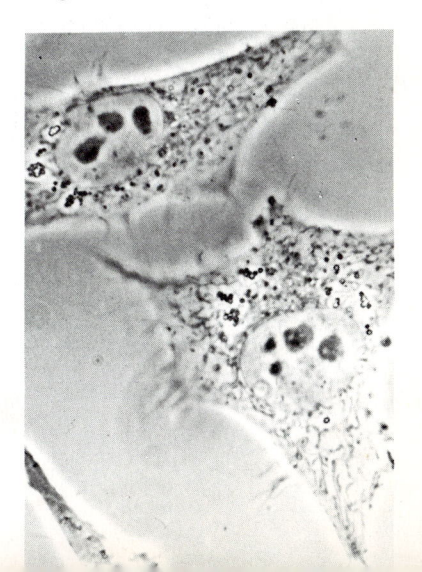

137

Life Inside Organisms

Laboratory Activity

How Many Tissues and Organs Can You Find in a Chicken Wing?

PURPOSE

A bird's wing is made up of groups of tissues and organs that work together to perform a job. In this investigation, you will dissect a chicken wing and look for the following tissues and organs: (1) skin, (2) fat, (3) muscle, (4) connective tissue, (5) tendon, (6) bone, (7) cartilage, (8) ligament, (9) blood vessel, (10) nerve.

MATERIALS

chicken wing tweezers
scissors paper plate

PROCEDURE

1. *Skin.* Before you begin the dissection, study the outside of the wing. Examine the weblike skin between the bones. Look for evidence that the skin was covered with feathers. Remove the skin from the largest bone and joint, as shown below. The following descriptions will help you to locate the tissues and organs in the chicken's wing.

2. *Connective Tissue.* This tissue resembles a flat, thin film or membrane. It is found in many places in the chicken's body. Connective tissue surrounds muscles and attaches the muscles to the underside of the skin.

3. *Muscle Tissue.* The pink-orange bundles of fiber found in the wing are skeletal muscles. Skeletal muscles are attached to bones. When the fibers expand or contract, they produce motion in the wing.

4. *Fatty Tissue.* Fat is stored in many places in the chicken's body. It can be recognized by its yellow color and greasy feeling. Fat stored on the underside of the skin helps to keep in the body's warmth. It also cushions and protects other body tissues.

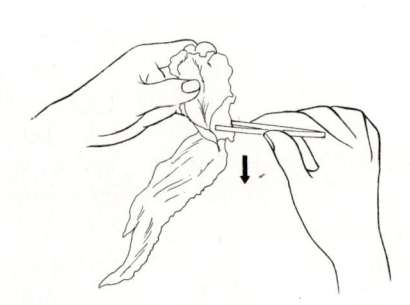

5. Tendon. Especially strong tissues attach muscles to bones. These shiny white tissues are called tendons.

6. Bone. Bones are hard tissues made up mostly of calcium minerals. There are hollow spaces inside many bones. Red blood cells are made by spongy tissue inside those spaces. The tissue that makes the red blood cells is called *marrow*.

7. Cartilage. This hard tissue—but not as hard as bone—is found at the ends of long bones. It is usually pearly white in color. Cartilage helps to prevent bones from grinding against each other.

8. Ligament. These strong white bands of tissue look like tendons. Ligaments connect two bones. They can best be located at a joint, where bones come together.

9. Blood Vessels. Arteries, veins, and capillaries are organs that carry blood throughout the body. Arteries have thicker walls than veins. Capillaries are too small to be seen without a microscope. Blood vessels are thin red tubes that you often see at the surface of muscles.

10. Nerve. Nerves are usually buried deep within an organism. They often lie close to long bones. Sometimes an artery, vein, and nerve are held together in one bundle by tissue wrapped around them. Nerves can be recognized by their white, threadlike appearance.

QUESTIONS

1. This diagram represents the dissection of a chicken wing as recorded by one student. Complete the diagram by naming the tissues and organs the student drew.

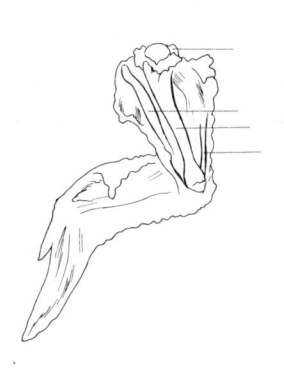

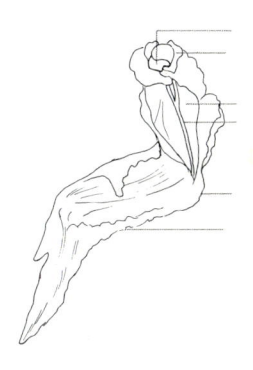

2. An organ system is a group of organs working together to perform a specific function. What five organ systems are present in the chicken wing?

3. Most chickens have short rounded wings. These are not suitable for flying long distances. Think of birds that spend much of their life in flight. How would you expect their wings to be different from those of a chicken?

Life Inside Organisms

LESSON REVIEW *(Think. There may be more than one answer.)*

1. Look at Figure 8.10 on page 137. From highest to lowest, the levels of organization within an organism such as you are
   a. population—organ—cell—tissue.
   b. system—organ—tissue—cell—organelle.
   c. organelle—tissue—cell—system—organ.
   d. population—community—ecosystem—biosphere—ecosphere.

2. The highest level of organization in Figure 8.10 is the
   a. individual.
   b. community.
   c. biosphere.
   d. ecosphere.

3. A population can live when
   a. cells within that population are dying.
   b. tissues within that population are dying.
   c. organs and systems within that population are dying.
   d. organisms within that population are dying.

4. Proof that cells, tissues, and organs do not have to die when an individual organism dies is shown by
   a. populations of rats that have lived for thousands of years.
   b. HeLa cells that now live, although the woman from whom they were taken has been dead many years.
   c. organs that now live after they have been taken from dead people.
   d. the fact that all populations will become extinct.

5. A system is made up of
   a. cells.
   b. tissues.
   c. organs.
   d. populations.

6. Within a living person,
   a. cells are dying all the time.
   b. tissue may die.
   c. organs can die.
   d. death of any part causes death of the individual.

7. Extinction refers to the death of
   a. organ systems.
   b. communities.
   c. populations.
   d. individuals.

KEY WORDS

organ system (p. 135)
levels of organization (p. 136)
extinction (p. 137)

Applying What You Have Learned

1. In a multicellular organism like yourself, life begins with one cell. What basic processes have to occur while the multicellular organism is developing?

2. Many people are willing to donate organs to other people who have lost their own by accident or disease. Why aren't organ transplants done more often?

3. If your body's rejection machinery was not working, you would not have to worry about such problems as hives, hay fever, reaction to bee stings, and allergies of all kinds. Would you want this? Why?

4. Arrange the following list of organization levels as they exist in nature. Start with the lowest level.
   a. organelle
   b. organ
   c. community
   d. tissue
   e. population
   f. system
   g. cell
   h. organism
   i. molecule
   j. biosphere
   k. atom

5. List the levels of organization for a single-celled organism.

6. Which level of organization is most likely to gain resistance to a pesticide?

7. Which function matches most closely with which organ?

Function	Organ
a. hearing	_____ stomach
b. walking	_____ kidney
c. digestion	_____ heart
d. waste removal	_____ brain
e. blood circulation	_____ foot
f. thinking	_____ ear

8. Match the organ(s) with each system.

Organ	System
a. skin	_____ digestive system
b. brain	_____ transport system
c. small intestine	_____ excretory system
d. white blood cells	_____ nervous system
e. spinal cord	_____ integumentary system
f. skull	_____ skeletal system
g. kidney	

Unit 4

Human Body Functions

Ask a person to describe an automobile to you, or the moon, or a baseball, or a play. You may get a strange look from that person. Who cannot describe an automobile? Who cannot describe the moon, or a baseball, or a play? Next, try asking two more questions. Where is your blood made? Why can't you live without your kidneys? You would find that most people know more about automobiles, sports, or plays than they know about their own bodies. No one will be likely to say that about you after you finish studying this unit.

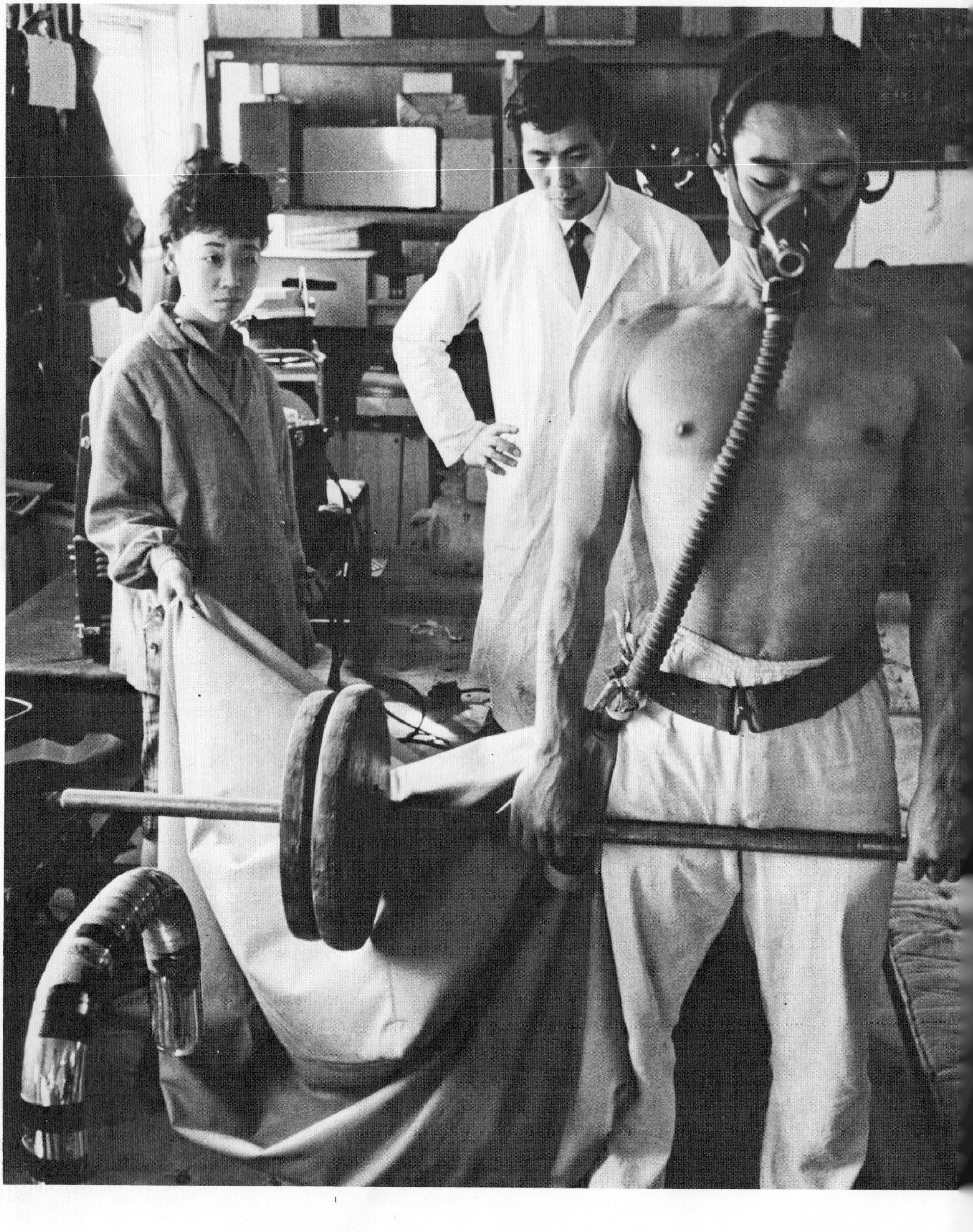

Figure 9.1: In preparation for the 1976 Olympic Games in Montreal, this Japanese weightlifter had his breathing capacity measured by two life scientists.

Chapter 9

DIGESTION, BREATHING, AND TRANSPORT

You have billions of cells in your body. They all need food and oxygen. They all have waste products that must be removed. You have three complete systems in your body that carry out these functions. These are the digestive, breathing, and transport systems. You will study each of these systems in this chapter.

DIGESTION

What is digestion? You cannot swallow a piece of food and expect your cells to use it the way it was when you ate it. The food has to be changed. **Digestion** is the process that changes food so that it can be used by your cells.

What changes does your food undergo? There are two basic changes. First, the food must be broken down into very small pieces. This process is called *mechanical digestion*. Second, the food must be changed chemically into molecules that your cells can use. This process is called *chemical digestion*.

digestion: the mechanical and chemical breakdown of food into nutrients the body can use

Human Body Functions

absorption:
the process by which digested food is taken into the bloodstream

digestive enzyme:
a chemical, usually a protein, that digests food

esophagus:
the organ that moves food from the mouth to the stomach

There is another process that is related to digestion. It is called **absorption.** This is the process by which digested food is taken into the bloodstream. A piece of food cannot be used by your cells until it is both digested and absorbed.

Digestion begins in the mouth. When you put food into your mouth, two things happen. You chew the food with your teeth. And the food is mixed with *saliva*, a watery fluid produced by glands in your jaws. Both processes—chewing and mixing with saliva—bring about mechanical digestion of the food. Thus, the food particles get smaller in size. The mixing with saliva also helps in the next process, which is swallowing.

Chemical digestion may also occur in your mouth. Saliva contains a **digestive enzyme** (eň′zīm). This is a chemical that breaks food down into smaller molecules. The digestive enzyme in your saliva is one that breaks down foods containing starch. Crackers, bread, and potatoes are examples. However, life scientists wonder how much digestion really takes place in your mouth. Certainly you have a digestive enzyme there. But food does not usually stay in your mouth long enough for the enzyme to go to work.

The esophagus moves food after it is swallowed. When you swallow, food moves from the back of your mouth into your throat. From there, it moves into a hoselike tube that is connected to your stomach. This tube is called the **esophagus** (ē sŏf′ə gəs).

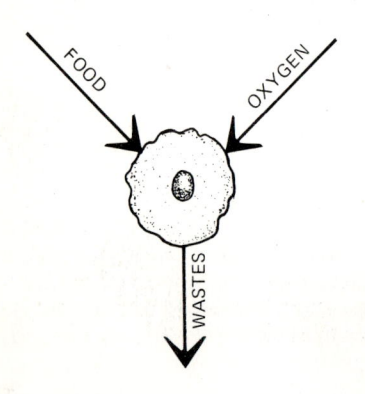

Figure 9.2: All cells need food. All cells need oxygen. All cells produce wastes that have to be removed.

Figure 9.3: Biting and chewing are the basic steps in the mechanical digestion of food that takes place in the mouth.

Digestion, Breathing, and Transport

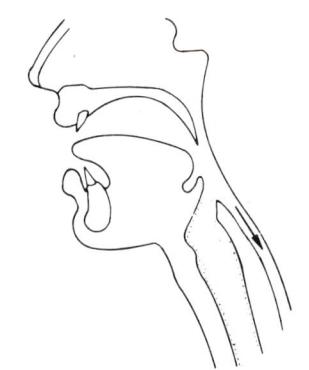

Figure 9.4: The esophagus does not have a digestive function. Food moves rapidly through the esophagus and into the stomach.

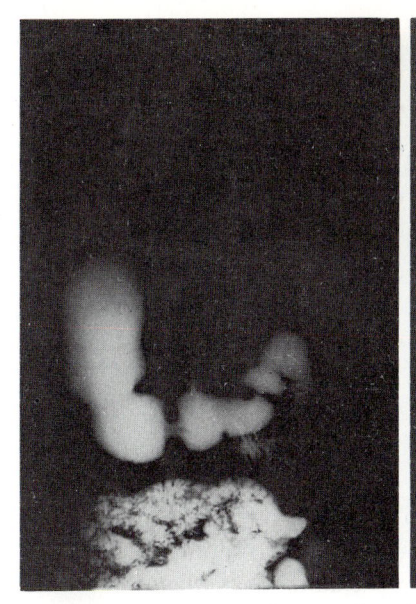

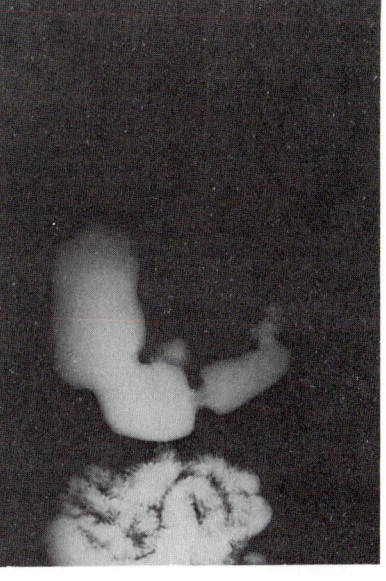

Figure 9.5: The stomach changes size when you eat. These X rays were taken of a woman lying on her right side. You see her nearly empty stomach on the left. Her stomach is fuller in the X ray on the right.

The esophagus has no digestive function. The food is in it only a short time. Bands of muscle cells in the wall of the esophagus push the food through it. Thus, food will get to your stomach even if you are weightless. That is why astronauts can eat in space. Back on Earth, the esophagus can transport food even if you are upside down!

The stomach stores and digests food. Eat a big meal and you can feel your **stomach.** That is because your stomach is elastic and stretches when you fill it. The elastic walls of your stomach contain bands of muscles. These muscles squeeze the food and make it mix with the liquids that are in the stomach. This squeezing process continues the mechanical digestion that was started in the mouth.

The liquids in the stomach play a part in digestion, too. Where do they come from? The stomach wall contains many glands. These glands produce digestive enzymes, hydrochloric acid, and **mucus** (myoo′kəs). The digestive enzymes of the stomach really begin the process of chemical digestion. (This assumes that little chemical digestion occurs in the mouth.) The hydrochloric acid aids the digestive enzymes in their work. The mucus is a fluid that keeps the food in a watery condition. It also lines the walls of your stomach and protects it from damage.

The small intestine is the main digestive organ. Food may stay in the stomach anywhere from 1 to 4 hours. Even-

stomach:
a main organ of digestion. It is a sac-like structure between the esophagus and small intestine.

mucus:
a slippery liquid produced by some glands and membranes in the body. It serves to moisten and protect the membranes.

Human Body Functions

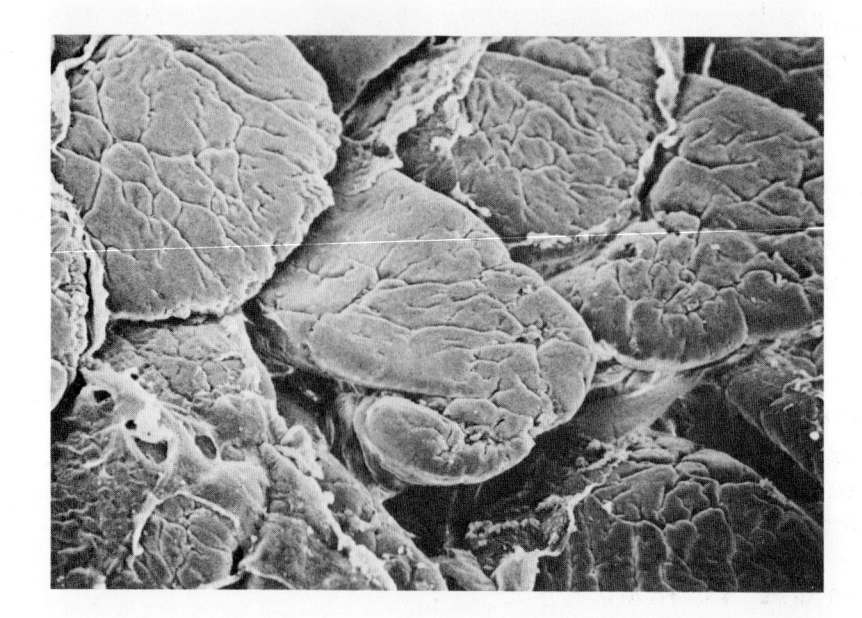

Figure 9.6: In this electron microscope view, human intestinal villi are magnified 800 times. This picture may give you a better idea of the surface area the villi offer for the absorption of digested food in the small intestine.

small intestine: the main organ of the digestive system. It is between the stomach and the large intestine. Most digestion and absorption occurs in it.

duodenum: the beginning portion of the small intestine

pancreas: an organ near the stomach that produces digestive enzymes. It also produces chemicals that help control the body's blood-sugar level.

bile: a digestive fluid that comes into the small intestine from the gall bladder

gall bladder: a saclike organ within the liver. It produces bile.

villi: very small fingerlike projections on the wall of the small intestine. Digested food is absorbed through the villi into the bloodstream.

tually it is squeezed out of the stomach and into the **small intestine.** The small intestine is your most important digestive organ. It is a tubelike organ about 7 meters (22 feet) long in an adult. It has many folds and twists. That is how it fits into your abdomen.

The first 20 centimeters (8 inches) or so of the small intestine is its most important region. This region is called the **duodenum** (doo′ə dē′nəm). It is here in the duodenum that most of the important digestive liquids are added to the food. Some of these are digestive enzymes produced in the wall of the duodenum. Other digestive enzymes enter from the **pancreas,** an organ near the duodenum. **Bile,** which aids in digestion of fats, comes into the duodenum from the **gall bladder.** The gall bladder is a saclike extension of the liver. (See Figure 9.7.)

The process of chemical digestion is completed as the food moves through the small intestine. The next process is absorption. Almost all of the digested food is also absorbed in the small intestine. This is where the digested food gets out of the digestive system.

You could understand the process of absorption better if you could look at the intestine wall with a microscope. You would see millions of tiny fingerlike things waving about. They are called **villi** (vĭl′lē). The villi greatly increase the area of the small intestine that is in contact with digested food. The digested food moves through the walls of the villi and into the bloodstream.

Water is absorbed in the large intestine. What is left after the digested food is absorbed in the small intestine? A good answer would be: waste materials and water. The waste

Digestion, Breathing, and Transport

materials consist mainly of parts of your food that could not be digested. These waste materials, along with the water, pass from the small intestine into the **large intestine.** The large intestine is also called the *colon* (**kō′lən**).

What happens in the colon? The main thing that happens is that much of the water is absorbed back into the body. Also, the waste materials are partly decomposed by the billions of bacteria that live in the colon. Normally, waste materials stay in the colon 24 to 30 hours. Then they are passed out of the body as **feces** (**fē′sēz**), the semisolid waste product of digestion.

large intestine:
the last part of the digestive system, also called the colon. Water is absorbed here and bacterial decay of waste takes place.

feces:
partly decomposed waste materials that pass out of the body after digestion has taken place. Feces are semisolid, much of the water having been absorbed in the large intestine.

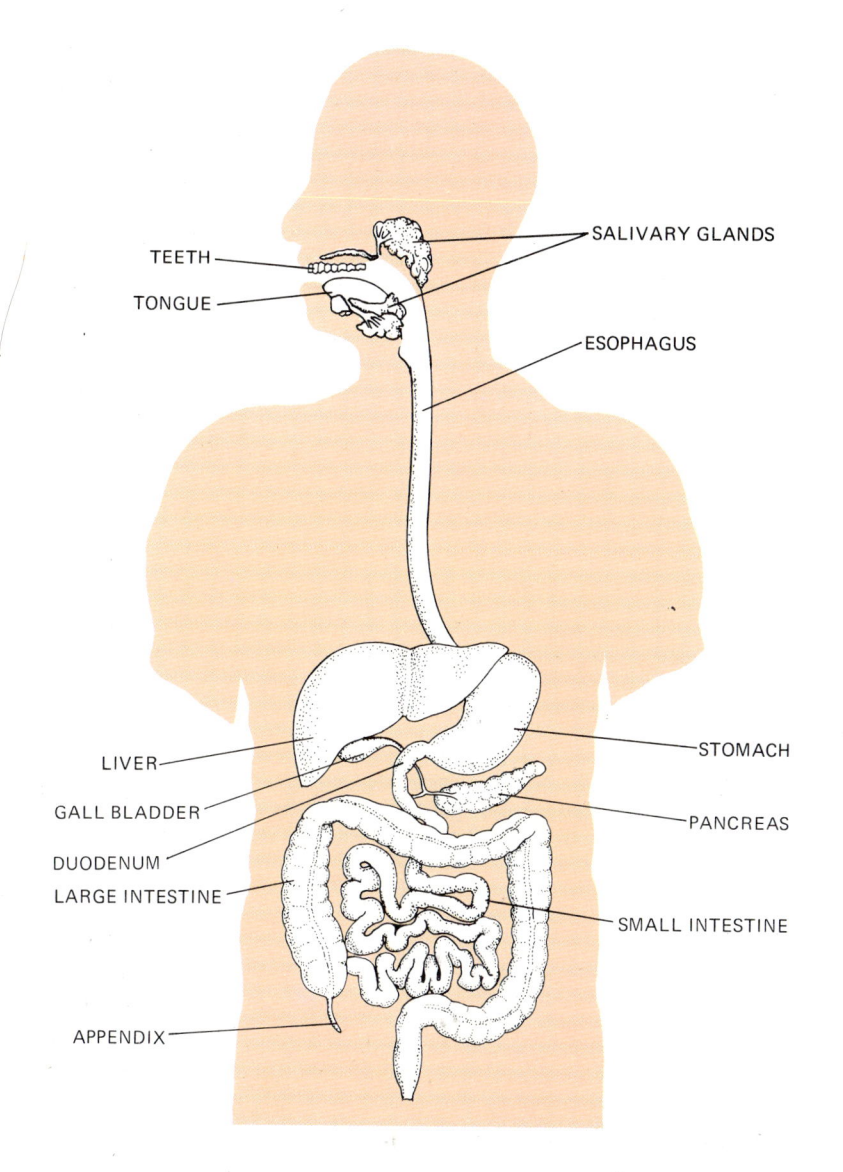

Figure 9.7: This diagram shows the entire digestive system. The appendix has no function, and appendicitis—an infection of this organ—is a troublesome disease.

149

Human Body Functions

Laboratory Activity

How Can Food Particles Pass Through All Those Walls?

PURPOSE

During digestion, the food you eat is broken down into nutrient molecules. The molecules are carried by the blood to nourish your body cells. To get to your body cells, the nutrients must pass through several membranes. (The life scientist would say that the particles "diffuse" through the membranes.) How is it possible for nutrients to pass out through the walls of the small intestine and in through the walls of the capillaries? How do these nutrients diffuse through the cell membrane and enter the cell? In this investigation, you will prepare a kind of membrane and determine if it is possible for food coloring and sugar to diffuse through it.

MATERIALS

cellophane
2 drinking glasses
test tubes
food coloring
teaspoon
sugar
4 rubber bands
Benedict's solution
test-tube holder
heat source

PROCEDURE

1. Soak two pieces of cellophane in a glass nearly filled with water for at least 1 hour.
2. Add water to each of two test tubes until they are three-fourths full. Add four drops of food coloring to one test tube. Dissolve some sugar in the second test tube.
3. Remove the cellophane from the water. Discard the water.
4. Cover the mouth of each test tube with a piece of cellophane. Secure the cellophane on each test tube. To do that, place one rubber band near the rim of the test tube. Place the second rubber band about 1.5 cm below the rim of the test tube, as shown in A. Be sure that the edge of the cellophane extends beyond both rubber bands.
5. Wipe the outside of each test tube.
6. Add water to each of two glasses until they are half full.

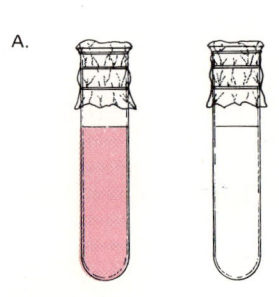

A.

B.   FOOD COLORING   SUGAR

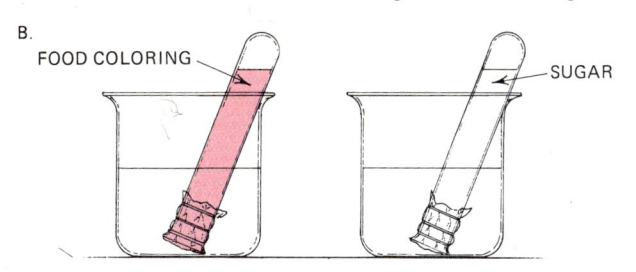

7. Turn one test tube upside down and prepare to lower it into the water in one of the glasses. Check the test tube for leakage. If the cellophane has been punctured, or is not securely fastened, liquid will leak through the membrane.
8. If leakage occurs, check the rubber bands to see if they are holding the cellophane in place. If leakage continues, you will have to use another piece of cellophane and repeat steps 4 through 7.
9. Lower the upside-down test tube into the water in the glass. Rest the test tube against the rim of the glass, as shown in B.
10. Repeat steps 7, 8, and 9 for the second test tube.
11. Let both test tubes remain undisturbed overnight.
12. Observe the materials the following day. What has happened to the food coloring?
13. Did sugar pass through the membrane? You can test for the presence of sugar by taste. But that may not be safe. Instead, perform the following test. Pour some water from the glass that may have sugar in it into a clean test tube. Add a few drops of Benedict's solution. Heat the test tube. (**Caution:** Follow your teacher's instructions. Take care not to burn yourself. *Do not boil the solution in the test tube.*) If sugar is present, the liquid will become a brick red color. Did sugar pass through the membrane?

QUESTIONS

1. What would happen if you used plastic food wrap instead of cellophane? Try it. What might be different about the cellophane that allows tiny particles to pass through?
2. Can other substances diffuse through the cellophane membranes? Would cornstarch pass through the cellophane? Plan an investigation to answer this question.
3. What do the results of these investigations tell you about the digestion of food?

Human Body Functions

LESSON REVIEW *(Think. There may be more than one answer.)*

1. Mechanical digestion occurs in
   a. the mouth.
   b. the stomach.
   c. the small intestine.
   d. the large intestine.

2. Chemical digestion occurs in
   a. the mouth.
   b. the stomach.
   c. the small intestine.
   d. the large intestine.

3. Absorption is the *only* process that occurs in
   a. the mouth.
   b. the stomach.
   c. the small intestine.
   d. the large intestine.

4. The duodenum is
   a. the most important region of the small intestine.
   b. the most important region of the large intestine.
   c. about as long as a football.
   d. a little longer than an unsharpened pencil.

5. In the large intestine,
   a. food is digested mechanically.
   b. food is digested chemically.
   c. food is absorbed.
   d. water is absorbed.

6. Which of the following terms best matches each phrase?
   a. digestion   h. duodenum
   b. stomach   i. pancreas
   c. absorption   j. bile
   d. mucus   k. gall bladder
   e. digestive enzyme   l. villi
   f. esophagus   m. large intestine
   g. small intestine   n. feces

   _____ chemical that breaks down food in the digestive system

   _____ breaking food down into nutrients the body can use

   _____ the movement of digested food into the bloodstream from the digestive system

   _____ organ where water is absorbed

   _____ fluid produced by the gall bladder that helps digest fats

   _____ first section of small intestine

   _____ tube between throat and stomach

   _____ moistens and protects membranes in the body

Digestion, Breathing, and Transport

_____ organ near the stomach that produces digestive enzymes

_____ projections into the small intestine that increase surface area for absorption

_____ the body's container that holds bile

_____ organ expands and stores food after eating

_____ organ where most digestion and all absorption of food occurs

_____ semisolid waste product of digestion

KEY WORDS

digestion (p. 145)
absorption (p. 146)
digestive enzyme (p. 146)
esophagus (p. 146)
stomach (p. 147)
mucus (p. 147)
small intestine (p. 148)

duodenum (p. 148)
pancreas (p. 148)
bile (p. 148)
gall bladder (p. 148)
villi (p. 148)
large intestine (p. 149)
feces (p. 149)

BREATHING

Gas exchange is the basic problem. All of the cells in your body need oxygen. All of your cells give off carbon dioxide, which must be removed from the body. Breathing is the process by which oxygen is taken into the body and carbon dioxide is removed. Sometimes other gaseous wastes are also removed during the breathing process.

What happens during breathing? The air that you breathe contains about 20 percent oxygen by volume. This air is *inhaled*, or taken into the body through the nose or mouth, or both. The nose is the better entrance. Hairs in the nose help remove dust and other particles. Special cells in the back of the nose have hairlike extensions that also clean the air. Air spaces in back of the nose, called **sinuses**, warm the air and add moisture if it is dry.

Air from the nose or mouth passes into the throat. From

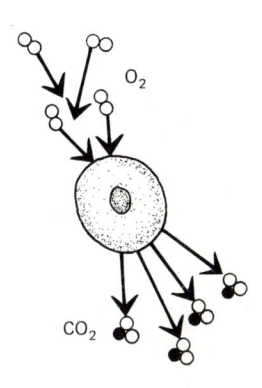

Figure 9.8: All your cells need oxygen molecules (O_2), and all of them give off carbon dioxide (CO_2) as a waste material.

sinus:
one of the air spaces behind the nose. The sinuses warm air that is breathed in, and add moisture to it if necessary.

153

Human Body Functions

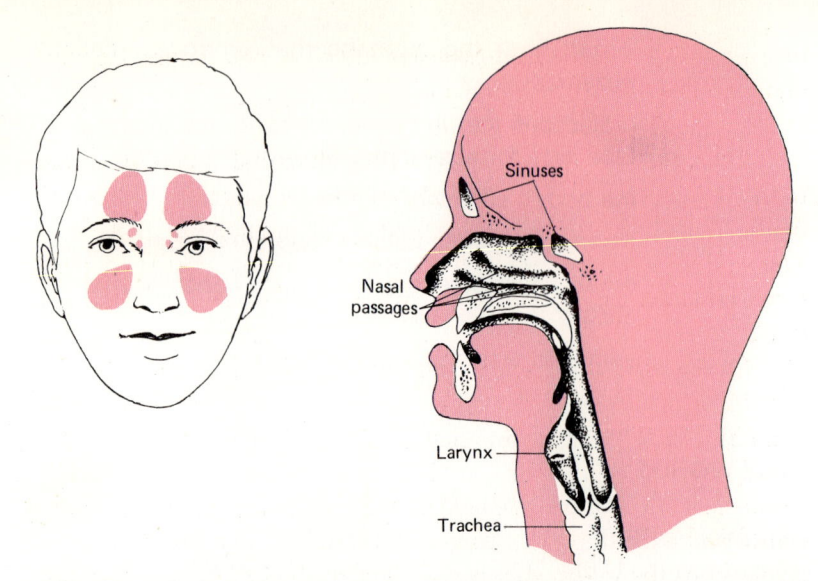

Figure 9.9: At the left is a diagram of the sinuses in the human skull. At the right is a side view of the air passages in the human skull. You can see that all air passages lead to the larynx and trachea.

larynx:
an organ between the throat and the trachea. The larynx contains the *vocal cords,* which are the basic source of speech production.

trachea:
the windpipe through which air moves between the larynx and the lungs

bronchial tubes:
the tubes in the lungs that transport air to and from the alveoli

lung:
one of two expandable organs in the chest in which gas exchange takes place. Carbon dioxide leaves the blood here and oxygen enters it.

there, it moves through a small opening in the **larynx,** or voice box. (The voice box is also called "Adam's apple.") This opening lets the air into the windpipe, or **trachea (trā′kē ə).** Feel just below your larynx. You can feel the stiff rings of tissue inside your trachea. The rings act like the wire in a car's radiator hose. They keep the trachea from collapsing.

Air passing through the trachea enters two **bronchial (brŏng′kē əl) tubes,** which are inside the **lungs.** These tubes branch and rebranch until they are very small. Each tube ends

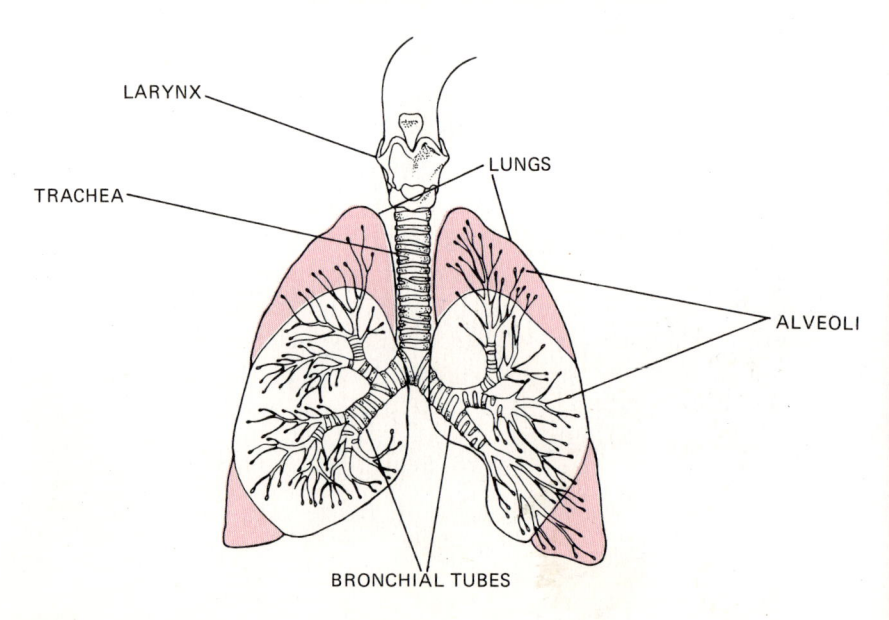

Figure 9.10: This cutaway diagram of the human breathing system shows the organs that are located in your neck and chest.

in a tiny air sac called an **alveolus** (ăl vē′ə ləs). There are very many of these alveoli (ăl vē′ə lī′).

alveolus: one of many small air sacs in the lungs through which oxygen and carbon dioxide are exchanged

Gas exchange occurs in the alveolus. The movement of oxygen and carbon dioxide to and from the blood occurs in the alveoli. The wall of each alveolus is only one cell thick. This cell is thin and flat. Just outside the alveolus wall is a capillary. This is a tiny blood vessel whose wall is also just one cell thick.

Oxygen passes through the alveolus wall and the capillary wall into the blood. After it is in the blood, it combines with **hemoglobin.** Hemoglobin is a red substance that is inside all red blood cells. It gives blood its red color. What does the hemoglobin do? If you were going to give hemoglobin a "job title" you would call it an "oxygen carrier." Hemoglobin transports oxygen throughout the body. When red blood cells carrying oxygen pass a tissue which is low in oxygen, hemoglobin releases its oxygen. Then the oxygen is used by the cells that need it.

hemoglobin: the oxygen-carrying substance in blood that gives it its red color

A tissue needing oxygen usually has a surplus of carbon dioxide that has to be removed. Most of the carbon dioxide is carried to the lungs in the blood **plasma.** Plasma is the liquid part of the blood. As blood flows by the alveolus, the plasma gives up carbon dioxide. The carbon dioxide passes through the capillary wall, then through the wall of the alveolus. After it is inside the lungs, it is exhaled, or breathed out.

plasma: the pale yellowish, liquid part of blood

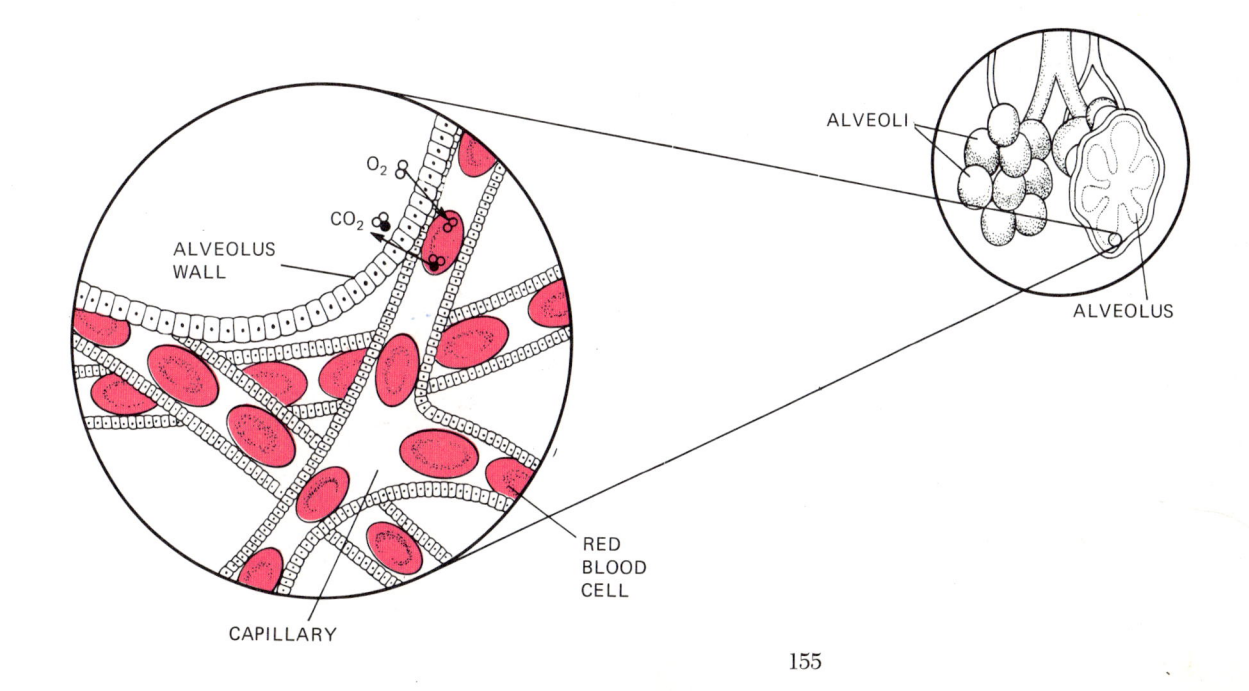

Figure 9.11: Actual gas exchange occurs in the alveolus. Note that the gases must move through two layers of cells.

155

Human Body Functions

diaphragm:
a large muscle below the lungs that aids breathing by moving up and down

emphysema:
a lung disease in which the elastic tissue, which normally helps push out air, has lost its "squeezing" ability

Figure 9.12: The emphysema patient often needs mechanical support in order for oxygen to enter the lungs. In this picture you see such a patient, who depends on a portable liquid oxygen system to help him get enough oxygen to live. The oxygen enters his nose through tubes held to his head.

Muscles and elastic tissue aid breathing. Your lungs could not do their work without muscles and elastic tissue. The muscles that aid breathing are in two main locations. Some are between your ribs. They help you expand your chest. One of the most important muscles is the **diaphragm** (dī′ə fram′). It is a large muscle below your lungs. The diaphragm moves up and down as you breathe—down as you inhale and up as you exhale.

Elastic tissue is found throughout your lungs. When you inhale, this elastic tissue is stretched. When you exhale, it is the squeezing of the elastic tissue that pushes out the air. Perhaps you have heard of the disease called **emphysema** (ĕm′fĭ sē′mə). It is a common disease, especially among smokers. People with this disease have lost the use of much of the elastic tissue in their lungs. In order to exhale, they have to use the squeezing power of their muscles instead. This require much more work. Just to breathe is a chore for such people, and their physical activity is limited.

LESSON REVIEW (Think. There may be more than one answer.)

1. During the breathing process,
    a. oxygen is released by your cells.
    b. carbon dioxide is absorbed by your cells.
    c. oxygen is taken into your body.
    d. carbon dioxide is released from your body.

2. If you did not have cartilage rings in your trachea,
    a. your trachea would close.
    b. your sinuses would dry out.
    c. your bronchial tubes would not branch.
    d. your alveoli would not open the larynx normally.

3. In the alveolus (Figure 9.11),
    a. oxygen must travel through three layers of cells to move from the lungs to the capillary.
    b. the red blood cells move from the capillary into the alveolus to pick up oxygen.
    c. carbon dioxide and oxygen are both present.
    d. there are red blood cells and capillaries.

4. During breathing,
    a. your diaphragm moves up as you inhale.
    b. your diaphragm moves up as you exhale.
    c. your diaphragm moves down as you inhale.
    d. your diaphragm moves down as you exhale.

5. Which of the following terms best matches each phrase?
   a. lung
   b. sinus
   c. larynx
   d. trachea
   e. bronchial tube
   f. alveolus
   g. hemoglobin
   h. plasma
   i. diaphragm
   j. emphysema

   _____ windpipe

   _____ a tiny sac in the lung that has a wall one cell in thickness

   _____ disease when elastic tissue in lungs is damaged

   _____ organ in which gas exchange takes place

   _____ red substance in blood that carries oxygen

   _____ voice box

   _____ a cavity in facial bones where air is warmed

   _____ muscle below the lungs that moves up and down during breathing

   _____ one of the passageways that connect the trachea with the alveoli

   _____ liquid part of the blood

KEY WORDS

sinus (p. 153)
larynx (p. 154)
trachea (p. 154)
bronchial tube (p. 154)
lung (p. 154)

alveolus (p. 155)
hemoglobin (p. 155)
plasma (p. 155)
diaphragm (p. 156)
emphysema (p. 156)

TRANSPORT

What is blood? About one-thirteenth of your body weight is blood. This liquid is the main carrier within the transport system. About half of the blood is *plasma*, a pale yellowish liquid. The other half of the blood is made up of solid materials, mostly cells. These solid materials are the red blood cells, the white blood cells, and the platelets.

What does blood do? We have already explained how blood transports oxygen and carbon dioxide. Blood has several other functions. It transports food to your cells, and it carries chemical regulators that you will study about in the next chapter. It helps keep your body temperature at an even level. In general, blood keeps the internal environment of your body the same everywhere.

Human Body Functions

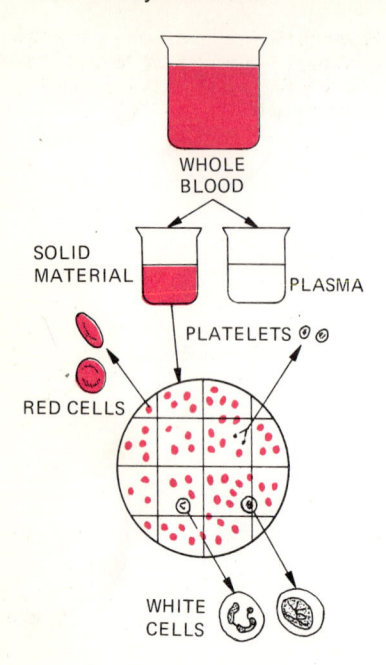

Figure 9.13: Whole blood contains roughly equal parts of solid material and plasma. The solid material is made up of red cells, white cells, and platelets. Which is most abundant?

Figure 9.14: Hemophilia is a disease that is inherited. A person with hemophilia does not have a normal blood clotting mechanism, so that any injury or other situation in which blood vessels are cut can lead to prolonged bleeding in the person. In this picture you see a young man who has hemophilia. He is shown with 121 units of blood clotting factors similar to those that were given to him after he had a tooth pulled. The blood clotting factors are one of a number of blood components that can be used from blood donors.

clotting:
the process by which proteins in blood plasma cause a network of fibers to form over a bleeding area

antibody:
any of several proteins in blood plasma that destroy certain disease-causing bacteria and viruses

red blood cell:
one of the solid parts of the blood. The red blood cells carry oxygen.

white blood cell:
one of the solid parts of the blood. The white blood cells help fight "invaders" of the body, such as bacteria and viruses.

platelet:
one of the solid parts of the blood. The platelets help in the clotting process.

heart:
a hollow muscle near the center of the chest. Its squeezing motions keep blood moving through the body.

The different parts of the blood have specific functions. For example, the blood plasma has proteins that do specific jobs. Some aid in the **clotting** process. This process helps you when something causes bleeding to occur. In the clotting process, a network of fibers forms a clot over the bleeding area. This seals off the blood flow.

Some of the other plasma proteins are called **antibodies.** Antibodies are a part of your rejection machinery. They fight and destroy foreign invaders that get into your body. Antibodies are the main reason you get over infectious diseases caused by bacteria and viruses.

The solid parts of your blood also have their specific jobs. The **red blood cells** carry oxygen. The **white blood cells** work along with the antibodies in fighting foreign invaders. Some of the white blood cells move out of the bloodstream and into the tissues. These white cells actually surround and eat invaders such as bacteria. The **platelets,** which are very small, aid in the clotting process.

The heart is the pump. Your blood is always in motion. That is because you have a **heart.** Your heart is near the

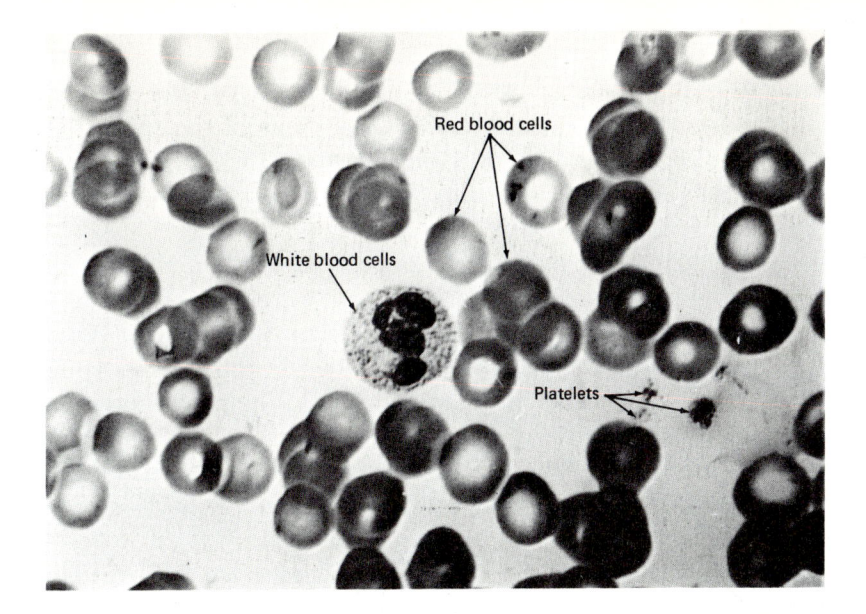

Figure 9.15: Red blood cells carry oxygen. Many of the white blood cells "eat" bacteria. Platelets help start the clotting process.

middle of your chest, but just slightly to your left side. It is a hollow muscle about the size of your fist.

Inside the heart, there are four little rooms, or chambers. The two upper chambers are called **atria** (ā′trē ə). The two lower chambers are called **ventricles.** Each atrium is connected to the ventricle below it by a **heart valve.** The valve opens and closes, letting blood flow through from atrium to ventricle. Blood does not flow the other way. There are no connecting valves between atria or between ventricles.

What you probably call a heart "beat" is the squeezing or contracting motion of the heart. The contraction of the heart muscle is what sets the blood in motion.

atria:
(singular, "atrium"), the two upper chambers of the heart
ventricles:
the two lower chambers of the heart
heart valve:
any of the valves in the heart that open and close to let blood flow, from an atrium into a ventricle, or out of a ventricle

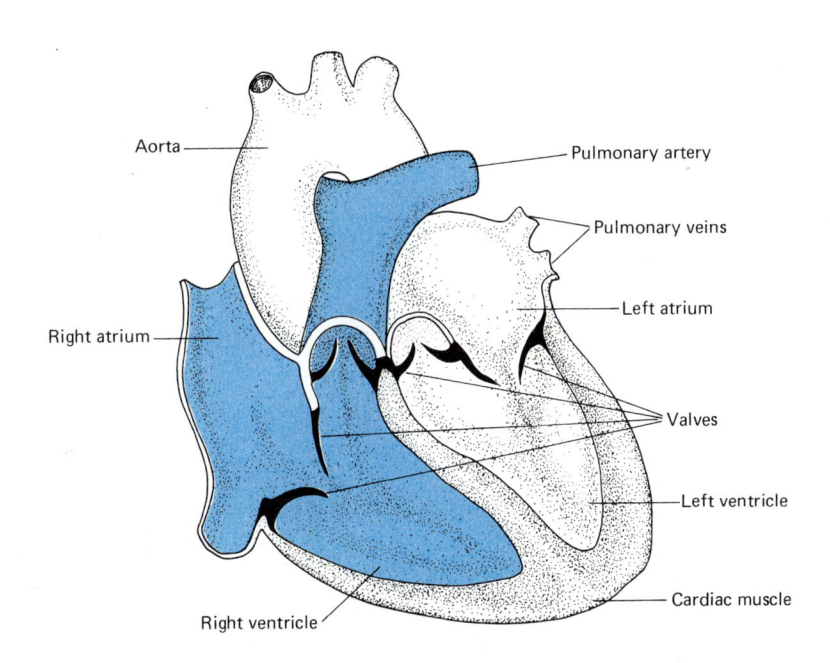

Figure 9.16: In this diagram of a human heart, the valves that control the flow of blood into and out of the heart are clearly seen. Figure out which valves are open and which are closed at the same time.

159

Human Body Functions

artery:
a blood vessel that carries blood *away from* the heart

heart rate:
the number of times the heart contracts, or beats, in one minute; also called the pulse

vein:
a blood vessel that carries blood *back to* the heart

capillary:
a blood vessel, one cell thick, through which blood passes from an artery to a vein

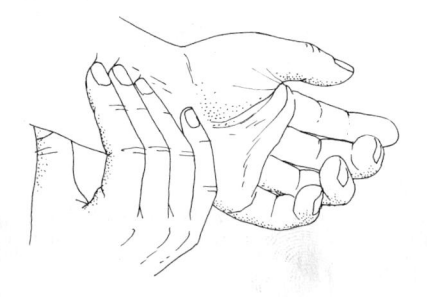

Figure 9.17: To take a pulse, place the first two or three fingers of the right hand firmly on the back of the wrist. Do not use the thumb—there is a pulse area in the thumb itself.

Arteries, veins, and capillaries carry the blood.
Blood vessel is the general name for any one of the tubes that the blood flows through. But there are specific kinds of blood vessels in the transport system. Each one has a different structure and function.

The **artery** is the kind of blood vessel that carries blood away from the heart. All arteries have a thick wall. This wall is elastic and will stretch when the blood is under pressure. When the heart contracts, the pressure of blood in the arteries rises. The artery walls respond by stretching. This helps reduce some of the pressure. Otherwise, too much pressure might break a small artery. When the heart muscle relaxes, the artery walls also relax. This motion of the artery wall squeezes the blood and helps push it along.

With the help of the diagram in Figure 9.17, you can feel the stretching and relaxing of an artery. You can also find out how often your heart contracts, or beats, in 1 minute. This is your pulse, or **heart rate.** The human heart rate varies from 60 to about 150 contractions per minute. Some of the things that affect the heart rate are your age, sex, exercise, and whether you are in good shape.

Veins are blood vessels that carry blood back to the heart. Veins have thin walls. You may be able to see some of your own veins by looking at the inside of your wrist. Veins are often found close to the skin. Arteries are usually much deeper in the body.

How does blood get from the arteries to the veins? That question was a mystery for many years. The riddle was solved when life scientists started using the microscope in their studies. Very thin blood vessels, called **capillaries,** were found connecting tiny arteries and veins. You have already learned about the capillary's role in the breathing process. You will recall that the wall of the capillary is only one cell thick.

Your capillaries leak. If you had a leak in the plumbing of your home, you might call a plumber. Do not call a doctor when we tell you that your capillaries are leaking. That is what they are supposed to do. A capillary is made up of cells that are rolled up into the shape of a tube. There are tiny spaces between these cells. Blood plasma leaks out of the capillary through

Figure 9.18: Capillaries are one cell layer thick. This diagram shows how the cells are arranged. Plasma leaks out of a capillary through the spaces between the cells.

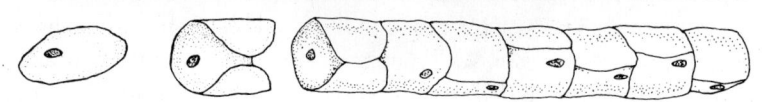

160

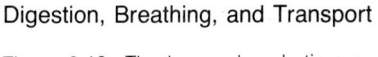

Figure 9.19: The human lymphatic system is a network of lymphatic vessels and lymph nodes.

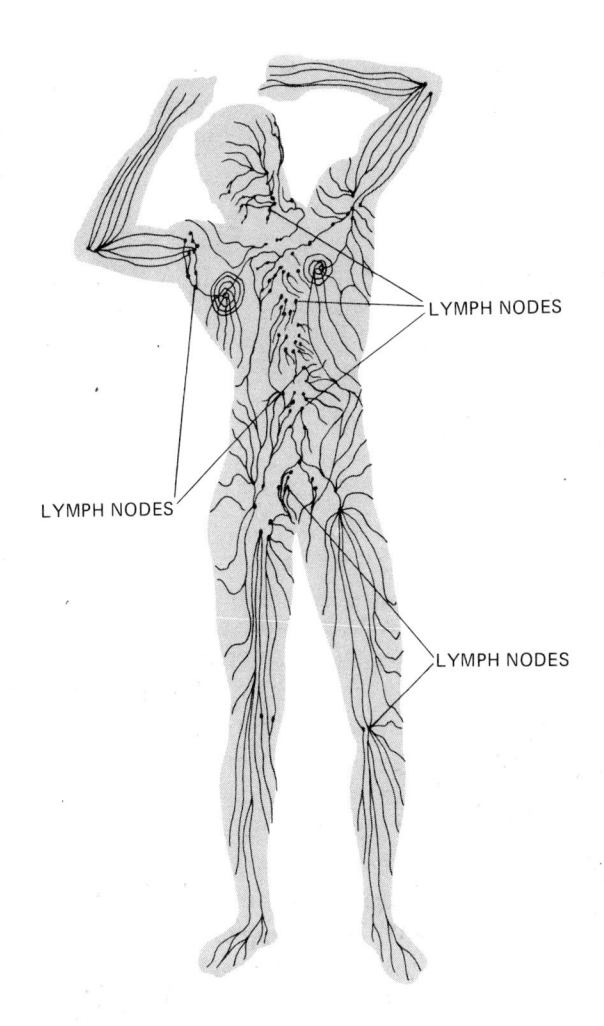

these spaces. The blood plasma flows into the tissues. That is the way nutrients are delivered to your cells.

Plasma gets back to the blood in two ways. If plasma leaks into the tissues, why aren't the tissues swollen with liquid? Actually, this can happen in certain diseases. But it does not happen normally. There are two reasons why. First, there is a two-way flow of plasma near the capillary. Plasma leaks out of the capillary. But plasma from the tissues also flows back into the capillary. Second, you have a **lymphatic system.** This system collects plasma and returns it to the blood.

What is the lymphatic system? It is mostly a network of tubes called lymphatic vessels. In many places throughout the body, the tubes widen out and form little reservoirs or storage areas. These storage areas are called **lymph nodes.** (Sometimes, you will also hear them called "lymph glands.") So far, we have been using the word "lymph" without defining

lymphatic system: a network of tubes that carries blood plasma from tissues back to the blood

lymph node: an enlarged area in the tubes of the lymphatic system where lymph is stored

Human Body Functions

Figure 9.20: The pulmonary circulation takes blood to the lungs and back to the heart.

PULMONARY CIRCULATION

LUNGS

PULMONARY VEINS

RIGHT ATRIUM

RIGHT VENTRICLE

LEFT ATRIUM

PULMONARY ARTERY

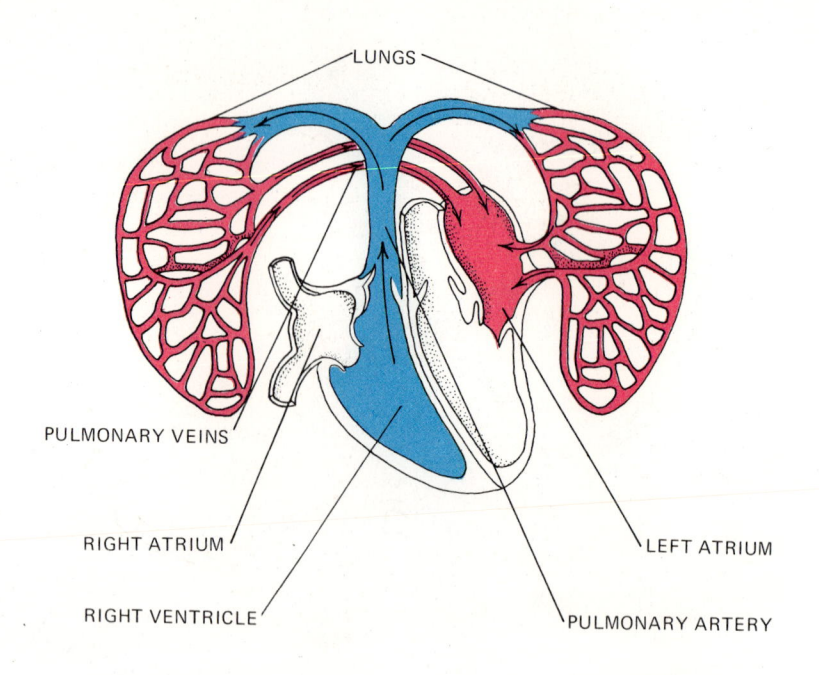

lymph:
a liquid that contains blood plasma from the tissues as well as white blood cells and antibodies

it. What is **lymph?** Actually, a good name for lymph is "used blood plasma." You have probably seen lymph. It is the clear liquid that drains from a broken blister.

Your lymphatic system has another important role. Much of the "combat" against foreign invaders goes on in the lymphatic system. Both antibodies and white blood cells do much of their "fighting" in the lymphatic system. Often you will know when such combat has occurred. There may be a greater than normal amount of lymph in your lymph nodes. This causes them to swell. If you have a sore throat, for example, you may be able to feel the swollen lymph nodes under your jaws.

Blood takes two circulatory routes. The human transport system is often called a "circulatory system." This is because your blood flows in circular routes. You can trace the blood's travels on the "road maps" in Figures 9.20 and 9.21.

pulmonary circulation:
the route the blood follows from the heart to the lungs and back to the heart

One route, called **pulmonary circulation,** takes blood from your heart, to the lungs, and back to the heart. The circle starts with blood that is brought to the heart by the veins. This blood is returning from your body. So it is low in oxygen and high in carbon dioxide. Large veins come together and empty this blood into the *right atrium.* From there, it flows through a valve into the *right ventricle.* From the right ventricle it is pumped out into the *pulmonary artery.* The pulmonary artery branches and takes the blood to each of the lungs. Oxygen and carbon dioxide

162

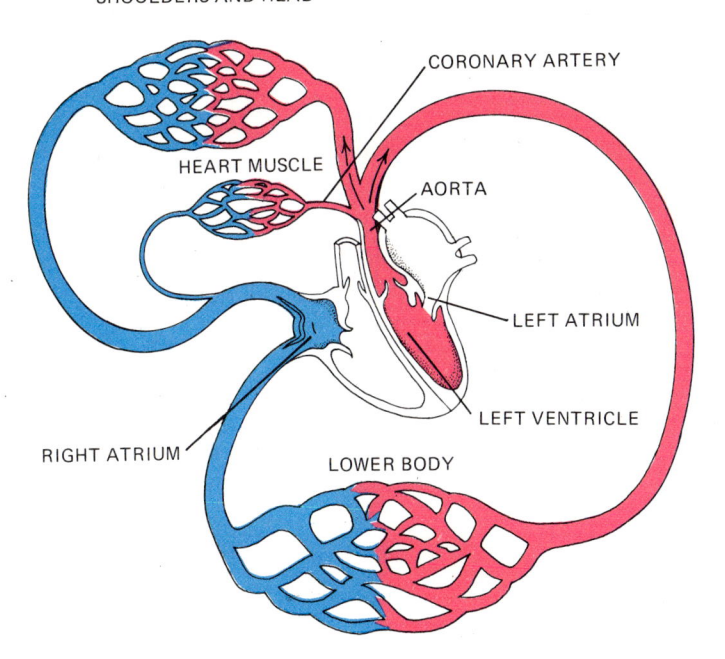

Digestion, Breathing, and Transport

Figure 9.21: The systemic circulation supplies blood to the entire body and takes blood back to the heart from all parts of the body.

are exchanged between the alveoli of the lungs and the blood in the capillaries. The blood is then collected into the *pulmonary veins.* These empty into the *left atrium.* What happens next? The blood is now ready to start its second circle tour.

The other circulatory route is called the **systemic circulation.** This trip takes blood to all parts of the body and back to the heart. We can start following the blood along this route where we left off above, in the left atrium. This is oxygen-rich blood from the lungs. From the left atrium, it flows through a valve into the *left ventricle.* The left ventricle pumps it out through the **aorta,** the largest artery in the body. The aorta gives off one branch just outside the heart. This is the *coronary artery.* The coronary artery supplies fresh blood to the heart muscle. The next branches of the aorta send blood to your shoulders and head. The aorta then arches over and continues down near your backbone. It sends off many branches that serve the lower part of your body. Eventually the blood moves into capillaries, and then into veins. The veins come together and bring blood back to the right atrium. Then another cycle begins. The blood moves to the right ventricle and out through the pulmonary artery.

How are wastes removed from the blood? Earlier in this chapter you learned that wastes are carried away from

systemic circulation: the route the blood follows from the heart to all parts of the body—except the lungs—and back to the heart

aorta: attached to the left ventricle, it is the largest artery in the body. It is the beginning of the systemic circulation outside the heart.

163

Human Body Functions

the cells by the blood. One of these wastes is carbon dioxide. You know what happens to that. It is released in the lungs and exhaled from the body. But what happens to other wastes? They too are removed from the blood. They are removed by a pair of bean-shaped organs called **kidneys**.

Your kidneys are located on each side of your backbone, just below your ribs. Large quantities of blood are always flowing through the kidneys. (See Figure 9.22.) Inside the kidney, much of the plasma is forced out of the blood and into thousands of small tubes. As the plasma flows through these tubes, special cells absorb it and return it to the blood. But these special cells do not take back the wastes. They pass on through the tubes. Eventually they pass out of the kidney and are stored in the **urinary bladder.** Later the wastes are eliminated from the body in the form of a liquid called **urine.**

kidney:
one of two bean-shaped organs, on either side of the lower backbone, which removes many waste materials from the blood

urinary bladder:
a saclike container connected to the kidneys that stores liquid wastes until they are passed out of the body as urine

urine:
liquid waste materials removed from the blood by the kidneys

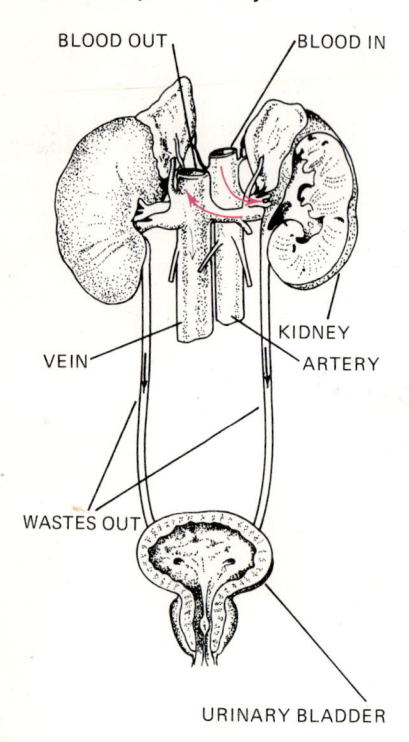

Figure 9.22: The two kidneys remove wastes from the blood. The wastes are concentrated in a solution called urine, which is stored in the urinary bladder until it is eliminated from the body.

LESSON REVIEW *(Think. There may be more than one answer.)*

1. The blood
    a. is about one-half plasma.
    b. is about one-fourth solid materials.
    c. is about one-fourth cells.
    d. is about three-fourths cells.

2. The blood
    a. helps regulate body temperature.
    b. has its own sealer that plugs up "leaks."
    c. carries hormones.
    d. determines whether we are friendly or unfriendly with people.

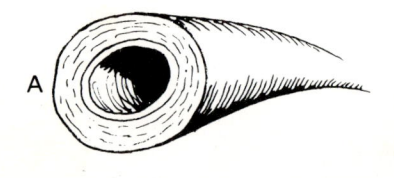

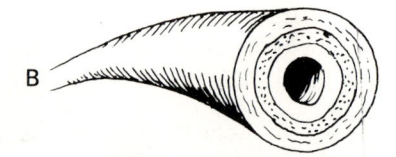

3. In the above illustration, blood vessel A
    a. is an artery.
    b. is a vein.
    c. is often found just under the skin.
    d. expands and stretches when the heart contracts.

4. In the above illustration, blood vessel B
   a. is an artery.
   b. is a vein.
   c. is usually found deep within the body.
   d. cannot stretch because of its thick wall.

5. Blood flows
   a. from the heart to the body to the lungs to the heart.
   b. from the heart to the lungs to the heart.
   c. from the heart to the body to the heart.
   d. from the heart to the lungs to the heart to the body to the heart.

6. After wastes are removed from the blood, they
   a. move from the kidneys to the urinary bladder.
   b. move from the urinary bladder to the kidneys.
   c. move from the capillaries to the lymph.
   d. move from the lymph to the capillaries.

7. Which of the following terms best matches each phrase?
   a. clotting
   b. antibody
   c. red blood cell
   d. white blood cell
   e. platelet
   f. heart
   g. atria
   h. ventricles
   i. heart valve
   j. artery
   k. heart rate
   l. vein
   m. capillary
   n. lymphatic system
   o. lymph nodes
   p. lymph
   q. pulmonary circulation
   r. systemic circulation
   s. aorta
   t. kidney
   u. urinary bladder
   v. urine

   _____ body in the blood that aids the clotting process

   _____ number of times the heart contracts in 1 minute

   _____ liquid waste processed by the kidney

   _____ largest artery in the body; main artery of systemic circulation

   _____ a protein in the blood that helps destroy foreign invaders

   _____ circulation that goes from the heart to the whole body and back to the heart

   _____ cell that carries oxygen

   _____ muscular pump for the transport system

   _____ circulation from the heart to the lungs and back to the heart

   _____ two lower chambers of the heart

   _____ system that transports plasma back to the blood

   _____ liquid that contains plasma, white blood cells, and antibodies

_____ any vessel that carries blood to the heart
_____ organ that cleanses the blood of wastes
_____ the body's storage container for urine
_____ process by which blood plasma changes into fibers
_____ functions as a gate between upper and lower chambers of the heart, and between vessels leaving the heart
_____ tiny blood vessel, one-cell thick, that serves body tissues
_____ any vessel that carries blood away from heart
_____ enlargements of lymph vessels where lymph is stored
_____ two upper chambers of the heart
_____ cells that combat disease

KEY WORDS

clotting (p. 158)
antibody (p. 158)
red blood cell (p. 158)
white blood cell (p. 158)
platelet (p. 158)
heart (p. 158)
atria (p. 159)
ventricles (p. 159)
heart valve (p. 159)
artery (p. 160)
heart rate (p. 160)

vein (p. 160)
capillary (p. 160)
lymphatic system (p. 161)
lymph node (p. 161)
lymph (p. 162)
pulmonary circulation (p. 162)
systemic circulation (p. 163)
aorta (p. 163)
kidney (p. 164)
urinary bladder (p. 164)
urine (p. 164)

Applying What You Have Learned

1. Villi increase the surface area of the small intestine. How does this help increase the amount of absorption?
2. If your heart contracts seventy times per minute, how many times would it contract in 1 day?
3. In what part of the small intestine would you expect to find most of the food absorption? Give a reason for your answer.
4. What causes food materials to move through the digestive tract?
5. Why is it better to have many small alveoli in the lung rather than just one large alveolus?
6. Why does air rush into the lungs when muscles in the chest expand the chest cavity?

Digestion, Breathing, and Transport

7. Why does exercising increase the rate of heartbeat and rate of breathing?
8. How is used blood plasma returned from the body cells to the blood?
9. How do the breathing and digestive systems depend upon the transport system?
10. How does the heart cause blood to circulate through the arteries and veins of the transport system?
11. What important exchange occurs between the alveoli and the capillaries outside them?

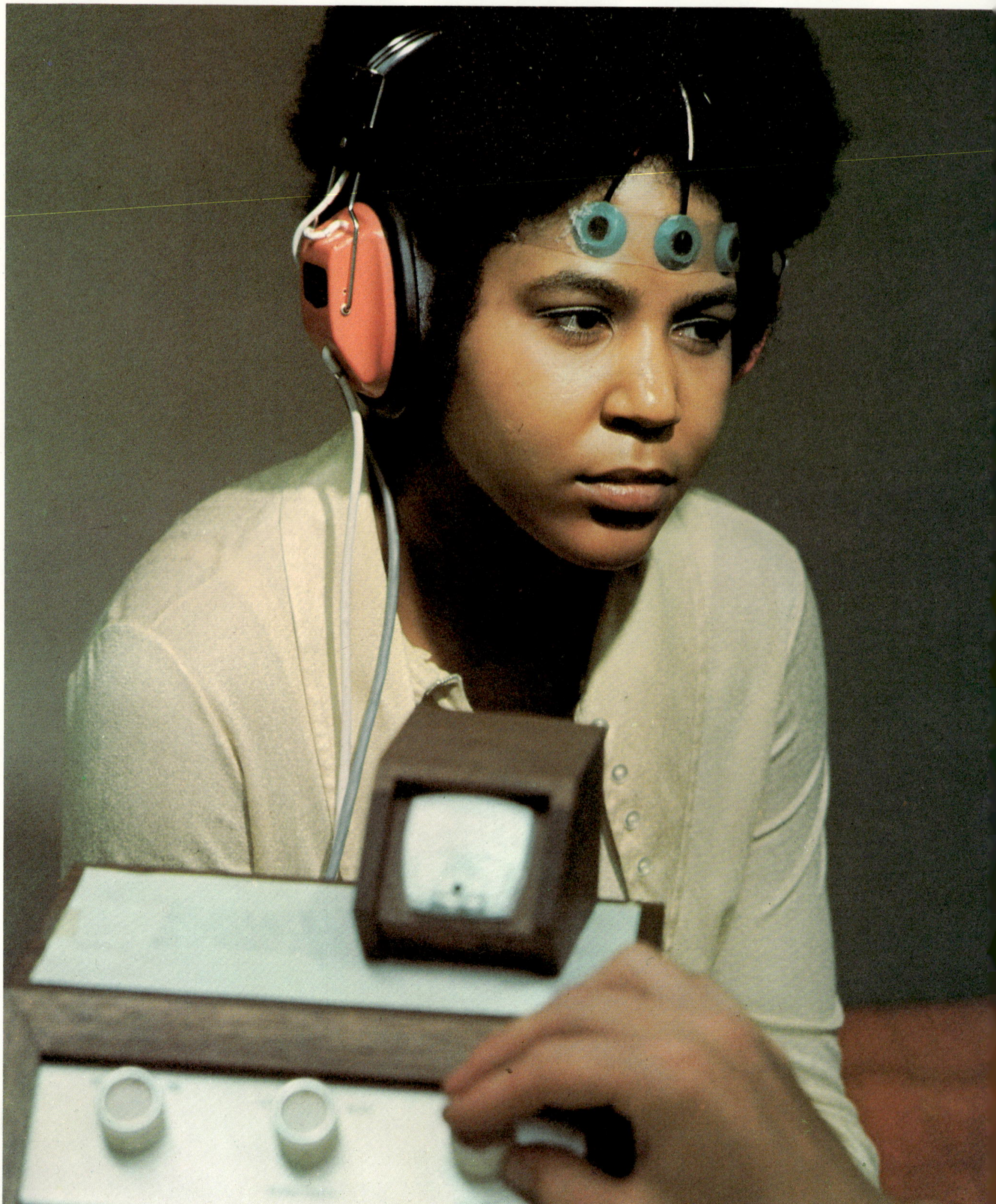

Figure 10.1: The woman is learning to reduce muscle tension. Electrical signals from her brain are recorded by the machine. As she learns to relax her muscles, the signals change. She uses the change in signals to know when she is controlling the tension in her muscles.

Chapter 10

CONTROL IN THE BODY

One of the most amazing things about your body is the way that all parts of it work together. Your cells, tissues, organs, and systems do just what they are supposed to do, when they are supposed to do it. We use the term "self-regulating" to describe the way the interior of your body is controlled. Self-regulating means that the interior of your body takes care of itself, automatically, without your doing anything. In this chapter, you will learn what causes the self-regulating systems to work. And you will study how one of the systems also lets you control the movements of your body.

HORMONE CONTROL

Hormones are chemical messengers. Have you ever been badly frightened, or very mad? If so, many things happened inside your body, things that you did not know about.

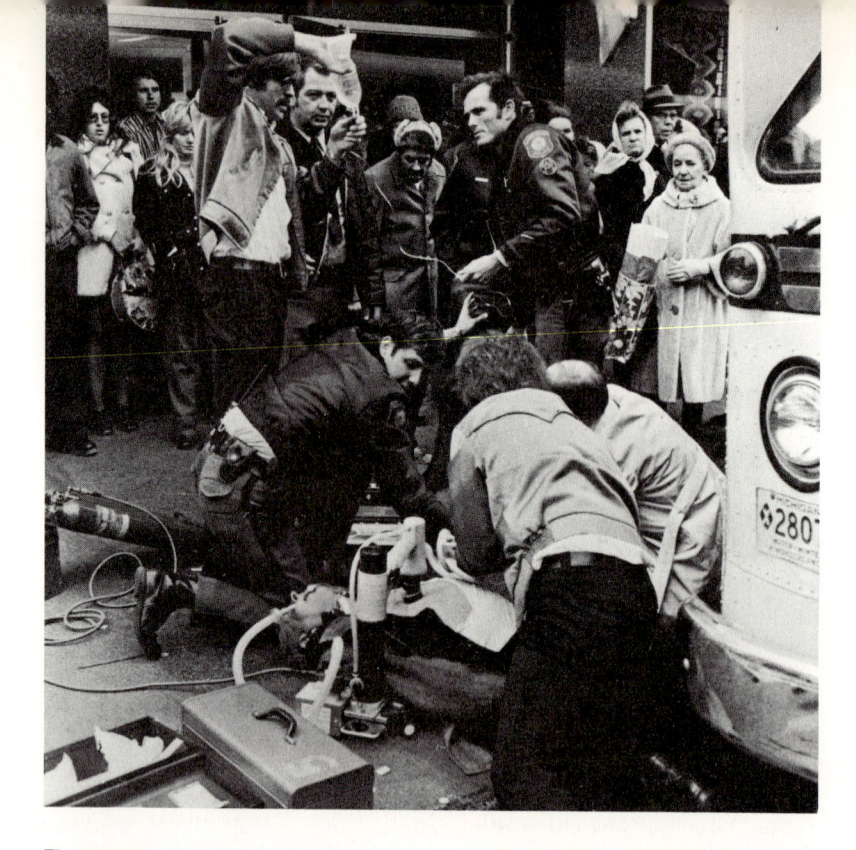

Figure 10.2: Emergency treatment—called cardiopulmonary resuscitation, or CPR—being given to a heart-attack victim at the spot where he fell on a city street. As part of such treatment, a doctor may direct, by radio, that adrenalin be injected to stimulate heart action. In recent years, on-the-spot CPR treatment of heart attack victims has saved many thousands of lives. Perhaps your community has a CPR program. Many U.S. communities now do.

For example, sugar that had been stored in your liver was suddenly released into your bloodstream. This sugar was used by your muscle cells. They used the sugar for quick energy. In this way, your body was made ready in case you needed to run or fight. This is an example of an automatic or self-regulating action inside your body. But what caused it?

adrenalin:
a hormone produced when a person is frightened or angry. It causes changes in the body that ready the body for "fight or flight."

adrenal gland:
one of two glands that produces adrenalin. Each adrenal gland is on top of one of the two kidneys in the body.

Adrenalin (ə drĕn′ə lĭn) caused the sugar to be released from your liver. Adrenalin is produced in two **adrenal** (ə drēn′əl) **glands,** one on the top of each of your kidneys. Say something stimulates or frightens you. Then adrenalin is released into the blood stream. The chemical is transported throughout the body. It acts as a messenger carrying instructions. The liver reacts to the adrenalin by releasing sugar into the blood. The heart reacts by pumping more blood. The arteries in your arms and legs increase in diameter. This allows more blood to flow to your arm and leg muscles. The muscles in your digestive system react by slowing down. All of these actions prepare your body for "fight or flight."

Adrenalin is an example of a **hormone.** A hormone can be thought of as a chemical messenger that travels in the bloodstream. A gland that produces a hormone is called an **endocrine** (ĕn′də krĭn) **gland.** All of the body's endocrine glands function as a system, the **endocrine system.**

The adrenal glands produce other hormones. Not all hormones ready your body for possible fight or flight. Hormones cause a wide variety of responses in your body. For example, the adrenal glands produce several other hormones besides adrenalin. And these other hormones help control a variety of functions. One of the best known of the adrenal hormones is **cortisone.** This hormone, along with others much like it, controls the use of water and minerals. These hormones also have an important effect on kidney function.

The thyroid and parathyroid are close together. You have two endocrine glands lying close together in your neck. But they produce hormones that have very different functions. The **thyroid** (thī′roid′) **gland** is the larger gland. It produces a hormone called **thyroxin** (thī rŏk′sĭn). Thyroxin stimulates cells to release more energy. In order for your thyroid to make thyroxin, it must have iodine. In many areas of the world, people get iodine from the water that they drink. But some regions have water low in iodine. Therefore, it is also added to table salt. "Iodized salt" is the label that you will see on salt with iodine added.

The **parathyroid gland** is thought of as one gland, but it is really four separate tiny glands. They are almost buried in the thyroid tissue. The hormone that they produce controls the use of calcium in your body. Calcium is needed in your blood for proper clotting. Your muscles must have it in order for them to contract. Your bones need it for the mineral that makes them hard. The parathyroid hormone causes the right amount of calcium to be at the right places in your body when it is needed.

Some hormones aid digestion. Both your stomach and duodenum produce hormones. The stomach's hormone causes it to produce digestive enzymes. That is, the hormone causes certain cells in the stomach wall to produce and release

Figure 10.3: This doctor is examining a woman with a greatly enlarged thyroid, or goiter. He may prescribe surgical removal of the diseased gland or radiation treatment to shrink it to more normal size.

hormone:
a chemical released within the body by a gland. Each hormone carries a specific chemical message. The message produces a specific bodily response.

endocrine gland:
any one of several glands that produce hormones

endocrine system:
one of ten organ systems in most animals. It is made up of glands that produce hormones, which are control "chemical messengers" of the body.

cortisone:
an adrenal hormone that helps to control the body's use of water and minerals

thyroid gland:
an endocrine gland located in front of and to both sides of the trachea. It produces thyroxin.

thyroxin:
a hormone produced by the thyroid gland. It contains iodine and stimulates the release of cellular energy.

parathyroid gland:
any of four small glands that are found within the thyroid gland. The parathyroid glands produce a hormone that controls the body's use of calcium.

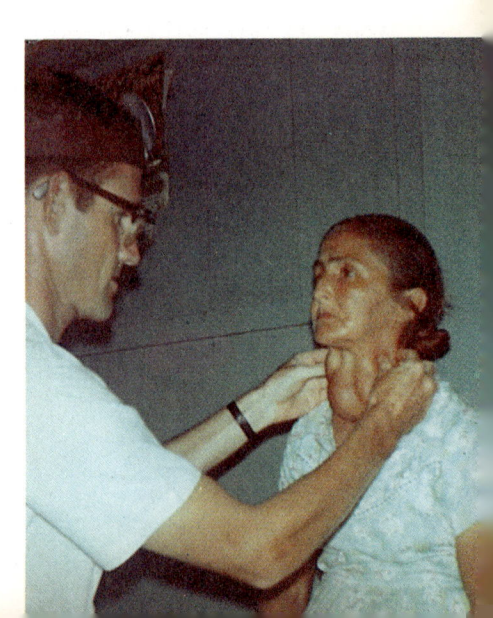

Human Body Functions

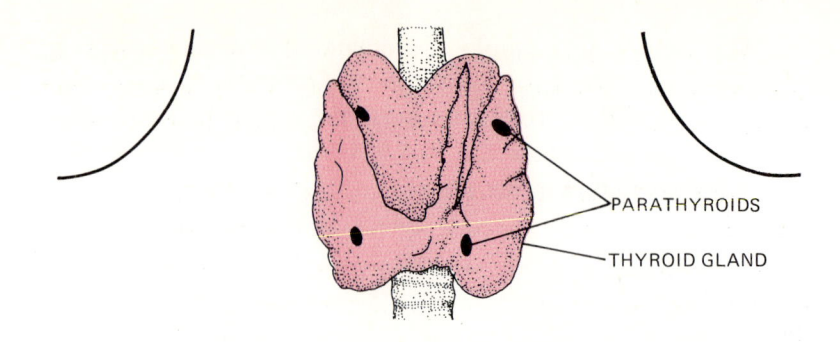

Figure 10.4: The thyroid and parathyroids are in the neck region. The typical number of parathyroids is four, but some people may have three, five, or six.

the enzymes. The hormone produced by the duodenum causes the pancreas to produce digestive enzymes. The same hormone causes the gall bladder to produce bile, which is also used in digestion.

The pancreas produces two hormones. The pancreas is an organ with two functions. We have already described how the pancreas produces digestive enzymes. The pancreas also produces two hormones. One of these hormones is well known. It is called **insulin.** Insulin causes excess sugar in the blood to be stored in the liver. It also causes cells to allow sugar to pass through their cell membranes. Some people have a pancreas that does not produce enough insulin. As a result, their blood sugar level is too high. This causes a type of disease called **diabetes.** People with diabetes often must buy insulin and inject it into their bodies. Such people learn how to use insulin to regulate their own blood sugar level.

insulin: the hormone produced by the pancreas that causes excess blood sugar to be stored in the liver

diabetes: a disease caused by a shortage of insulin in the body. That results in excess sugar in the blood, which, if uncontrolled, can lead to very serious problems.

Figure 10.5: Diabetes is one of the most serious of diseases among children. In this picture, a visiting nurse is teaching a mother and her young, diabetic son the proper way to inject insulin for the control of the disease. The nurse will also instruct the mother in the proper diet her son should follow to help keep the amount of his blood sugar under control.

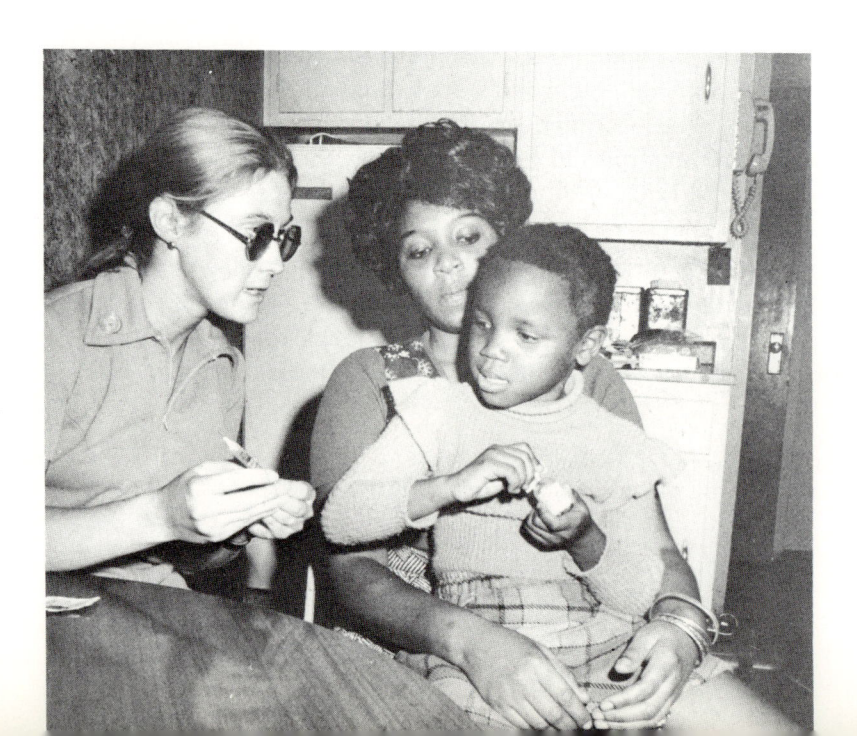

The second hormone produced by the pancreas has the opposite effect of insulin. It causes sugar to be released from storage in the liver. In this way, it acts much like adrenalin.

Your body has a "master" endocrine gland. There is a small gland just below your brain, right in the middle of your head. This gland is the "master" endocrine gland and is called the **pituitary gland.** To see why the pituitary gland is called the master endocrine gland, all you need is a list of its "jobs." It produces a hormone that causes the thyroid gland to produce thyroxin. It produces another hormone that causes the adrenal glands to produce their hormones. Other pituitary hormones cause reproductive organs to produce their hormones. (You will study these reproductive hormones in Chapter 12.) Now you see why the pituitary is called the master endocrine gland. Its hormones control most of the other endocrine glands.

The pituitary also produces hormones that affect specific body functions. One of the best known is the **growth hormone.** This hormone affects cells that grow and divide in the bones and muscles. Too little of the hormone can cause an individual to be a dwarf. Too much of the hormone can cause the opposite—a giant. Growth hormone is now available for use by doctors. With this hormone, they can sometimes prevent a young person from becoming a dwarf.

pituitary gland:
the "master" endocrine gland located below the brain. Pituitary hormones control most of the other endocrine glands.

growth hormone:
a pituitary hormone that controls the growth and division of cells in the bones and muscles

---

LESSON REVIEW *(Think. There may be more than one answer.)*

1. Hormones
    a. travel in the bloodstream.
    b. are digestive enzymes.
    c. are produced by endocrine glands.
    d. are obtained in the food that we eat.

2. Hormones are produced by the
    a. pancreas and thyroid gland.
    b. brain and spinal cord.
    c. adrenal glands and parathyroid glands.
    d. stomach and pituitary gland.

3. The endocrine gland that controls many other endocrine glands
    a. is the pituitary gland.
    b. is the thyroid gland.
    c. is a reproductive organ.
    d. is an adrenal gland.

Human Body Functions

4. Which of the following terms best matches each phrase?
   a. adrenalin
   b. adrenal gland
   c. hormone
   d. endocrine gland
   e. endocrine system
   f. cortisone
   g. thyroid gland
   h. thyroxin
   i. parathyroid gland
   j. insulin
   k. diabetes
   l. pituitary gland
   m. growth hormone

   _____ any chemical "message" carried in the bloodstream

   _____ adrenal hormone that controls the use of water and minerals

   _____ hormone produced by the pituitary gland

   _____ disease caused when the pancreas fails to produce insulin

   _____ master endocrine gland

   _____ gland on top of a kidney

   _____ system produces all hormones

   _____ gland that controls blood calcium

   _____ large endocrine gland in the neck

   _____ hormone that causes excess blood sugar to be stored in liver

   _____ any gland that produces hormones

   _____ hormone that causes cells to release more energy

   _____ hormone that causes stored sugar to be released into bloodstream

KEY WORDS

adrenalin (p. 170)
adrenal gland (p. 170)
hormone (p. 171)
endocrine gland (p. 171)
endocrine system (p. 171)
cortisone (p. 171)
thyroid gland (p. 171)

thyroxin (p. 171)
parathyroid gland (p. 171)
insulin (p. 172)
diabetes (p. 172)
pituitary gland (p. 173)
growth hormone (p. 173)

NERVE CONTROL

The nervous system works faster. Picture in your mind some of the cells in the adrenal glands that produce adrenalin. Suppose some of these cells were to grow long branches that reached out and connected to the liver. Then when you became angry or frightened, the ends of these cells would place adrenalin right into the liver cells. What might you say about such a system?

You do not have long, branching adrenal cells in your body. But you do have a system that works much like the one that is described above. This is your nervous system. It is more effective than the endocrine system in one main way: It works faster.

The nerve cell is the key unit in the nervous system. Nerve cells have long branches. Some have branches that are more than 1 meter (more than 3.3 feet) long in humans. These cells form a network throughout your body. Messages travel very fast along the nerve cell branches. Some of the longer nerve cells send messages at the rate of 100 meters (110 yards) per second. That is like going the length of a football field in the time you could say "one hundred one"! That is much faster than hormones travel in your bloodstream.

The nerve cell message is actually a type of electrical *impulse*. When the electrical impulse reaches the end of a nerve cell branch, it causes a chemical to be released. (Interestingly, one chemical that may be released is adrenalin.) This chemical may bridge the tiny gap between the ends of two nerve cell branches. In this way, the impulse can move on from one nerve cell to another. Sometimes the end of a nerve cell is in contact with other kinds of cells. This happens in muscles and in certain glands. In muscle, a nerve branch ending may release a chemical that causes the muscle cells to contract. Other nerve cell branches may cause glands to release chemicals. For example, impulses to the salivary gland cells cause them to release saliva.

Nerve cells are organized. The nerve cells in your body are organized into tissues and organs. When groups of nerve cells form a separate tissue, this structure is called a **ganglion** (plural, "ganglia"). You have ganglia scattered throughout your chest and abdominal region.

nerve cell:
a cell that carries electrical "messages" throughout the body. The messages produce specific body responses.

ganglion:
a tissue made up of groups of nerve cells

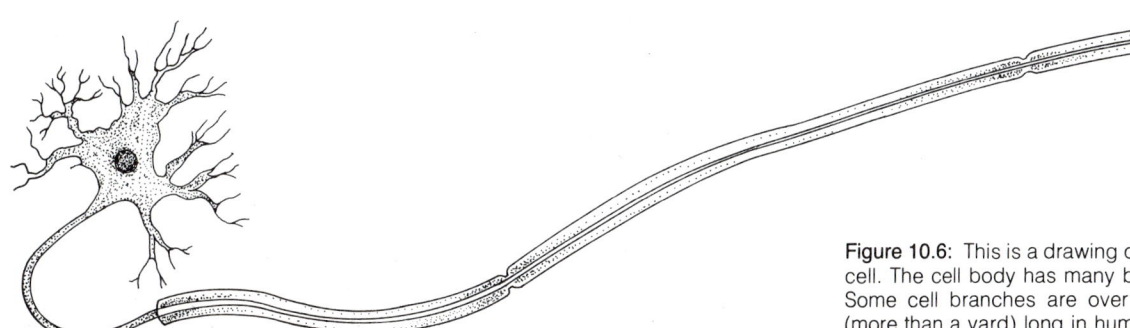

Figure 10.6: This is a drawing of a nerve cell. The cell body has many branches. Some cell branches are over a meter (more than a yard) long in humans.

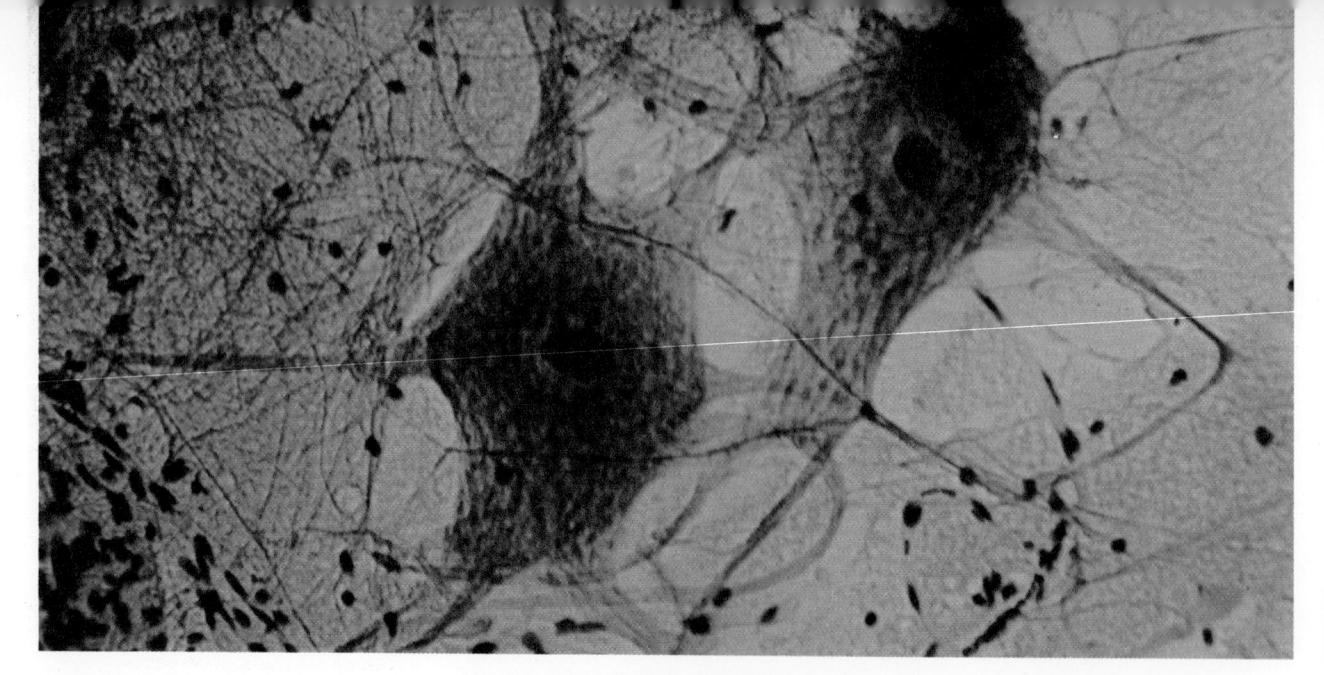

Figure 10.7: A section through nerve tissue shows a massive network of nerve cell branches. In this section you can also see two nerve cell bodies. (Courtesy Carolina Biological Supply Co.)

Most of the nerve cells in your body are concentrated in the brain and the spinal cord. The brain has several regions. Each region of the brain has its own special structure and function. The spinal cord is a long organ. It connects with your brain and extends down your back. It is enclosed and protected by the bones of your spine. The spinal cord is also connected, by nerve cells, to all parts of your body.

Nerves are like telephone cables. It is a good idea to think of a nerve cell as having two main parts. One part is the **nerve cell body.** This is the central part of the cell, including the nucleus and most of the cell's organelles. The other main part of the nerve cell is the **nerve cell branch.** This extends out, away from the cell body.

Now you are ready to understand the structure of a nerve. Picture in your mind a group of nerve cell bodies in some region

nerve cell body: the central part of a nerve cell. It contains the nucleus and most of the organelles.

nerve cell branch: the long part of a nerve cell that extends away from the nerve cell body

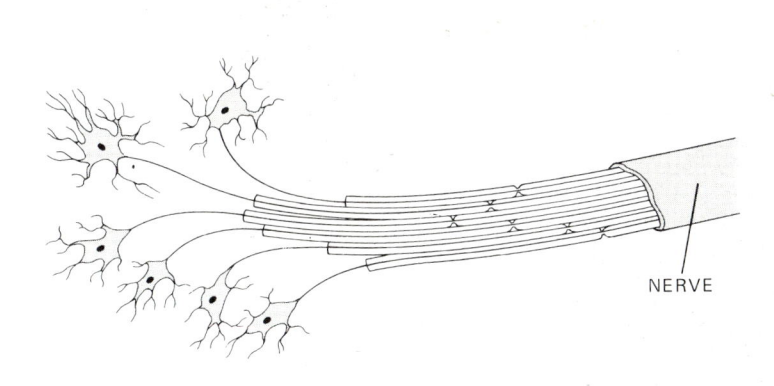

Figure 10.8: A nerve is a bundle of nerve cell branches. Note the similarity between a nerve and a telephone cable.

Figure 10.9: If you could cut across a nerve and look at the ends, they would look like this telephone cable.

of the brain. Next, picture the long branches of the different cells coming together as you see in Figure 10.8. Finally, picture all of these branches enclosed in a sheath. This bundle of branches is a **nerve**. As you can see by comparing Figures 10.8 and 10.9, a nerve closely resembles a telephone cable. And like a telephone cable, a nerve carries messages as electrical impulses. Remember, a nerve is not one cell. A nerve is a bundle of individual nerve cell branches.

Nerve control is both voluntary and involuntary. At the beginning of this chapter we said that your internal body control is self-regulating and automatic. You saw that you have no control over the hormones that exert control in your body. Likewise, you have no control over the way your nervous system regulates your body processes. This type of automatic nervous control is called **involuntary control**.

Your nervous system also lets your body respond to the environment. If somebody throws a rock at you, you will duck or try not to get hit. Your nervous system allows you to make this response. But the response is voluntary. That is, you purposely controlled your action. You could have let the rock hit you. In this case, nervous control is called **voluntary control**.

nerve:
a bundle of nerve cell branches enclosed in a sheath

involuntary control:
control in the endocrine system and part of the nervous system that is automatic

voluntary control:
control in part of the nervous system that is not automatic. The control is brought about by conscious thought.

Human Body Functions

Laboratory Activity

How Do Your Eyes Adjust to Bright Light?

PURPOSE

Some actions take place without your having to think about them. Such actions are called *reflex* actions. Your nervous system controls them automatically. You have no control over them. In this investigation, you will observe a reflex action of your eye.

MATERIALS

2 or 3 partners   chair
bright flashlight   handkerchief

PROCEDURE

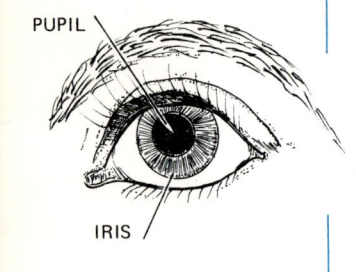

1. You will work with one or more partners in this investigation. Have one partner hold a lighted flashlight. Ask the other partner to sit in a chair to face the flashlight. The seated person is the *subject* whose actions are to be studied.
2. The subject should close one eye, or use a hand to cover one eye. Ask the subject to look toward the flashlight, but not directly into its light.
3. Hold a folded handkerchief in front of the subject's open eye. After a few minutes, remove the handkerchief. Observe the eye (see diagram). How did the pupil and the iris change when the handkerchief was removed? Open both eyes.
4. Repeat the procedure above. This time have the subject keep both eyes open. Cover one of the subject's eyes with a folded handkerchief. Observe the size of the pupil in the uncovered eye.
5. Continue to watch the pupil of the uncovered eye. Quickly remove the handkerchief from the covered eye. Describe the pupil and the iris of the *uncovered* eye.
6. Ask the subject if she or he was aware of the changes in the iris and pupil of the eye.
7. Repeat the investigation with a different partner sitting in the chair. Record your results. Compare your results with the results that were obtained by other groups in the class.

QUESTIONS

1. What happens to the pupil when the eye is exposed to bright light?
2. There are two sets of muscles in the iris of the eye. One set is circular. One set is radial. What happens to these muscles when the pupil gets larger? When the pupil gets smaller?
3. Explain what happened to the pupil of the uncovered eye in step 5.

Most of your muscles are voluntary. You can control them. The workings of your internal organs are involuntary. You cannot control them. Keep in mind this basic difference between the endocrine system and the nervous system. The endocrine system is completely automatic. The nervous system is both voluntary and involuntary.

LESSON REVIEW *(Think. There may be more than one answer.)*

1. The nervous system is more effective than the endocrine system because
    a. it produces stronger hormones.
    b. it can work longer.
    c. it can work faster.
    d. it can work all over the body.

2. There are nerve cells in your body that are
    a. longer than an unsharpened pencil.
    b. longer than a football.
    c. longer than a football field.
    d. longer than your leg.

3. At the end of nerve cell branches,
    a. there may be another nerve cell.
    b. a chemical may be produced.
    c. there are sharp points that cause muscles to react when they are touched.
    d. there are small electric generators.

4. The function of nerve cell bodies is
    a. to receive nerve impulses.
    b. to analyze nerve impulses.
    c. to transfer nerve impulses.
    d. to produce hormones.

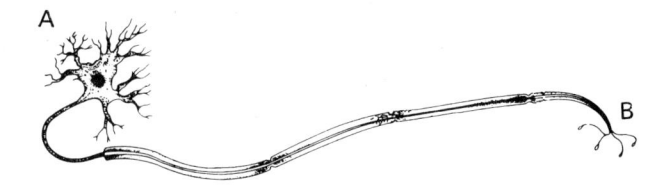

5. The above structure
    a. is a nerve cell.
    b. is a nerve.
    c. moves nerve impulses from A to B.
    d. moves nerve impulses from B to A.

Human Body Functions

6. Which of the following terms best matches each phrase?
   a. nerve cell
   b. ganglion
   c. nerve cell body
   d. nerve cell branch
   e. nerve
   f. involuntary control
   g. voluntary control

   _____ extensions from a nerve cell body

   _____ when a muscle cannot be controlled consciously

   _____ main part of a nerve cell, where the nucleus is located

   _____ cluster of nerve cells

   _____ bundle of nerve cell branches

   _____ when a muscle can be controlled consciously

   _____ specialized cell of the nervous system

KEY WORDS

nerve cell (p. 175)
ganglion (p. 175)
nerve cell body (p. 176)
nerve cell branch (p. 176)

nerve (p. 177)
involuntary control (p. 177)
voluntary control (p. 177)

BRAIN CONTROL

brain:
a large mass of nerve tissue enclosed in the skull. The brain interprets sensory input, directs body functions and activities, and is the center of thought and feeling.

The brain is the main control center. Nerves have been compared to telephone cables. The **brain** can be compared to a central telephone switching station. A girl in Ohio wants to

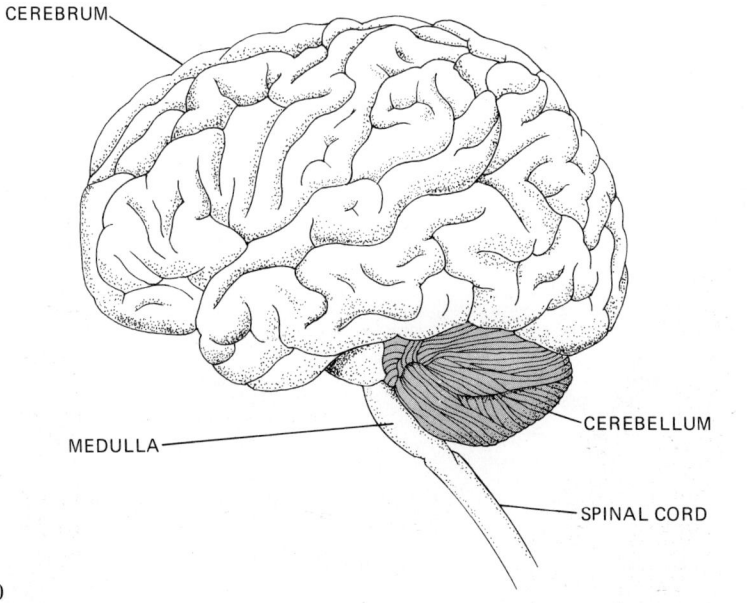

Figure 10.10: This is a diagram of the human brain.

180

call her Uncle Joe in Texas. She dials a number on her telephone. Signals go to a central switching station. A computer switches the signal through millions of possible cables. But the signal goes out on just the right cable. There is a loud ring in Uncle Joe's house in Texas. He jumps out of his chair to answer the phone.

About the same thing happens when a person sees a big mean dog coming after him or her. Signals go to the brain through nerves. The brain works like a computer—only much more effectively. The signals are switched throughout the brain. Each bit of information is interpreted: "There is a dog." "It's a big, mean dog." "The big, mean dog is coming after me." All this switching and interpreting occurs in a fraction of a second. Then signals are sent out of the brain through another set of nerves. These signals go to the leg muscles. In an instant, the brain has taken information and caused the person to go running down the street.

The brain has three main regions. One main region of the brain is called the **medulla** (mə **dŭl′**ə). Doctors often use the term "brain stem" to describe the medulla. It is at the base of the brain, just above the spinal cord. The medulla interprets information coming from the internal organs of the body, such as the lungs and heart. Information from the bloodstream may come to the medulla when muscles need more oxygen. If so, the medulla sends out signals that cause the lungs to breathe faster. It may also send out information to cause the heart to beat faster. That will speed up the flow of oxygen-rich blood to the muscles.

The second main region of the brain is the **cerebellum** (sĕr′ə **bĕl′**əm). It interprets and sends out information that keeps muscles coordinated. Think of a tightrope walker walking between two tall buildings. A gust of wind tilts the person. This information is carried by nerves to the cerebellum. Signals from the cerebellum have to be sent to the muscles so that the muscles

medulla:
a major region of the brain, also called the brain stem. Located at the base of the brain and connected to the spinal cord, it controls many internal organs.

cerebellum:
a major region of the brain. It controls and coordinates muscular responses.

Figure 10.11: The twin towers of the World Trade Center in New York City are 405 meters (1350 feet) tall. They are the second-tallest buildings in the world. In the summer of 1974, illegally and as a publicity stunt, a French tightrope walker strung cable between the towers and walked from one to the other.

Human Body Functions

will react properly. But how much should one leg muscle contract? And how much should another leg muscle relax? And how should the muscles of the back and shoulders relax or contract? The slightest mistake would mean disaster. But the cerebellum comes through (we hope). Just the right amount of information goes to each muscle. Some contract and others relax. And the tightrope walker regains balance.

The third main region of the brain is the **cerebrum** (sĕr′ə brəm). This is the largest region. Also, it is the region where all thought occurs. Most information from a person's environment is interpreted in the cerebrum. The taste of food is interpreted in one region of the cerebrum. Odors, sounds, and sights are interpreted in other regions. Also, the cerebrum is the center of memory and imagination. That is one way humans differ from all other animals. The cerebrum of the human brain is much larger in proportion to the rest of the body than is that of other animals. This helps explain why humans and not chimpanzees can write poetry and collect information from Mars.

cerebrum: the large upper layer of the brain. It is the brain region where all thought occurs and where input from the senses is interpreted.

The spinal cord is a switching center. The **spinal cord** is a thick, long body of nerve tissues. It is enclosed by bones called **vertebrae** (vûr′tə brē). These are small, irregular-shaped bones that make up the *spine*, or backbone.

The spinal cord receives information from parts of the body below the head. This information is usually sent to the brain, interpreted, and sent back through the spinal cord and then to the body.

Sometimes information from the body does not go all the way to the brain to be interpreted. Sometimes information comes to the spinal cord and is sent directly to another part of the body. The knee-jerk reflex is an example. Sit on a table with

spinal cord: a thick, long body of nerve tissues that extends from the base of the brain, branching to form smaller nerves that serve various parts of the body

vertebrae: the bones of the spine, or backbone

Figure 10.12: This diagram maps some of the specialized regions of the cerebrum. Damage to any of these regions will affect the function that it controls.

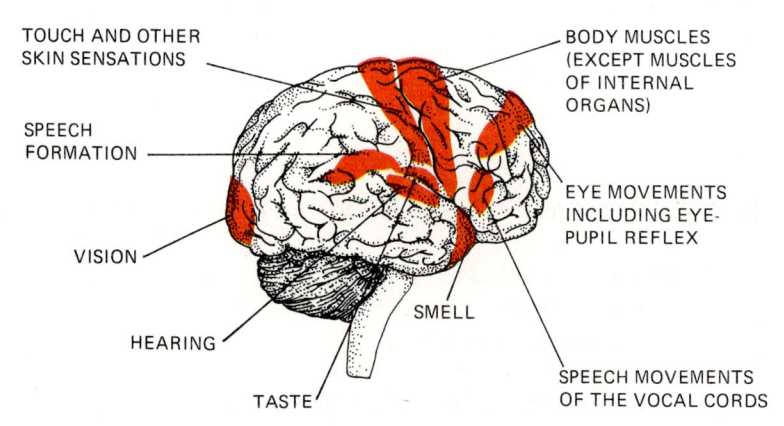

Control in the Body

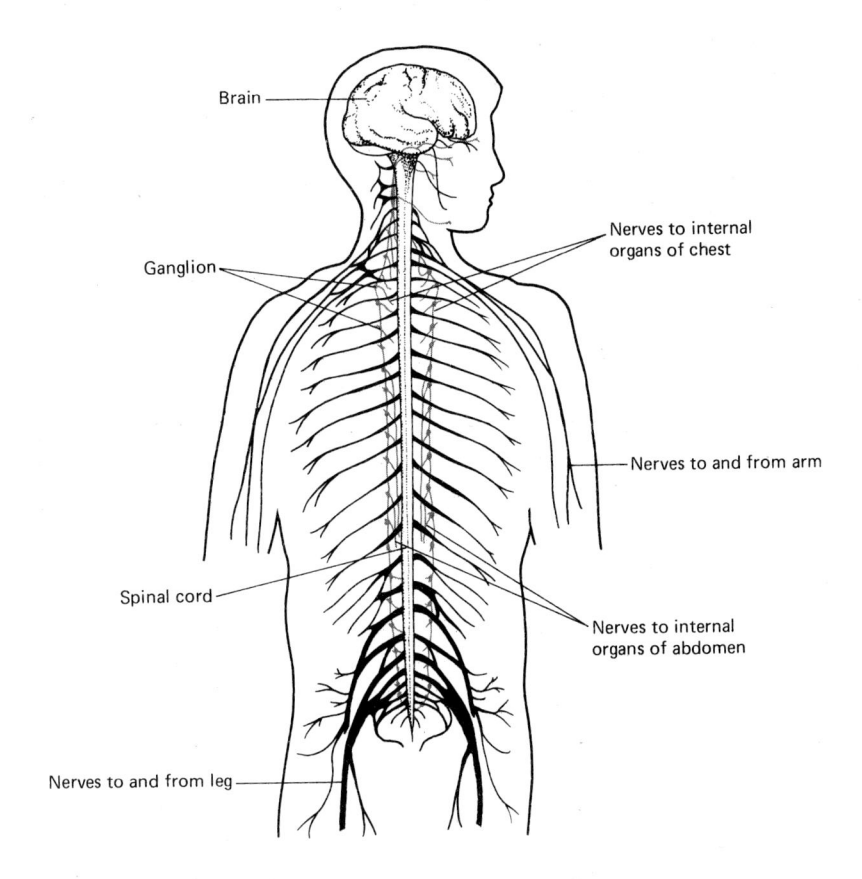

Brain
Nerves to internal organs of chest
Ganglion
Nerves to and from arm
Spinal cord
Nerves to internal organs of abdomen
Nerves to and from leg

Figure 10.13: This is a diagram of the central nervous system and some of its branches. Impulses sent over one part (black) are under conscious control and cause muscles to contract or relax. Impulses sent over another part (gray) are not under conscious control. These control heartbeat and other involuntary body actions.

a leg hanging loosely. Tap the region just below the knee. The lower leg will kick automatically. If you are relaxed, you cannot stop this movement. It is involuntary. In this case, the information goes only from your knee to your spinal cord. Signals are sent from the spinal cord directly to your leg muscles, causing them to respond.

LESSON REVIEW *(Think. There may be more than one answer.)*

1. The brain works like
    a. a computer.
    b. a telephone cable.
    c. a television set.
    d. a central telephone switching station.

183

Human Body Functions

2. The main control center of the nervous system is the
   a. spinal cord.
   b. endocrine system.
   c. medulla.
   d. brain.
3. The medulla
   a. is located at the top of the brain.
   b. controls many internal organs.
   c. is located at the base of the brain.
   d. is the largest region of the brain.
4. The cerebellum
   a. is the largest region of the brain.
   b. is located just above the spinal cord.
   c. controls and coordinates muscles.
   d. is the center of thought and feelings.
5. The cerebrum
   a. is the largest region of the brain.
   b. is between the cerebellum and medulla.
   c. interprets sensory input.
   d. is the center of thought and feelings.
6. The spinal cord
   a. acts as a switching station between the brain and other parts of the body.
   b. controls some reflexes directly.
   c. sometimes sends messages from one part of the body to another, without involving the brain.
   d. is enclosed by vertebrae.
7. Which of the following terms best matches each phrase?
   a. brain   d. cerebrum
   b. medulla   e. spinal cord
   c. cerebellum   f. vertebrae

   _____ irregular bones that protect the spinal cord

   _____ the center of thought and feelings

   _____ is made up of three main parts

   _____ a thick, long body of nerve tissue

   _____ is attached to the spinal cord

   _____ controls muscle action

KEY WORDS

brain (p. 180)
medulla (p. 181)
cerebellum (p. 181)

cerebrum (p. 182)
spinal cord (p. 182)
vertebrae (p. 182)

Applying What You Have Learned

1. What are two systems that control and regulate body functions?
2. Why can the nervous system react faster than the endocrine system?
3. What will happen if the nerve that controls a certain gland or muscle becomes injured and cannot transmit or carry an impulse?
4. Indicate which of the following body actions are involuntary and which are voluntary. Could any of these actions be controlled in both ways?
   a. blinking your eyelids
   b. throwing a rock
   c. walking home from school
   d. secreting digestive enzymes
   e. reacting to an electrical shock
   f. coughing
   g. writing
   h. heartbeat
5. What system helps the endocrine system and distributes hormones throughout the body?
6. What system detects changes and activity in the body's internal and external environments?
7. How is the purpose of nerve endings in the eyes and ears different from the purpose of those in muscles and glands?
8. For which two different systems does the pancreas work? Describe its function in each.

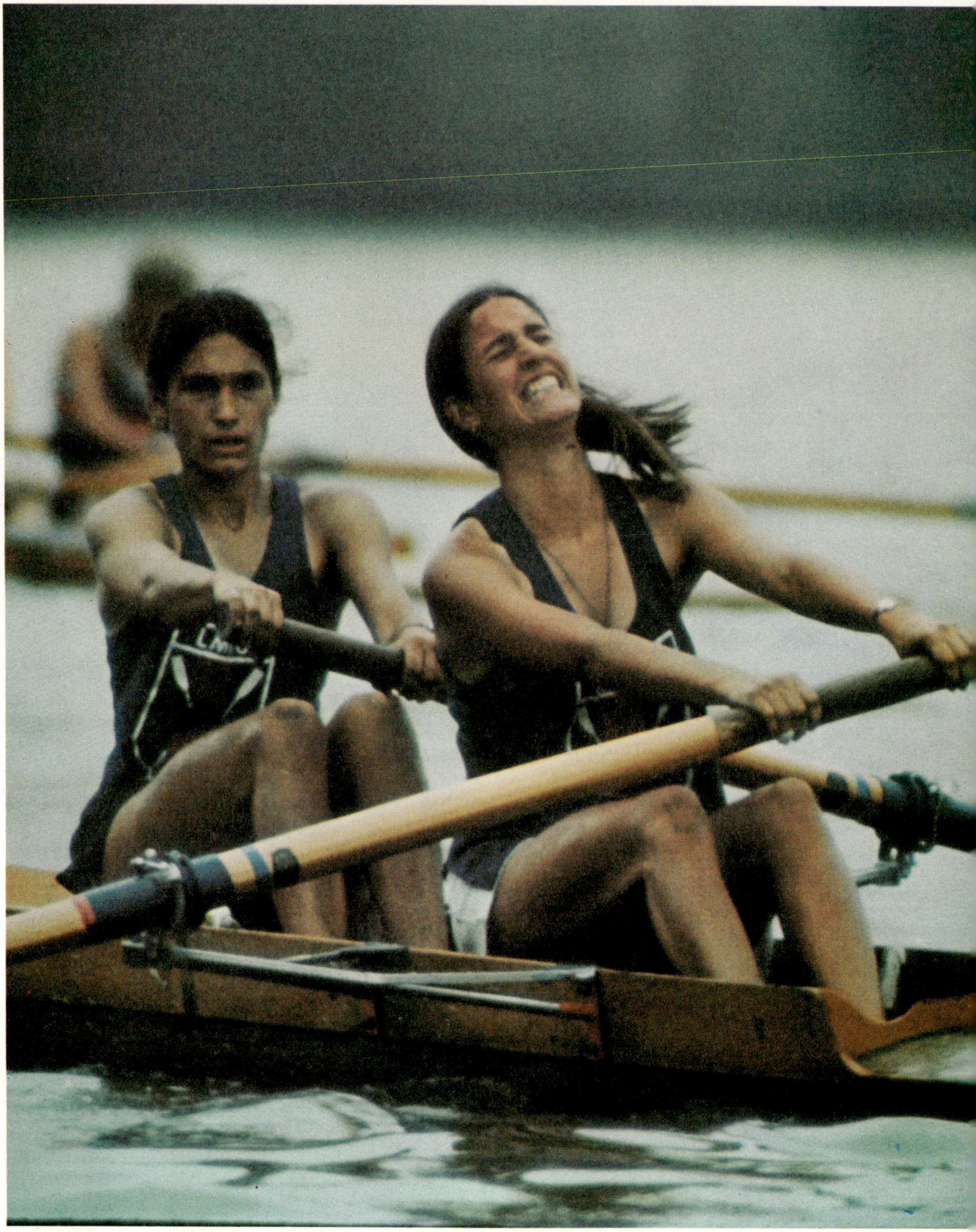

Figure 11.1: You use many muscles when you row a boat. For you to compete in rowing races, your back, shoulder, and leg muscles must be strong and well developed.

Chapter 11

SKIN, MUSCLES, AND BONES

The skin, or **integument** (ĭn tĕg′yə mənt), covers your body and has several important functions. More than half of your body weight is made up of muscles and bones. In this chapter, the integumentary system, the muscular system, and the skeletal system are discussed.

integument:
the skin; an organ of the integumentary system

THE INTEGUMENTARY SYSTEM

The skin has two main layers. The entire human body is covered with skin, which is the main organ of the integumentary system. The skin itself has two main layers.

The outer layer of the skin is called the **epidermis** (ĕp′ĭ dûr′mĭs). It, in turn, is made up of layers of dry, tough cells. The skin is always shedding the outer layers of the epidermis. We see this as dry, flaky skin, or as dandruff when it comes from our scalp. This shedding is normal, despite advertisements that tell you otherwise.

epidermis:
the outer of two layers that make up the skin

Human Body Functions

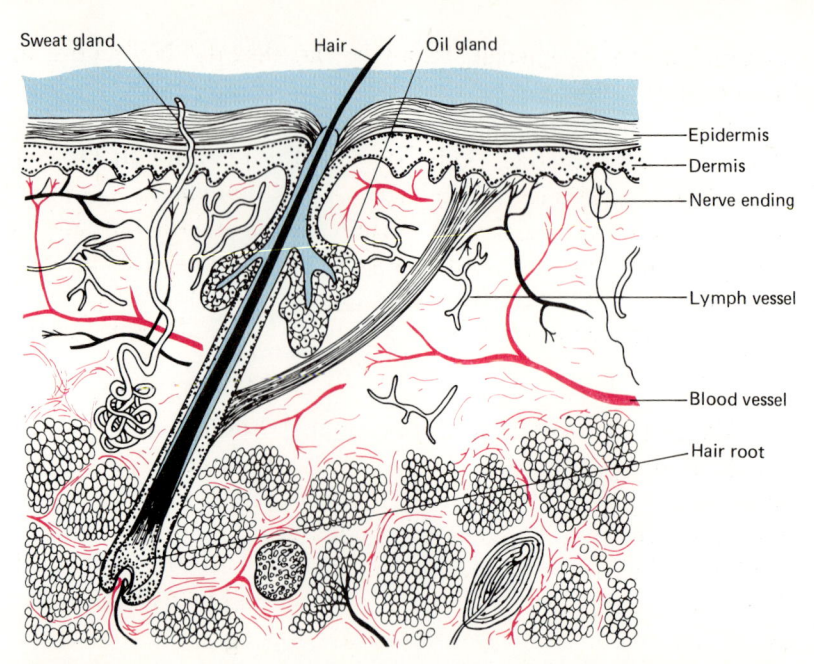

Figure 11.2: This cross section of the skin shows the different layers and glands. The small circles at the base of the skin represent a layer of fat cells usually found under the skin.

callus:
a thickened area of the epidermis

dermis:
the inner of two layers that make up the skin

The epidermis does not have capillaries. Therefore, it has no blood. Nor does it have nerve endings. For these reasons, the epidermis can be cut without bleeding or causing pain.

We can cause thick layers of epidermis to develop. Any part of the epidermis that receives extra rubbing or wear will thicken and cause a **callus** (kal′əs). A carpenter will develop calluses on the inside of the hands. A guitar player will develop them on the tips of the fingers. Look at the middle finger of your writing hand. Look at the part pressed on by the pen or pencil when you write. You may see a callus, or "writer's bump," there.

The inner layer of skin is called the **dermis.** It, in turn, has several specialized layers of cells. One produces new cells to replace those being lost by the epidermis. This layer can also produce new cells to replace those destroyed by cuts and other minor injuries. Another layer of the dermis produces brown pigment, which gives the skin its color. Also in the dermis are nerve endings, small blood vessels, and bands of connective tissue. The connective tissue gives the skin its strength and elasticity.

The integumentary system has organs besides the skin. Nails and hair are two other organs of the integumentary system. Nails on the fingers and toes are a hard protein material made by special tissues in the dermis. Nails help to protect the fingers and toes.

Hair covers most of the body. It grows out of a pit in the skin called a **follicle** (fŏl′ĭ kəl). At the bottom of the follicle, special cells divide and cause the hair itself to grow longer. Near the follicle is a small muscle. At certain times—when the skin is cold or when you are feeling strong emotion—these muscles contract. This causes the hair to stand on end. It also causes small bumps to appear on the skin. Many people call them "goose bumps."

follicle:
the tiny sac in the skin from which a hair grows

There are two types of glands in the skin. One type of gland in the skin produces oil. It is called a **sebaceous** (sĭ bā′shəs) **gland.** (This name is derived from *sebum* (sē′bəm), one name for the oil.) Usually the sebaceous glands are next to the hair follicles and secrete oil directly into the follicle. The oil functions to keep the hair from getting dry and brittle. Also, the oil on the skin helps prevent excessive flaking of the epidermis and drying.

sebaceous gland:
one of the tiny glands in the skin that secretes oil into a hair follicle

The second type of gland in the skin is the **sweat gland.** It produces sweat, which is water with a small amount of dissolved salt. Sweat has an even smaller amount of a waste product called *urea* (yoo rē′ə). The breakdown of urea by microorganisms on the skin produces the odors that we call "body odor."

sweat gland:
one of the tiny glands in the skin that secretes sweat (perspiration) through openings, or pores, in the skin

The skin has three main functions. One main function of the skin is to protect the body. The skin prevents the tissues below it from being injured. It also prevents them from drying out, since the skin is almost waterproof. In addition, it keeps foreign matter such as microorganisms and chemicals from contacting the inner tissues.

The second function of the skin is to help regulate body temperature. The sweat glands play an important role in this function. When the body has excess heat, the sweat glands go into action. Sweat covers the skin. When the sweat evaporates, the body is cooled. Blood vessels in the skin also help to control heat loss. When the body has excess heat, more blood flows through the skin. Some of the excess heat carried by the blood is lost to the outside. When the body is cold, a much smaller quantity of blood flows through the skin. This conserves body heat.

The third function of the skin is to serve as a sense organ. Nerve endings in the dermis give us a sense of touch. With our eyes closed, we can feel most objects and identify them because of these nerve endings. Other specialized nerve endings in the dermis allow us to sense pain, heat, and cold.

Human Body Functions

LESSON REVIEW *(Think. There may be more than one answer.)*

1. The integumentary system includes
   a. skin.
   b. nails.
   c. hair.
   d. sweat glands.

2. The skin
   a. protects the body.
   b. produces hair.
   c. regulates body temperature.
   d. acts as a sense organ.

3. A hair
   a. is made of urea.
   b. cannot stand on end.
   c. grows out of a sebaceous gland.
   d. grows from its base.

4. A knife cut of the skin that stops before hitting muscle
   a. can be felt in the epidermis.
   b. can be felt in the dermis.
   c. is too shallow to become infected.
   d. might heal without a visible scar.

5. Glands in the skin
   a. cause dandruff.
   b. produce sweat.
   c. produce oil.
   d. act as sense organs.

6. Which of the following terms best matches each phrase?
   a. integument   e. follicle
   b. epidermis   f. sebaceous gland
   c. callus   g. sweat gland
   d. dermis

   _____ secretes perspiration

   _____ tiny sac from which a hair grows

   _____ a thickened area of the outer skin layer

   _____ the inner of two skin layers

   _____ the skin

   _____ causes oily hair

   _____ skin layer without capillaries

KEY WORDS

integument (p. 187)
epidermis (p. 187)
callus (p. 188)
dermis (p. 188)

follicle (p. 189)
sebaceous gland (p. 189)
sweat gland (p. 189)

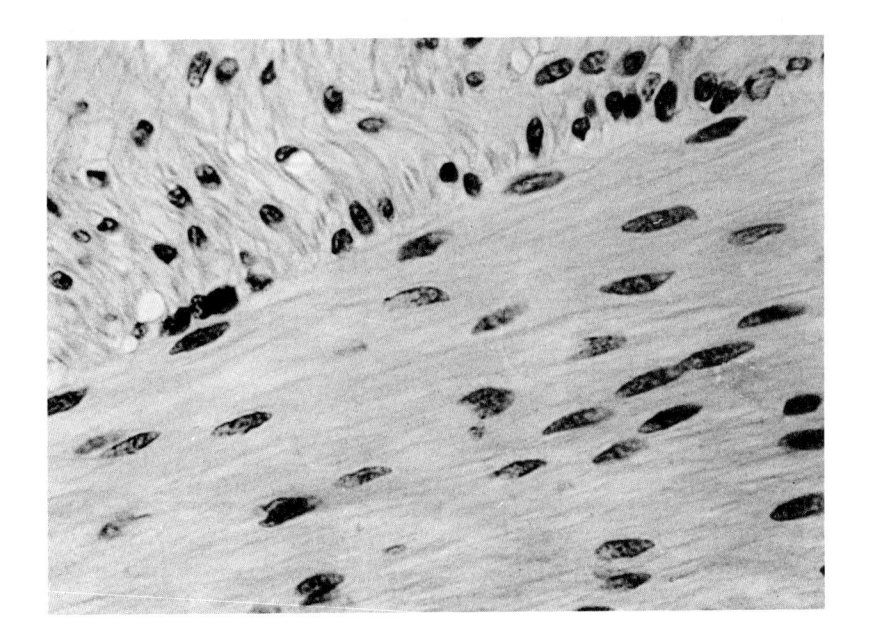

Figure 11.3: Two layers of smooth muscle tissue from the human digestive system can be seen in this microscopic picture. The tissue layer at the right of the view contracts and expands in an up-and-down direction. The tissue layer at the left contracts and expands in a circular direction.

THE MUSCULAR SYSTEM

There are three kinds of muscles in the human body. One kind of muscle is found only in the internal organs of the body. This is called **smooth muscle.** It is found, for example, in the walls of the digestive organs. The largest mass of smooth muscle is found in the female **uterus** (yōō′ter əs). This hollow organ, where the human **embryo** (ĕm′brē ō′) and **fetus** (fē′təs) develop, is almost entirely made up of smooth muscle.

Smooth muscle is made up of long thin cells that overlap and form layers. This muscle contracts more slowly than the other types of muscle in the body. Also, its range of movement is larger. The stomach and uterus are both hollow organs with smooth muscle in their walls. The muscle in these organs may relax and expand to three times or more its normal size. But it also may contract and cause the stomach to be a small tube and the uterus to be an almost solid organ.

One of the main characteristics of smooth muscle action is that it is *involuntary*. That means conscious thinking cannot cause it to relax or contract. It is controlled by the spinal cord, the medulla of the brain, and by other parts of the nervous system.

Cardiac muscle is found in the heart. Another type of muscle is **cardiac muscle.** It is the muscle that makes the heart work, and is not found anywhere else in the body.

Cardiac muscle is like smooth muscle in one way. Its action is involuntary. You have no conscious control over your

smooth muscle:
one of several types of muscles. Smooth muscles are found only in internal organs.

uterus:
a thick, hollow organ in females. The human embryo and fetus develop within the uterus. Also known as the womb (wōōm).

cardiac muscle:
one of several types of muscle. Cardiac muscle is found only in the heart.

Human Body Functions

Figure 11.4: In this microscopic view of cardiac muscle tissue, you can see how the muscle fibers lie alongside one another.

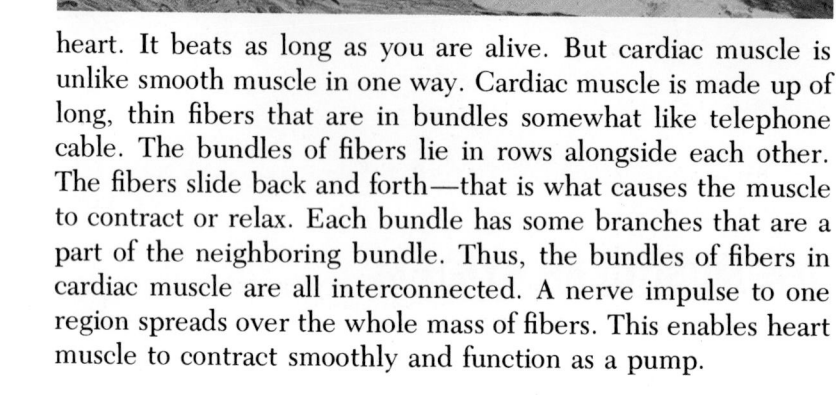

heart. It beats as long as you are alive. But cardiac muscle is unlike smooth muscle in one way. Cardiac muscle is made up of long, thin fibers that are in bundles somewhat like telephone cable. The bundles of fibers lie in rows alongside each other. The fibers slide back and forth—that is what causes the muscle to contract or relax. Each bundle has some branches that are a part of the neighboring bundle. Thus, the bundles of fibers in cardiac muscle are all interconnected. A nerve impulse to one region spreads over the whole mass of fibers. This enables heart muscle to contract smoothly and function as a pump.

Skeletal muscle moves bones. A third type of muscle is called **skeletal muscle.** Most skeletal muscle is attached to bones. By contracting and relaxing, it causes the bones to move. This allows movement of the body and its limbs.

Skeletal muscle is *voluntary* muscle. It responds to your conscious thoughts. Like cardiac muscle, it is made up of bundles of fibers that lie alongside each other. But there is one

skeletal muscle: one of several types of muscle. The skeletal muscles control all outside-the-body movements.

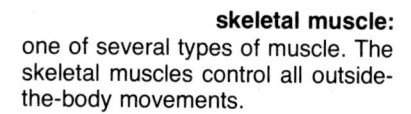

Figure 11.5: The muscles of the body have similarities and they have differences.

A COMPARISON OF THREE MUSCLE TYPES

Smooth Muscle	Cardiac Muscle	Skeletal Muscle
1. long slender cells 2. found in internal organs 3. involuntary 4. contracts more slowly than other muscles	1. bundles of fibers that are interconnected 2. found only in the heart 3. involuntary	1. bundles of fibers that are insulated from each other 2. attached to skeleton and cause movements of bones 3. voluntary 4. capable of quicker and more precise movements than other muscles

Skin, Muscles, and Bones

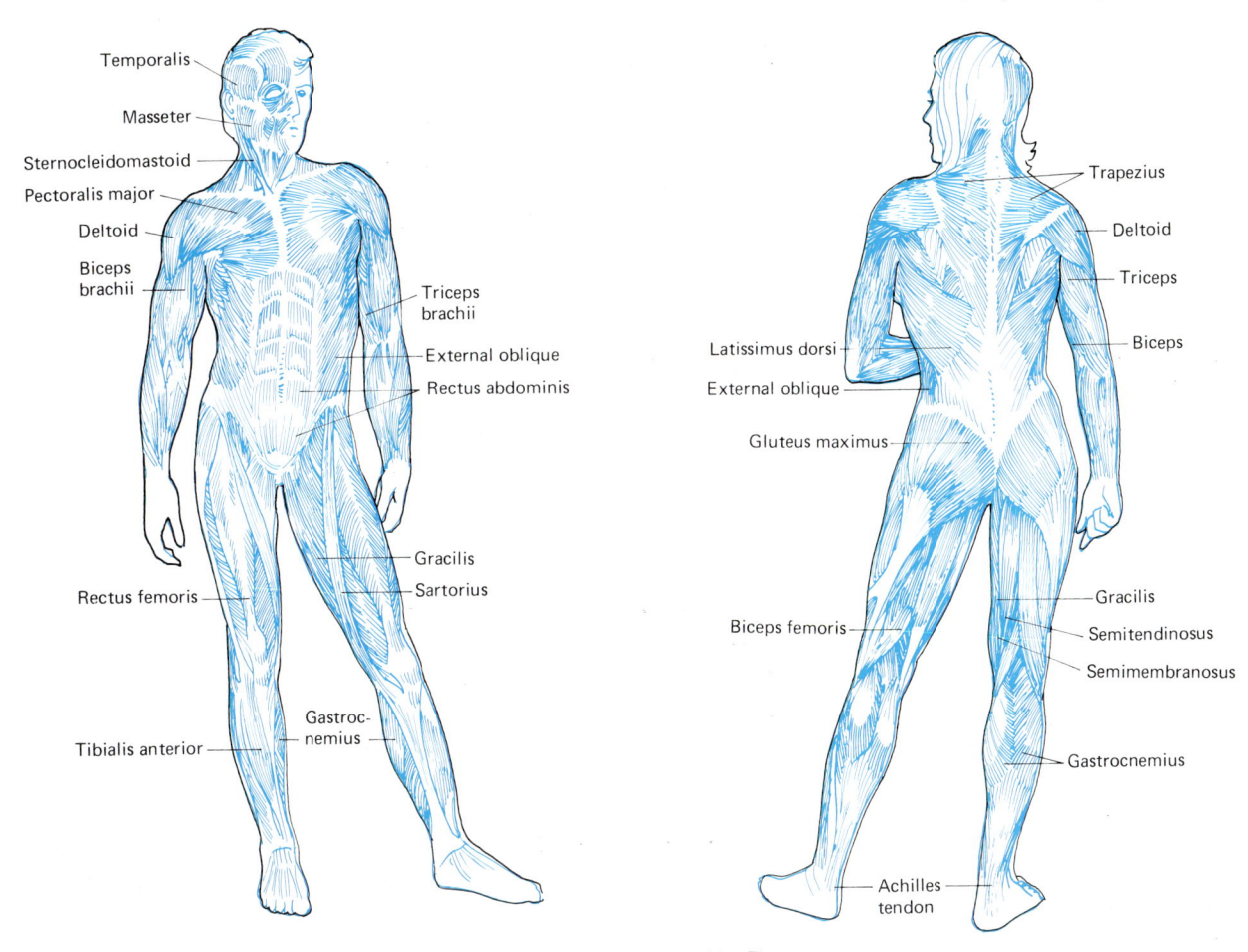

Figure 11.6: These diagrams show the major skeletal muscles below the skin. There are many other skeletal muscles that lie deeper within the body.

difference. The cardiac muscles have bundles of fibers that are branched. That is not true of skeletal muscles. Each bundle of fibers in skeletal muscles is surrounded by a membrane. Thus, the fiber bundles are separated from each other, like wires in telephone cable. Nerve impulses have to reach each individual bundle of fibers before they will relax or contract. This is a complicated system, but it pays off in efficiency. Your skeletal muscles can make quicker and more exact movements than cardiac or smooth muscles.

Skeletal muscles can be conditioned. The size of the fibers in skeletal muscle can change. If the skeletal muscles are used often, the fibers increase in size. If the skeletal muscles are not used, the fibers decrease in size. If injuries cut the nerves to skeletal muscle, a tragic thing happens. Because the muscle cannot contract without nerve stimulation, muscles without nerves get smaller and smaller and shrink away.

Figure 11.7: These high school cross-country runners are nearing the end of a race. How have they trained for the race? What purposes were served by their training?

muscle tone: the condition that keeps a muscle ready for action. It is observed as a slight, but constant, contraction.

Most people are aware of the fact that exercise can make skeletal muscles larger. But exercise also increases **muscle tone**. This is a condition of a resting skeletal muscle. It determines the muscle's readiness to do work. A muscle with good tone can respond quickly and exactly. In other words, it is more likely to do what you want it to do. Also, it will do more work without getting fatigued.

What is muscle fatigue? Everyone knows about "tired muscles." Tired muscles do not work when you want them to do so.

Skeletal muscles, like all muscles, need energy before they can do work. They get their energy by breaking down sugar and starch molecules. Aside from energy, one of the products is called lactic acid. The lactic acid acts like a muscle poison. It is what causes your muscles to become fatigued. If enough lactic acid builds up in the muscles, they will eventually stop all movement.

As you might guess, you have a way to get rid of the lactic acid that builds up in your muscles. In the presence of oxygen, much of the lactic acid that forms is converted back into sugar and starch. Energy is stored in the process. But note the problem. Oxygen is needed to get rid of the lactic acid. When you exercise, your body cannot get oxygen to your muscles fast enough to get rid of the lactic acid. That is why you eventually get tired. So you must rest. And while you rest, more and more oxygen is carried to your muscles. The lactic acid gradually disappears and you gradually get over your tired muscles.

Oxygen debt is the price of exercise. We use the term **oxygen debt** to describe the lack of oxygen in skeletal muscles that have been working. You go into oxygen debt when you exercise. You pay back the debt when you rest. Now you can see how athletic training serves two purposes. It increases the tone of the skeletal muscles. This gives the athlete more control over them. Also, it increases the efficiency of the breathing and transport systems that bring oxygen to the muscles.

Imagine a marathon runner in a 26-mile race. That person cannot afford to build up a very large oxygen debt. Such an athlete must have supereffective breathing and transport systems. Years of training are required to develop such a system.

oxygen debt: a condition in which skeletal muscle cells have less oxygen in them after muscular work has been done

---

LESSON REVIEW (Think. There may be more than one answer.)

1. Smooth muscle
    a. consists of long, overlapping cells.
    b. is found in the heart and other internal organs.
    c. contracts slowly.
    d. is controlled by conscious thinking.

2. Cardiac muscle
    a. consists of interconnected bundles of thin fibers.
    b. is found in the heart and other internal organs.
    c. works like a pump.
    d. is controlled by conscious thinking.

3. Skeletal muscle
    a. consists of separate bundles of thin fibers.
    b. attaches to bones.
    c. contracts precisely and slowly.
    d. is controlled by conscious thought.

Human Body Functions

4. Exercise of skeletal muscles
    a. increases their size.
    b. decreases their tone.
    c. makes them react involuntarily.
    d. makes them react quicker.

5. Lactic acid
    a. fatigues skeletal muscle.
    b. fatigues cardiac muscle.
    c. is broken down by oxygen.
    d. affects sprinters more than distance runners.

6. Which of the following terms best matches each phrase?
    a. smooth muscle   d. skeletal muscle
    b. uterus   e. muscle tone
    c. cardiac muscle   f. oxygen debt

    _____ keeps a muscle ready for action

    _____ a thick, hollow organ made up almost entirely of smooth muscle

    _____ found only in the heart

    _____ what muscles have after they have worked

    _____ controls all outside-the-body movement

    _____ found in all internal organs except the heart

KEY WORDS

smooth muscle (p. 191)   skeletal muscle (p. 192)
uterus (p. 191)   muscle tone (p. 194)
cardiac muscle (p. 191)   oxygen debt (p. 195)

THE SKELETAL SYSTEM

The skeletal system is more than bones. It is common to think of the term "skeleton" when the skeletal system is mentioned. A human skeleton, which is the bony remains of a person, was once a part of a skeletal system. But there is more to the system than dry bones. Living bone has soft tissue on the inside. That tissue is called **marrow** (măr'ō). Some of this marrow (in the flatbones of adults) produces red and white blood cells.

marrow: soft tissue in certain bones. The marrow produces blood cells.

Skin, Muscles, and Bones

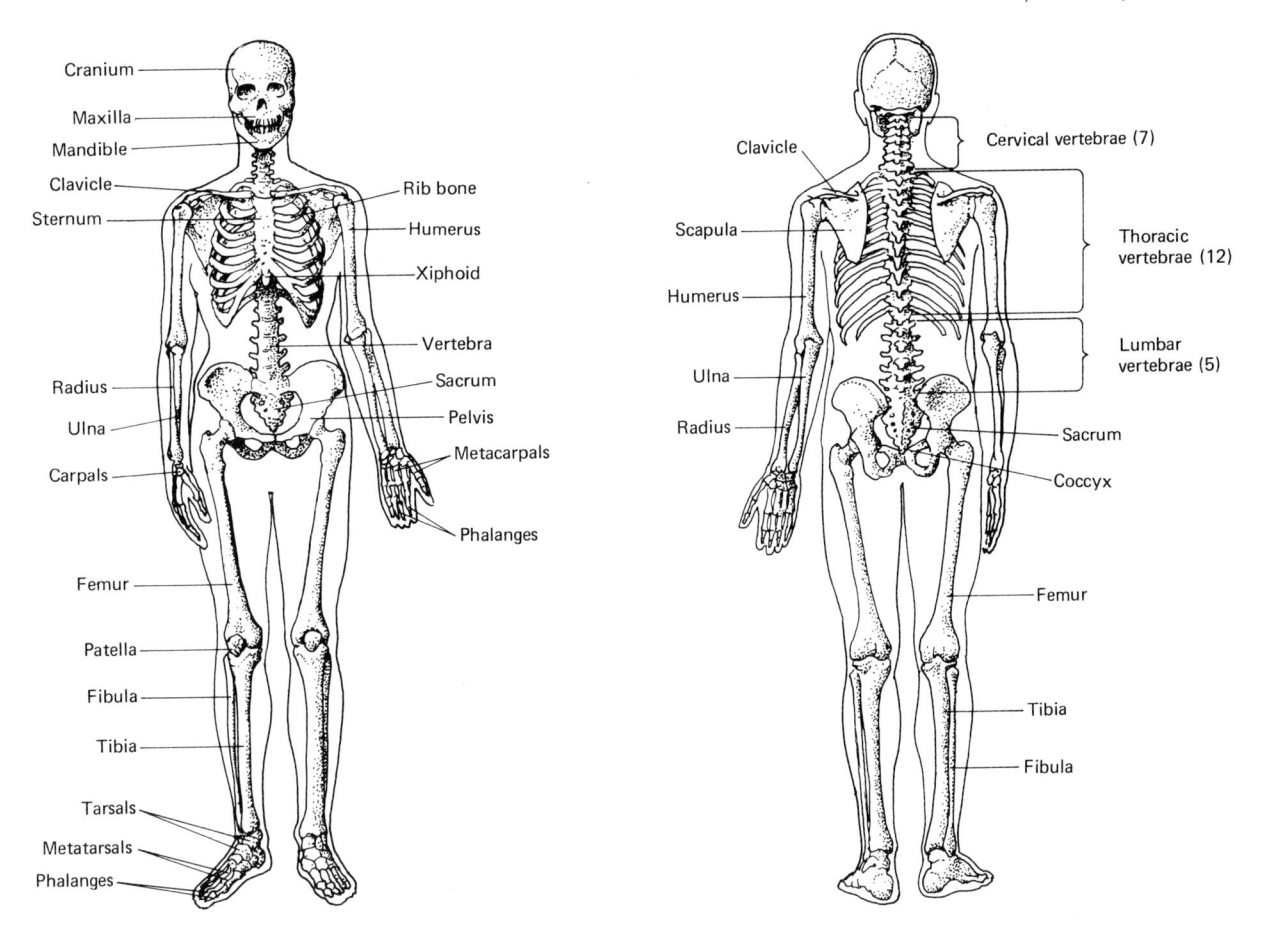

Figure 11.8: These diagrams show the major bones of the human body. In all, there are usually 206 bones present in an adult human.

Many parts of the skeletal system, especially in young people, are made up of **cartilage.** This is a semisolid material. You see it as the flexible cap on the end of a chicken drumstick. You can feel cartilage in your nose and in your ear. The skeletal system of a newborn baby is mostly cartilage. Over the years most of that cartilage gradually changes to bone.

Connective tissue also is a part of the skeletal system. Some connective tissue forms a tough membrane that surrounds the bones. Other connective tissue forms into tough white bands of material that hold bones together. These bands of connective tissue are called **ligaments.** Bands of connective tissue that attach skeletal muscles to the bones are called **tendons.**

There are joints in the skeletal system. A place where two or more bones come together and usually are connected is called a **joint.** In some cases, joints do not move. There

cartilage:
a semihard material that may develop into bone. Some parts of the body are formed mainly from cartilage, such as the nose and the ears.

connective tissue:
tissue that holds internal body parts together

ligament:
a band of connective tissue that holds bones together

tendon:
a band of connective tissue that attaches muscles to bones

joint:
the point at which bones come together or are connected

197

Human Body Functions

are some joints in your skull like that. Other joints move slightly, like those in the middle of your foot and the palm of your hand. Other joints, such as your rib joints, are fixed in one place but are elastic and can stretch. Finally, there are movable joints, such as those in your elbows and knees. The greatest number of ligaments is found in these joints. These are the joints you are most likely to injure in contact sports, such as football.

What are the functions of the skeletal system? One function of the skeletal system is obvious to anyone. It provides a place for your muscles to attach so that you can move.

A second function is support. Your spine and legs support your body and help you keep an upright posture.

A third function of the skeletal system is protection. The hollow backbones—the vertebrae—surround and protect the spinal cord. Your ribs provide a protective case around your lungs and heart. Your skull encloses and protects your brain.

The fourth function of the skeletal system has already been mentioned. Some of the marrow in the bones produces red and white blood cells.

Figure 11.9: At the left, you see an actual human hip joint. Note the "ball and socket" arrangement where the femur fits into the pelvis. At the right, you see an artificial hip joint. Each year, thousands of people, especially older ones, have to have one or both hip joints replaced as a result of accidents or disease.

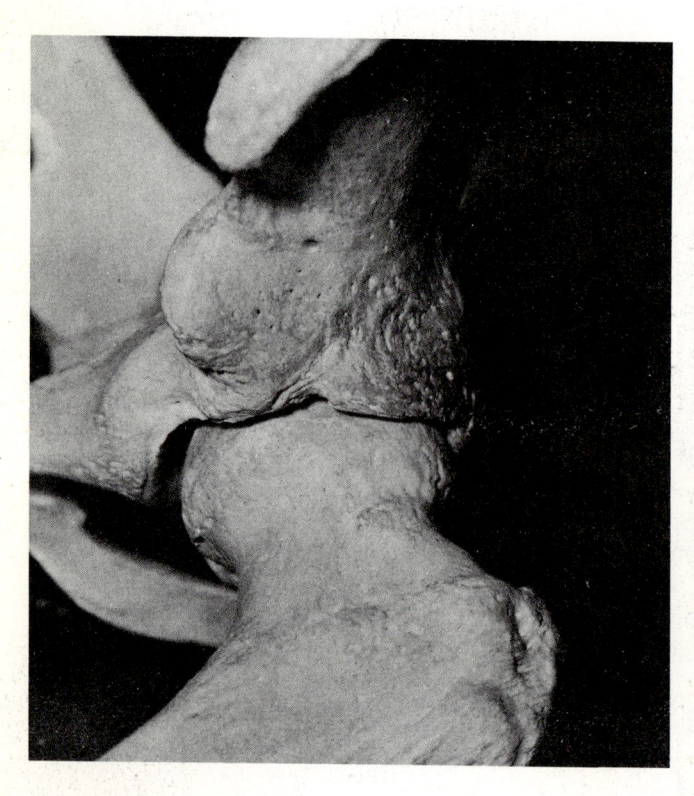

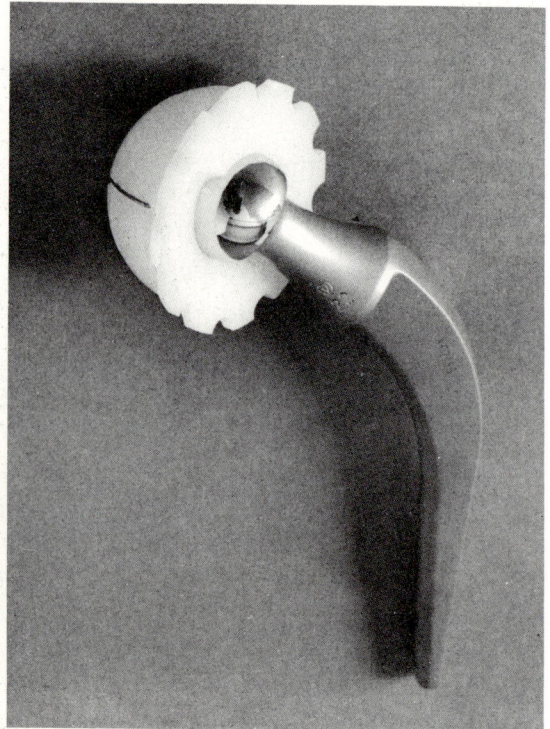

Career Spotlight

Although he or she never scores a point, the *athletic trainer* is one of the most important members of a team. Playing any sport includes the risk of injury. It is the athletic trainer's job to deal with this problem. The trainer works to prevent avoidable injuries. Educating team players in the proper care and use of their muscles is a big part of that. The trainer also gives immediate first aid when an injury occurs. Then the trainer follows the orders of a doctor to help the injured athlete return to good health.

One of the trainer's most important jobs is to keep minor injuries from becoming worse. Often, an athlete can continue playing with a minor injury if it is properly cared for. The trainer tapes, wraps, and pads these minor injuries. She or he may also use physical therapy to ease the pain of sprains or pulled muscles and ligaments.

To be a trainer, you must meet standards set by the National Athletic Trainer's Association, or NATA. One way to do that is to take a special college program. Another is to work with a "certified" trainer to learn what you need to know. In either case, you must then pass the NATA Certification Examination. Almost all colleges hire trainers for their varsity and intramural sports teams. Soon, it may become law that all high schools do the same. Job opportunities for women and men as athletic trainers seem very good as a result.

Human Body Functions

Laboratory Activity

Do You Know Your Own Body?

PURPOSE

Because we see them every day, we often overlook the most amazing parts of our world. In this investigation, you will take a fresh look at your own body. You will explore skin, joints, and muscles.

MATERIALS

dissection microscope rubbing alcohol
sharp probes cotton

PROCEDURE

1. Place your hand on the stage of the dissection microscope. Explore your hand under low power. When you discover something particularly interesting, zero in with higher magnification. Record your observations.
2. Under the microscope, gently prod your skin with a probe. Test your skin for stretch. Find calluses, scars, and places where epidermis is flaking away. Record your observations.
3. Under the microscope, examine your fingerprints. Record your observations.
4. Disinfect an area of dermis on a hand or finger. Disinfect a probe. Use rubbing alcohol to do this. Under the microscope, expose your dermis in the disinfected area by scraping gently through the epidermis with the disinfected probe. Also observe freckles and other pigment in the dermis. Record what you observe.
5. Explore your fingernails under the microscope at high power. Examine how they grow from within the finger. Gently scratch your fingernail with a sharp probe. Record your observations.
6. Under the microscope, at high magnification, explore the hair growing on your hand. Search for different colors and lengths. Notice the follicle out of which a hair grows. Record your observations.
7. You will not see an oil gland under the microscope, but you can see what it does. Pull a few long hairs out of your head. Select one with a white glob on the end and observe it under the microscope. What do you observe?
8. Locate a blackhead on your hand and observe it under the microscope. This is no more than a sweat or sebaceous gland plugged with dirt. What do you observe?
9. Explore the joints of your finger, wrist, elbow, chest, and neck. Place your fingers on each joint. Feel the connective tissue slide as the joint moves. Notice how each type of joint moves differently from other types. Record your observations.

Skin, Muscles, and Bones

10. Place your little finger in the "starting position" as shown at the left in the diagram. Extend it straight out, as shown at the right. This is one complete "time." How many times can you extend it in 60 seconds? Record your data.

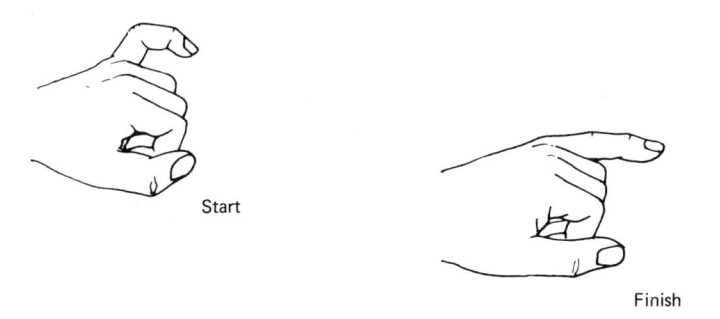

Start

Finish

11. Organize a finger tournament for 1 week from today. Choose official timekeepers and judges. Design a workout schedule to get your little finger into shape. Be careful not to strain it during training. The diagram shows how a tournament was organized for eight competitors. Student F beat students E, G, and D to win the tournament.

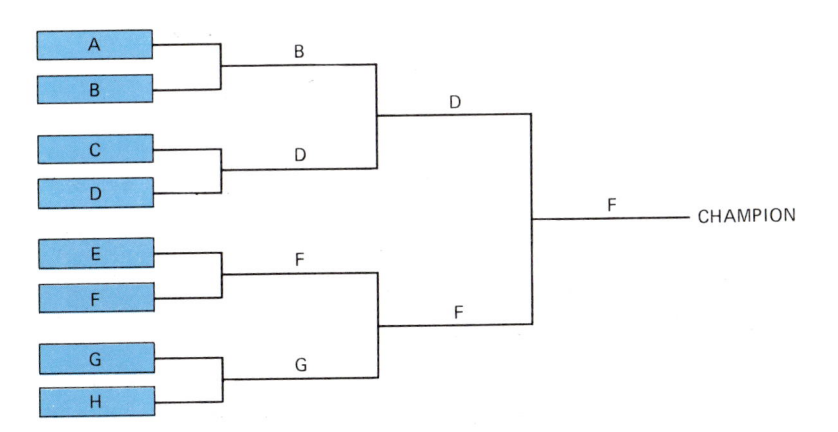

QUESTIONS

1. Of what other material does your skin remind you?
2. Find out what function your fingerprints serve.
3. Why are fingernails sometimes a place where infectious microorganisms flourish?
4. What was the white substance you observed at the end of a hair? How did it get there?
5. Why does brushing make your hair shine?
6. Why do joints move only in certain directions? Find out what a "double joint" is.
7. You may have experienced fatigue and lack of coordination when you "ran the finger race" in step 10. Why was that?

Human Body Functions

LESSON REVIEW *(Think. There may be more than one answer.)*

1. The skeletal system includes
   a. bones.
   b. marrow.
   c. cartilage.
   d. ligaments.

2. A bone
   a. is alive.
   b. may produce blood cells.
   c. is joined to another bone by ligaments.
   d. is eventually replaced by cartilage.

3. A joint
   a. is a place where bones come together.
   b. may not move at all.
   c. often contains tendons.
   d. often contains ligaments.

4. Connective tissue
   a. connects nerves to bone marrow.
   b. holds bones together.
   c. holds muscle to bone.
   d. is a type of muscle.

5. The skeletal system
   a. protects the brain and spinal cord.
   b. produces red and white blood cells.
   c. supports the body.
   d. allows us to move.

6. Which of the following terms best matches each phrase?
   a. marrow   d. ligament
   b. cartilage   e. tendon
   c. connective tissue f. joint

   _____ a place at which bones come together
   _____ tissue in certain bones that produces blood cells
   _____ holds internal body parts together
   _____ connective tissue that holds muscles to bones
   _____ early stage in the development of bone
   _____ connective tissue that holds bones together

KEY WORDS

marrow (p. 196)   ligament (p. 197)
cartilage (p. 197)   tendon (p. 197)
connective tissue (p. 197)   joint (p. 197)

Applying What You Have Learned

1. Why is it a benefit to humans that our epidermis flakes off?
2. Design a covering for a robot that will perform all the functions of the human skin.
3. Certain tiny organisms live on the surface of the human integumentary system. Pretend you are their size. Describe the landscape you would see as vividly as you can. Be sure to include all the organs that make up the integument.
4. Discuss reasons why most Americans seem so anxious to disguise their body odor and sweat.
5. Why is smooth muscle involuntary in its action?
6. After years of training, how would the heart of a distance runner look and function differently from the heart of a sprinter?
7. A female athlete seldom develops large skeletal muscles. Why is it still advantageous for her to train?
8. Why do young children, who seem to be falling down all the time, seldom break bones?
9. If your skeleton were on the outside, as it is with insects, how might your life be different?

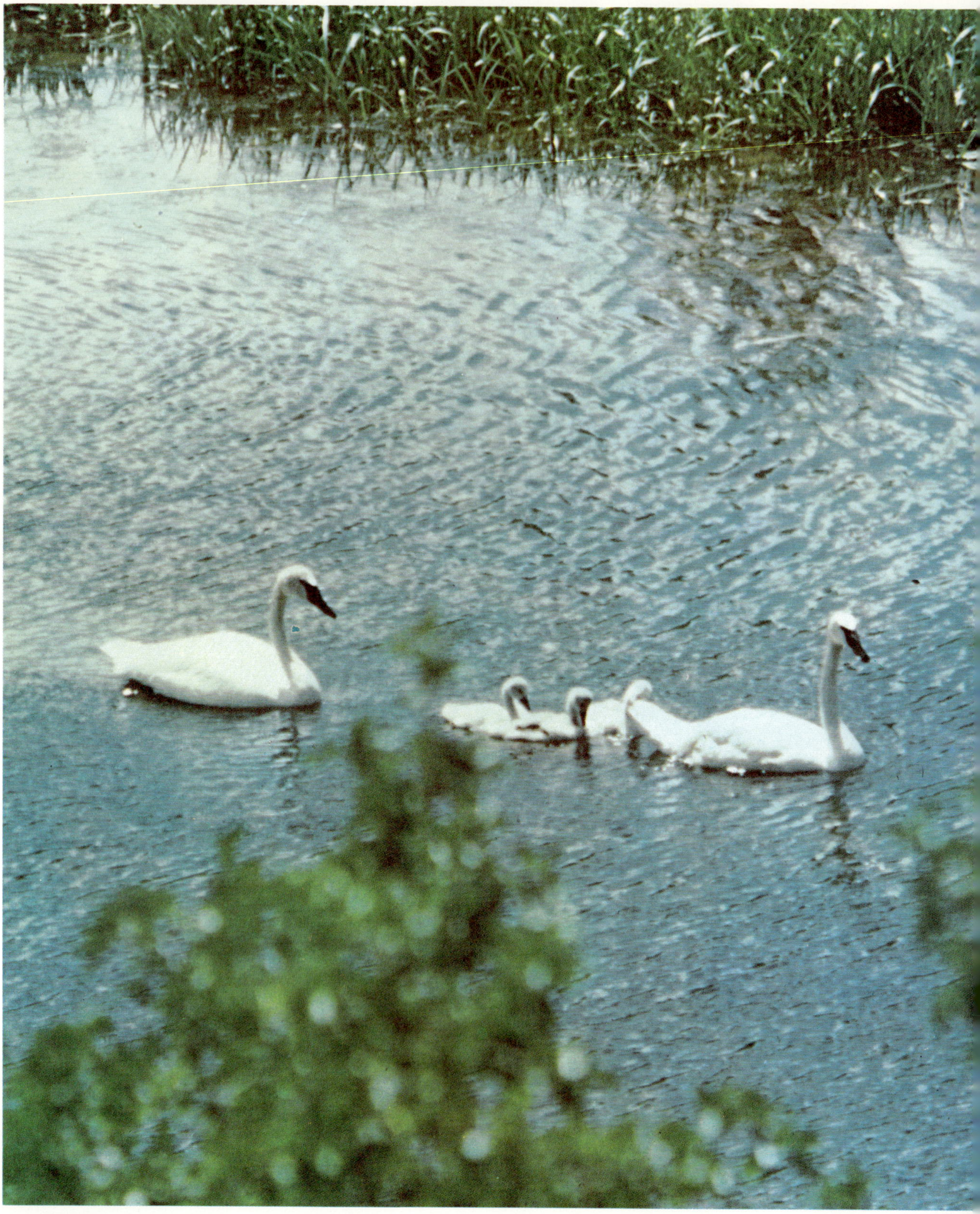

Figure 12.1: The reproductive pattern of the trumpeter swan produces offspring unlike either parent.

Chapter **12**

REPRODUCTION

Reproduction means "making more of the same." Trees reproduce when they make more trees. Elephants reproduce when they make more elephants. Humans reproduce when they make more humans. This chapter is mostly about human reproduction—not about the reproduction of trees or elephants. However, much of what you learn will help you understand how all other organisms carry out this function.

THE NATURE OF THE PROCESS

Cells can reproduce themselves. To understand the process of reproduction, you have to think about cells. Most cells can reproduce themselves. Because of this, tissues, organs, systems, individuals, populations, and communities can be reproduced.

How does a cell make another cell? Strictly speaking, it does not. A cell reproduces by dividing into two cells. Each

Human Body Functions

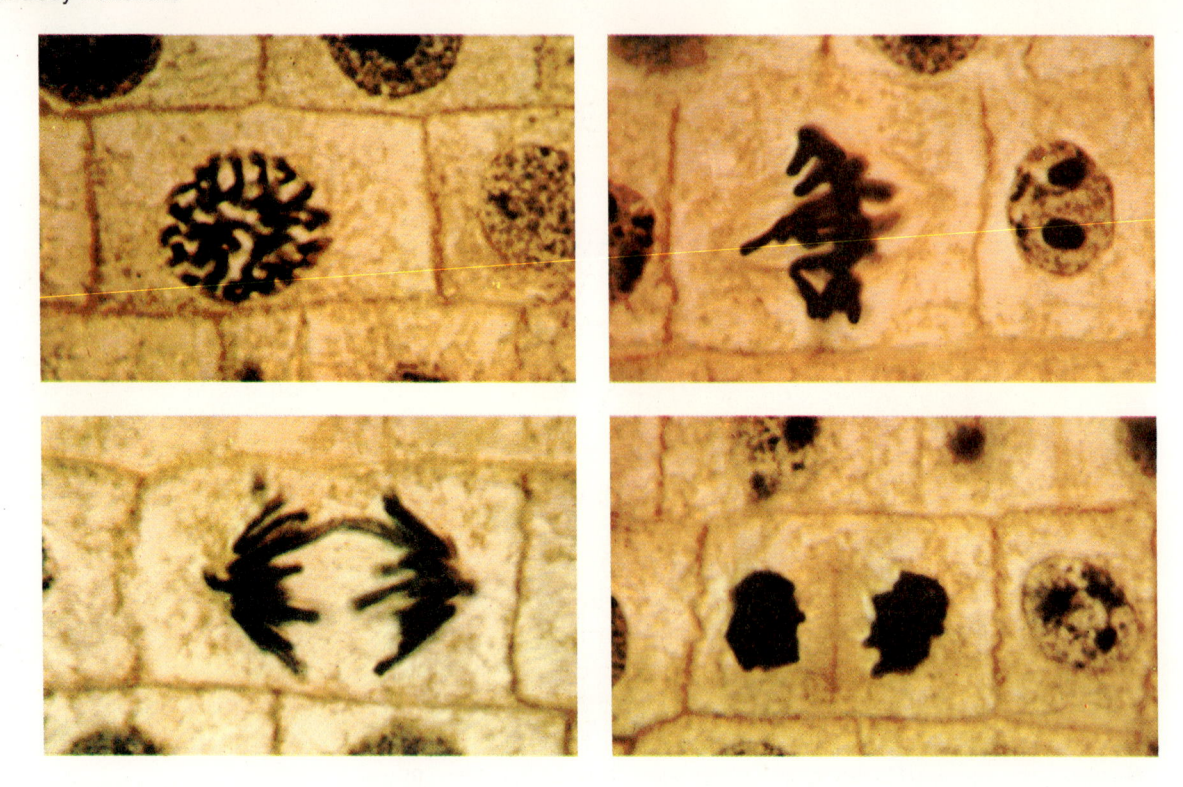

Figure 12.2: These photographs show the major events that occur when an onion cell divides. At the upper left, the nucleus has changed form. It no longer looks solid. The material of the nucleus has condensed into bodies called chromosomes, which have duplicated themselves at this point. Since the genes are in the chromosomes, they too have been duplicated. At the upper right, the chromosomes have gathered in the middle of the cell. At the lower left, the two complete sets of chromosomes are moving to opposite ends of the cell. At the lower right, cell division is almost completed. The chromosomes have lost their shape and one cell is almost divided into two. Both "daughter" cells will have the same chromosomes and genes as the "parent" cell. (Courtesy Carolina Biological Supply Co.)

cell division:
a process by which cells reproduce themselves by splitting apart

"daughter" cell is half as large as the "parent" cell. Study the pictures and caption in Figure 12.2. Notice that before the cell divides, the important parts are duplicated. Then one of each duplicated part goes to each daughter cell. This is an important fact about **cell division.** It guarantees that each daughter cell will be like the parent cell.

Cells can grow. Imagine what would happen if the cell division process were to continue for a long time. Eventually you would have a large number of very small cells. And the cells would keep getting smaller every time a division occurred.

Cell division is not the only process that occurs during reproduction. Growth also occurs. Most growth occurs inside cells. Each cell takes in materials from its environment. These serve as raw materials that the cell uses in making the parts that it needs. The cell membrane also grows larger. This provides

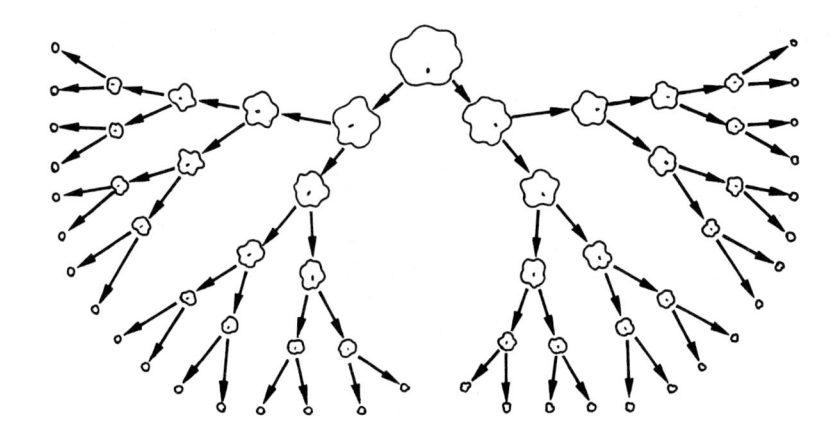

Figure 12.3: Unless cell growth occurred during reproduction, the cells resulting from cell division would be very small. Both processes—cell division and cell growth—take place.

more room inside the cell. Eventually a cell will stop growing. No one yet knows why.

Cells can specialize. You started out your life as one cell. That cell divided and redivided. Eventually, through cell division and growth, your body was made up of trillions of cells. Now stop and think: Are your cells all alike? No. They are not. You have many different kinds of cells that are organized into tissues, organs, and systems. Therefore, something else must have happened during your development besides cell division and growth.

A third process did occur during your development. Shortly after cell division began, some of the cells began to specialize. Some became blood cells. Others became skin cells, muscle cells, nerve cells, and so on. As soon as they were formed, the specialized cells began to organize themselves into tissues. Then the tissues formed organs. Eventually complete systems were formed.

What causes a cell to change its shape and function and become specialized? Life scientists would very much like to know the answer to that question. Cancer cells are cells that have changed their shape and function. If life scientists knew what causes normal cells to specialize, they might learn what causes normal cells to become cancer cells.

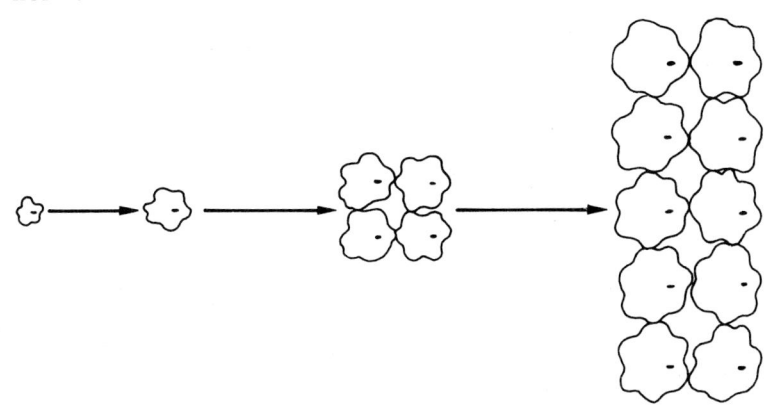

Figure 12.4: All cells in the body are not alike. They are specialized to make up different kinds of tissues.

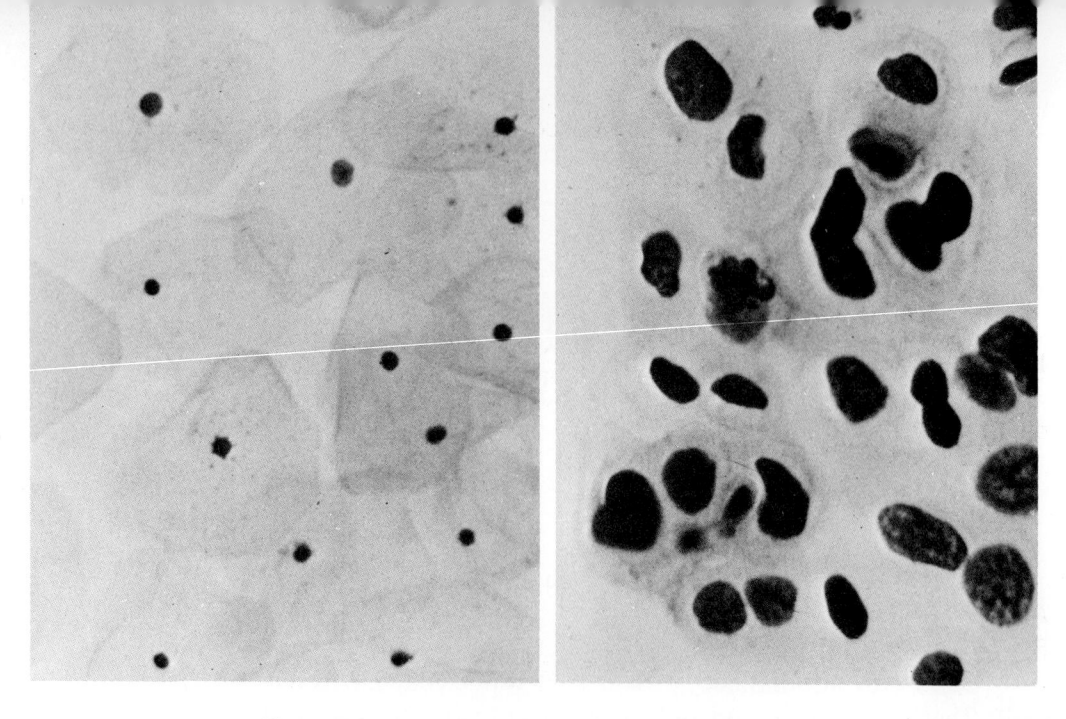

Figure 12.5: The cells at the left are from a normal human cervix. The cells at the right are from a woman with cancer of the cervix.

to that question. Cancer cells are cells that have changed their shape and function. If life scientists knew what causes normal cells to specialize, they might learn what causes normal cells to become cancer cells.

LESSON REVIEW *(Think. There may be more than one answer.)*

1. The reproduction of tissues, organs, systems, individuals, and populations is made possible because
    a. cells can make hormones.
    b. cells can reproduce themselves.
    c. cells are all alike.
    d. cells are all different.

2. The cells in your body are
    a. all alike.
    b. all different.
    c. specialized.
    d. alike if they are in the same tissue.

3. The processes that occur during the development of a multi-cellular organism are
    a. cell division.
    b. cell digestion.
    c. cell growth.
    d. cell specialization.

4. The process by which cells reproduce themselves is called
    a. cell division.
    b. cell digestion.
    c. cell growth.
    d. cell specialization.

KEY WORD

cell division (p. 206)

ASEXUAL AND SEXUAL REPRODUCTION

Asexual reproduction is simple and fast. Organisms use two different methods of reproduction. One is asexual reproduction, and the other is sexual reproduction. Let us consider first the simpler and faster method: **asexual reproduction.** Here is the way that it usually works. A cell, or a group of cells, is released from an organism's body. That cell or group of cells starts to divide, grow, and specialize. The result is another organism just like the parent from which the cells came.

Black bread mold is shown at the left in Figure 12.7. At the right of the figure, you see a microscopic view of a spore case that has broken open. The tiny round objects that are being released are **spores.** A spore is an example of a single cell, protected by a coating, that is released by a parent. The spore cell will develop into a new mold if it lands in a favorable environment.

asexual reproduction: a process by which some organisms reproduce their own kind. In asexual reproduction, cells divide and grow to form another organism.

spore: a reproductive cell that develops into an organism

Figure 12.6: An individual that results from asexual reproduction has the same genes as the parent.

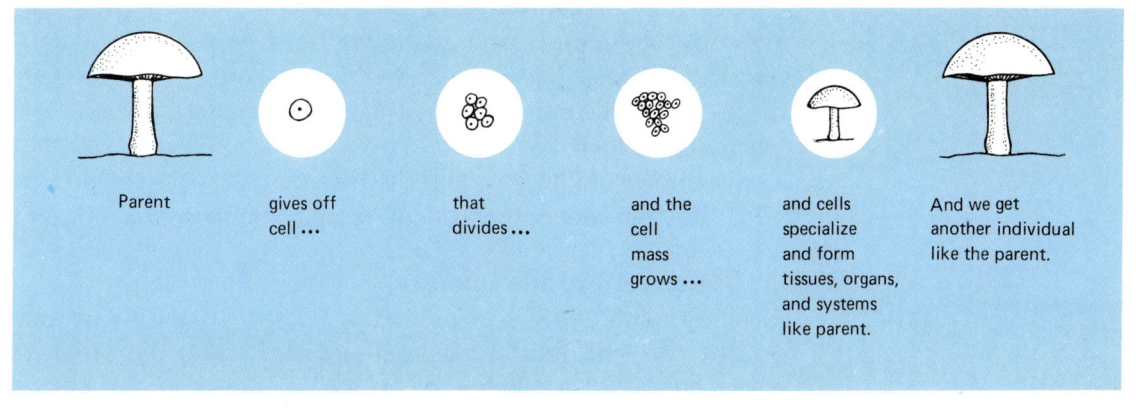

Parent — gives off cell... — that divides... — and the cell mass grows... — and cells specialize and form tissues, organs, and systems like parent. — And we get another individual like the parent.

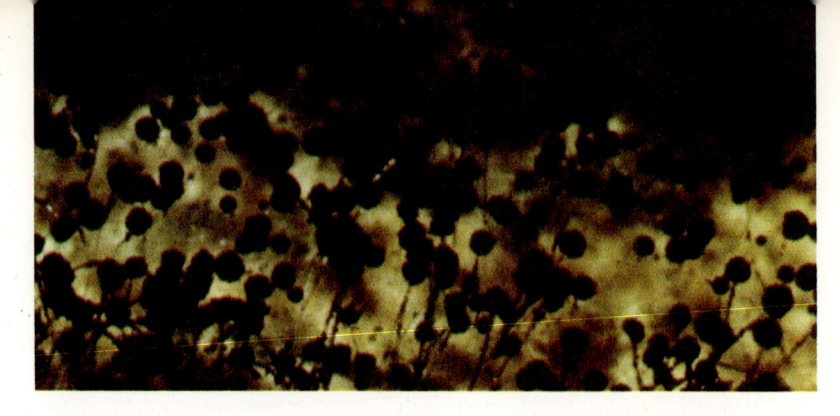

Figure 12.7: Black bread mold gets its name from the many black spore cases that it produces. The rest of the mold is white.

sexual reproduction: a process by which organisms reproduce their own kind. In sexual reproduction, a male reproductive cell unites with a female reproductive cell.

fertilization: in sexual reproduction, the uniting of a male reproductive cell with a female reproductive cell

sperm cell: a male reproductive cell

egg cell: a female reproductive cell

fertilized egg cell: the single cell formed when a male reproductive cell unites with a female reproductive cell

genes: units in cell nuclei that control the ways in which an individual will develop. Genes contain instructions for building all body proteins.

Sexual reproduction is more complicated. The second way that organisms reproduce is by **sexual reproduction.** Here is a typical way this method works. One organism, called the *male*, will produce a special reproductive cell. The *female* organism will also produce a special reproductive cell. These two cells must come together and form one cell. The process of the two reproductive cells coming together to form one cell is called **fertilization.**

In many simple organisms that live in the water, the male and female reproductive cells are much alike. When released, they swim around in the water until they collide. Fertilization, therefore, is an accidental process for these simple organisms.

Larger and more complex organisms usually produce reproductive cells of two different types. The cell produced by the male usually can swim through a liquid. Such a cell is called a **sperm cell.** The reproductive cell produced by the female, called an **egg cell,** is usually much larger than the sperm cell. Seldom is the egg cell capable of swimming. For fertilization to occur, the sperm cell must swim to the egg cell. During fertilization, it bores its way through the membrane of the egg cell and becomes a part of it. The single cell that results is called a **fertilized egg cell.**

Sexual reproduction is the more common method. Asexual reproduction is simpler and faster than sexual reproduction. Yet, when one compares the methods that different organisms use to reproduce, a strange fact appears. Or at least it seems strange at first glance. Asexual reproduction is not the more common method of reproduction. **Sexual reproduction is by far the more common method found in nature.**

Genes enter the picture. Why does nature favor sexual reproduction over a simpler and faster method? To answer that question, we must bring some strangers into the picture: **genes** (jēnz). Perhaps genes are not total strangers to you. Many

Reproduction

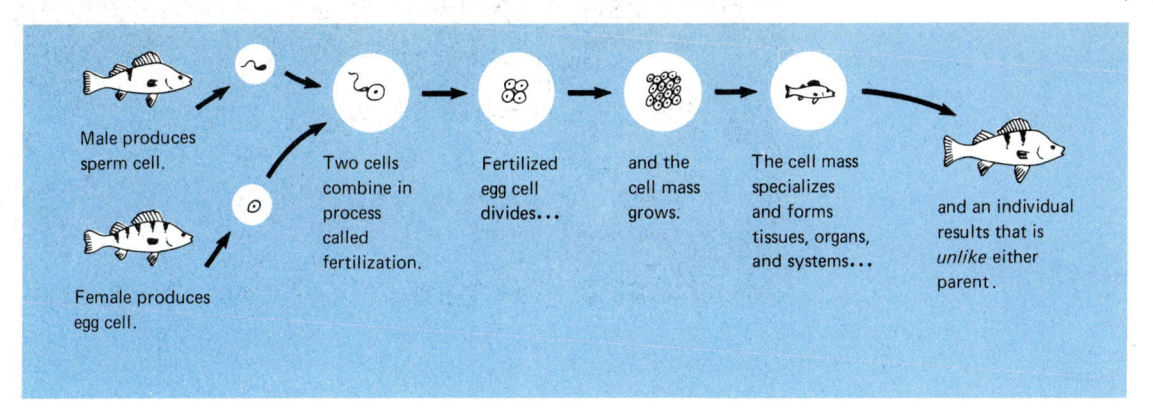

Figure 12.8: An individual that results from sexual reproduction has one set of genes from each parent. For this reason, the offspring is not exactly like the parents.

people know that they are the units that control the **heredity** of individuals. Genes determine what an individual may develop into. A giraffe is a giraffe and not an elephant because of its genes. And one giraffe is different from another giraffe because of its genes. You will study more about genes in the next chapter. Here is what you need to know now. Genes are passed on to the offspring during the reproductive process. What might this mean to the survival of a population?

heredity: the physical traits passed on to an individual from its parents. Heredity is controlled by the individual's genes, half of which come from its "father," the other half from its "mother."

Sexual reproduction is a safety mechanism. Suppose a population reproduces only by asexual methods. This means that all of the individuals will be much alike. This will be so because they all have about the same genes. Now suppose a drastic change occurs in the environment of such a population. Suppose the drastic change is bad for one individual. Won't it be bad for all individuals? Suppose the change causes the death of one individual. Isn't it likely that the same change might kill all the individuals?

Now consider what happens within a population that reproduces sexually. A male reproductive cell has a set of genes from one parent. A female reproductive cell has a set of genes from

(a) Population that reproduces asexually— members have similar genes.

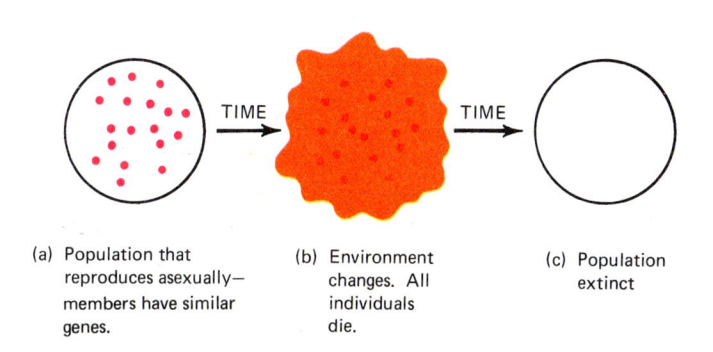

(b) Environment changes. All individuals die.

(c) Population extinct

Figure 12.9: When the organisms in a population are alike, any change in the environment that is bad for one will be bad for all.

Laboratory Activity

How Fast Does Yeast Reproduce Asexually?

PURPOSE

Dried yeast grains are made up of many tiny yeast plants. These plants grow and reproduce asexually when placed in a sugar solution. How fast can asexual reproduction be? In this investigation, you will observe the rate at which yeast reproduces under favorable conditions.

MATERIALS

2 glasses	paper clip
2 teaspoons	plastic wrap
sugar	2 rubber bands
yeast	desk lamp

PROCEDURE

1. Fill two glasses half full of water. Add 2 teaspoons of sugar to each glass. Stir until the sugar is dissolved. Wash and dry the spoon.
2. Prepare labels reading "yeast" and "control." Tape the "yeast" label on one glass, placed below the water level. Turn the label so the lettering faces in. Tape the "control" label to the second glass in a similar way.
3. Take ½ teaspoon of sugar solution from the glass labeled "yeast." Carefully place the spoon on the desk top. Place a grain of dried yeast in the sugar solution in the spoon. Using a paper clip, mash the yeast grain until it is completely mixed with the sugar solution in the spoon.
4. Pour the yeast–sugar water mixture back into the glass labeled "yeast." Stir.
5. With a clean, dry spoon, stir the sugar solution in the glass labeled "control." No yeast is added to the water in this glass.
6. Cover the mouth of each glass with plastic wrap. Secure each wrap with a rubber band.
7. Place the glasses close to, but not touching, the bulb on a desk lamp. Turn the bulb on.
8. Read the labels on both glasses. To do this, look through the solutions to see the label facing inward on the far side. Note how clearly you can read each label.
9. Set the glasses aside. Observe the glasses 24 hours later. Stir each solution. Be sure to use a different spoon to stir each solution. Look at the labels on both glasses. Repeat this procedure for several days. Record what you observe each time.

QUESTIONS

1. What is the purpose of the "control" sugar solution?
2. Why was sugar added to the water in both glasses? Would the yeast multiply in water to which no sugar was added?
3. Why do you think the desk lamp is used in step 7. How would you test your idea?
4. Why are different spoons used in step 9?

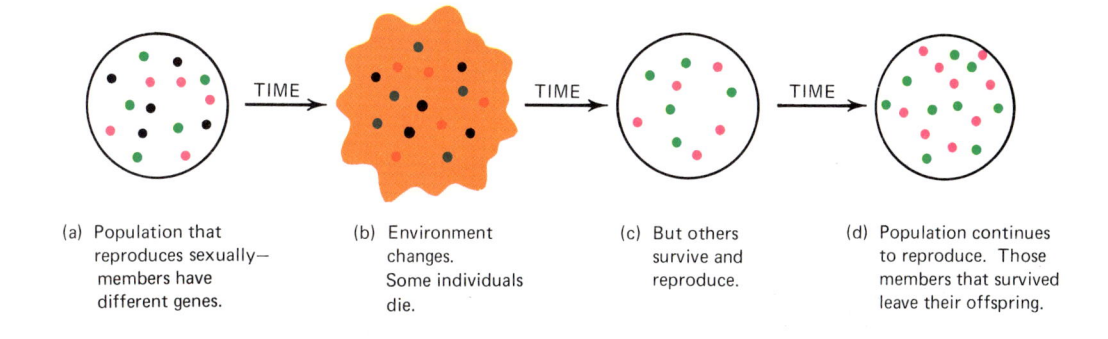

| (a) Population that reproduces sexually— members have different genes. | (b) Environment changes. Some individuals die. | (c) But others survive and reproduce. | (d) Population continues to reproduce. Those members that survived leave their offspring. |

Figure 12.10: All of the individuals are different in a sexually reproducing population. A change in the environment that is bad for some is not likely to be bad for all. This kind of population can usually "rebound" and survive in a changing environment.

another parent. The individual that results from the fertilization of the two cells has genes from two different parents. These genes produce a completely different individual. The individual that results from sexual reproduction is almost certain to be unlike either parent.

Suppose a drastic change occurs in the environment of a sexually reproducing population. The change that is bad for one individual may not be bad for another. It might even be helpful. A change may kill one member of the population. It is not nearly so likely to kill all the others. Sexual reproduction guarantees that there will be a wide variety of individuals in a population.

Sexual reproduction can be considered a safety mechanism in a population. Changes are always taking place in the environment of a population. Some of them are drastic. A cold spell, a dry spell, an earthquake, and a flood—these are just a few examples. To survive such disasters, a population must be able to bounce back. The ability to bounce back is much greater in a population with a variety of individuals. In most populations, sexual reproduction insures that there will always be a variety of individuals.

LESSON REVIEW *(Think. There may be more than one answer.)*

1. An individual exactly like its parents has been
    a. reproduced from a fertilized egg.
    b. reproduced by sperm cells.
    c. reproduced asexually.
    d. reproduced sexually.

Human Body Functions

2. An individual not exactly like either parent has been
    a. reproduced from a fertilized egg.
    b. reproduced by sperm cells.
    c. reproduced asexually.
    d. reproduced sexually.
3. Sexual reproduction
    a. is faster than asexual reproduction.
    b. is much more commonly used by populations.
    c. guarantees variety among the individuals in a population.
    d. is slower than asexual reproduction.
4. In an environment that is always changing,
    a. a population of identical individuals is more likely to survive.
    b. a population of different individuals is more likely to survive.
    c. asexual reproduction is a safety mechanism.
    d. sexual reproduction is a safety mechanism.
5. Which of the following terms best matches each phrase?
    a. asexual reproduction   f. egg cell
    b. spore   g. fertilized egg cell
    c. sexual reproduction   h. genes
    d. fertilization   i. heredity
    e. sperm cell

    _____ the uniting of a sperm cell and an egg cell

    _____ a process that produces individuals like the parent

    _____ a reproductive cell that swims

    _____ product after sperm cell and egg cell are joined

    _____ a process that produces individuals unlike either parent

    _____ the reproductive cell that develops into an organism

    _____ units that control heredity

    _____ a reproductive cell that does not swim

    _____ physical traits passed on by parents to children

KEY WORDS

asexual reproduction (p. 209)   egg cell (p. 210)
spore (p. 209)   fertilized egg cell (p. 210)
sexual reproduction (p. 210)   genes (p. 210)
fertilization (p. 210)   heredity (p. 211)
sperm cell (p. 210)

Reproduction

THE HUMAN REPRODUCTIVE PROCESS

Human beings reproduce only sexually. Are you a reader of science fiction stories? If so, you may have heard of *cloning* (**klō**′ning) people. This means reproducing people asexually. A fictional technique for doing this requires one of your cells—for example, a cell from the inside of your cheek. By some means, such a cell then develops into another person exactly like you. You may or may not like the idea of this sort of reproduction. But life scientists are predicting that such a technique could be worked out sometime in the future.

Right now there is only one way that human beings can reproduce. That is by sexual reproduction. The male has an entire reproductive system. So does the female.

The male system produces sperm. The organs of the male reproductive system are shown in Figure 12.11. The key function of the male system is to produce sperm cells—called *sperm*. The sperm are produced in the **testes** (**těs**′tēz), both of which hang below the body in a saclike pouch called the **scrotum**.

There are three other main organs of the male reproductive system. The **seminal vesicles** (**věs**′ĭ kəlz) produce a white liquid called **semen**. The semen provides a liquid environment for the

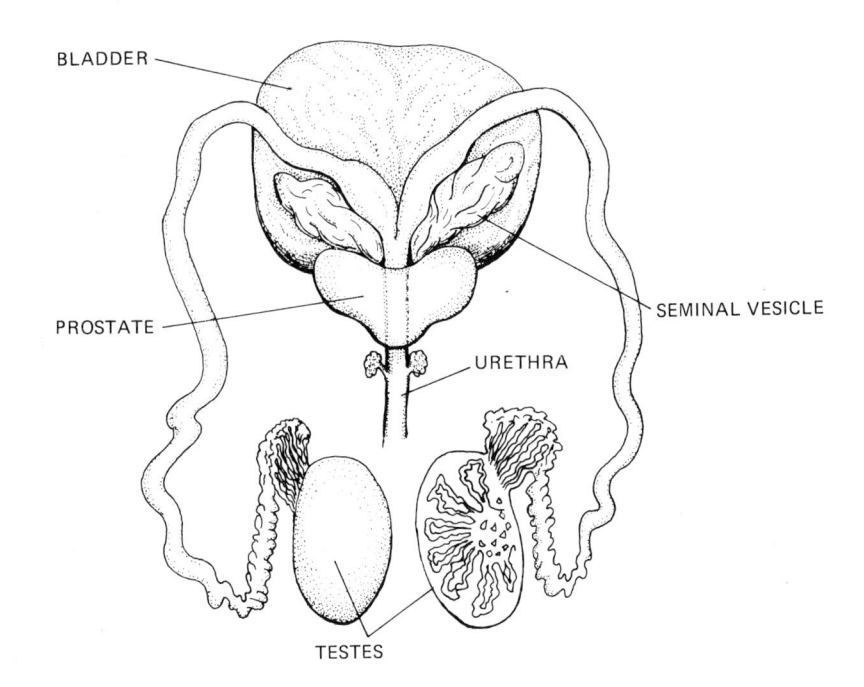

testes:
male reproductive organs, also called testicles; usually occur as a pair and produce sperm cells

scrotum:
the saclike pouch that houses the male animal's testes

seminal vesicles:
male reproductive organs that produce semen

semen:
a white liquid produced by the male reproductive system. It provides a liquid for sperm.

Figure 12.11: The male reproductive system.

Human Body Functions

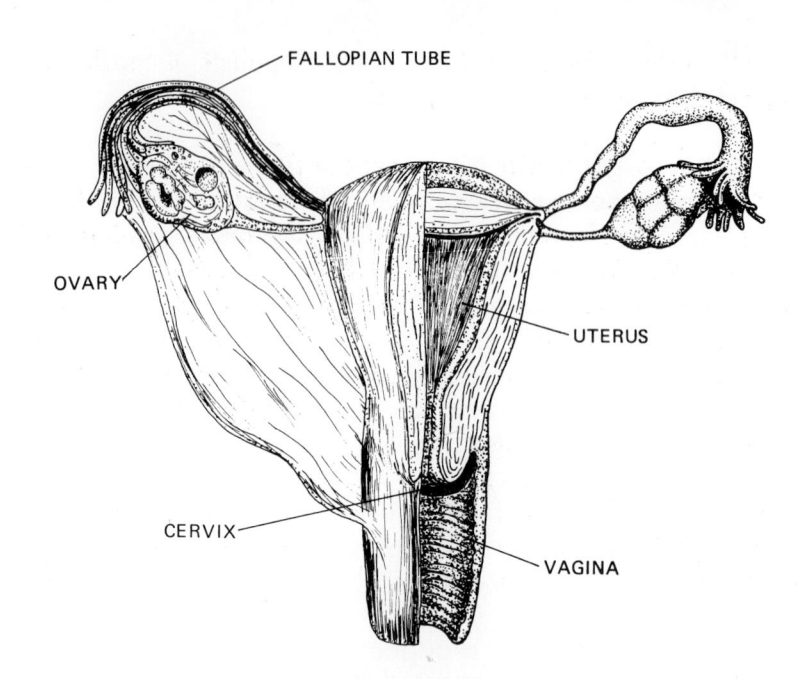

Figure 12.12: The female reproductive system.

prostate:
a male reproductive organ that produces semen. Strong contractions of the prostate eject sperm and semen through the penis.

penis:
the external organ of the male reproductive system, for the ejection of sperm and semen. It is also an organ of the male excretory system, for the passage of urine.

coitus:
in sexual reproduction, the process by which sperm and semen are introduced by the male into the female reproductive system

testosterone:
a hormone produced by the testes. It is largely responsible for the onset of sexual maturity in males.

ovary:
the female reproductive organ that produces egg cells

estrogen:
a hormone produced by the ovaries. It is largely responsible for the onset of sexual maturity in females.

sperm. The **prostate** (prŏs′tāt) has two functions. It produces semen. And it has a set of muscles that can contract strongly and eject sperm and semen through the penis. The **penis** is used during **coitus** (kō ī′təs), the process by which sperm and semen are introduced into the female reproductive system.

The testes have another important role besides the production of sperm. They contain special cells that produce a hormone called **testosterone** (tĕs tŏs′tə rōn). This hormone is first produced during the male's early teenage years. It causes the male to develop certain sex characteristics. These include the growth of body and facial hair and a deepening of the voice.

The female system has three functions. The female reproductive system is shown in Figure 12.12. One of its functions is to produce egg cells, or "eggs." A second function is to provide the proper environment for the first 9 months of development of the human being. Third, by producing milk, the female reproductive system can provide food for the human baby. This can last from the baby's birth up to age 18 months or older.

The ovaries produce eggs and estrogen. The **ovaries** are small glands about the size of walnuts. Normally they alternate in producing eggs, releasing them at a rate of one about every 28 days. The ovaries also produce **estrogen** (ĕs′trə jən). This is one of the main female hormones. It is responsible for causing the female to develop her special sex characteristics. Like testosterone in the male, estrogen production begins during the early teenage years. It causes development of the breasts,

Reproduction

growth of body hair, and internal development of the female reproductive system.

Other organs aid the reproductive process. The egg is produced on the outside of either one of the ovaries. When completely developed, the egg falls off the ovary and into the mouth of the **Fallopian tube.** From there, it moves through the tube and eventually reaches the uterus. If the egg is fertilized by a sperm, the fertilized egg will become attached to the uterus wall. (By the time it attaches, it will have undergone several divisions.) The uterus will house the developing human being for about 9 months.

The uterus connects at its lower end to the **vagina** (və jī′nə). This tubelike structure expands when the baby is ready to be born. At birth, the baby passes through the vagina and out of its mother's body. The vagina is called the birth canal during the process.

The female has a menstrual cycle. In the sexually mature male, sperm are produced all the time, by the billions. The female system works very differently. Egg release is accompanied by a 28-day series of events called the **menstrual cycle.** There are three main events in the menstrual cycle:

1. A thick blood-rich tissue grows on the inside wall of the uterus. This provides a suitable environment in which a fertilized egg could develop.
2. One of the ovaries releases an egg. This process is called *ovulation* (ŏ′vyə lā shən).
3. If the egg is not fertilized, the uterus sheds the blood-rich lining. The lining is discharged through the vagina over a *menstrual period* of a few days.

Fallopian tube: either of two tubes that are part of the female reproductive system. Each leads into the uterus and serves as a path for an egg cell, which drops into it from an ovary.

vagina: an organ of the female reproductive system. It serves as a path to the uterus and Fallopian tubes for sperm cells ejected into it. It also serves as a birth canal through which a baby is born from its mother.

menstrual cycle: a series of events in a sexually mature female. It includes preparation of the uterus for a potential fertilized egg. That is followed by ovulation. Then, when fertilization does not take place, vaginal discharge during the menstrual period completes the cycle.

Figure 12.13: This diagram shows the three main events in the menstrual cycle: (a) buildup of the uterine lining, (b) ovulation, and (c) shedding of the lining—the time of menstrual discharge.

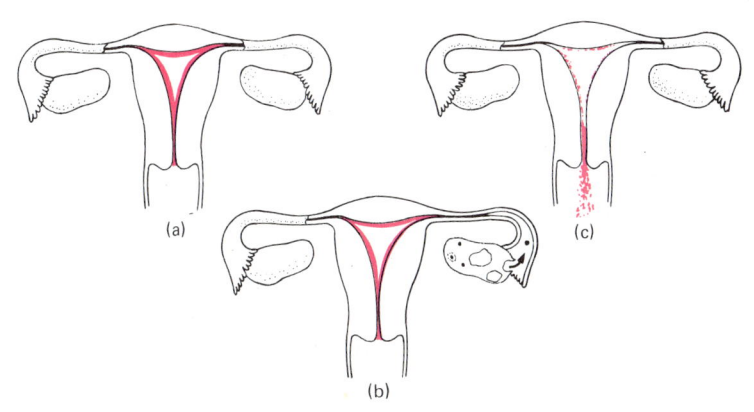

217

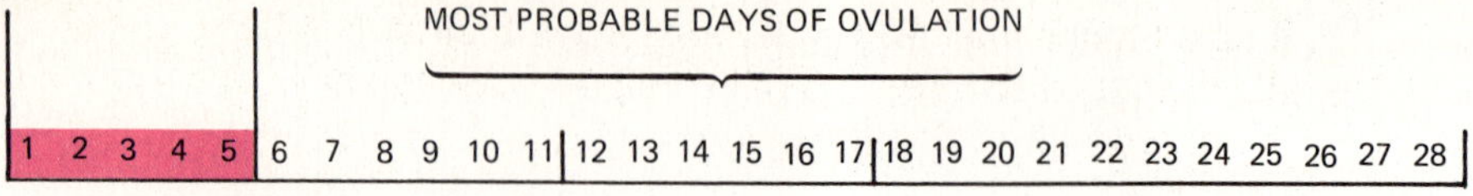

Human Body Functions

```
|1 2 3 4 5|6 7 8 9 10 11|12 13 14 15 16 17|18 19 20 21 22 23 24 25 26 27 28|
                              MOST PROBABLE DAYS OF OVULATION
```

Figure 12.14: The menstrual cycle is often a regular 28-day cycle. However, it is important to remember that both the length of the cycle and the timing of the events in it can be quite irregular.

The diagram in Figure 12.14 is a calendar of the menstrual cycle, which is about 28 days long. The menstrual period occurs during the first few days of the cycle (usually days 3 through 5). During the following days, the inside wall of the uterus grows thicker. At the same time, an egg is maturing in the ovary. Ovulation usually occurs on or about the fourteenth day. If the egg is not fertilized, the wall of the uterus breaks down. This happens near the twenty-eighth day. Then the cycle is repeated. A normal cycle may take somewhat more or less than 28 days.

The menstrual cycle can be very irregular. In fact, it is quite normal for the teenage female to have an irregular cycle. Also, even when there is a 28-day cycle, the time of ovulation may vary. Thus, fertilization could occur at just about any point in the menstrual cycle.

LESSON REVIEW *(Think. There may be more than one answer.)*

1. Sperm are produced by
    a. the prostate.
    b. the seminal vesicles.
    c. the testes.
    d. the penis.

2. The female reproductive system
    a. produces sperm.
    b. produces eggs.
    c. houses the developing child for about 9 months.
    d. produces testosterone.

3. In the female reproductive system,
    a. eggs are produced by the uterus.
    b. a fertilized egg attaches to the wall of the uterus.
    c. fertilization usually takes place in the vagina.
    d. development of the embryo usually takes place in a Fallopian tube.

4. During a 28-day menstrual cycle,
    a. ovulation is most likely to occur on the eighth day.
    b. the egg matures during days 1 through 13.

Reproduction

      c. the uterine wall thickens during days 6 through 14.
      d. vaginal discharge usually occurs during days 24 through 28.
5. The menstrual cycle
    a. always occurs during 28 days.
    b. usually occurs during 28 days.
    c. is sometimes very irregular.
    d. is normally irregular for teenage girls.
6. Which of the following terms best matches each phrase?
    a. testes   h. testosterone
    b. seminal vesicles i. ovary
    c. scrotum   j. estrogen
    d. semen   k. Fallopian tube
    e. prostate   l. vagina
    f. penis   m. menstrual cycle
    g. coitus

    ____ external male reproductive organ

    ____ liquid environment for sperm

    ____ saclike housing for the testes

    ____ usual place where the egg cell is fertilized by a sperm cell

    ____ produce sperm cells

    ____ process by which sperm and semen are introduced into the female reproductive system

    ____ strongly contracting male reproductive organ

    ____ only function is to produce semen

    ____ hormone responsible for male sex characteristics

    ____ occurs in sexually mature females

    ____ hormone responsible for sexual maturity in females

    ____ serves as the birth canal

    ____ produces the egg cell

KEY WORDS

testes (p. 215)
scrotum (p. 215)
seminal vesicles (p. 215)
semen (p. 215)
prostate (p. 216)
penis (p. 216)
coitus (p. 216)

testosterone (p. 216)
ovary (p. 216)
estrogen (p. 216)
Fallopian tube (p. 217)
vagina (p. 217)
menstrual cycle (p. 217)

GROWTH, BIRTH, AND EARLY DEVELOPMENT

pregnancy
the time during which the uterus houses a developing organism

embryo:
the earliest stage in the development of organisms reproduced sexually

The embryo stage lasts 2 months. Pregnancy, or the period of time from fertilization to birth, is about 9 months. The first 2 months of this period is called the **embryo** stage. The term "embryo" is the name given to all organisms during their early stages of development.

What happens to the embryo after it starts to develop? First, it becomes attached to the wall of the uterus. It does this by growing fingerlike structures. These embed themselves in the wall of the uterus. This triggers the growth of a large mass of tissue in the uterus. That new growth of tissue is called the **placenta**.

placenta:
a large mass of tissue that attaches the embryo to the uterus. It is the place where materials are exchanged between them. It also produces hormones needed by the mother.

umbilical cord:
a flexible extension of tissue that joins the embryo to the placenta. It is filled with blood vessels. These transport blood from the embryo to the placenta and back.

The placenta has three basic functions. First, it attaches to the embryo, in time, through a long, flexible extension. That extension is called the **umbilical cord.** The second function of the placenta is to produce hormones that the mother needs during pregnancy. Finally, the placenta and umbilical cord serve as a transport and exchange system.

Blood from the embryo flows in an artery through the umbilical cord. It divides into capillaries in the placenta. Blood from the mother also flows in an artery into the placenta. In the placenta, the blood vessels are close together. An exchange of materials can take place. Waste products leave the embryo's blood and pass through membranes into the mother's blood. At the same time, oxygen and other nutrients pass from the mother's blood, across membranes into the embryo's blood. The embryo's capillaries join to form a vein. The enriched and purified blood flows in this vein through the umbilical cord and back to the embryo. The cord is attached to the embryo at its *navel.*

During the first 2 months, the embryo develops all of the vital organs. It also develops membranes that enclose it like a sack. Fluids surround the embryo. The fluids provide a protective cushion during the entire 9-month period of development. This allows the mother to lead an active and normal life without endangering the life of her unborn child.

fetus:
an unborn baby from about the third month of development in the uterus until birth. During the first 2 months of prebirth development, the unborn child is called an embryo.

The fetus develops into a baby. After 2 months, the developing embryo is called by a new name: **fetus** (fē't'əs). The name means "offspring." There is a good reason for this change in name. The development pattern changes after the first 2 months. From this time on, growth, rather than specialization, is the main process of development. At the end of the 2-month period, the embryo is about 3 centimeters (about 1

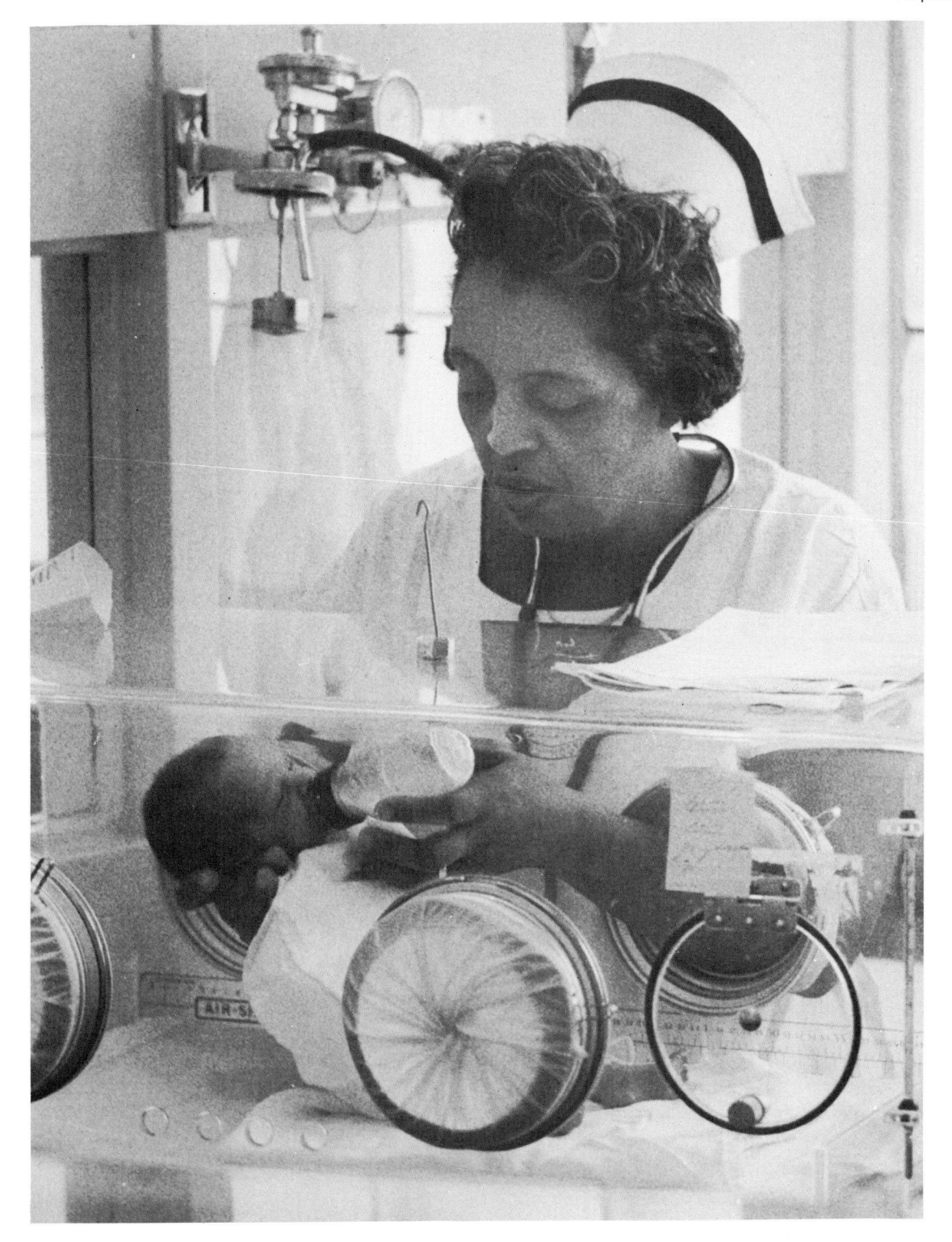

Figure 12-15: A baby born before the full 9-month period is a premature baby. Premature babies often require special life-support equipment to provide a healthy artificial environment until they are able to live normally.

Human Body Functions

inch) long. At the end of 9 months, the fetus may be over 50 centimeters (more than 20 inches) long.

Labor causes birth. The birth process begins when the uterus starts to contract. (Remember, the uterus is a muscle. In fact, it is the most powerful muscle in a woman's body.) Usually the first contractions break the membranes that surround the fetus. Then a rhythmic set of contractions begins. They are infrequent at first. Then they occur more often. These rhythmic contractions are called **labor.**

Before labor, the opening of the uterus, called the **cervix,** is almost closed. Now it must expand to a diameter of about 10 centimeters (about 4 inches). The squeezing action of the uterus forces the baby's head through the opening of the uterus and then out through the birth canal. During the process, the pressure on the baby's head sometimes causes it to become misshaped. That is no problem, however, because the skull bones are soft and will reshape themselves.

After the baby is born, the uterus keeps contracting until the placenta is expelled. The expelled placenta, along with the membranes connected to it, is called the **afterbirth.**

What are the needs of a newborn baby? A newborn baby needs food, clothing, shelter—and care. Everybody knows that. But what specific kinds of care? There is no "guidebook" that has all of the answers. In fact, life scientists who study such matters are still seeking answers to some very basic questions.

Here is an example of the type of question being asked: Should a baby be held? If so, how much? Nobody knows for sure. It is known that during the last months of development, the fetus hears its mother's heartbeat. It is also known that mothers, whether right-handed or left-handed, tend to hold their babies on the left side. Is that an instinctive way to allow the baby to hear a heartbeat? One doctor thought so. He played a tape-recorded heartbeat in a nursery where babies were kept during the first few days after birth. Normally babies lose some weight right after birth. Most of the babies that heard the heartbeat either gained weight or stayed the same.

Here is another related question: What happens if a baby is not held? A life scientist who experimented with rhesus (rē′səs) monkeys gave us a clue. He raised infant monkeys without their mothers. Dummies were used in place of the real

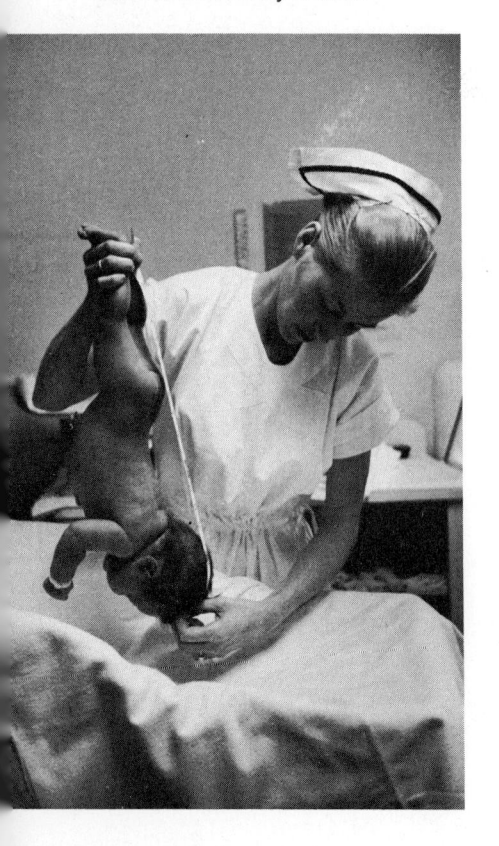

labor: rhythmic contractions of the uterus that force a baby out of its mother at the time of birth

cervix: the opening of the uterus

afterbirth: the placenta and the membranes attached to it that are expelled from the uterus after a baby is born

Figure 12-16: At the end of the 9-month period in the uterus, the embryo has developed into a "newborn." The baby may be more than 50 centimeters (20 inches) in length and can average about 3 kilograms (6½ pounds) in weight.

Figure 12.17: A baby rhesus monkey clutches its terry-cloth "mother." This "mother" seemed to be better than a bare wire frame.

mothers. The dummies were made in several ways. Some were made of wire mesh. Some were covered with terry cloth to make them soft. Some were given faces; others were not. Some had bottles with nipples for nursing attached; some did not. Some of the dummies were heated so as to be warm to the touch. Others were left unheated.

One of the findings of the experiments done with the infant monkeys and the dummy mothers was quite surprising. Feeding was *not* the most important function of the dummies to the infants. Of far greater importance was a dummy that could serve as a source of comfort and attachment for the infants. The baby monkeys most often ran to the terry-cloth dummies. They developed the strongest attachment to them.

The baby monkeys raised with dummy mothers did not grow to normal adulthood. In particular, they did not show normal behavior to their own babies when they became mothers themselves. In fact, some abused their infants so severely that the infants had to be removed from the mothers for safety.

What do these experiments tell us about the two questions? Baby monkeys are not human babies. And a terry-cloth "person" is not a real person. But a life scientist might say this much: These experiments suggest that holding a baby is important. Maybe it even makes a difference how the baby is held.

Human Body Functions

LESSON REVIEW *(Think. There may be more than one answer.)*

1. The placenta
   a. attaches the embryo to the ovary.
   b. produces hormones.
   c. forms a junction between the bloodstreams of the mother and the embryo.
   d. is another name for the embryo.

2. The process of cell and tissue specialization
   a. occurs mostly in the embryo.
   b. occurs mostly in the fetus.
   c. occurs equally in the embryo and the fetus.
   d. does not occur until a baby is born.

3. During labor,
   a. the cervix contracts, the fetus expands, and the uterus gets smaller.
   b. the fetus contracts and the uterus expands.
   c. the uterus contracts and the cervix expands.
   d. the fetus expands and the uterus gets smaller.

4. Judging from limited research, it seems probable that babies
   a. need a mother who wears terry cloth.
   b. need to be held.
   c. should not be held.
   d. need to be left alone.

5. Which of the following terms best matches each phrase?
   a. pregnancy   e. fetus
   b. embryo   f. labor
   c. placenta   g. cervix
   d. umbilical cord   h. afterbirth

   _____ muscle contractions of the uterus during the birth process

   _____ expelled by the uterus after a baby is born

   _____ tissue that holds the embryo and the fetus to the uterus

   _____ unborn baby from about the third month until birth

   _____ tissue that holds the embryo and fetus to the placenta

   _____ opening between the uterus and the vagina

   _____ time when a uterus is housing a developing organism

   _____ earliest stage in the development of an organism reproduced sexually

KEY WORDS

pregnancy (p. 220)　　　　fetus (p. 220)
embryo (p. 220)　　　　　 labor (p. 222)
placenta (p. 220)　　　　　cervix (p. 222)
umbilical cord (p. 220)　　afterbirth (p. 222)

Applying What You Have Learned

1. What are three ways that a cell changes during development?
2. Describe a population that reproduces by asexual reproduction.
3. Describe a population that reproduces by sexual reproduction.

Questions 4 and 5 are based on this paragraph:

   A freshwater pond may contain several populations of single-celled organisms. Some of these can reproduce by both sexual and asexual reproduction. If the environment offers good growing conditions, the organisms will reproduce by cell division and cell growth. When the environment changes and becomes unfavorable, two similar organisms will unite and combine their gene material. This union produces a new cell that is unlike either of the two parent cells. This new combination of gene material may produce a cell that is able to adapt to unfavorable changes in the environment.

4. When do these single-celled organisms reproduce by asexual reproduction?
5. How and when do these single-celled organisms reproduce by sexual reproduction?
6. A population of single-celled organisms that can reproduce asexually has an advantage over those that cannot. Why?
7. What safety mechanism allows these populations to bounce back from unfavorable changes in the environment?
8. What guarantees variety in the populations of single-celled organisms?
9. Describe the monthly menstrual cycle. Describe how the cycle will be affected if fertilization occurs.
10. How is development in the embryo stage different from development in the fetal stage?

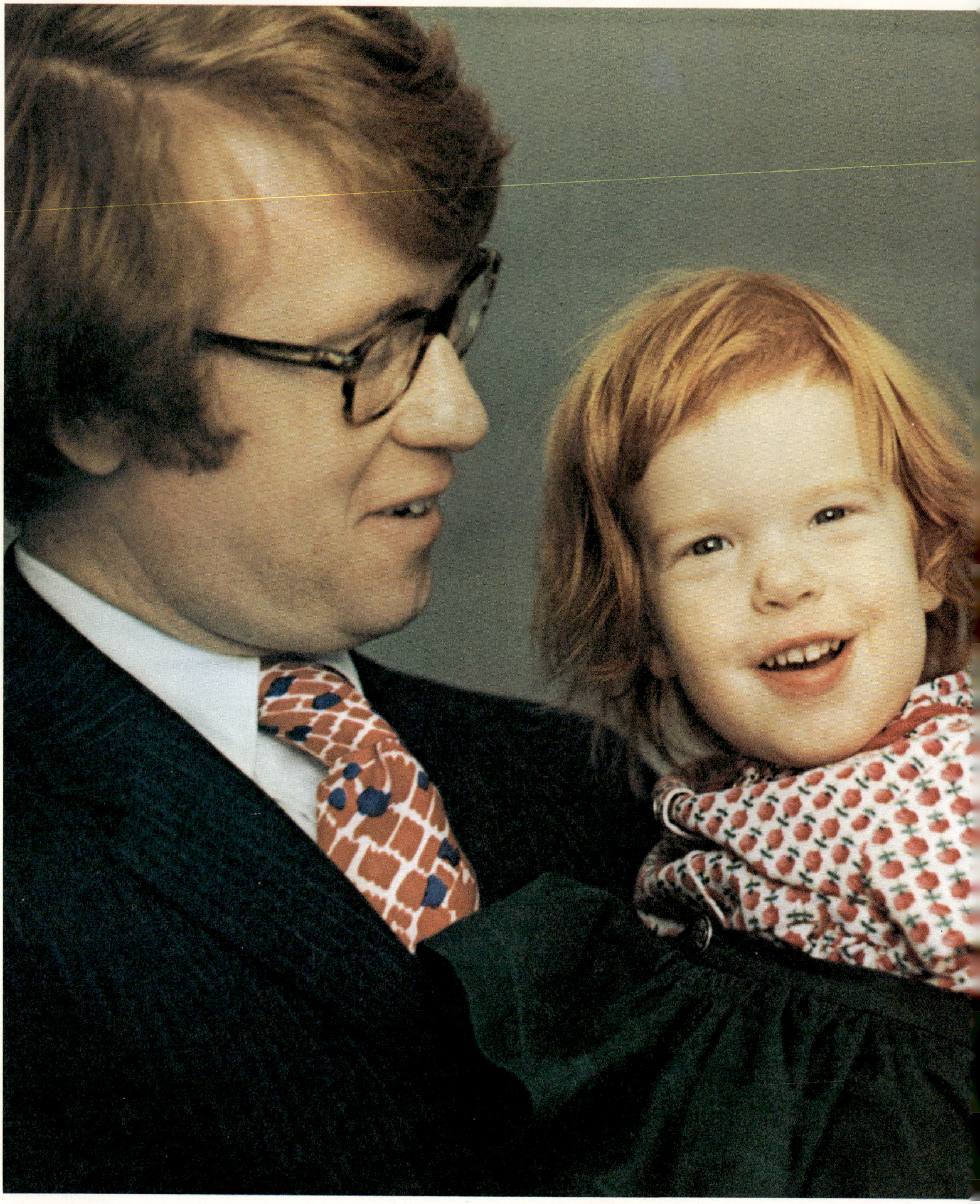

Figure 13.1: A child inherits its looks from its mother and father. Some of the personal characteristics this girl inherited from her father are easily observed.

Chapter 13

GENES AND HEREDITY

Almost everyone knows something about heredity. For example, people may say that you have got your father's eyes, or that you have your mother's hair. Such statements point out that personal characteristics, or **traits,** are passed on from generation to generation. But what determines a trait such as brown eyes or black hair? Or blue eyes or blonde hair? Dark skin or light skin? How are traits passed on? Can a trait be changed? These are just a few of the questions that will be discussed in this chapter.

trait: a physical characteristic, such as hair or eye color, that may be inherited

GENES

What are genes? Genes are what you inherit from your parents. You get two sets of them, one set from your father and one set from your mother. The father's sperm cell contains one set of his genes. The mother's egg cell contains one set of

Human Body Functions

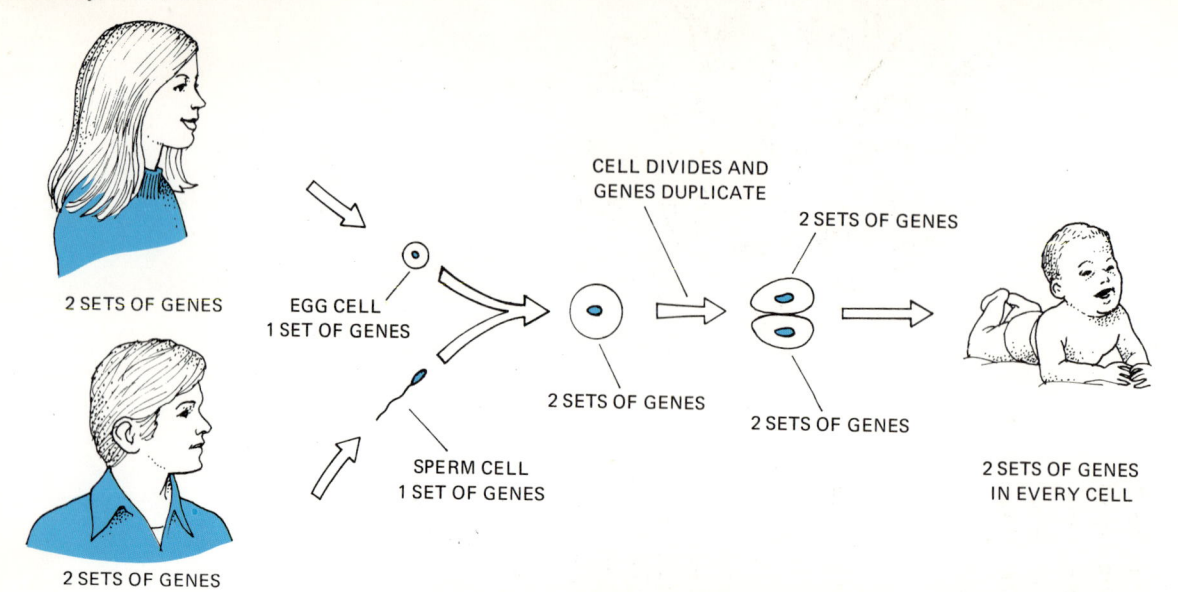

Figure 13.2: A child's genes are inherited from its parents. Every cell in the body has two sets of genes, one set from each parent.

chromosome: one of the small bodies in a cell's nucleus that contains DNA tied to protein

her genes. During fertilization, the sperm unites with the egg cell. The two sets of genes are then in one cell.

Where are the genes? They are in the nucleus.* Where in the nucleus? In the **chromosomes**. What are chromosomes? They are small bodies that are visible during the time that the cell is dividing. The chromosomes contain molecules of DNA. The genes are in the molecules of DNA.

When the fertilized egg cell first divides, the chromosomes are duplicated. As a result, so are all the DNA and all the genes. Thus, the two cells that result each have identical sets of genes. And every time these cells divide, their sets of genes are always duplicated. By the time an individual is completely developed, every cell has the same two sets of genes. Later, when the individual is old enough to produce sperm or egg cells, there is a change in this pattern. Egg cells and sperm get only one set of genes each.

What do your genes do? Suppose you had a set of instructions to build a model airplane or to sew a dress. Now suppose someone asked you: What do these instructions do? You could not answer because the instructions themselves do not do anything. They are used by whoever is building a model airplane or sewing a dress.

Likewise, one cannot answer the question: What do your

---

*Very small amounts of genetic material have been discovered outside the nucleus.

Genes and Heredity

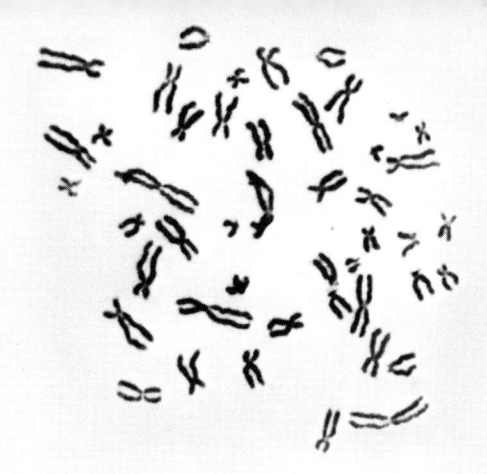

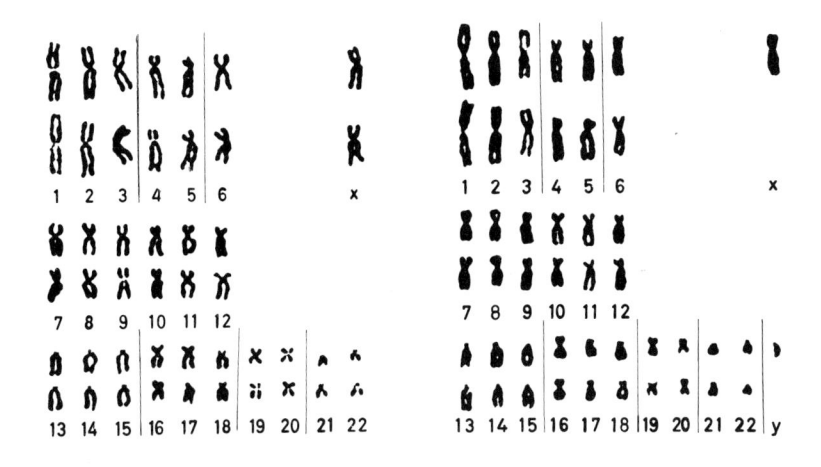

Figure 13.3: This photomicrograph (top) shows the 23 pairs of chromosomes of a human being. The chromosomes are in the "double-strand" stage—they have been duplicated and are about to separate into two identical sets. In the two lower photos, the chromosome "snapshots" of a female (left) and a male (right) have been rearranged into pairs for study by life scientists. (Top photo, courtesy Carolina Biological Supply Co.)

genes do? Your genes are two sets of instructions. Instructions for what? Your genes contain instructions for building all the different proteins that are in your body.

What are proteins?   Proteins are molecules. Like all molecules, they are made up of atoms. Most proteins are very large molecules, made up of very many atoms. What should you know about proteins? Proteins are very important parts of any living organism—of yourself, for example.

Many of the proteins in your body are like building blocks. They are like the bricks and stones that make up a building. Your hair, muscles, and skin are examples of these materials. They are all made up of protein.

Most of the other proteins in your body are **enzymes**. Enzymes are the "doers" in your body. They control almost all your

enzyme:
any one of many proteins that affect chemical changes in the body

229

Human Body Functions

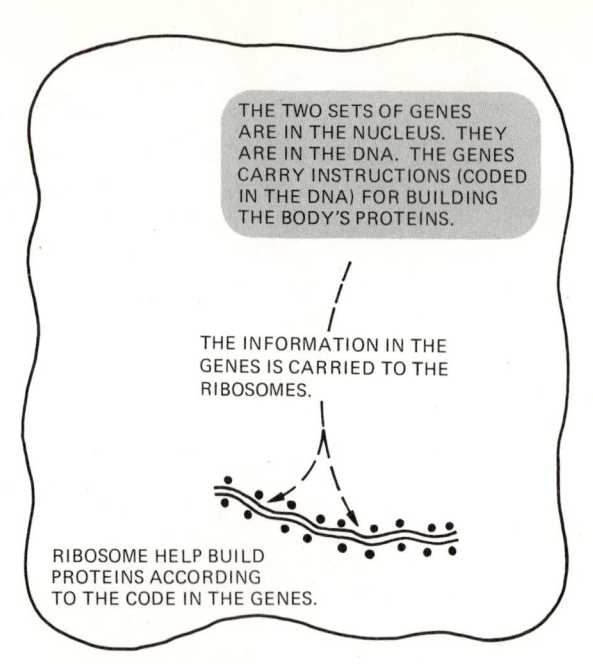

> THE TWO SETS OF GENES ARE IN THE NUCLEUS. THEY ARE IN THE DNA. THE GENES CARRY INSTRUCTIONS (CODED IN THE DNA) FOR BUILDING THE BODY'S PROTEINS.
>
> THE INFORMATION IN THE GENES IS CARRIED TO THE RIBOSOMES.
>
> RIBOSOME HELP BUILD PROTEINS ACCORDING TO THE CODE IN THE GENES.

Figure 13.4: The genes in the nucleus are contained in the DNA. Each section of the DNA molecule is a gene that controls the making of a single protein.

Figure 13.6: The genes in DNA provide the instructions for building all the different proteins that control and carry out the body's functions and activities.

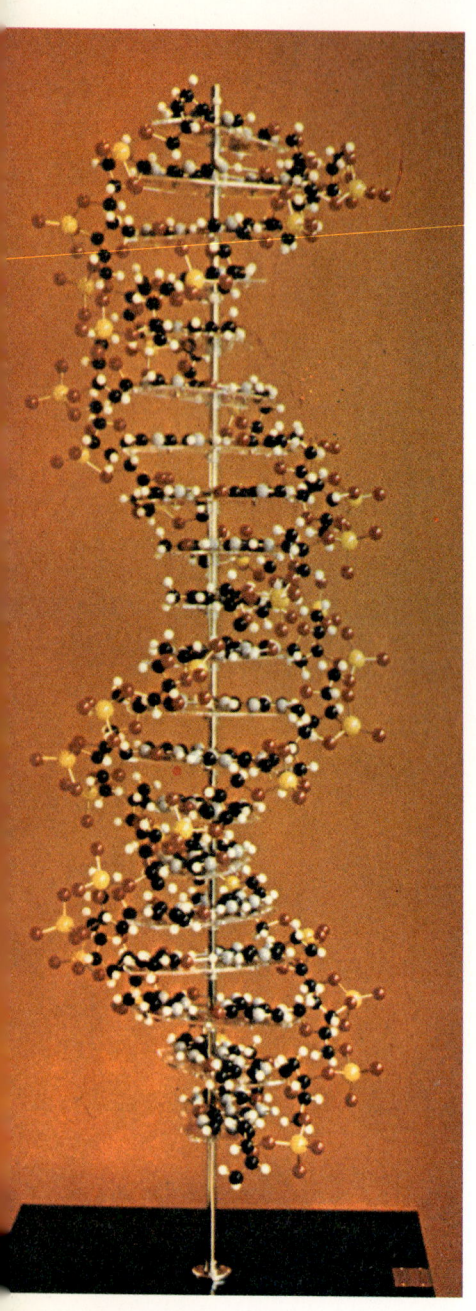

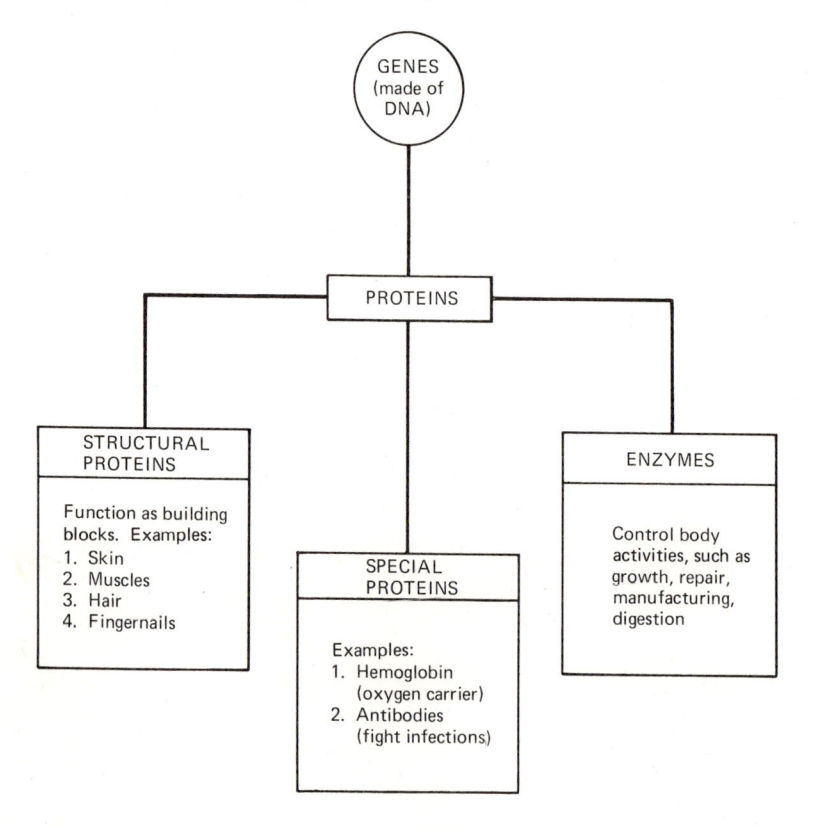

```
         GENES
        (made of
          DNA)
            |
         PROTEINS
      /     |      \
STRUCTURAL  SPECIAL  ENZYMES
PROTEINS    PROTEINS
```

STRUCTURAL PROTEINS

Function as building blocks. Examples:
1. Skin
2. Muscles
3. Hair
4. Fingernails

SPECIAL PROTEINS

Examples:
1. Hemoglobin (oxygen carrier)
2. Antibodies (fight infections)

ENZYMES

Control body activities, such as growth, repair, manufacturing, digestion

Figure 13.5: The atoms in the DNA molecule are arranged in a double helix pattern, where each strand complements the other chemically. When the cell divides, the two strands come apart, and a new complement is formed for each strand.

Genes and Heredity

body's activities. They help to build, they help to tear down, and they regulate.

Here is another way to think about proteins. Picture one of your cells as a factory. The factory itself is made up mostly of proteins. Most of the molecules that are made there are proteins. And almost all the "workers" and "supervisors" are proteins! Now you can understand why the instructions in your genes are so important. All the proteins in your body were built from these plans.

LESSON REVIEW *(Think. There may be more than one answer.)*

1. Each cell that lives on the inside of your cheek contains
    a. a set of genes from either of your parents.
    b. a set of genes from your father only.
    c. a set of genes from your mother only.
    d. a set of genes from your mother and a set of genes from your father.

2. Which of the following are correct statements?
    a. There are instructions in the proteins and chromosomes for manufacturing all the enzymes found in the nucleus.
    b. The genes are in the chromosomes, which are in the nucleus.
    c. There are instructions in the enzymes for manufacturing all the genes that are in the chromosomes.
    d. Proteins and enzymes are made from instructions in the genes, which are in the nucleus.

3. A sperm cell has
    a. one set of genes.
    b. two sets of genes.
    c. a set of genes from your father.
    d. a set of genes from your mother and a set of genes from your father.

4. Which of the following terms best matches each phrase?
    a. trait
    b. chromosome
    c. enzyme

    ____ body in the nucleus where DNA is stored

    ____ a personal characteristic

    ____ one of a group of proteins that control, build, and tear down the body

---

KEY WORDS

trait (p. 227)
chromosome (p. 228)
enzyme (p. 229)

THE GENE POOL

What happens to genes? You have two sets of genes. Who used them before you did? That is an easy question to answer. Your mother and father used them. Who used them before that—and long before that? You cannot answer that question. You do not know all the people who have used your genes. They have been used by thousands of individuals. They will probably continue to be used in future years, perhaps by thousands of individuals.

What happens to genes? Genes in a population can be compared to a deck of playing cards. The playing cards in a deck are used over and over. So are genes. For each new game the card player gets a new hand of cards. Each new individual gets a new combination of genes. Between each card game, the cards are shuffled. Sexual reproduction in a population causes the genes to be reshuffled for each new generation.

Each population has a gene pool. The total of all the genes in a population is called a **gene pool**. A gene pool is somewhat like a huge deck of playing cards. It is how one population differs from another. A population of rabbits will have some of the same genes as a population of elephants. But a rabbit population will have some genes that no other population has. So will an elephant population.

Each of the individuals in a population uses two sets of genes from the gene pool. If these individuals reproduce, their offspring will use the same genes after they die. Over the years, a great many individuals use genes from the population's gene pool. A gene pool lasts for thousands or millions of years. What could cause the change of a gene pool?

gene pool: the total of all genes in a population

Figure 13.7: A gene pool changes slowly, over many years, as gene mutations build up in the population. As a gene pool changes, the population changes. This process is called evolution.

EVOLUTION = ANY CHANGE IN A GENE POOL

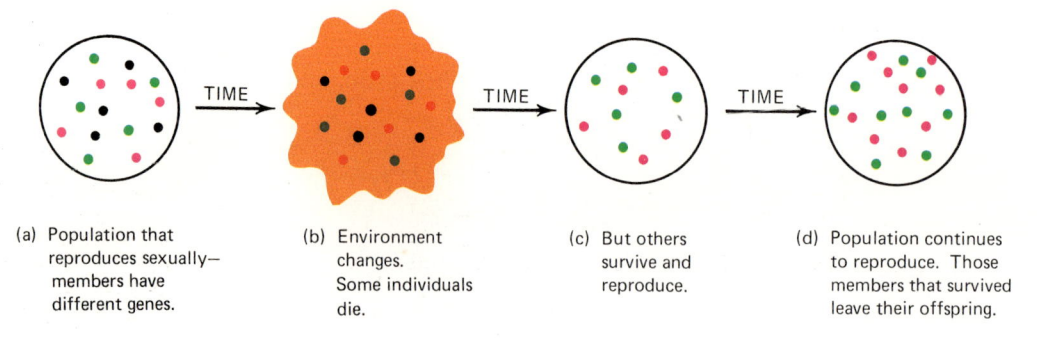

(a) Population that reproduces sexually—members have different genes.

(b) Environment changes. Some individuals die.

(c) But others survive and reproduce.

(d) Population continues to reproduce. Those members that survived leave their offspring.

Genes and Heredity

Genes can mutate. Do genes always remain the same? Are all the genes that you are using now the same as they were 2000 generations ago? The answer is no. Genes can change and they do change. When they change they are said to *mutate*. The name for a change in a gene is a **mutation.**

What happens when a gene mutates? Remember, a gene is a set of instructions for building a protein, or a part of a protein. If a gene mutates, the instructions for building are changed. The protein does not get built the same way that it did before. Now the question is: Will the protein be able to function as it did before?

Life scientists have much more to learn about gene mutations. But the evidence that they have now suggests that most mutated genes do not remain in the gene pool. In other words, the proteins that are built from these changed genes do not function properly. The individual using the mutated gene either fails to live or fails to reproduce. If the individual does live and does manage to reproduce, the offspring may fail to live or reproduce. Most mutated genes do not survive more than a few generations.

mutation:
any change in a gene. It may lead to a change in an organism.

What happens when mutated genes do survive? When mutated genes do remain in the gene pool, the gene pool is no longer the same as it was before. Therefore, the population is not the same as it was before. The population has changed. The life scientist says that the population has *evolved*. The process of change in a population is called **evolution.**

Most evolution is **adaptive evolution.** This means that most changes in a population improve its chances of surviving in its environment. As an example, suppose that a mutated gene gave a grasshopper a stronger flight muscle. That grasshopper might have a better chance to survive and to leave offspring. Some of

evolution:
the process of change in a population as its gene pool changes
adaptive evolution:
a change in the gene pool of a population that better suits it to survive in its environment

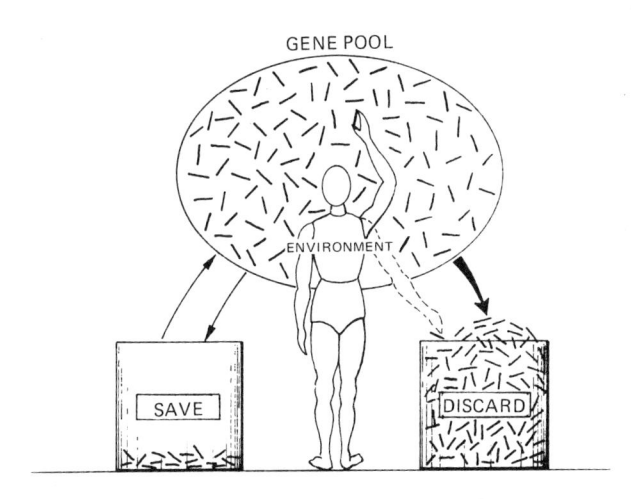

THE FATE OF MUTATED GENES

Figure 13.8: When a gene mutates, the instructions are changed. The protein that is built differs in some way from the original. Usually this is harmful to the individual, and most mutations do not remain in the gene pool.

233

Human Body Functions

Figure 13.9: This fruit fly cannot fly. It has mutated genes that produce useless wings. This fly could not survive in nature. It can exist only in genetics laboratories. (Courtesy Carolina Biological Supply Co.)

adaptation: a trait that particularly helps an organism to survive in its environment

the offspring would inherit the mutated gene. Their offspring might have a better chance to survive and to reproduce. After many years, sexual reproduction could scatter this new gene throughout the population. Then we could say that the population had evolved. We might also be able to say that the evolution was adaptive. A population of grasshoppers that are stronger fliers might have a better chance of surviving in its environment. The stronger flight muscle that resulted from the mutated gene would be called an **adaptation.**

LESSON REVIEW *(Think. There may be more than one answer.)*

1. Your genes
    a. have only been used once—by you!
    b. were used by your mother and father.
    c. were used by your grandmothers and grandfathers.
    d. were used by thousands of people.

2. A population of chipmunks
    a. has a gene pool.
    b. has genes that no other population has.
    c. has genes that other populations have.
    d. has genes that will be used by thousands or millions of chipmunks.

3. A mutated gene
    a. is a gene that has been changed.
    b. will cause a different protein to be built.
    c. is likely to create new individuals in a population.
    d. usually does not survive in a gene pool.

4. A mutated gene that is adaptive
   a. will be more likely to survive in a gene pool.
   b. will give individuals with that gene a better chance of surviving and leaving offspring.
   c. cannot be spread through a population by sexual reproduction.
   d. is not likely to survive in a gene pool.

5. Which of the following terms best matches each phrase?
   a. gene pool
   b. mutation
   c. evolution
   d. adaptive evolution
   e. adaptation

   _____ all the genes of a population

   _____ change in a population

   _____ change that improves chances of surviving

   _____ change in a gene

   _____ a trait that helps an organism to survive in its environment

KEY WORDS

gene pool (p. 232)
mutation (p. 233)
evolution (p. 233)

adaptive evolution (p. 233)
adaptation (p. 234)

PREDICTING INHERITANCE

Human traits are difficult to predict accurately. Newly married couples often wonder what their children will be like. Here are typical questions they ask. "I'm blond and my husband is black-haired. What color hair will our children have?" "I have blue eyes and my wife has brown eyes. What color eyes will our children have?" "My husband is 188 centimeters (6 feet 2 inches) tall and I am 157.5 centimeters (5 feet 2 inches) tall. How tall will our children be?"

It is impossible for the life scientist to make accurate predictions about most human traits. Here is why. For traits such as hair color, eye color, and tallness, many different genes are involved. Genes from both parents are at work. Sometimes a gene from one parent will dominate a gene from the other parent. But many times, genes from both parents will have an effect. This results in a blend or mixture, which gives a trait that may be like or unlike either or both parents.

Human Body Functions

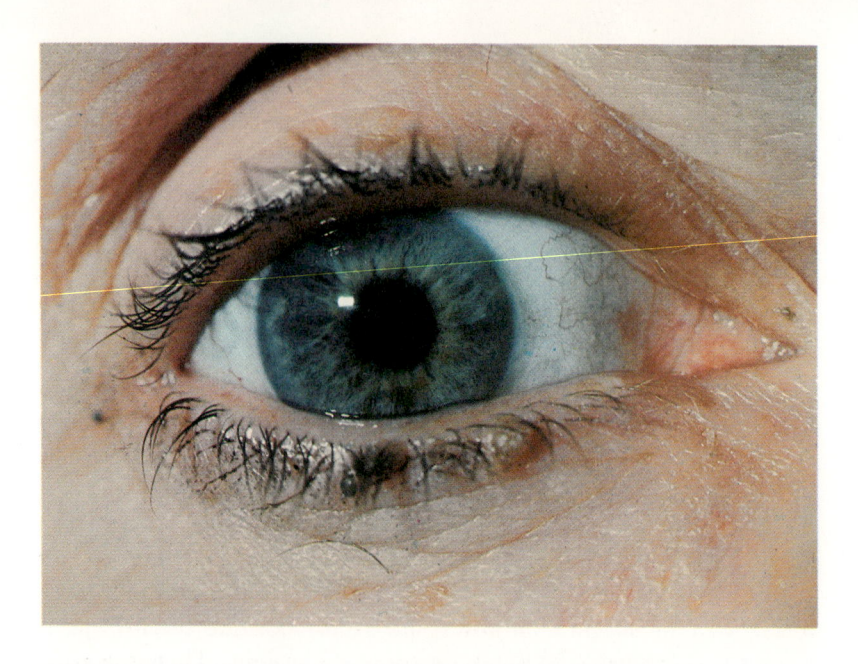

Figure 13.10: This closeup of a hazel eye shows the multicolored nature of this eye color. Each different color is the result of a genetic pattern in the individual's DNA.

ABO blood type:
a way to classify blood based on the type of protein on the surface of red blood cells

dominant gene:
a gene that most strongly determines a trait that is the result of two or more genes

Human ABO blood type can be predicted. There are a few human traits that can be predicted. The human ABO blood type is one of them. **ABO blood type** refers to the type of protein that is on the surface of your red blood cells. Your ABO blood type is an important matter if you happen to need blood as the result of an injury or illness. A transfusion of the wrong ABO blood type can be fatal.

There are three different genes that may determine the ABO blood type. We can call these genes A, B, and o. Note that we use capital letters for A and B, and a small letter for o. This is because both the A and B genes are **dominant genes.** That is, both A and B dominate the o gene when it is present.

You have two of the three genes that produce the ABO blood type. You got one from your father and one from your mother. There are six possible pairs that you can have: AA, Ao, BB, Bo, AB, and oo. If you have AA or Ao, you have type A blood. (Remember: A or B dominates o.) If you have BB or Bo, you have type B blood. If you have AB, you have type AB blood. If you have oo, you have type O blood.

The big question: genes or environment? What will determine how tall you will grow? Will it be the genes you received from your parents? Or will it be determined by the food and vitamins that you have eaten, or will eat? If your height is determined by your genes, we could say that genes determine tallness. If your height is determined by food and vitamins, we

236

could say that the environment determines tallness. Which is right?

The question about tallness is just one of many that we could ask. What determines personality: genes or environment? What determines intelligence: genes or environment? Such questions cannot yet be answered by the life scientist. But there is one general statement that can be made. **Genes determine what an individual *may* become; environment determines what an individual *will* become.**

Think of your inheritance as an inborn ability. You did inherit the ability to grow to a certain height. But how tall you will grow depends upon things in your environment, such as food. You may have inherited an ability to sing. But whether you will sing will be decided by factors in your environment. You may have inherited an ability to learn rapidly. But whether you will learn rapidly will be decided by factors in your environment. For most traits, neither the genes nor the environment is more important. All traits require the genes. That is a fact. But many traits need a favorable environment in which the genes can be used.

Figure 13.11: Although the genetic pattern determines certain traits, the way these traits show up is due to the environment. These trees grow this way because of harsh prevailing winds.

Laboratory Activity

What ABO Blood Type Genes Could You Have?

PURPOSE

There are four different blood types, A, B, AB, and O. Your combination of two of the three genes A, B, and o determines your ABO blood type. In this investigation, you will determine your ABO blood type.

MATERIALS

microscope slide   cotton
felt-tip pen   alcohol
anti-A serum   disposable lancets
anti-B serum   toothpicks

PROCEDURE

1. With a felt-tip pen, draw two circles on a clean, dry slide. Each circle should be about the size of a quarter. Label one circle "Anti-A." Label the second circle "Anti-B."

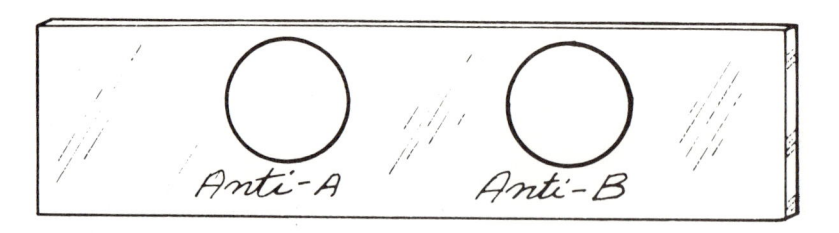

2. Place one drop of anti-A serum in the circle labeled "Anti-A." Place one drop of the anti-B serum in the circle labeled "Anti-B."
3. Wash your hands with soap and water. Soak a ball of cotton in alcohol.
4. Gently press your thumb against the first joint of your middle finger. This pressure will cause the blood to stay near the tip of your middle finger.
5. Wipe the tip of your finger with the alcohol-soaked cotton.
6. Open the package containing the sterile lancet. Be careful not to touch the point of the lancet.
7. Prick the tip of your finger with the lancet. Dispose of the used lancet so that it cannot be used again by another person. If you have to prick your finger again later in the investigation, *use a fresh lancet.*
8. Wipe off the first drop of blood with the alcohol-soaked cotton. Remove the second drop of blood with the wide end of a toothpick. Transfer the blood to the anti-A serum. Stir the serum with the toothpick.
9. Using a second toothpick, remove another drop of blood, and transfer it to the anti-B serum. Stir the serum with the toothpick.

10. Wipe the tip of your finger with the alcohol-soaked cotton. Hold the cotton on your finger tip until the bleeding stops.
11. Look at both circles on the slide. Wait several minutes to see if a reaction occurs in either circle. Compare your results with the chart above. If you see blood clumps in the circle labeled "Anti-A," your blood type is A. If you see blood clumps in the circle labeled "Anti-B," your blood type is B. If clumping occurred in both circles, your blood type is AB. If there is no clumping in either circle, your blood type is O. What is your ABO blood type?

BLOOD TYPE	SERUM Anti-A	SERUM Anti-B
A	clumping	no clumping
B	no clumping	clumping
AB	clumping	clumping
O	no clumping	no clumping

QUESTIONS

1. What is the frequency of each blood type in your class? Record the blood type of each student in the class. What percentage of the students have type A blood? Type B? Type AB? Type O? You may want to tabulate your results on a chart similar to this one.

Blood type	Number of students	Percentage of the class
A		
B		
AB		
O		

2. Blood is not only labeled by its ABO type. It is also labeled as "Rh negative" or "Rh positive." Find out what you can about the Rh factor in blood. What is its genetic significance?

Human Body Functions

LESSON REVIEW *(Think. There may be more than one answer.)*

1. Most human traits
   a. can be predicted accurately.
   b. cannot be predicted accurately.
   c. are caused by one gene dominating another.
   d. are caused by the action of both parents' genes.

2. A person with type A blood
   a. could have a mother with type A or O blood.
   b. could have a father with type A or O blood.
   c. could have a mother with type B blood and a father with type O blood.
   d. could have a mother with type O blood and a father with type O blood.

3. The ability to roll one's tongue (as shown in the figure) is controlled by a dominant gene *R*. The nondominant gene is *r*. If you cannot roll your tongue,
   a. your mother could have the two genes RR.
   b. your mother could have the two genes Rr.
   c. your father could have the two genes RR.
   d. your father could have the two genes Rr.

4. If you
   a. can roll your tongue, you may have the two genes RR.
   b. can roll your tongue, you may have the two genes Rr.
   c. cannot roll your tongue, you may have the two genes Rr.
   d. cannot roll your tongue, you have the two genes rr.

5. A child is mentally retarded because its mother could not obtain enough protein when she was pregnant. This is an example of how
   a. genes determine what an individual *may* become.
   b. genes determine what an individual *will* become.
   c. environment determines what an individual *may* become.
   d. environment determines what an individual *will* become.

KEY WORDS

ABO blood type (p. 236)   dominant gene (p. 236)

Applying What You Have Learned

Questions 1 and 2 refer to the following paragraph:
  The chemical pesticide DDT was developed to help control insect populations. A few genes in some insect gene pools may adapt individuals to be resistant to DDT. Suppose DDT is sprayed on an insect population that has DDT-resistant members.

1. Will the number of insects with genes that led to DDT resistance increase, decrease, or remain the same? Why?

Genes and Heredity

2. Will further use of DDT be as effective in killing this insect population? Why?
3. Explain why different combinations of gene sets occur within a population.
4. A dominant gene for a specific trait is inherited along with a nondominant gene for the same trait. Which gene's set of building instructions will be used to build the specific proteins?

From what you have learned about blood type genes, complete questions 5 and 6 in the following table. Use the example as an aid to solving questions 5 and 6. Work out your answers and record them on a separate piece of paper.

	Example		Question 5		Question 6	
Parents' Blood Types	Father A	Mother B	Father AB	Mother O	Father A	Mother A
Possible gene combinations in parents	AA and Ao	BB and Bo				
Possible egg and sperm genes	A or o	B or o				
Possible gene combinations in offspring	AB, Bo, Ao, or oo					
Blood types not possible from these parents	None					
Possible offspring blood type	AB, A, B, or O					

7. Explain how nonadaptive mutations may get eliminated from a population.

Questions 8 through 10 refer to the following paragraph:

For many years, a population of gray moths has lived in an area where there are many birch trees with gray bark. The moths spend much of the day resting on the trunks of these trees. Their natural enemies are birds that catch them while they are resting. About 100 years ago, coal-burning factories began to pollute the area. The bark on the trees became darker and darker over the years. Eventually it was almost black.

8. Over a period of 100 years the population of gray moths changed into a population of black moths. Explain how this could have occurred.
9. Would the change in the moth population be an adaptive change? Explain your answer.
10. Pollution in the area is now decreasing. The bark of the birch trees is getting lighter. How do you think the moth population will adapt to that change?

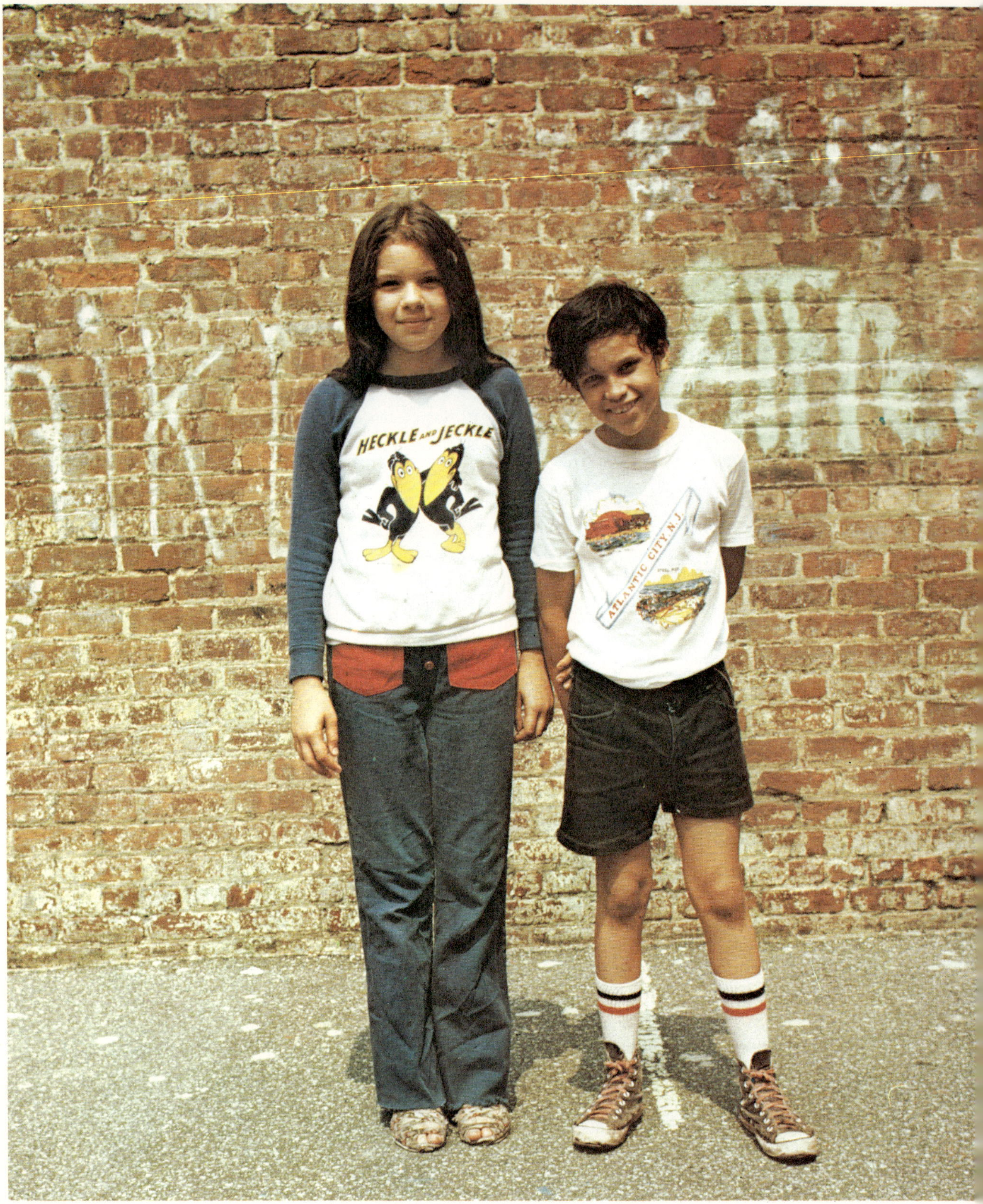
HECKLE AND JECKLE
ATLANTIC CITY N.J.

Figure 14.1: This boy and girl are both about age 12. The boy was taller than the girl 6 years ago. Who is likely to be taller 6 years from now?

Chapter 14

GROWTH AND DEVELOPMENT

Since you were born, your body has probably multiplied in weight at least ten times. Your body length is about three times what it was at birth. Your face looks different. Your voice is different. Even the internal parts of your body, such as your skeleton, muscles, and brain, are different.

How have these changes in your body come about? What caused them? What changes can you expect to occur in your body from now on? What problems might you meet? These are the general questions that will be answered in this chapter.

HOW GROWTH OCCURS

What is growth? *Growth* means to "get larger." The ability to grow is not limited to living things. Nonliving things can grow also. Icicles can grow. Rivers can grow. Mountains and volcanoes can grow.

Human Body Functions

There is a great difference in the way that nonliving and living things grow. Most nonliving things grow by an addition process. Materials are added on. An example is the way a large snowball is made for a snowman. You start with a little ball, and you roll it in the snow. It grows larger and larger. It gets larger because snow is added to the outside.

You did not grow the way a snowball grows. However, the materials that caused your **growth** did come from outside of your body. The difference is that you took the growth materials inside. Your body broke them down into smaller pieces. Then it rearranged them. Your growth took place inside your body.

growth: the process by which something gets larger

Your growth occurred in three ways. Like all living things, your body is made up of cells and the products that cells manufacture. Growth occurs because of your cells and their products.

One way that growth occurs is by *cell division*. Growth occurs because two cells take up more room than one cell.

The second way that growth occurs is by *cell enlargement*. Cells take in materials from their environment. Some of these materials are kept inside and used as building materials. The result is larger cells.

The third way that growth occurs is by the *accumulation of cell products*. Some bones grow in this way. Bone cells manufacture calcium minerals. These pass out of the cell and then harden. A bone grows because of the accumulation of these calcium minerals.

Development is growth and change of form. In the years you have lived, your body has done more than grow. If all you did was get larger you would look like a huge baby. Your body has also developed. **Development** is the process by

development: the process by which the body grows and changes form

Figure 14.2: Your body grows in three ways: (a) by cell division, (b) by cell enlargement, and (c) by accumulation of cell products.

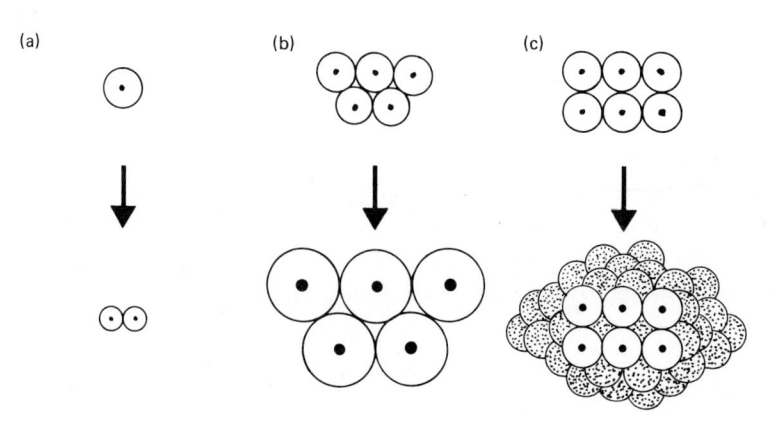

244

which your body grows and changes form. A change in form means a change in shape and proportion. You can see an example of how form changes in Figure 14.3. Note how the baby's head is large in proportion to the size of its body. During the process of human development, the head gradually becomes smaller in proportion to the rest of the body. The head does grow larger, but the rest of the body grows much faster.

Figure 14.3: During development before birth, the head grows faster than the body. After birth, the body grows faster than the head. A child's head is proportionately larger than an adult's head.

How do you grow taller? When you were born, much of your skeleton was made of cartilage. During the first year of your life, the cartilage in your skeleton began to change to bone. This process continued throughout your childhood.

Now you have just a little cartilage left in your bones. In the long bones, such as those in your arms and legs, there is a layer of cartilage near each end. (See Figure 14.5.) New bone is being made on each side of these cartilage layers. That makes the bone longer and you taller. You will stop growing when the cartilage layers change to bone.

How do your muscles grow? You probably have fewer muscle cells now than you had when you were born. Does that surprise you? The reason is that after a certain time muscle cells do not reproduce themselves. That time occurred before you were born. Since your birth, a few of your muscle cells have probably died.

Then how have your muscles grown? They have grown by cell enlargement. Materials have accumulated inside the cells

Human Body Functions

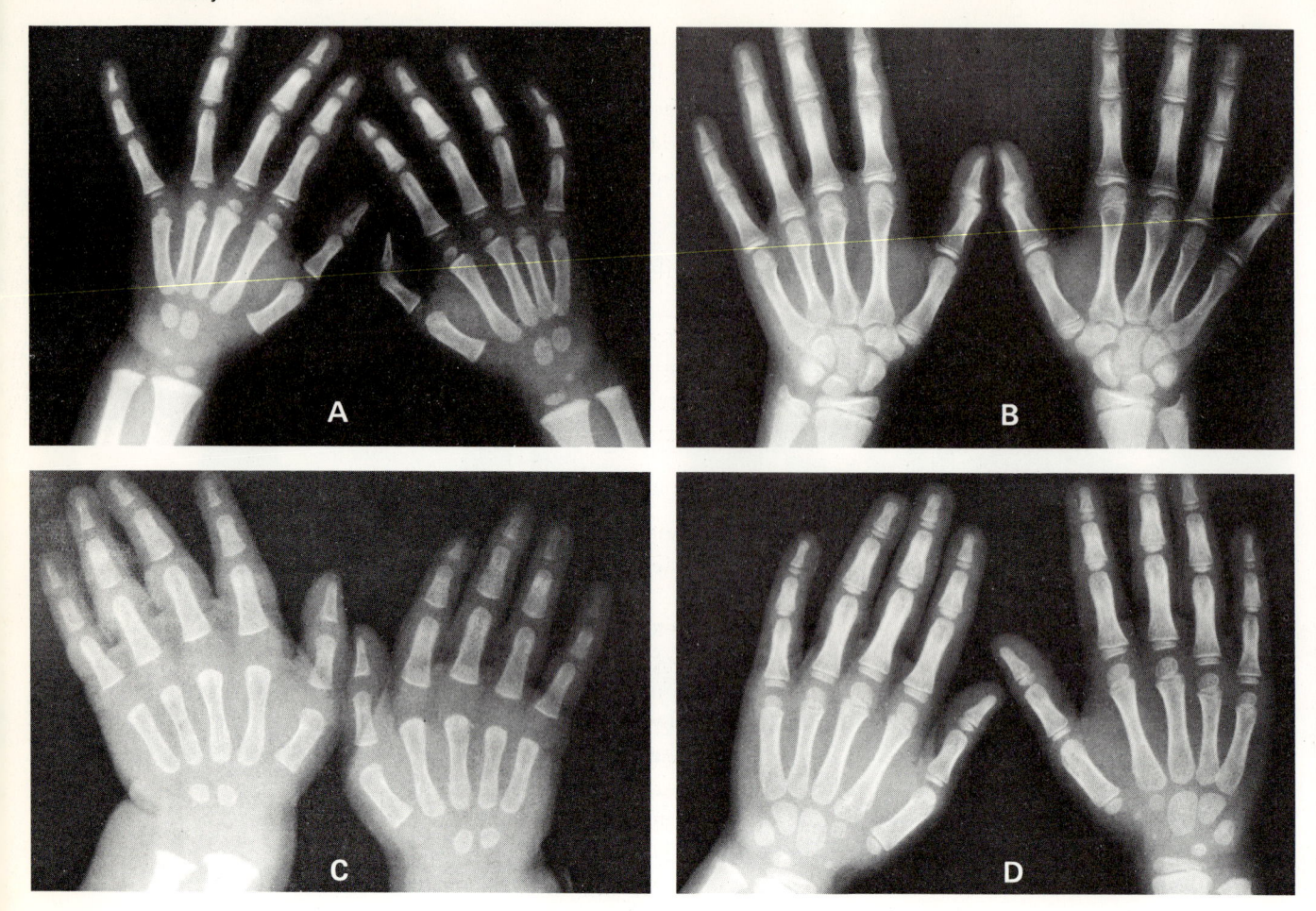

Figure 14.4: As a human being gets older, the amount of cartilage decreases and the amount of bone increases. The x rays above were taken of hands at ages 1, 4, 7, and 10. Match the x ray with the proper age.

Figure 14.5: These X rays show the knees of an adolescent (left) and an adult (right). Note the cartilage layer at the ends of the long bones. In the adolescent, this layer has not changed to bone—new bone can still be made, and the legs can thus grow longer.

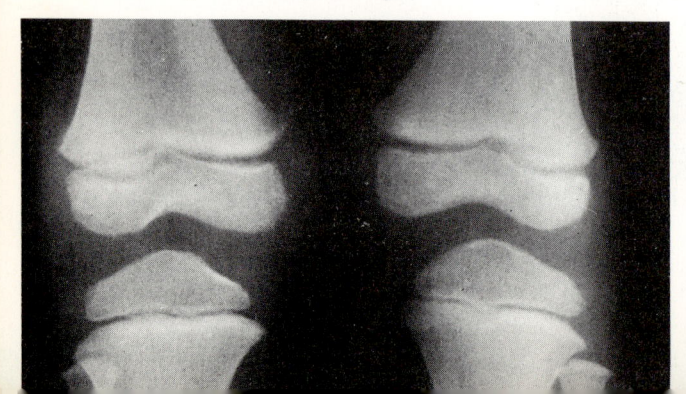

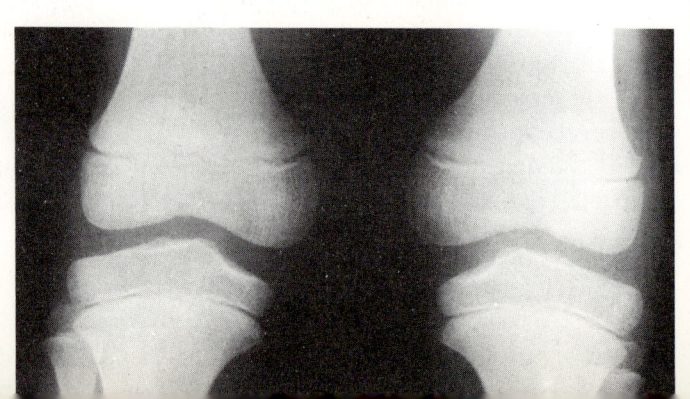

and caused them to get larger. This accumulation process is greatly stimulated when the muscle cells have to work. For example, if you do hard exercise regularly, your muscles will grow larger. (See Figure 14.6)

The opposite is also true. If you stop all exercise, your muscles will shrink. What does a leg or arm that has been in a cast for 2 or 3 months look like?

Growth and Development

How does your brain grow? Your brain is larger now than when you were born. But it has fewer cells. Brain cells are like muscle cells. They do not reproduce themselves. You have a certain number at birth. From that time on, the number decreases steadily. Many brain cells die every day. It is a good thing that you have far more than you need!

Your brain cells are larger now than they were at birth. That is why your brain is larger. It grew by cell enlargement. Also, many of your brain cells have changed in form. As you have aged, your brain cells have developed many branches. These branches reach out and intertwine with the branches of other cells. It is the branching of the brain cells that allows them to carry out many of their amazing functions.

What causes growth? Many things cause growth. First, there are the two sets of genes that you inherited from your parents. They contain all of the "instructions" for your growth.

As you learned in Chapter 10, hormones also control growth. Thyroxin from the thyroid gland stimulates growth. And your pituitary gland—the "master gland"—produces a very powerful growth hormone. It stimulates growth of your whole body, but it is most effective on your "long bones." These are the bones of your arms and legs.

A proper diet is also a very important requirement for growth. Most of the new material that is used for growth is protein material. Therefore, a good supply of protein in the diet is absolutely necessary. Your body cannot make proteins out of other kinds of food. If the protein is not in your diet, there will be no protein for growth materials.

Figure 14.6: These foods are rich in protein—the body-building material. Without protein, you cannot grow and develop properly.

Human Body Functions

LESSON REVIEW *(Think. There may be more than one answer.)*

1. Human growth
    a. occurs the same way as crystal growth.
    b. occurs the same way as river growth.
    c. occurs the same way as a snowball's growth.
    d. occurs inside the body.

2. A bone grows mostly by
    a. cell division.
    b. cell enlargement.
    c. accumulation of cell products.
    d. all of the above equally.

3. The human brain
    a. grows after birth.
    b. does not grow after birth.
    c. reproduces more cells after birth.
    d. loses cells after birth.

4. Muscle cells
    a. reproduce if they are exercised.
    b. get larger if they are exercised.
    c. get smaller if they become inactive.
    d. decrease in number after birth.

5. Body growth is aided by
    a. a thyroid hormone.
    b. a pituitary hormone.
    c. a diet rich in foods like bread and potatoes.
    d. a diet rich in foods like beans and meat.

6. Human development
    a. includes growth.
    b. does not include growth.
    c. includes changes in the body's form.
    d. begins before birth.

KEY WORDS

growth (p. 244)
development (p. 244)

STAGES OF GROWTH AND DEVELOPMENT

Everyone has an internal clock. Some babies walk when they are 9 months old. Others do not walk until they are 16 months old. Some children lose their front teeth at 5 years. Others lose them at 7 years. Some boys start a period of rapid

Figure 14.7: Every person has a unique rate of development. Some grow faster than others at a particular time. Some start later than others. Most people grow rapidly during adolescence.

growth at 13. Others begin at 15. There is a good reason for these differences in the rate of development. Everyone has a development regulator that works like a clock. The rate of the clock is probably determined by your genes and set before birth. The result is a development rate that is unique in each person. You are developing at a rate that is unlike that of any other person.

Is there any way to control your internal clock? In other words, suppose you are a boy and you are "scheduled" to start growing a beard at 15 years, 3 months. Is there any way to speed up this process, or slow it down? There may be a way, but life scientists have not discovered it yet. In general, life scientists know very little about the regulators that control the rate of development.

Childhood development is slow and steady. The first 12 years of life are usually called the "childhood years." This is a period of slow and steady development. During the first 4 years, boys and girls develop at about the same rate. From age 11 to 12, most girls are taller and heavier than most boys.

Adolescence brings rapid growth. The years from about 12 to 18 are the years of **adolescence.** These are difficult years for many young people because of the rapid development that occurs. Other than the fetal period before birth, the adolescent period is the time of the greatest development.

adolescence:
roughly the period between 12 and 18 years of age. It is noted for its rapid development and as the time of sexual maturation.

249

Human Body Functions

Figure 14.8: Can you tell the face of the adolescent from the face of the same person as an adult in each of these pairs of photos? How? What characteristics give a person's face the appearance of an adult?

What happens during adolescence? Many things. Growth is spectacular during this period. Most girls start a period of rapid growth between the ages of 11 and 12. And at age 16, most of their growth is completed. Boys start later on their growth spurt. Some boys may start as early as 12½. Many are growing rapidly at 13½. But quite a number of boys are 14½ or 15 before they really start their rapid growth. Boys have usually completed most of their growth by age 18.

Puberty also arrives during adolescence. Adolescence is also the time of puberty. **Puberty** is the time of sexual maturity, when reproduction is possible. Puberty is brought on mainly by the hormone testosterone in males and by the hormone estrogen in females. Both of these hormones also cause the development of **secondary sexual characteristics.** These include the growth of body hair, beard, and a lowering of the voice in males. Some body hair also grows on the female. However, the most important changes in the female are the development of breasts and a widening of the hips. Both of these changes prepare the female for childbirth. Puberty for the female is also when menstruation begins.

Adolescence brings changes in body form. The face of a man is not the same as the face of a boy. And the face of a woman is not the same as the face of a girl. The changes that occur in facial characteristics take place during adolescence. In general, young faces are rounded. Mature faces have features that are more angular, or sharper. Usually there are changes in the forehead, nose, and chin. They generally become more prominent. There are also little shifts in muscle and fatty tissue

puberty:
the time of sexual maturity in humans, when reproduction first becomes possible

secondary sexual characteristic:
a developmental change that appears at puberty

under the skin. These changes all help create the "adult" look. (See Figure 14.8.)

The rest of the body also undergoes changes in form at different times, so that most of the time the body form is out of proportion. The feet start growing first, then the legs, arms, hips, chest, and shoulders. The trunk, which is between the hips and chest, is often the last part of the body to grow.

Adolescents usually have a muscle problem. Adolescents usually have a problem with their muscles. By the time a young person is 15 or 16, he or she usually has adult-size sets of bones and muscles. With these larger muscles, the adolescent experiences a gain in strength. But the young person does not gain nearly as much strength as will be gained later. This lag in strength creates a problem if the adolescent tries to perform physical work like an adult. It also creates a problem if an adult expects an adolescent to "live up to her or his size." Either way, the teenager can end up with muscle tears, strains, or even a **hernia**. A hernia is a tear in one or more of the abdominal muscles—the muscles between the chest and the legs.

hernia:
a tear in one or more of the abdominal muscles

Adolescents do not grow proportionately. Teenagers are often very sensitive about their appearance. This can create problems if they do not know what to expect during adolescence. For example, it was mentioned earlier that the feet are the first part of the body to start growing rapidly. Imagine how some teenagers feel when they notice that their feet have grown way out of proportion. Others worry because their legs have grown long and out of proportion to their trunk. Adolescence is truly an "awkward age," and there is no way to change the clock that is creating the problems.

The slow developer has a special problem. The slow developer is the one who usually suffers the most during adolescence. And the sufferer is much more likely to be a boy than a girl. The problem is worse if the boy does not understand that he must live with a "slow clock."

Albert Einstein and Winston Churchill are two famous examples of slow developers. Einstein, who was one of the world's greatest geniuses, was quite old before he started talking. His parents thought he was retarded. Churchill, who was a great statesman and writer, was always much smaller than his classmates. His teachers thought that he was dull. He even failed several grades when he was in school. Both men leaped way ahead of their age group when they reached their late teens and early twenties. Thousands of outstanding persons have had similar experiences.

Human Body Functions

Laboratory Activity

How Do They Grow?

PURPOSE Young people grow. But they do not grow at the same rates. Age makes a difference. The sex of the individual makes a difference. Each person has his or her own "internal clock." That makes a difference, too. In this investigation, you will look at growth patterns and see how they are influenced by a person's age and sex.

MATERIALS
string
tape measure or meter stick
graph paper
scissors

PROCEDURE

1. Measure the heights of five persons, 18 years of age and under. Record their ages, sexes, and heights. You will pool your data with your classmates. Be sure to include as many different age groups as possible within the class data. Be sure that two people in the class do not measure the same person. If you use a tape measure marked in inches, convert the measurements to centimeters. (To convert, multiply by the factor 2.5 cm/in.)
2. It may not be convenient for you to carry a ruler. Then here is one way you can measure a person's height. Carry with you a few meters of string and some folded pieces of masking tape. When you are ready to measure someone, ask the person to stand shoeless, back to the wall. Use a pencil eraser to mark the wall at the level of the top of the person's head. Stretch a piece of string between the floor and the mark on the wall. (You may have to tape one end of the string.) Cut the string. Prepare a label with the name, age, and sex of the person you have measured. Attach the label to the string with masking tape. Bring each of the strings back to the class and measure the lengths.
3. If you measure the height of a baby, be sure to measure when the baby's legs are extended.
4. Tabulate the data from all students on classroom charts similar to those shown. You probably want to copy the classroom charts in your notebook, too.

A.

HEIGHT IN CENTIMETERS

MALE LIST HEIGHTS	AGE 0-1	1	2	3	4	5	6	7	8	9	10	11	12	13	14	15	16	17	18

252

Growth and Development

B.

FEMALE LIST HEIGHTS

	HEIGHT IN CENTIMETERS																		
AGE 0-1	1	2	3	4	5	6	7	8	9	10	11	12	13	14	15	16	17	18

5. Average the heights recorded in each box. Prepare a chart (as shown) to show the average heights, for each age and both sexes, of all the people who were measured by your class.
6. Draw a graph to show the class results. Use the same graph to show results for females and males.

C.

	AVERAGE HEIGHT IN CENTIMETERS																		
AGE	0-1	1	2	3	4	5	6	7	8	9	10	11	12	13	14	15	16	17	18
MALE																			
FEMALE																			

QUESTIONS

Using the graph, answer the following questions:
1. At what ages are boys taller than girls?
2. When are girls taller than boys?
3. When are boys and girls the same height?
4. During what years does the growth rate slow down for both sexes?

Human Body Functions

Figure 14.9: Albert Einstein (left) and Winston Churchill (right) were slow developers. Both of them caught up with and surpassed their age group when they were in their late teens and early twenties. Both men soon became famous —and helped change the history of the world.

The biggest problem for the slow developer is relations with other people. Teenagers sometimes become so unhappy with themselves that they begin showing off to get needed attention. Because they seem less grown-up and less good-looking, they may feel left out when among their friends. Typically, they are bossy, rebellious, or aggressive. They feel the need to be liked by friends and family. Yet they often do things that cause them to be disliked. This creates unhappiness. It also leads to conflict with friends, family, teachers, and society.

LESSON REVIEW *(Think. There may be more than one answer.)*

1. A 7-year-old girl will often be
    a. taller than a boy of the same age.
    b. shorter than a boy of the same age.
    c. about the same height as a boy of the same age.
    d. taller or shorter than a boy of the same age.

2. A 12-year-old girl will often
    a. be taller than a 12-year-old boy.
    b. be shorter than a 12-year-old boy.
    c. have a body that is proportional.
    d. have a body that is not proportional because some parts are more developed than others.

3. During adolescence,
    a. the growth rate stops.
    b. the growth rate slows down.
    c. females begin to menstruate.
    d. secondary sex characteristics begin to develop.
4. Generally, during the adolescent period,
    a. girls' feet and legs often grow faster than the rest of their body.
    b. all girls and boys grow at the same time and at the same rate.
    c. slow developers will be slow at other things throughout their lives.
    d. slow developers may try to prove their maturity by attention-getting devices.
5. Which of the following terms best matches each phrase?
    a. adolescence   c. secondary sexual characteristic
    b. puberty   d. hernia

    _____ time of sexual maturity

    _____ time of most rapid development after birth

    _____ body hair

    _____ a tear in an abdominal muscle

KEY WORDS

adolescence (p. 249)   secondary sexual characteristic
puberty (p. 250)   (p. 250)
                                hernia (p. 251)

Applying What You Have Learned

1. Describe three ways in which growth occurs.
2. How is the growth of a brain cell similar to the growth of a muscle cell?
3. Can cells shrink as well as grow? Explain.
4. List three things that help determine your growth rate.
5. How is bone different from cartilage?
6. Giants grow to large sizes because of too much growth hormone. What do you think would happen if a person did not receive enough growth hormone?
7. Why is adolescence called an "awkward age"?
8. Consider the food you have eaten today. List some of the protein foods you have had.
9. Do other living things go through periods of growth and development before reaching maturity? Give an example.
10. What is one of the problems that a slow developer may have to overcome?

Unit 5

Challenges to Human Survival

A little bug fell in a pie.
"'Tis a pity," said a fly
who was strolling by.
"But who will know?
You're not too big,
and we all must die."

"I know that's true,"
said the bug to the fly.
"But a pie isn't the place
for me to die.
Bail me out
and let me live like a man,
with a cigarette cough
and Nikoban.
Where I can walk in smog
or on oil in the sea—
and can kill myself
with dignity."

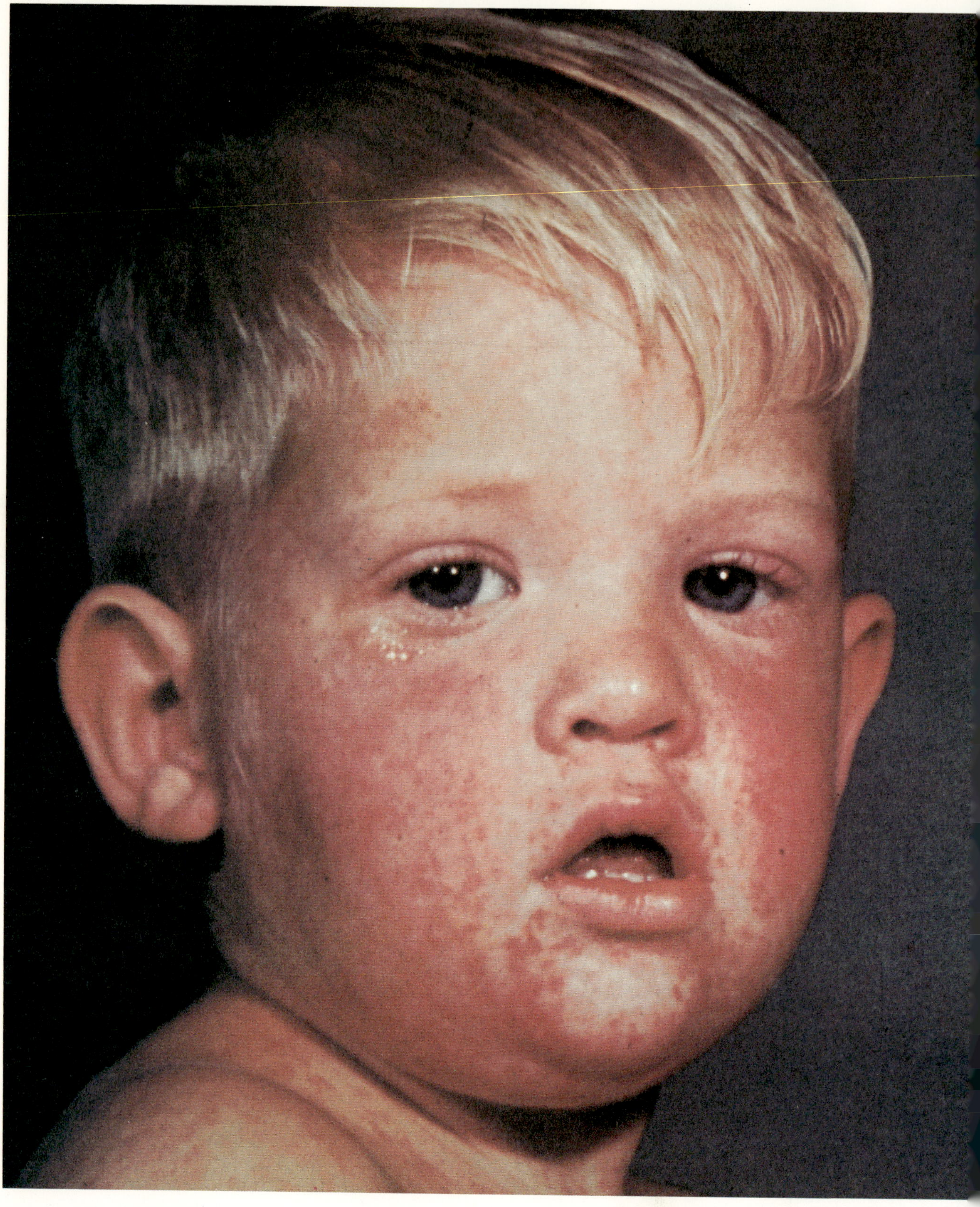

Figure 15.1: Measles was once a common childhood disease in the United States. As you will learn later in this chapter, a way has been found to prevent this disease.

Chapter 15

DISEASE

Disease is one of your enemies. This enemy can make your life unpleasant. It can also cripple you. Sooner or later, a disease of some kind will probably cause your death. Every day of your life, you must use various weapons to combat this enemy. What weapons? Knowledge is a key weapon. Correct action is another. You will learn how to use both when you finish this chapter.

WHAT CAUSES DISEASE?

Foreign invaders cause disease. Many organisms in your environment can cause you to have a disease. Most of these organisms are small or microscopic in size. Some are bacteria.

Bacteria are microscopic single-celled organisms. Most bacteria are harmless. Most of them function as decomposers

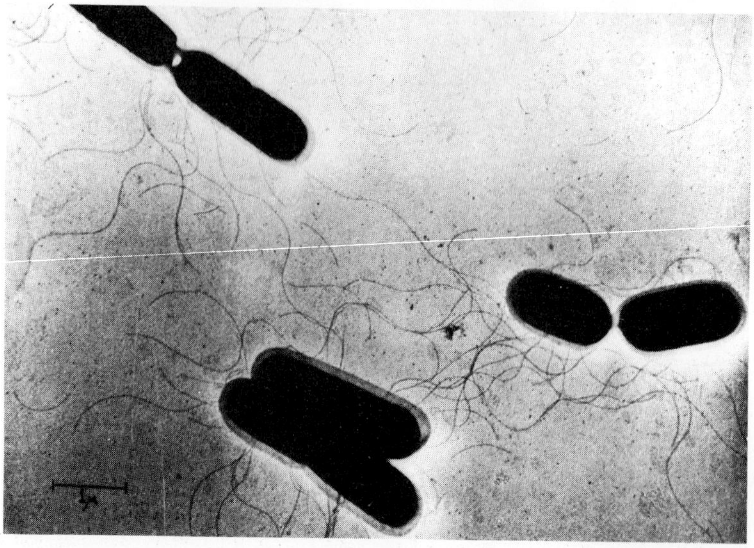

Challenges to Human Survival

Figure 15.2: These are Clostridium tetani bacteria, the organisms that can cause tetanus, or "lockjaw," in a deep wound. Such injuries should be seen by a doctor, for tetanus can be fatal if not treated.

in our ecosystems. But a few are capable of causing disease in humans. To do so, they must get inside the body, and conditions must be right for them to multiply.

Clostridium tetani (klŏs trĭd′ē əm tĕt′nī) is an example of a bacterium that can cause human disease. (Do not try to memorize its name.) It causes *tetanus*, which is sometimes called "lockjaw." The soil contains large numbers of these bacteria. Any kind of a wound that gets dirt in it is also likely to have these bacteria. Will they cause tetanus, a disease that is often fatal? Maybe. Much depends upon the wound. The tetanus bacteria multiply only in the *absence of oxygen*. Deep wounds or puncture wounds, such as might be caused by a nail, knife, or bullet, are the most dangerous. Tetanus bacteria are not likely to multiply in shallow wounds where they are closer to the air. Of course, there are other disease-causing bacteria that do multiply in shallow wounds!

fungi: simple, plantlike organisms that lack chloroplasts. The singular is fungus (fŭng′əs).

Fungi cause ringworm and athlete's foot.

The **fungi** (fŭn′jī) are common organisms in terrestrial ecosystems. Molds and mushrooms are familiar examples. Many fungi are microscopic in size. Most of them function as decomposers. A few cause human disease. Such diseases are generally not as common, or as dangerous, as those caused by bacteria. You will learn more about fungi in Chapter 19.

Ringworm is a common fungus disease. *Athlete's foot* is another. Both are diseases of the skin. Both are caused by microscopic molds. With ringworm, the mold grows in a circular pattern, causing an irritation of the skin. This disease is common in children, and it usually affects the scalp. Athlete's foot is more common in adults. The mold usually grows in the moist areas between the toes. Neither of the two diseases is considered serious.

Viruses cause many common diseases. Viruses are not true organisms. You will learn more about their characteristics in Chapter 18. Right now, there are two things that you should know about viruses. First, they are very, very small—much smaller than bacteria. Second, they have the ability to reproduce themselves when they are inside living cells.

The ability to reproduce themselves inside living cells is the reason viruses cause disease. When they reproduce inside a cell, they take their building blocks from the cell. This is like burning all the wood in a house in order to keep warm. The house ends up being destroyed. So does the cell in which the viruses are multiplying. Eventually the cell will break open and a hundred or more viruses will spill out. These viruses will invade other cells and continue to multiply. Each cell in which they multiply ends up as a dead cell. A virus disease results when a group of cells in a tissue is destroyed by multiplying viruses.

What are some virus diseases? There are many. Some of the common ones include mumps, measles, chickenpox, influenza ("flu"), and the common cold.

Have you ever had fever sores? These are caused by a type of virus called a **latent virus.** "Latent" means that it "acts, or is capable of acting later." The latent virus that causes fever sores is common in the cells of many people. But it is usually inactive. This means that it is not reproducing and destroying cells—and thus not causing the sores. Under certain conditions, the

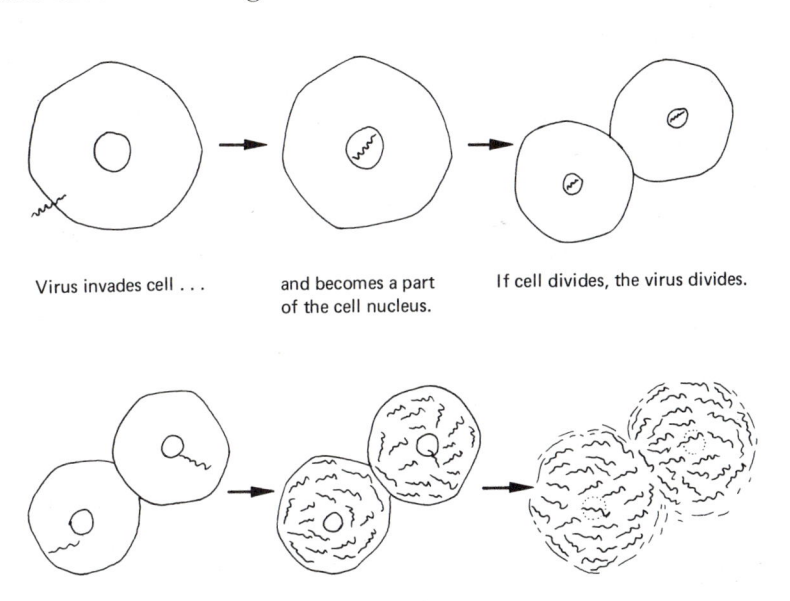

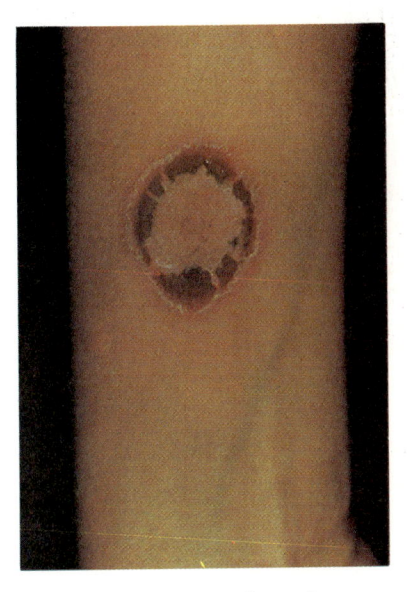

Figure 15.3: "Ringworm" got its name from the typical pattern of the skin irritation seen in this common fungus disease.

latent virus:
an inactive virus in a cell. It may at some other time begin to reproduce and destroy the cell.

Virus invades cell . . .   and becomes a part of the cell nucleus.   If cell divides, the virus divides.

Stress—virus moves out into cytoplasm . . .   reproduces more of itself . . .   and the cell is destroyed.

Figure 15.4: A latent virus can remain inside a cell without doing any damage until something causes it to become active. Then the virus takes over the cell and reproduces.

parasite:
an organism that lives at the expense of another organism

viruses can become active. They can begin to multiply and destroy cells. What conditions can cause this? Life scientists are not certain about all of them. Stress conditions are usually said to be the cause. Another disease, such as a cold, can trigger the viruses into action. A nervous condition may also cause them to start multiplying.

Worm parasite diseases are declining. There are several kinds of worm **parasites** that can multiply and cause damage in people. Some of them, such as the roundworm and hookworm, used to be fairly common in warm, moist regions of the United States. Eggs of the roundworm were common in the soil. They gained entry to the body with dirt taken into the mouth. In these same areas, tiny hookworms were often in moist soil. People in such areas used to go barefoot more than they do now. The worms would enter their bodies by boring through the bottoms of their feet. The wearing of shoes, and cleaner health habits, have helped cut down on the disease caused by these two worms.

The *trichina* (trĭ kī′nə) *worm* can also cause disease. This worm lives in the muscles of various animals. Pigs sometimes get infected with this worm because of their eating habits. If a person eats undercooked pork from an infected pig, he or she too can become infected. The safest preventive measure is to cook all pork thoroughly. Thorough cooking kills the worm.

Swimmer's itch is a bothersome disease in some areas. It is caused by a small worm. This worm lives part of its life in the body of a snail. It then leaves the snail and migrates through the water of a pond or lake. When a person swims in such a body of water, the worms attach themselves to the skin. Then they

Figure 15.5: These worm parasites can cause disease. This enlarged photo shows Ascaris (ăs′kə rəs), a roundworm that is common in soil. (Courtesy Carolina Biological Supply Co.)

try to bore through the skin and enter the body. They cannot do it. But in the process of trying, they cause an uncomfortable skin irritation.

Some diseases are inherited. Up to now, we have discussed diseases caused by an agent that invades the body. A **genetic disease** is another completely different type of disease. Genetic diseases are caused by faulty genes. As you have learned, genes contain information. This information allows the body to build all necessary proteins. A faulty gene is one that causes a faulty protein, or no protein at all, to be built. A faulty protein, or no protein, will cause an abnormal structure or function—a disease. Because there are so many different genes and proteins, there are also many possible genetic diseases. Some genetic diseases can be controlled (but not cured) with special diets or medicines.

Hemophilia was one of the first genetic diseases to be understood by life scientists. Individuals with this disease are called "bleeders." They lack the proper proteins to cause their blood to clot. After an injury, their blood just keeps flowing. Such people must always avoid activities that are likely to cause injury. Bleeding caused by injury can often be stopped by putting in another person's blood that has the proper proteins.

Improper diet causes deficiency diseases. A proper diet must include a wide variety of nutrients. Among these are carbohydrates, fats, minerals, vitamins, and protein. The lack of any one of these can cause a **deficiency disease.**

Carbohydrates (sugars and starches) and fats are seldom in short supply, and almost everybody gets enough of them. Minerals, even in trace amounts, are needed for a variety of purposes. Generally, they are not very likely to be missing in one's food. The lack of vitamins or proteins is usually the problem in a deficiency disease.

Table 15.1 lists the vitamins needed in the human diet. Also in the table are the foods that contain the vitamins, and the deficiency diseases that can result from a lack of them. In countries where people are educated to the need of vitamins, such diseases are fairly rare. (See Figure 15.8.)

Protein deficiency disease is tragic. Protein deficiency disease is a worldwide problem, and it is getting worse! Most foods that are rich in protein are expensive. Such foods are also in short supply in many parts of the world. Poor people, wherever they live, are most likely to suffer from some degree of protein deficiency disease. Most of the people in our world are poor people.

genetic disease: any of several diseases, caused by a faulty gene, that can be inherited

deficiency disease: any of several diseases caused by a lack of an essential nutrient

VITAMIN FUNCTIONS, SOURCES, AND DEFICIENCY DISEASES

Vitamin	Needed for	Found in	Deficiency disease or condition
Vitamin A	Vision in dim light Growth Healthy mucous membranes	Liver, kidney, egg yolk Butter, margarine, cheese Yellow fruit (cantaloupe) Green and yellow vegetables	Night blindness Poor growth
Vitamin B_1 (thiamine)	Normal digestion Healthy nervous system Growth	Fish and poultry Meat and eggs Beans and peas (soybeans) Whole-grain cereals Enriched bread	Beriberi (incomplete digestion, nervous disorders) Loss of weight Poor growth
Vitamin B_2 (riboflavin)	Healthy skin and mouth Oxygen use by cells Eye functioning Growth	Meat, liver, and kidney Milk and eggs Green vegetables, soybeans Fish and poultry Whole-grain cereals	Tongue inflammation Premature aging Dim vision, oversensitivity to bright light Poor growth
Vitamin B_6 (pyridoxine)	Healthy skin Growth	Whole-grain cereals, nuts Meat, liver, egg yolk	Skin inflammations Poor growth
Niacin (nicotinic acid)	Healthy skin and nervous system Functioning of digestive system	Meat and liver Fish, poultry, milk Peanut butter Whole-grain cereals Tomatoes and potatoes	Pellagra (mental disorders, digestive problems, sore tongue, skin inflammations)
Vitamin B_{12}	Formation of red blood cells	Liver and meat Green vegetables	Pernicious anemia (reduced number of red blood cells)
Vitamin C (ascorbic acid)	Healing of wounds, broken bones, and bruises Strong blood vessels Healthy gums	Citrus fruit (orange, lemon, grapefruit) Other fruit Leafy vegetables Tomatoes, green peppers	Scurvy (tendency to bruise easily, bleeding in joints, failure of wounds to heal) Sore gums
Vitamin D	Use of calcium in building bones and teeth Growth	(Produced in skin when exposed to sunlight) Milk with vitamin D added Eggs and liver	Rickets (soft bones which cause bowlegs; poor tooth development)
Vitamin E (tocopherol)	Healthy reproductive system	Leafy vegetables, seed oils Milk and butter	Not determined
Vitamin K	Normal blood clotting	(Produced in digestive tract) Green vegetables, tomatoes	Hemorrhaging (excessive bleeding)

Figure 15.6: Vitamins are essential parts of the human diet. A shortage of one or more vitamins can cause one of the deficiency diseases.

How does protein deficiency disease affect the body? In many ways, most of which are very bad. All structures and functions of the body depend upon the presence of proteins. When proteins are not in the diet, they cannot be made by the body. Figure 15.7 shows one visible result of protein deficiency disease. Such children are not fat. They have potbellies because

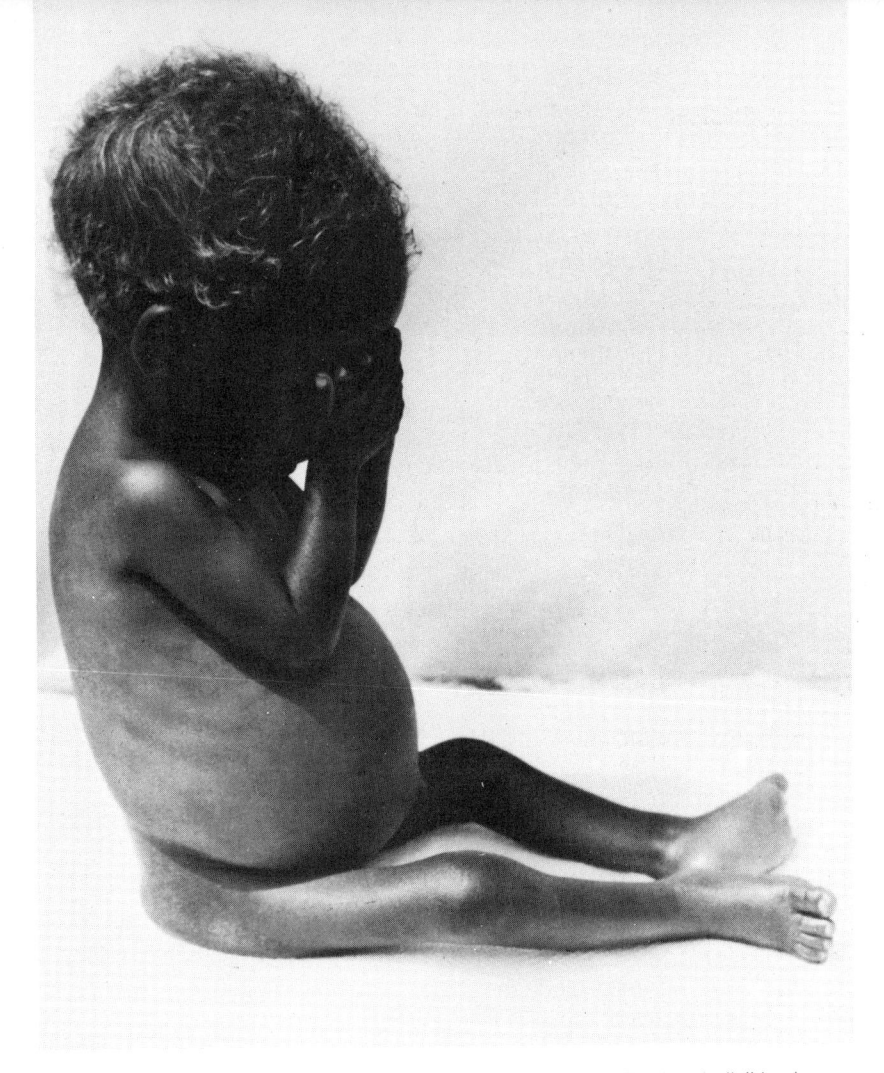

Figure 15.7: This child is suffering from protein deficiency. The "potbelly" is characteristic of this disease. Many children in the underdeveloped areas of the world are victims of this disease.

their muscles have not developed properly. The swelling is caused by the intestines pushing out against thin muscles.

Probably the most tragic effect of protein deficiency disease is brain damage. In the last chapter, you learned that brain cells are all formed before birth. They do not multiply after that time. The pregnant mother must furnish the proteins necessary for the proper number of brain cells to form. A mother without enough protein in her diet can cause her child to be born with fewer than normal brain cells. Then, if the newborn child also lacks protein in the diet, the brain cells do not develop properly. The result is a child with an improperly developed brain and a lower than normal intelligence. The tragedy is that there seems to be no way of correcting the problem after the child reaches about 4 years of age.

Figure 15.8: This girl in Indonesia has rickets, a deficiency disease caused by a lack of vitamin D. This bowlegged condition is now rare in the United States, largely because vitamin D is added to milk.

Challenges to Human Survival

LESSON REVIEW *(Think. There may be more than one answer.)*

1. Most bacteria
    a. cause disease.
    b. are harmless.
    c. are decomposers in ecosystems.
    d. cause tetanus, or "lockjaw."

2. Most fungi
    a. cause disease.
    b. are harmless.
    c. are decomposers in ecosystems.
    d. cause ringworm or athlete's foot.

3. In the diagram, what is most likely being depicted?
    a. A fungus has caused ringworm.
    b. Bacteria have multiplied outside a cell.
    c. A virus has multiplied.
    d. A trichina worm has invaded a cell.

4. Worm parasite diseases
    a. are increasing in the United States.
    b. are decreasing in the United States.
    c. can be caused when the parasites enter the mouth.
    d. can be caused when the parasites bore through the skin.

5. Foods containing protein
    a. are the most expensive foods.
    b. are the world's most abundant foods.
    c. are primarily needed for energy.
    d. are needed for proper brain development in developing babies and young children.

6. Which of the following terms best matches each phrase?
    a. fungi   d. genetic disease
    b. latent virus   e. deficiency disease
    c. parasite

    _____ an organism that lives at the expense of another
    _____ any disease that is inherited
    _____ virus that is inside a cell but is not reproducing
    _____ are mostly decomposers
    _____ disease caused by lack of nutrients

Disease

KEY WORDS

fungi (p. 260)
latent virus (p. 261)
parasite (p. 262)

genetic disease (p. 263)
deficiency disease (p. 263)

HOW ARE DISEASES TREATED AND PREVENTED?

The body combats invaders in two ways. Most foreign organisms that invade your body are doomed to die. That is because your body has two very good ways of destroying them. They are destroyed by phagocytes (făg'ə sīt) and antibodies.

The term **phagocyte** means "eating cell." This name has been given to some of the white blood cells that circulate in your bloodstream. They actually do eat bacteria and other small invaders. Somehow phagocytes are attracted to any site where bacteria are multiplying. Your body also responds by making many more phagocytes when there is an infection. This fact will help a doctor know if you have an infection. He can take a sample of your blood and count the white blood cells. The count may be higher than normal. If so, you probably have an infection and your body is trying to combat it.

As you learned in Chapter 9, *antibodies* are special proteins that circulate in your blood. Somehow they destroy invaders such as bacteria and viruses. Life scientists know how some of them function. The method of action of others is a complete mystery.

Antibiotics kill bacteria. Many organisms produce chemicals that kill other organisms, or at least stop them from growing. People have made good use of this fact. We have put

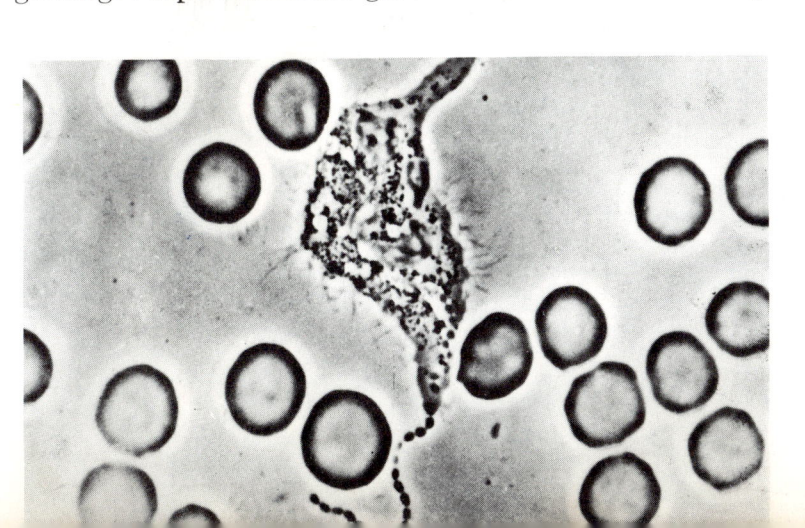

phagocyte:
a type of white blood cell that attacks and digests some disease-causing organisms

Figure 15.9: This white blood cell is attacking and eating invading bacteria. Such "eating cells," or phagocytes, circulate constantly in your bloodstream.

Challenges to Human Survival

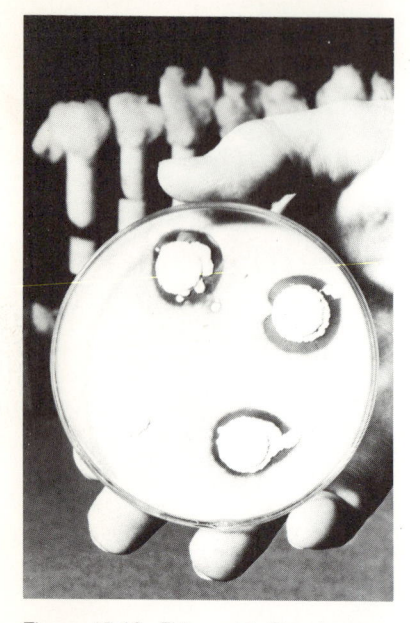

Figure 15.10: This mold, Streptomyces rimosus (strĕp tō mī′ sēs rə mō′ səs), is the source of an antibiotic. The clear space around the mold cultures shows the region where the mold has killed off the bacteria.

some of these chemicals to work for us. They are called antibiotics.

Penicillin is an example of an antibiotic. It is a chemical produced by a mold. Penicillin kills certain kinds of bacteria. It will kill these bacteria inside your body. And it will not destroy any of your own cells in the process. This fact is important. There are many antibiotics in nature. But there are few that will kill disease organisms without causing other damage in your body.

Antibiotics are most effective against bacteria. They have very little if any effect on viruses and fungi. They have no effect on worm parasites. We badly need an effective virus antibiotic. Maybe you can help find one someday.

Surgery cures a variety of diseases. Surgery is the process of changing the body by some physical method. It is usually used to cure a disease. (Some people use surgery to have a physical imperfection repaired, or to repair the scars of an accident; others use it to make themselves more attractive.) A surgeon will remove an infected *appendix*. That is a way to cure *appendicitis*, an infection of that organ. Some kinds of disease may be cured when a surgeon replaces a diseased heart

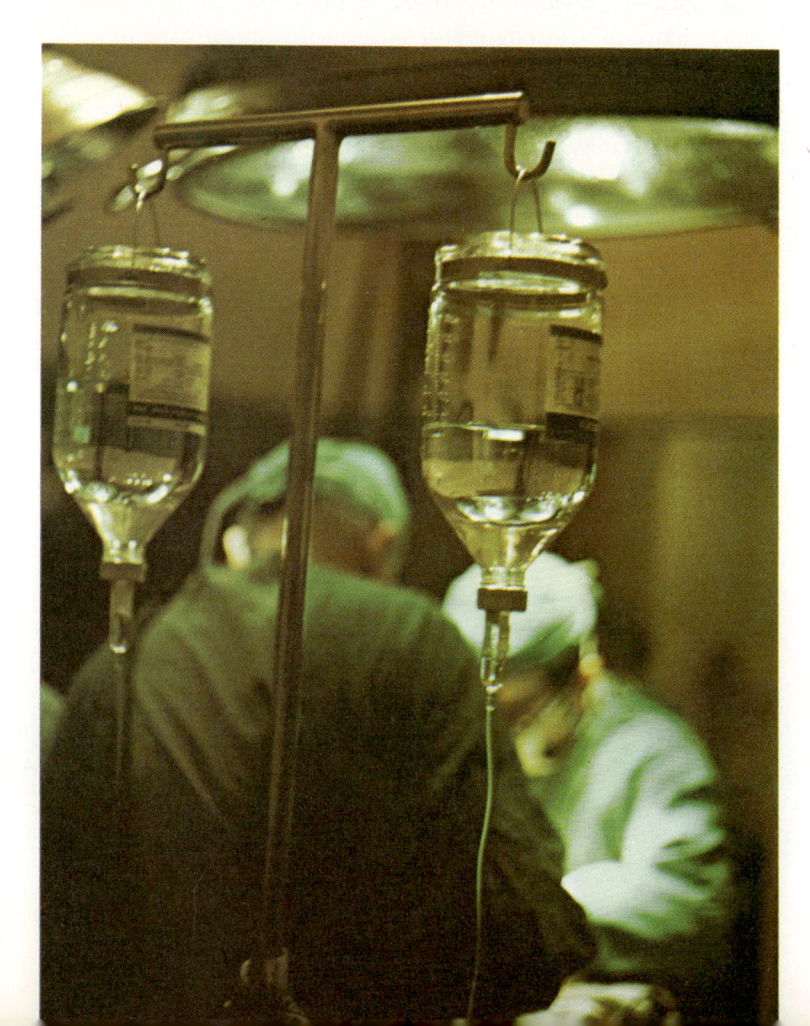

Figure 15.11: Modern surgery is able to cure or correct many diseases. Medical scientists are doing research in many areas of corrective surgery. The transplantation of entire organs or the implantation of artificial ones is one such area of research.

Figure 15.12: Measles has become almost unknown as a childhood disease in the United States since the development of a vaccine that creates immunity against the disease. Mass vaccination of children has been found an effective way to protect against measles.

valve with an artificial one. The surgical transplanting of kidneys has saved the lives of thousands of people with incurable kidney disease.

A disease may create immunity. If you have had diseases like measles and mumps, you are not likely to have them again. Here is the reason why. When you had the disease, your body manufactured antibodies to fight it. Eventually the antibodies destroyed the invaders, and that ended the disease. Now those same antibodies are still in your bloodstream—or at least some of them are. If the viruses that cause measles or mumps enter your body again, now they will not have a chance to multiply. The antibodies in your bloodstream will destroy them. If this is the case, you are said to have an **immunity** to these diseases.

How long will immunity last after you have had a disease? This depends upon the disease. Diseases such as measles and chickenpox usually create a lasting immunity. That is, you will probably be immune for the rest of your life. For other diseases, the immunity may last only a short time. Influenza is an example of such a disease. For other diseases such as the common cold, there is little if any immunity after the disease. Life scientists do not know why some antibodies remain in the body and protect it while others do not.

We can create immunity. Let us suppose we are going to do an experiment. We will take a tube of viruses that cause some human disease and we will kill them. Then we will inject the dead viruses into some person. Could we fool that

immunity: protection against a disease by antibodies produced in the blood in reaction to the disease or by immunization

person's antibody-making machinery? Could we make that machinery act as if live viruses had invaded the body? If so, that person would start making antibodies. After a while, the individual would have enough antibodies in the blood to destroy live viruses if they entered the body. We could make that person immune to the disease.

The preceding experiment has been done many times, and with good results. We have been very successful in fooling the human body's antibody-making machinery. The usual method for doing this is to inject dead viruses—as in our experiment. That stimulates the individual to produce antibodies, which remain and give immunity. If you are immune to whooping cough, polio, or measles, that is probably the way you became immune. Any process in which we create immunity to a disease is called **immunization.**

Another way to create immunity is to inject a live virus that is similar to the disease virus. This live virus causes only a mild reaction in the body. But the antibody-making machinery is fooled. Antibodies against the disease virus are made. You are probably immune to the disease called smallpox. If so, this immunity was created by injecting live cowpox viruses into your arm or leg. (Look for the scar caused by the cowpox viruses.) The cowpox virus is like the smallpox virus, and it causes the body's antibody-making machinery to go into action. The antibodies that result will destroy the smallpox virus if it invades your body later.

We can also create immunity to diseases that are caused by bacteria. We do it in the same way by fooling the body's antibody-making machinery.

immunization: the process of causing the body to produce antibodies by injecting it with dead or weakened disease-causing viruses or bacteria. Sometimes immunization involves the injection of antibodies from another animal.

Figure 15.13: This laboratory technician is injecting fertilized chicken eggs with flu virus. After the virus reproduces—at the expense of the chicken embryo—it is withdrawn from the egg and processed into flu vaccine.

Laboratory Activity

Discovery: How Life Scientists Can Create Immunity to Disease

INTRODUCTION

Less than 100 years ago, life scientists disagreed strongly about the causes of disease. Many life scientists were then studying the strange creatures (many were bacteria) that the microscope revealed. Some of them suspected, and later proved, that diseases were caused by those tiny microorganisms. They were often called "microbes," and disease-causing microorganisms came to be called "germs." Doctors could prevent smallpox—by using a vaccine (immunization) to infect a person with cowpox virus—but did not know how this worked. (It was a similar virus creating immunity. But nobody had ever seen a virus.) Louis Pasteur (1822–1895) was investigating microbes that cause disease. His discovery of the way to immunize was part accident and part genius. The exact events are uncertain, but Paul de Kruif reconstructed them in the following excerpt from the book *Microbe Hunters* (Harbrace Paperbound Library, Harcourt, Brace & World, New York, 1953).

A READING

In 1880, Pasteur was playing with the very tiny microbe that kills chickens with a malady known as chicken cholera. . . . Pasteur was the first microbe hunter to grow it pure, in a soup that he cooked for it from chicken meat. And after he had watched these dancing points multiply into millions in a few hours, he let fall the smallest part of a drop of this bug-swarming broth onto a crumb of bread—and fed this bread to a chicken. In a few hours the unfortunate beast stopped clucking and refused to eat, her feathers ruffled until she looked like a fluffy ball, and the next day Pasteur came in to find the bird tottering, its eyes shut in a kind of invincible drowsiness that turned quickly into death.

Challenges to Human Survival

Roux and Chamberland nursed these terrible wee microbes along carefully; day after day they dipped a clean platinum needle into a bottle of chicken broth that teemed with germs and then carefully shook the same still-wet needle into a fresh flask of soup that held no microbe at all—so day after day these transplantations went on—always with new myriads of germs growing from the few that had come in on the moistened needle. The benches of the laboratory became cluttered with abandoned cultures, some of them weeks old. "We'll have to clean this mess up to-morrow," thought Pasteur.

Then the god of good accidents whispered in his ear, and Pasteur said to Roux: "We know the chicken cholera microbes are still alive in this bottle . . . they're several weeks old, it is true . . . but just try shooting a few drops of this old cultivation into some chickens. . . ."

Roux followed these directions and the chickens promptly got sick, turned drowsy, lost their customary lively frivolousness. But next morning, when Pasteur came into the laboratory looking for these birds, to put them on the post-mortem board—he was sure they would be dead—he found them perfectly happy and gay!

"This is strange," pondered Pasteur, "always before this the microbes from our cultivations have killed twenty chickens out of twenty. . . ." But the time for discovery was not yet, and next day, after these strangely recovered chickens had been put in charge of the caretaker, Pasteur and his family and Roux and Chamberland went off on their summer vacations. They forgot about those birds. . . .

But at last one day Pasteur told the laboratory servant: "Bring up some healthy birds, new chickens, and get them ready for inoculation."

"But we only have a couple of unused chickens left, Mr. Pasteur—remember, you used the last ones before you went away—you injected the old cultures into them, and they got sick but didn't die?"

. . . "Well, all right, bring up what new chickens you have left—and let's have a couple of those used ones too—the ones that had the cholera but got better. . . ."

The squawking birds were brought up. The assistant shot the soup with its myriads of germs into the breast muscles of the chickens—into the new ones, *and into the ones that had got better!* Roux and Chamberland came into the laboratory next morning—Pasteur was always there an hour or so ahead of them—they heard the muffled voice of their master shouting to them from the animal room below stairs: "Roux, Chamberland, come down here—hurry!"

They found him pacing up and down before the chicken cages. "Look!" said Pasteur. "The new birds we shot yesterday—they're dead all right, as they ought to be. . . . But now see these chickens that recovered after we shot them with the old cultures last month. . . . They got the same murderous dose yesterday—but look at them—they have resisted the virulent dose perfectly . . . they are gay . . . they are eating!"

Roux and Chamberland were puzzled for a moment.

Then Pasteur raved: "But don't you see what this means? Everything is found! Now I have found out how to make a beast a little sick—just a little sick so that he will get better, from a disease. . . . All we have to do is to let our virulent microbes grow old in their bottles . . . instead of planting them into new ones every day. . . . When the microbes age, they get tame . . . they give the chicken the disease . . . but only a little of it . . . and when she gets better she can stand all the vicious virulent microbes in the world. . . . This is our chance—this is my most remarkable discovery—this is a *vaccine* I've discovered, much more sure, more scientific than the one for smallpox where no one has seen the germ. . . . We'll apply this to anthrax too . . . to all virulent diseases. . . . We will save lives . . . !"

A lesser man than Pasteur might have done this same accidental experiment—for this was no test planned by the human brain—a lesser man might have done it and would have spent years trying to explain to himself the mystery of it, but Pasteur, stumbling on this chance protection of a couple of miserable chickens, saw at once a new way of guarding living things against virulent germs, of saving men from death.

. . . . Hurriedly Pasteur and Roux and Chamberland set out to confirm the first chance observation they had made. They let virulent chicken cholera microbes grow old in their bottles of broth; they inoculated these enfeebled bugs into dozens of healthy chickens—which promptly got sick, but as quickly recovered. Then triumphantly, a few days later, they watched these birds—these *vaccinated* chickens—tolerate murderous injections of millions of microbes, enough to kill a dozen new birds who were not immune.

So it was that Pasteur, ingeniously, turned microbes against themselves. He tamed them first, and then he strangely used them for wonderful protective weapons against the assaults of their own kind.

DISCUSSION

Pasteur had discovered a way to create immunity: Use "weakened" bacteria. (See his method diagrammed below.) This immunization method can be used with both bacteria and viruses. Pasteur used heat to weaken bacteria and develop a vaccine for *anthrax,* a disease of sheep and human beings. And he went on to develop a vaccine for *rabies,* though he never saw the "germ" (actually a virus)—also using heat to weaken the germ. Life scientists later used dead viruses and other techniques to create immunity. But the basic method Pasteur discovered has been used to develop immunizations against diseases that once killed millions of children and adults.

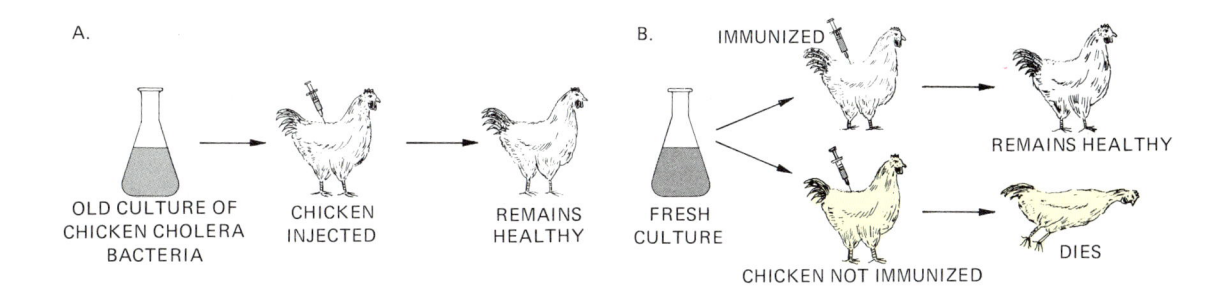

A.
OLD CULTURE OF CHICKEN CHOLERA BACTERIA → CHICKEN INJECTED → REMAINS HEALTHY

B.
FRESH CULTURE → IMMUNIZED → REMAINS HEALTHY
            → CHICKEN NOT IMMUNIZED → DIES

Challenges to Human Survival

Another organism's antibodies can be used. Sometimes it is not possible to inject live or dead microorganisms into a person's body in order to create immunity. If that is the case, life scientists have found another way to do it. They inject the microorganisms into an animal such as a horse. Then they take blood from the horse and remove the antibodies that the horse developed. Those antibodies may then be injected into a person. That will give the person immunity. However, this sort of immunity usually does not last as long as that produced by other methods. This method is used to give temporary immunity to such diseases as measles, mumps, diphtheria, and tetanus.

Killer number one is yet to be conquered. Millions of people are killed or crippled every year by diseases that cannot be cured. One such disease is **atherosclerosis** (ă thə rə sklə rō′sĭs). This is considered a type of heart disease. It kills more people in the United States than any other disease.

atherosclerosis: a disease of the arteries in which fats build up in the arterial walls

Atherosclerosis is a disease of the arteries, the elastic vessels that carry blood from the heart. Fat slowly builds up in the walls of the arteries. This fat gradually plugs up the openings inside the arteries. The heart has to work harder to push the blood through the smaller tubes. But a bigger problem may develop. The blood may form small clots when it is forced through cramped spaces. These clots can cause death or severe damage in the body. For example, a clot in an artery serving the brain causes a **stroke**. The stroke may result in all or part of the brain being destroyed. If any of the coronary arteries that serve the heart muscle are blocked, some of the heart muscle may die. When it does, a **coronary heart attack** occurs. (This is often simply called a "coronary.") If enough of the heart muscle is destroyed, the heart will stop beating and the individual will die.

stroke: the death of brain tissue served by an artery that is clogged by a blood clot. Paralysis, loss of speech, or death usually results.

coronary heart attack: the death of heart tissue served by an artery that is clogged by fat or a blood clot

Figure 15.14: This coronary artery (left) shows a large opening that allows the blood to flow freely. This artery (right) is plugged by fatty deposits and is almost closed down.

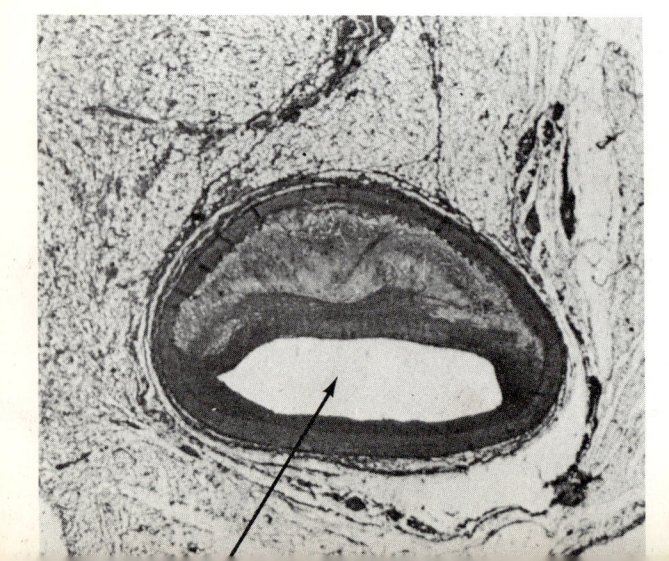

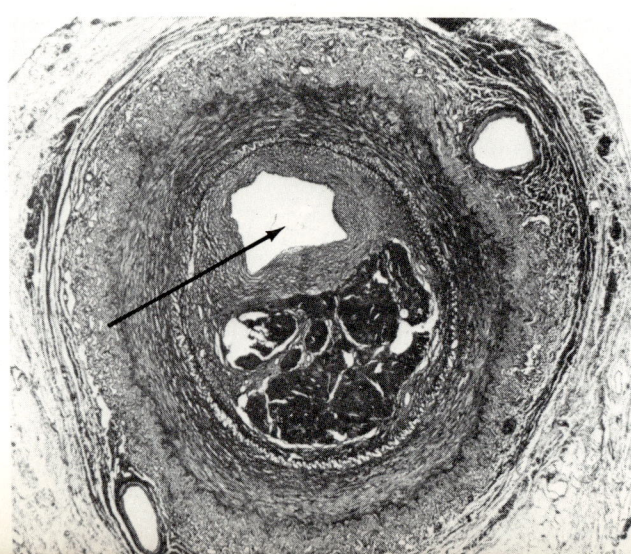

Disease

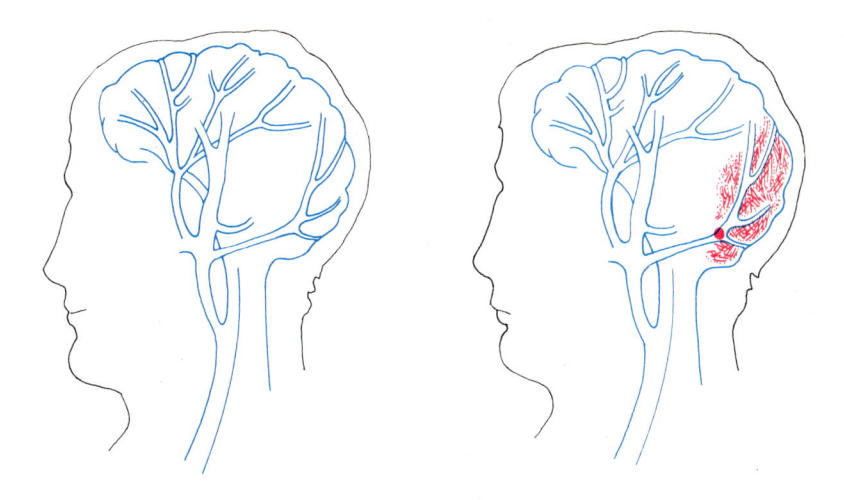

Figure 15.15: The picture shows a stroke. A clot plugs up an artery to part of the brain. That portion of the brain tissue dies and those functions it controls are destroyed.

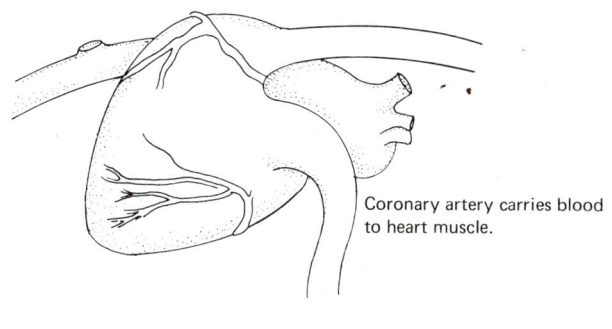

Coronary artery carries blood to heart muscle.

(a) NORMAL.

(b) "CORONARY." A branch of the coronary artery is plugged, causing the death of the heart muscle not being served by blood.

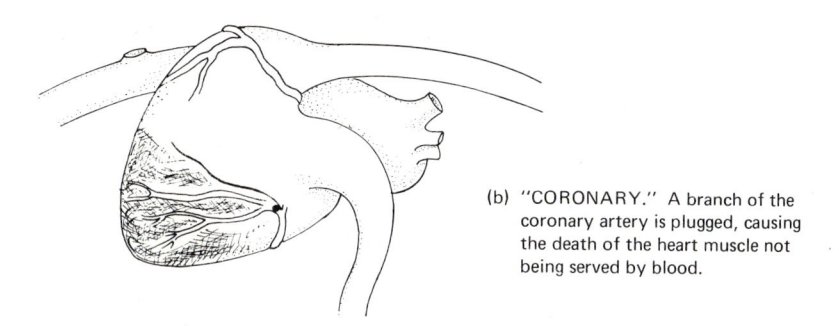

Figure 15.16: If the coronary artery is plugged, part of the heart muscle may die. If it dies, a person will suffer a coronary heart attack, which may result in death.

What causes atherosclerosis? Doctors and life scientists are not exactly sure why fat begins to build up in the arteries of a person. But they do know of five factors that seem

275

Challenges to Human Survival

to speed up the process. One factor is the diet of a person. A diet high in animal fats is likely to be dangerous over a long period of time. Overweight people are much more likely to die of atherosclerosis. A second factor is stress. People who are under pressure and who are "uptight" seem to be more likely to develop the disease. A third factor is lack of exercise. People who are active physically are less likely to have the disease than those who are not. The fourth factor is related to the sex of the individual. Males suffer from the disease more than females. Finally, a fifth factor is smoking, which doctors have not yet been able to explain. Heavy smokers are much more likely to die from coronary heart attacks than nonsmokers.

LESSON REVIEW *(Think. There may be more than one answer.)*

1. Foreign invaders in your body may be destroyed by
    a. certain types of white blood cells.
    b. certain types of red blood cells.
    c. phagocytes.
    d. antibodies.

2. Antibiotics
    a. are produced by organisms.
    b. are most effective against viruses.
    c. have little or no effect on bacteria.
    d. have little or no effect upon diseases such as mumps, measles, and colds.

3. If dead viruses are injected into your body,
    a. they can cause a disease.
    b. they may cause antibodies against that virus to be produced.
    c. they might give a person the ability to kill live viruses of the same kind if they enter the body.
    d. they might cause a person to be immune to a disease.

4. Live viruses are injected to create immunity for
    a. diphtheria.
    b. measles.
    c. cowpox.
    d. smallpox.

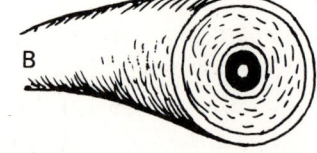

5. Figure A is a cross-sectional diagram of a normal artery. Figure B is probably
    a. the cross section of an artery from a patient with atherosclerosis.

b. the cross section of an artery almost plugged up with fatty deposits.
c. the cross section of an artery from a patient who died of a coronary heart attack or stroke.
d. the cross section of a vein.

6. Which of the following terms best matches each phrase?
   a. phagocyte   d. atherosclerosis
   b. immunity   e. stroke
   c. immunization   f. coronary heart attack

   _____ condition that occurs when some branch of the heart's artery becomes blocked

   _____ natural protection from a disease

   _____ condition that occurs when there is a blockage of blood vessels in the brain

   _____ cell that eats bacteria

   _____ process that creates immunity

   _____ disease caused by the buildup of fat in arteries

KEY WORDS

phagocyte (p. 267)
immunity (p. 269)
immunization (p. 270)

atherosclerosis (p. 274)
stroke (p. 274)
coronary heart attack (p. 274)

Applying What You Have Learned

1. Briefly describe three ways in which immunity to a disease may be developed.
2. What does immunization mean?
3. How are genetic diseases passed on to other people?
4. What are some results of severe protein deficiency?
5. Atherosclerosis has not always been the number one killer. What are some reasons why atherosclerosis is so common today?
6. Explain what happens to cause a stroke.
7. Explain what happens to cause a "coronary."
8. Describe how a single virus organism can eventually be the cause of a disease.
9. List one or more diseases for each of the following:
   a. fungus   c. bacteria   e. genetic diseases
   b. virus   d. diet deficiencies
10. Antibiotics are important in curing diseases caused by bacteria. What may happen if some of the bacteria treated with antibiotics survive?

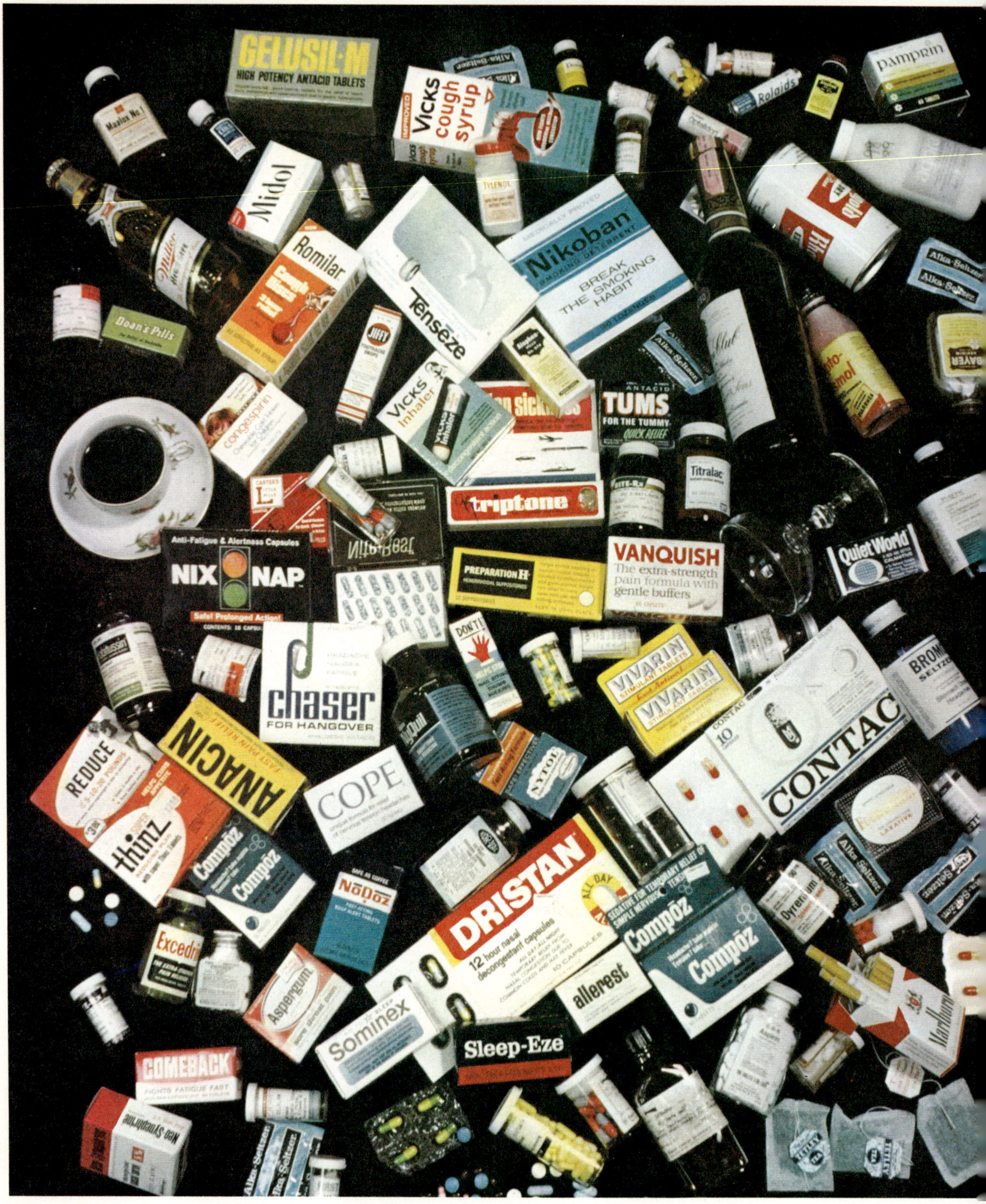
GELUSIL-M
HIGH POTENCY ANTACID TABLETS
Maalox No. 1
Vicks cough syrup
Midol
Rolaids
Pamprin
Tylenol
Romilar
Tenseze
Nikoban
BREAK THE SMOKING HABIT
Miller
Doan's Pills
Alka-Seltzer
Jiffy
Vicks Inhaler
TUMS
FOR THE TUMMY
QUICK RELIEF
BAYER
congespirin
Titralac
triptone
Anti-Fatigue & Alertness Capsules
NIX NAP
Safe! Prolonged Action!
PREPARATION H
VANQUISH
The extra-strength pain formula with gentle buffers
Quiet World
DON'T
chaser
FOR HANGOVER
VIVARIN
STIMULANT TABLETS
CONTAC
BROMO SELTZER
REDUCE
ANACIN
thinz
COPE
NYTOL
Compōz
NoDoz
DRISTAN
12 hour nasal decongestant capsules
ALL DAY
Excedrin
Aspergum
Sominex
allerest
Sleep-Eze
COMEBACK
FIGHTS FATIGUE FAST
Neo-Synephrine
Marlboro

Figure 16.1: These are some common drugs that people use. All of them can be purchased legally. Most of them can be purchased in "drugstores."

Chapter **16**

PEOPLE AGAINST THEMSELVES

Disease organisms are not the only enemies that threaten your survival. The Red Baron, the famous World War I flying ace, was not killed by a disease organism. Neither was Janis Joplin, who was a famous singer.

What did kill the Red Baron? Bullets. What did kill Janis Joplin? Drugs. Are bullets and drugs enemies of people—like disease organisms? Do they threaten your survival? You may not think so after you finish this chapter.

THE REAL ENEMY

Placebos tell us something about ourselves. Ask any doctor to tell you about **placebos** (plə sē′bō). She or he may frown for a minute. You may be asked how you know about them. But then you may see a smile come to the doctor's face. Most doctors are fascinated with the subject of placebos.

placebo:
a fake drug or treatment used as a control in medical experiments. Often a placebo brings about a "cure" because the patient thinks it is real.

Challenges to Human Survival

What are placebos? An example will help you understand them. Suppose six people come to a doctor's office. They are all complaining of a bad pain somewhere. Suppose the doctor gives three patients a pain-relieving drug. Then the other three patients are given a pill that looks and tastes like the pain-relieving drug. This fake drug would be a placebo. What results could the doctor expect?

It would be difficult to predict what would happen to the first three patients. People react differently to drugs. Most doctors would probably expect two of the three to get some relief from their pain. What about the three patients who took the placebo? The results for these are easier to predict. Careful scientific studies have shown that the placebo will probably give one of the three patients relief from the pain.

The mind can protect or destroy. The behavior of human beings under hypnosis reveals the great power of the human mind. Dentists are able to perform all kinds of usually painful work on patients by hypnotizing them. Patients under hypnosis are told that they will not experience pain—and they do not.

The mind can even protect an individual from damage. In a famous experiment, forty volunteers allowed their arms to be burned. Then they allowed themselves to be hypnotized. They were told that the same amount of heat would again be applied to their arms. They were also told that their left arms would not be damaged by the heat, but that their right arms would. Then identical amounts of heat were applied to each of the volunteers' arms. Thirty out of the forty volunteers suffered little damage to their left arms. All of them were burned as before on their right arms.

The mind can also destroy an individual. *Ulcers* are open sores on the stomach or small intestine. Millions of people suffer

Figure 16.2: For years "civilized" people scoffed at the techniques of witch doctors and medicine men. Now we suspect that they can have healing power when they can control the patient's mental attitude.

from ulcers. Medical scientists believe that many people have ulcers because of their mental attitudes. Worry may be one of the chief causes of ulcers.

Ulcers are not the only disease that can be created by the mind. In any large hospital, doctors can point out patients with a variety of diseases partly or wholly caused by the mind. Doctors will also tell you that many patients die without any known disease. These patients die simply because they have lost their desire to live.

An extreme case of mental self-destruction occurs among the natives of central Australia. An individual may be punished by the tribe in a ceremony called "pointing the bone." If the tribe's medicine man points a bone at a man, that man believes himself to be doomed. The man may fall to the ground and start moaning as if he were in pain. Eventually, he crawls away and ignores the rest of his tribe. He stops eating. He just sits and waits for the death that eventually comes.

The mind can be a disease agent. In the last chapter, you learned about viruses, bacteria, and other disease agents. What you have read so far in this chapter is designed to alert you to another disease agent: your own mind. Your mind is your greatest gift. But you must also look at it as a possible enemy that can destroy you just as effectively as viruses or bacteria.

Here is an example. There is a type of heart disease caused by poor circulation of blood in the heart muscle. It is a sign of atherosclerosis. The disease is called *angina pectoris* (ăn jī′nə pĕk′tə rĭs). It causes severe pain in the chest. The pain gets much worse when the patient exercises. The disease can be detected with an electrocardiograph. This is a machine that makes a picture of the electrical waves produced by the heart muscle. (Remember that nerve impulses are electrical in nature.)

Heart surgery can sometimes cure the disease. One doctor has become quite famous because of a special method he uses to treat angina pectoris. He told several patients with this disease that they needed heart surgery. When the patients agreed, the doctor went through the whole surgical process— except for one thing. He did not cut into the chest cavity. Instead he made a cut in the skin so that the patients thought they had had heart surgery. The results: rapid improvement of the patients. One patient had been in extra bad condition. Before the "operation," he could exercise only 4 minutes before severe pain stopped him. The electrocardiograph also showed evidence of his disease. After the "operation," the same patient was able to exercise ten minutes without pain. The electrocardiograph also showed a more normal picture.

LESSON REVIEW *(Think. There may be more than one answer.)*

1. Pill A has a chemical in it that is a known pain reliever. Pill B looks and tastes like A but has no pain reliever in it.
    a. Pill A could be used as a placebo.
    b. Pill B could be used as a placebo.
    c. Pill A could be used to relieve the pain of a patient.
    d. Pill B could be used to relieve the pain of a patient.

2. In an experiment, volunteers had burning heat applied to their arms after being hypnotized. This experiment demonstrated
    a. how the mind can cause damage to other parts of the body.
    b. how the mind can protect other parts of the body from damage.
    c. that heat will always damage the skin.
    d. that heat will repair skin that is already damaged.

3. The doctor who did fake surgery on his heart patients proved
    a. that fake surgery can be an effective placebo.
    b. that heart patients really do not have a disease.
    c. that fake surgery cannot heal patients who have a true disease.
    d. that the mind can be powerful enough to cause the rapid improvement of a serious disease.

KEY WORD

placebo (p. 279)

THE DRUG PROBLEM

What is the drug problem? Suppose a teenage boy drives a car around a curve at 80 miles per hour. Suppose the car rolls out of control and the boy is killed. What killed the boy? Some would say that the car killed him. Others might say that the curve in the road killed him. Few would say that the boy killed himself.

This same kind of thinking can be applied to the so-called drug problem that you have heard about. Everywhere teenagers hear about the evils and dangers of drugs. "Drugs are killing our youth" is a cry that is often heard. This statement shows the wrong kind of thinking, for two reasons. First of all, drugs are talked about as though they were monsters that are alive and roaming the Earth. You would think that these monsters were hiding behind every corner just waiting to kill people. Drugs are not alive. Drugs do not run around finding people

Figure 16.3: There are many places where alcoholics can get help to learn to control themselves in the absence of alcohol.

and killing them. Neither do automobiles or bullets. People can kill themselves or other people with automobiles. People can kill themselves or other people with bullets. People can kill themselves or other people with drugs. The teenage boy killed himself with an automobile. The Red Baron was killed by another man with bullets. Janis Joplin killed herself with drugs.

The statement "Drugs are killing our youth" shows poor thinking for a second reason. This statement is almost always made by parents and older people. And, in fact, *they are the ones who are killing themselves with drugs.* And they are doing so at an amazing speed. For example, there is a special group of about 5 million Americans. They would, no doubt, be amazed at an Australian native who dies after being pointed at with a bone. Yet these same 5 million Americans are slowly killing themselves. They are called **alcoholics.** They use large quantities of the drug called alcohol every day of their lives. They know that the drug is wrecking their lives and that it will eventually kill them.

The problem with alcohol is very great. By causing accidents and diseases, alcohol is sometimes claimed to be the "number three killer" in America—after heart disease and cancer. That statement hides the real truth. Here is a more accurate statement: Self-destruction with alcohol may be the number three cause of death in America.

Here is another example. Many millions of adults use the drugs that are in tobacco cigarettes. The dangers of these drugs are well advertised. But listen carefully to the antismoking messages that you hear. Here, for example, is what a Commissioner of Health for the State of New York has said: "No other single factor kills so many Americans as cigarette smoking. . . . Bullets, germs,

alcoholic:
a person with alcoholism. That is a disease that makes the alcoholic dependent on drinks made with grain alcohol. The only successful cure for alcoholism is self-control. The alcoholic *must not drink* beverages containing grain alcohol.

Challenges to Human Survival

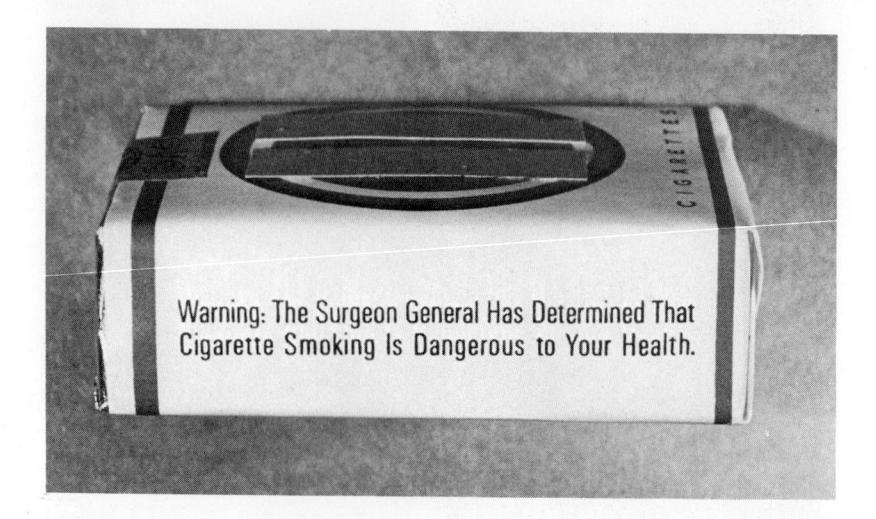

Figure 16.4: The evidence is on the cigarette pack for everyone to see. Self-destruction with this drug cannot be labeled as "accidental."

and viruses are killers; but for Americans, cigarettes are more deadly than any of them. No single known lethal agent is as deadly as the cigarette." Is that really an accurate statement? How would you revise the statement to make it more accurate?

We have a human problem. We do not have a drug problem in America. We have a human problem. We have far too many human beings who are destroying themselves. Much of this self-destruction is being done with drugs.

Are people trying to kill themselves? Usually not. Then why are they doing it? There are several reasons. In this chapter, we can only try to eliminate one of the reasons: ignorance. Some people did not know enough about drugs before they started to use them. They learned about them after it was too late. This could be called "accidental self-destruction." Such accidents should not happen. They cannot happen to you if you have the correct knowledge about drugs.

Drugs are not good or bad. People kill themselves with automobiles. People injure themselves with automobiles. Yet you cannot say that automobiles are bad. It all depends upon how they are used. The same kind of thinking should be applied to drugs. This is the first fact that you should learn about them. People can use drugs and get a variety of results. Lives can be saved with them. Pain can be relieved with them. They can be used for stimulation or relaxation. They can cause injury. They can cause death. Drugs cannot be labeled "good" or "bad."

Drugs are different and so are individuals. There are thousands of different drugs that are sold and used in our modern society. Try counting just the laxatives, cold tablets, and diet medicines in one drugstore.

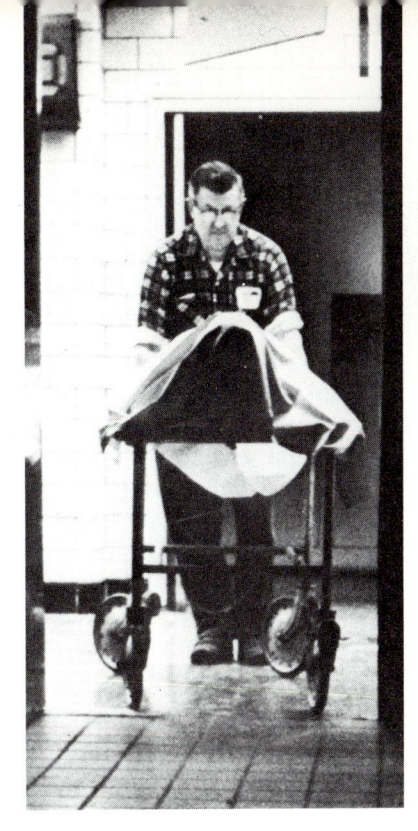

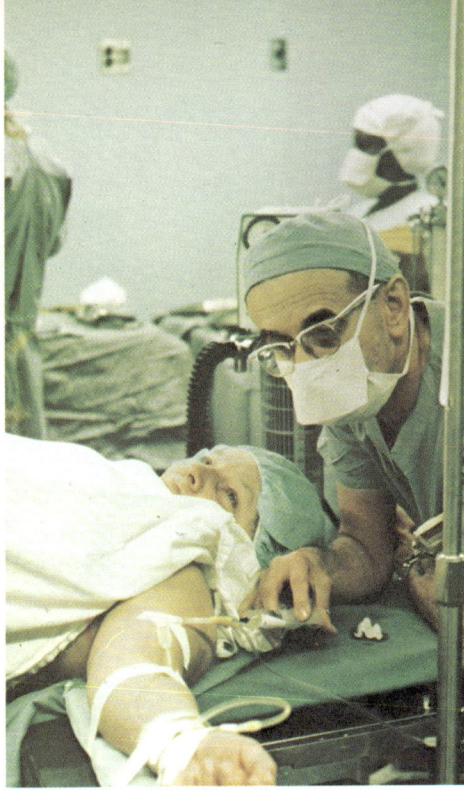

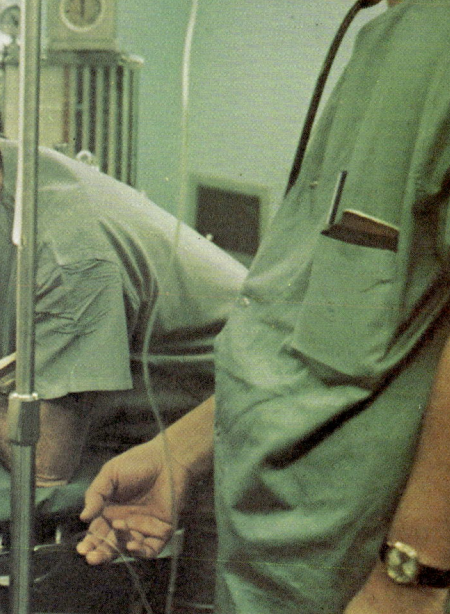

Each individual is different. Thus, it is not always possible to know how a person will react to a drug. Penicillin has been called the "wonder drug" because of the way it cured certain diseases. It is also a fact that some people can have very bad reactions to penicillin. A few people have been killed by this drug.

Figure 16.5: Barbiturates illustrate well the dual role of drugs in our society. The person on the left committed suicide with them. The person on the right is undergoing pain-free, life-saving surgery because of them.

Figure 16.6: In a modern drugstore, hundreds or thousands of drugs are available to anyone who wants to use or misuse them.

Challenges to Human Survival

Laboratory Activity

Do You Live in a "Drug Society"?

PURPOSE

Magazines and newspapers are rich sources of information about drugs. Advertisements tell you why you should buy and use a specific drug. News articles report on the effects of drugs. Editorials present different viewpoints about different drugs. In this investigation, you will examine how newspapers and magazines deal with drugs. This can help you to better understand drugs and society's involvement with drugs.

MATERIALS

newspapers
magazines
scissors
envelopes

PROCEDURE

1. Discuss with your class the meaning of the term "drugs." You know that alcoholic beverages and tobacco are considered drugs. Medications prescribed by a physician, and many nonprescription medications, such as aspirin and cough syrup, are drugs too. What about cooking wine, coffee, mouthwash, rubbing alcohol, beauty preparations, and deodorant? How will you classify marijuana? LSD? Set limitations on the definition of "drugs" in the group discussion.

2. Look through magazines and newspapers for advertisements of drugs. Should you include advertisements that picture drugs even if the advertisement is not for that particular drug? For example, suppose an advertisement for a travel agency shows a picture of a restaurant in which a bottle of liquor is shown on the table. Decide whether your class should include this advertisement as an advertisement about drugs.

3. Should news articles written about drugs be included? Editorials? What newspapers and magazines will you review? You may want to work in groups of several students. Be sure to review a wide range of magazines and newspapers.

4. Collect newspapers and magazines. If you borrow the publication be sure to ask permission to cut pages from it.

5. Look through the publications. When you find material on drugs, clip it from the magazine or newspaper. Write the name of the publication in one corner of the torn-out page. Continue your search. If you are in doubt about whether an article is appropriate, take it to class and discuss it with other students. Someone else may have the same question.

6. Arrange your clippings into groups such as alcoholic beverages, tobacco, headache remedies, cough cures, complexion creams, deodorants, cooking supplies, and weight-loss aids. These classifications will make the class discussion more systematic and orderly.

7. Take your organized clippings to school. In a classroom discussion, present the advertisements you were able to collect. How many different headache relievers were found? How many types of cigarettes were advertised? What really unusual drugs were you able to uncover? Who was able to find the most advertisements? From what types of publications did most of the ads in each category come?

QUESTIONS

1. Estimate how much money was spent on drug advertisements in each of several publications.
2. How many different brands of headache remedies are available in the drugstore? How many cough remedies?
3. What percentage of the advertisements in each of several magazines or newspapers were advertisements for drugs?

Challenges to Human Survival

tranquilizer:
a drug prescribed to calm the nerves or to relieve anxiety. Two such drugs—Librium and Valium—are the most heavily prescribed drugs in the world.

What is drug abuse? If you want to start an argument, get some people together. It does not matter what their backgrounds are—doctor, lawyer, barber, or baker. Just ask them to explain the term "drug abuse." Within a few minutes, you should have a good argument going.

Why do people argue about drug abuse? There are two main reasons. The first is that people tend to be dishonest with themselves. Here is an example. A busy father gets a call at the office. It is his wife. The school principal called her and said that their son John was caught smoking a marijuana cigarette. The father tells her to get hold of herself—have a drink, and he will be home as soon as possible. The father lights a cigarette and starts making phone calls. He lights another and clears his desk. He smokes two more on the way home. Meanwhile, the wife took his advice. She had a martini. She had been drinking coffee and chain-smoking cigarettes since the principal's call. The martini had little effect on her nerves. She fixed another, and also took one of her **tranquilizers.** The father arrived home desperate for a scotch and water. He also fixed his wife another martini. He gulped down one drink and fixed another. Together they sat on the sofa, in a cloud of smoke, with drinks in their hands, and talked about John's "drug abuse." These parents probably would not think or admit that *they* might be guilty of drug abuse. Unintentional dishonesty of this kind feeds many arguments.

The second main reason for arguments about drug abuse is a lack of knowledge. Most drugs are complex molecules. The human body is extremely complex. It is fantastically difficult to learn what happens to a drug once it gets inside the body. Does it affect the brain cells? If so, what does it do to them? Does it damage chromosomes? Does it cause gene mutations? These are just a few of the hundreds of questions that a drug user might ask. But years and years of research are required to answer such questions. People want facts now—right now! But facts about many drugs are not available. They will not be available for many years, and no amount of money can buy them. Then what do people do? They argue. And they argue for what they want to believe.

People do become dependent upon drugs. Some facts are known about drugs. One of them is that a drug user can become dependent on the drug. In other words, the user can reach a point where he or she is uncomfortable or unhappy without the drug. Dependence can create a great many problems for the user.

How drug-dependent do people become? There is a wide range of drug dependence. There is the "mild desire" type of dependence. The user thinks about the drug and wants the drug.

Figure 16.7: The first step on the road to prison for many people was taken when they became addicted to drugs.

But that type of user can stop thinking about it and do other things if she or he chooses. There is a "strong desire" or "craving" type of dependence. This user badly wants the drug. The person thinks about it if he or she does not have it. It is difficult for this type of user to think about other things. She or he cannot do other things when the drug is not available. This person will sometimes change normal behavior and behave differently until the drug is obtained.

An extreme type of drug dependence is called **addiction**. A person addicted to a drug has several characteristics. This person is totally involved with the drug. Throughout the waking hours, the person's thoughts are on the drug. Life habits are built around the drug. If there is a choice, the drug will probably be chosen over family, friends, or work responsibility. If the drug is one that is illegal or expensive, the problem is more severe. The addicted person can become a criminal or an outcast from society.

Can drug dependency be broken? The answer to this question is yes. However, you must keep two facts in mind. First, the addicted person must want the dependency to be broken. Drugs do not force themselves upon anybody. A person has to want to use a drug in order to use it. A person also has to

addiction: the strongest type of drug dependence. A person in this condition cannot function without the habit-forming drug.

289

Challenges to Human Survival

physical dependence: a condition in which a person's body loses its ability to function normally without a drug

want to stop using a drug in order to stop being dependent upon it.

Here is the second fact that you should know about breaking drug dependency. Some drugs can produce **physical dependence.** This means that an addicted person has more than a mental desire for the drug. Other parts of the body react badly when the drug is not taken. There may be pains and cramps in the digestive system. Muscles may be uncontrollable. The person may become unconscious. The person may even die.

Back to the question: Can drug dependency be broken? The answer is yes if (1) the addicted person wants to break the dependency, and (2) the person can stand the body's reactions to the absence of the drug if he or she is physically dependent upon it. Medical help is usually available for people who want to break a physical dependence upon a drug. The medical help can ease the body reactions when the drug is no longer used. However, there is still a problem. The "medical help" is usually some other drug. This other drug may create a new dependence.

LESSON REVIEW *(Think. There may be more than one answer.)*

1. The two posters shown above
    a. give true messages. Alcohol and other drugs do kill people.
    b. are not completely true. Drugs are always killers, while alcohol usually is not.
    c. give false messages. Neither alcohol nor drugs kill people.
    d. give false messages. If people are going to die from alcohol or drugs, they have to kill themselves.
2. Drugs can
    a. save lives.
    b. relieve pain.
    c. cause pleasure and relaxation.
    d. cause injury or death.

3. Which one of the following statements best describes the definition of "drug abuse"?
   a. Drug abuse is the overuse of any drug.
   b. Drug abuse is the heavy smoking of cigarettes.
   c. Drug abuse can only be defined by a doctor.
   d. Drug abuse is defined by people according to their personal views.
4. Drug dependency can be broken
   a. if substitute drugs are made available to ease the pain of withdrawing.
   b. if the drug-dependent person wants the dependency broken.
   c. if the drug-dependent person can tolerate the withdrawal.
   d. if the drug-dependent person is not totally addicted to the drug.
5. Which of the following terms best matches each phrase?
   a. alcoholic   c. addiction
   b. tranquilizer d. physical dependence

   ____ a condition in which the body feels the need for a drug

   ____ a person addicted to alcohol

   ____ an extreme type of drug dependence

   ____ a drug prescribed to relieve anxiety

KEY WORDS

alcoholic (p. 283)
tranquilizer (p. 288)

addiction (p. 289)
physical dependence (p. 290)

PROBLEM DRUGS

What are problem drugs? There are two ways to look at problem drugs. One way is from the drug user's point of view. A drug that causes damage to the user's mind or body could be called a "problem drug." A drug that the user can become dependent upon can also be a problem drug. A third type of problem drug is the illegal drug. The illegal drug may or may not cause physical damage or dependency. The simple fact that the drug is illegal is enough to create serious problems for the user.

The other way to look at a problem drug is from the viewpoint of the nonuser. There is always this question: Do drug users have the right to destroy themselves if that is what they

Challenges to Human Survival

want? The nonuser of drugs usually says no. Consider a male alcoholic, who is married and has children. Because of his drug use, other people must work to support his family. Because of his drug use, the children are without a father at times when they need one. The wife is without a husband when she needs one. Certainly the wife and family of this man would call alcohol a "problem" drug even if the user would not.

The rest of this chapter is about certain drugs that most users and nonusers would call problem drugs.

Alcohol is a social drug.

Millions and millions of people use alcohol. They use it at parties, picnics, and at other social events. The drug is taken in liquid beverages. Beer, wine, whiskey, gin, brandy, and vodka are the most common of these. The effects of alcohol on the human body depend much upon the quantity used.

Even in small quantities, alcohol slows down the activities of the central nervous system. For this reason, alcohol belongs to a group of drugs called **central nervous system depressants.** These depressants can be very dangerous. The central nervous system controls such vital body activities as breathing and heart rate. An accidental overdose of alcohol can depress these activities and cause death. However, accidental death from alcohol is not common. It usually occurs only to young and inexperienced users of the drug.

central nervous system depressant: a drug that affects the function of the brain and the spinal cord. The effect is for body reflexes and functions to slow down.

Some people can become physically dependent upon alcohol. As mentioned earlier, this type of addict is called an alcoholic. Alcoholics use large quantities of alcohol every day. Such quantities cause damage to the whole body. The liver is also damaged by a hardening process called *cirrhosis* (sĭ rō′sĭs). Cirrhosis is often what causes the death of alcoholics.

withdrawal: the step-by-step removal of a drug from an addicted person. It is usually accompanied by severe physical and

Withdrawal is the process by which drug users stop using a drug. Withdrawal from alcohol addiction usually creates illness and other bad effects for the addict. Sudden withdrawal can cause death. There are a few special hospitals and clinics where alcoholics can get personal help with their problem. Many more such places are needed because there are more alcohol addicts than all the other drug addicts put together.

Tobacco is a new problem.

Tobacco has been chewed, sniffed, and smoked for over 300 years. During most of that time, a few doctors and life scientists suspected that it caused body damage. But they had no proof. So for many years, tobacco was not considered a problem. Now we know better. Tobacco smoke contains a number of dangerous chemicals. These include carbon monoxide, nicotine, and a whole family of chemicals called tars. Doctors and life scientists have proven

Figure 16.8: Most people have seen television reruns of the old Perry Mason show. William Talman (shown) played the role of the district attorney who was always defeated by Perry Mason. William Talman is now dead. He died of lung cancer. Before he died he asked to appear on television to bring a special message to people. As you see him in this picture, Mr. Talman was saying: "He (Perry Mason) used to beat my brains out on TV every week for about 10 years. . . . You know, I didn't really mind losing those courtroom battles. But I'm in a battle right now that I don't want to lose at all because if I lose it, it means losing my wife and kids. I've got lung cancer. So take some advice about smoking—and losing—from someone who's been doing both for years. If you haven't smoked—don't start. If you do smoke—quit. Don't be a loser."

that the chemicals in tobacco smoke can cause cancer and emphysema. Lung cancer, for example, is twenty times more common for tobacco smokers than for nonsmokers. It has also been discovered that tobacco use is linked in some way to coronary heart disease. Male smokers between 45 and 54 years of age have more than three times the death rate from heart attacks than nonsmokers do. The heavy use of tobacco is not even a gamble. Repeated studies have shown that heavy smoking always causes some kind of damage to the body.

Millions of people use tobacco. They use it knowing that it will damage their bodies. They know it probably will cause them to have an earlier death. Why do they use it? Part of the reason is that tobacco use creates a type of dependence. It is not a strong physical type of dependence like alcohol addiction. But many people—doctors and scientists included—who have quit smoking will tell you that withdrawal is seldom easy.

The opioid user faces serious problems. There is a family of drugs that comes from the poppy plant. It includes

Challenges to Human Survival

THIS IS A DRUG ADDICT BEING COOL.

DON'T JOIN THE LIVING DEAD.
The Mayor's Narcotics Control Council

Figure 16.9: This type of poster may be having some helpful results in persuading young people to avoid addicting drugs.

opioid:
opium, or any drug made from opium, which comes from a particular poppy plant

opium and drugs made from it. These include heroin, morphine, and codeine (kō′dēn), all of which are called **opioids** (ō′pē oid). These drugs have been used for many years by doctors to relieve pain. That is the only legal way they can be used. However, these drugs are also sold illegally. They are used for purposes other than to relieve pain. These drugs relieve hunger, sex drives, and feelings of anger, fear, and worry. Heroin is the opioid most often used for such purposes.

The opioid user faces serious problems. First of all, the use of opioids often leads to a serious type of physical addiction. This type of addiction is much like a living nightmare for many people. This is partly because the drugs are illegal and very expensive. The cost of being addicted to heroin—the most common opioid used—may average over $50 *a day!* Finding money to support this kind of addiction can easily lead to criminal acts.

Withdrawal from opioids is difficult, but it can be done. There are drugs that can be used to ease the pains of withdrawal. One of the biggest problems for the ex-user is to adjust to a new kind of life. The user's whole life had been built around the drug. Most of the user's friends were other users. It is not always easy to change one's way of life. It is not always easy to find new friends. These facts help explain why many ex-users walk right back into the nightmare by returning to the drug.

People against Themselves

Barbiturates are depressants. The **barbiturates** (bär bĭch′ər ĭt) and related drugs make up a whole family of drugs. They have a number of medical uses. These are based upon their *sedative action*. A sedative is a *depressant*, a chemical that slows down the activities of the central nervous system. Barbiturates affect the central nervous system much the same as alcohol does. Doctors prescribe certain barbiturates as sleeping pills. Certain others are prescribed for persons with nervous disorders. A few of the barbiturates are used as general anesthetics. General anesthetics are used to make patients unconscious during surgery. If you ever have to undergo surgery, you will be very thankful that they exist.

Barbiturates are also sold illegally. They are used for a variety of reasons. Some persons use them for their sedative action. Others use them with stimulating drugs. Alcoholics use them to quiet their nerves during periods of withdrawal.

There are two reasons why these drugs are a problem. First, their use can create a strong physical addiction, and all the problems that go with it. Second, it is easy to get an overdose that will cause death. **The combination of a barbiturate and alcohol can be extremely dangerous.**

Withdrawal by long-term barbiturate users is a painful process. The user may suffer cramps, nausea, and vomiting. In some cases, there are seizures and convulsions—both violent reactions of the body. Sudden withdrawal after long-term use can cause death. Medical supervision is always recommended.

Amphetamines are stimulants. The name **amphetamine** (ăm fĕt′ə mēn′) is a general name that is applied to several drugs with similar action. All of them stimulate the central nervous system. Their effects are somewhat opposite those of alcohol and barbiturates.

The amphetamines are sometimes bought illegally. Some people who use them illegally are looking for the stimulating effects of the drug. Overdosage is common among such users. Such overdoses can cause personality changes. Typical is the **paranoid** (păr′ə noid′) reaction. The paranoid person believes that "everyone is out to get me." Such a person can be dangerous. She or he may use physical violence in an attempt to get even. Personality changes caused by amphetamines are temporary and disappear after the person stops using them.

Amphetamines do not create the same strong type of physical addiction as barbiturates, opioids, and alcohol do. However, users can develop a level of dependence that will cause them to feel a steady need for the drug. This will cause them to seek the drug and use it as much as possible.

barbiturate:
one of a group of chemically related compounds used mainly as central nervous system depressants or to produce sleep

Figure 16.10: A lethal equation: Barbiturates plus alcohol equals death.

amphetamine:
one of a family of drugs that stimulates, or peps up, the user. Amphetamines can create addiction.

paranoid:
relating to a mental disorder in which a person feels persecuted or has imagined fears

Challenges to Human Survival

psychedelic drug: any drug that causes a change in normal perception. The strongest reaction to such a drug is a hallucina-

The psychedelics change perception. Perception is the awareness you have through your senses. You perceive by seeing, hearing, tasting, smelling, and feeling. **Psychedelic** (sī'kĭ dĕl'ĭk) **drugs** cause changes in a person's perception. The changes may be mild. Things seen by the individual may be brighter or more colorful. The changes can also be drastic. Sights may be heard; sounds may be seen. In some cases, the psychedelics cause wild or weird perceptions that may be worse than nightmares. These are called hallucinations.

Psychedelic drugs have little if any medical value. They are illegal drugs. LSD is about the most powerful of the psychedelic drugs. Marijuana, which is mildly psychedelic, is the most common.

Why are psychedelic drugs a problem? There are several reasons. First of all, their effect upon the mind is drastic. Any person can react very badly to potent psychedelics like LSD. Numerous accidents and suicides have occurred during or after LSD "trips." LSD effects can occur long after use of the drug has stopped. These flashbacks are thought to cause serious depression, paranoid behavior, or fear reactions.

What do psychedelic drugs do to the mind and body? Thousands of doctors and life scientists would like to know the answer to that question! Right now they have very few facts that they can use to give an answer. The problem is that the psychedelic drugs have been in use for only a short time in America and Europe. Researchers have not had enough time or experience with psychedelic drug users. As yet, none of the psychedelic drugs is definitely known to cause body damage. But remember, this was also the case with tobacco just a few years ago. Many scientists are very suspicious of LSD. Doctors warn all pregnant women to avoid LSD, because of the strong possibility that it may cause deformed babies. Some studies have also shown that people can develop personality changes after long-term use of LSD or marijuana. The changes include a loss of interest in work, study, and personal appearance. Some users also show undesirable behavior patterns such as the paranoid behavior discussed earlier.

Figure 16.11: This is a cannabis plant. It is the source of marijuana and hashish.

A last word about drugs. America has been called a "drug society." That is a good name for us. Just look in your home medicine cabinet. Count the number of drugs, and you will probably see why. Americans look at the drugstore as a place to buy an answer for all of their problems. If we have a pain, we buy a drug to kill the pain. If we are too fat, we buy a drug that will make us slim. If we are too skinny, we buy a drug that will make us fat. If we are ugly, we buy a drug that will make us beautiful. If we smell bad, we buy a drug that will make us smell good. The examples could go on and on.

You are in the process of entering our "drug society." You will be surrounded by drugs all your life. Advertisements will be selling drugs to you. Friends will be trying to give drugs to you. Some of these will be prescription drugs—drugs that have been legally prescribed by a doctor. (That does not make them legal for you!) Others will be drugs that have been purchased illegally. The fact that a drug is illegal does not mean that it will not be available. As long as someone wants to pay for a drug, there will always be someone on a street corner selling it. Law enforcement is difficult when many of the people of a country want to buy a drug.

So now is a good time to ask yourself some questions. What kind of a drug user will I be? Am I the type of individual who might look to drugs for answers to my problems? Am I going to use more or fewer drugs than my parents? Am I going to learn about the drugs I permit to enter my body? An honest answer to these questions may alert you to a possible enemy: yourself.

LESSON REVIEW *(Think. There may be more than one answer.)*

1. Alcohol
    a. is a central nervous system depressant.
    b. is a central nervous system stimulant.
    c. can cause death by overdose.
    d. can cause strong addiction.

2. Tobacco smoking
    a. has been linked to lung cancer.
    b. has been linked to the mental disease of severe depression.
    c. has been linked to coronary heart disease.
    d. causes a severe type of addiction.

3. Opioids
    a. relieve hunger, pain, and the sex drive.
    b. are costly when purchased illegally.
    c. cause a severe type of addiction.
    d. cause an addiction that can never be broken.

4. Barbiturates
    a. are central nervous system depressants.
    b. are central nervous system stimulants.
    c. can cause a severe type of addiction.
    d. are deadly because they can so easily be overdosed, especially when taken with alcohol.

Challenges to Human Survival

5. Amphetamines
    a. are central nervous system depressants.
    b. are central nervous system stimulants.
    c. cause a permanent paranoid personality in abusers.
    d. cause an addiction similar to alcohol, opioids, and barbiturates.

6. Psychedelic drugs
    a. cause changes in perception.
    b. have value only in medicine.
    c. are suspected of causing serious health problems.
    d. and their effects on the body have not been thoroughly studied by medical researchers.

7. Which of the following terms best matches each phrase?
    a. central nervous system depressant
    b. withdrawal   e. amphetamine
    c. opioid   f. paranoid
    d. barbiturate   g. psychedelic drug

    _____ heroin is one

    _____ a central nervous system stimulant

    _____ deadly when mixed with alcohol

    _____ it can change perception

    _____ affects the brain and spinal cord

    _____ the process of stopping drug usage

    _____ relating to a mental state in which a person thinks that others are out to get him or her.

KEY WORDS

central nervous system depressant (p. 292)
withdrawal (p. 292)
opioid (p. 294)

barbiturate (p. 295)
amphetamine (p. 295)
paranoid (p. 295)
psychedelic drug (p. 296)

Applying What You Have Learned

1. Pick out a drug that people consider safe. This should be a drug that is common in most homes. Think of a way to argue successfully that this drug could be dangerous, not safe.

2. What is wrong with the following statement: "Bullets, germs, viruses, and cigarettes are killers"?

3. Alcohol is one of the most common drugs used by people. Yet many people who use it argue strongly that they are not drug users. Why do you think this is true?
4. Many opioid users cure their addiction and then become users again. Why might this happen?
5. Suppose it should be discovered that some widely used drug causes gene mutations. Then you hear a user of that drug make this statement: "I think I should be able to take any drug that I want. It's my own business if I want to destroy my body." Use what you learned about the gene pool in Chapter 13 to argue against this person's point of view.

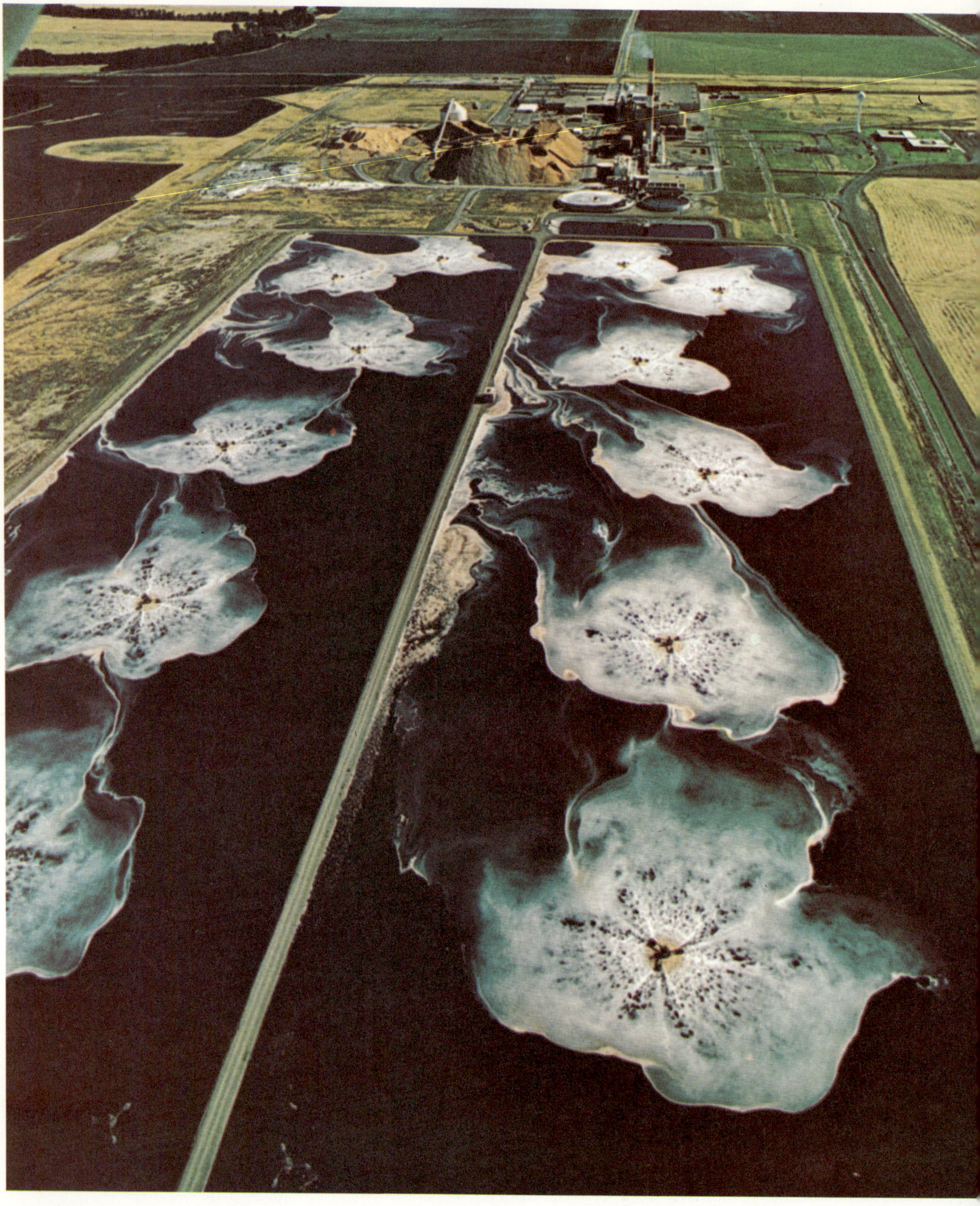

Figure 17-1: Aerators are used in this pollution-control center. Aeration adds oxygen to the water by exposing a constant spray of water droplets to the air. This process also removes carbon dioxide, hydrogen sulfide, and taste-producing gases or vapors.

Chapter 17

POLLUTION OF THE ENVIRONMENT

The title of the last chapter was "People against Themselves." The title of this chapter could be "People against Everybody." The person who pollutes the environment has turned against other people and all other forms of life. What is pollution? How do we pollute? What does pollution do to you? These are a few of the questions that you will find answered in this chapter.

WHAT IS POLLUTION?

Pollution is a relative term. When is something polluted? How much is something polluted? These questions will not be answered the same by everybody. That is what is meant by the statement: **Pollution** is a relative term. How much something is polluted depends upon who is making the decision. A man who lives next to a garbage dump may have fifteen

pollution:
the products of human activity that harm—or destroy—the quality of the physical environment

301

Challenges to Human Survival

old car bodies in his yard. He might not think that his environment is polluted. You might disagree. A woman rakes leaves from her lawn and burns them. She might not think she is polluting the air. You might disagree. Water you might not want to drink may be just right to cool the furnaces of a steel mill. Yet water you might drink may have so many minerals in it that it would damage the boilers of an electric power company.

Because pollution is a relative term, pollution is also a controversial term. *Controversial* means that people will argue about it. There are many things that can affect such arguments. Suppose you live near a steel mill that pumps tons of smoke and gas into the air. Is that steel mill polluting the air? You might say yes. But what if your job was in that steel mill working at a furnace or in an office? What if you had to help support your family and had no money but your paychecks from that mill? Then would you think that the steel mill was polluting the air? Of course you cannot answer that question. You do not know what it is like to be in that situation. But you can see why a worker would defend the steel mill and its wastes. So you can understand why the wastes of the steel mill make a controversial pollution problem.

What is pollution? We have said that pollution is a relative term. We have said that pollution is a controversial term. But what is pollution? Pollution can be defined as the products of human activities that harm or destroy the quality of the environment.

How can the quality of the environment be harmed or destroyed? There are many ways. Pollution could be anything that affects the health of people. Pollution could be anything that upsets the delicate balance of an ecosystem. Pollution could be anything that destroys the beauty of the environment. Pollution could be anything that ruins the recreational use of the environment. What other ways are there that pollution can destroy the quality of the environment?

Pollution can affect or destroy a person's health. Many years ago, a city used to be a very dangerous place to live. Deadly diseases spread through city populations like wildfire. Thousands died in great plagues. Why? Mainly because of pollution. Wastes of all kinds were dumped in the streets and gutters. These wastes contained many kinds of disease organisms. From the streets and gutters they spread everywhere. The drinking water often became polluted and spread the killer disease organisms. This disease problem was ended by modern sewer systems and water treatment plants.

But people in cities are still dying—sometimes by the

Pollution of the Environment

Figure 17.2: The Great Plague of London was due to pollution. Wastes containing disease organisms were dumped in the streets, where they could spread to the whole population.

thousands. In December 1952, a 5-day **smog** covered London, England. At least 4000 deaths were probably caused by that smog. In 1963, a similar smog condition existed over New York City. The deaths of 405 people were blamed on that smog.

Death is the most severe result of pollution. But disease is also an undesirable result. Pollution can spread organisms that cause disease. Pollution can also be the direct cause of disease. Examples of both will be explained later in this chapter.

smog: a heavily polluted condition of the air. It usually is made up of smoke particles and poisonous sulfur and nitrogen gases.

Pollution can upset the balance in ecosystems. There are many ways that pollution can affect the balance in ecosystems. You have already learned about DDT pollution and how the "top dogs" in some ecosystems are being killed. This leaves the food web without an important group of consumers. As a result, the other populations will change. The new ecosystem may or may not be desirable from the human point of view.

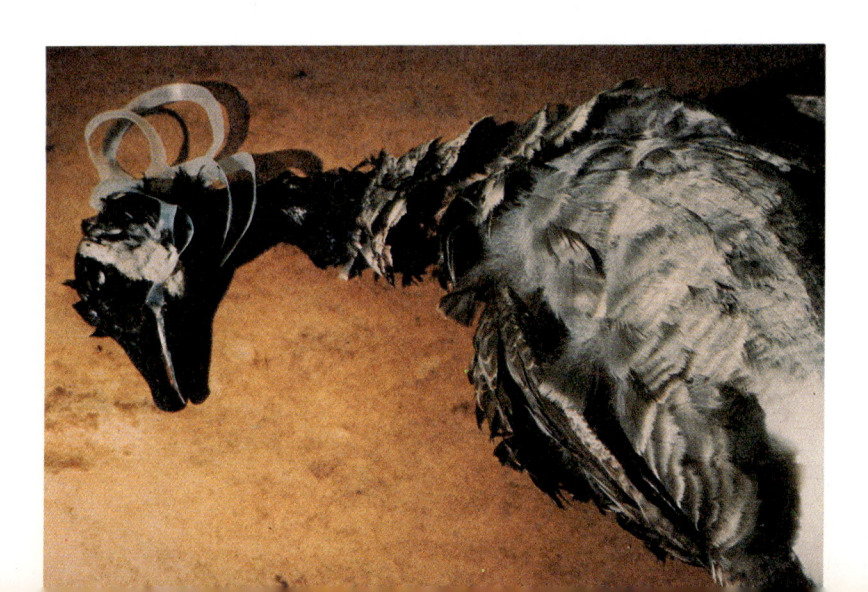

Figure 17.3: This Canadian goose was found dead on the shore of a lake in Ashley National Forest, Utah. It somehow got caught in the "six-pack" plastic holder and strangled. Who do you think should be blamed for the death of this once beautiful bird?

FIGURE 17-4: Recycling not only rids our landscape of unsightly junk, but it also allows many materials to be reused. First, waste materials, such as newspaper and glass, have to be collected (left). Then machines or chemicals can help to prepare the materials for reuse. The newspaper being recycled (right) will be mixed with water to create pulp. The pulp will then be used to make new paper products.

Lake Erie is one of the Great Lakes. This lake is an ecosystem. This ecosystem was nearly destroyed by pollution. A few years ago, Lake Erie produced tons of fish for commercial fishers. Thousands of people enjoyed living near the lake. Many thousands more enjoyed the swimming and boating the lake offered. Cities and industries had always dumped their wastes in the lake. After World War II, the areas bordering the lake grew rapidly. And these areas continued to dump their wastes in the lake. The result was a disaster for the people who lived near the lake. Instead of a body of water, the lake was once called a "chemical tank." People realized what was happening before the lake was lost forever. Pollution controls were established. Slowly the lake is recovering.

Pollution can destroy scenic beauty. There are hundreds of ways pollution can destroy the scenic beauty of the environment. Who can see a sunset when smog covers a city? Who can enjoy a river with trash floating in it? Who can enjoy a scenic highway trip when old car bodies litter the countryside? Who can enjoy a picnic surrounded by broken glass and beer cans?

Why have we let pollution destroy the beauty of our environment? One reason has to do with our **values**—what we think is most important. For years financial value has always been rated higher than scenic value. A grassy meadow was more valuable when a shopping center was built upon it. Marshlands

values: those things a person or society believes to be important

had more value when they were drained and used for housing subdivisions. Forests of oak and hickory trees had more value when open pit mines were dug in them.

It is important for us to have shopping centers, houses, and products of mines. It is also important for us to have scenic meadows, marshlands, and forests of oak and hickory. What may or may not destroy our scenic beauty will always be controversial—and a matter of values.

LESSON REVIEW *(Think. There may be more than one answer.)*

1. The term "pollution" can be described as
    a. relevant.
    b. relative.
    c. controversial.
    d. decisive.

2. Which of the following could be examples of pollution?
    a. A student drops a gum wrapper in the hall of the school.
    b. A farmer burns the weeds that are plugging up an irrigation ditch.
    c. A husband tells his wife "Happy Birthday" with a large billboard along a city expressway.
    d. A music fan on a crowded beach turns a radio up to full volume.

3. Pollution can destroy
    a. individuals.
    b. populations.
    c. communities.
    d. ecosystems.

4. Smog
    a. occurs naturally.
    b. may occur where there is heavy traffic.
    c. may occur where there is heavy industry.
    d. may kill.

KEY WORDS

pollution (p. 301) values (p. 304)
smog (p. 303)

WATER POLLUTION

The nature of the problem is people. Human population is increasing rapidly. Human activities are increasing rapidly. Yet the amount of water on the Earth remains the same. That just about explains the problem of water pollution today.

People have always used rivers for sewers. We have always used lakes and oceans for dumping grounds. Only recently have these practices created problems. Rivers and other bodies of water can handle small amounts of pollution. Many of our waste products can be broken down and recycled without affecting the aquatic ecosystems. But there is a limit. Lake Erie and some other bodies of water have almost passed that limit. A great many other bodies of water are near the critical stage. Why?

Decomposers use oxygen. Many things will decompose when they are dumped in a large body of water. These include household garbage, body wastes, and wastes from pulp mills, canneries, and packing plants. Years ago, this was what usually happened. The wastes were attacked by decomposer bacteria. The by-products enriched the water, and the life in the water benefited. Gaseous wastes passed back to the atmosphere. However, there were two problems. One of the problems was usually corrected.

Decomposer bacteria use oxygen. The greater the number of them, the more oxygen they use. There are always great numbers of decomposer bacteria in water where wastes are dumped. They use up much of the oxygen that is dissolved in the water. That creates a problem, especially if you are a fish such as a trout that needs a good supply of oxygen. This problem used to solve itself in rivers and large bodies of water. Moving

Pollution of the Environment

Figure 17.5: Decomposer bacteria multiply when wastes are dumped into a stream or lake. They use up oxygen. If the decomposers multiply too rapidly, they use up most of the available oxygen and kill the fish.

water picks up oxygen from the atmosphere and puts it in solution. A tumbling river, with falls and white water, replaces lost oxygen quite rapidly. The water in a large lake circulates. Water that is low in oxygen in one region will be replaced with water rich in oxygen from another.

Disease organisms lurk in polluted water. A second problem in polluted water does not correct itself. There are several possible disease organisms in human wastes. If they get into a river or lake, they are not easily destroyed. They may be in the water for long periods of time, ready to cause diseases such as hepatitis and typhoid. Such water can be treated with chemicals like chlorine that will kill these organisms. That solves the problem for those who would drink the water. But swimmers, and others who use the water for recreation, are still in danger.

Rapid decomposition speeds eutrophication. Eutrophication (yōō trəf′ĭ kā′shĭn) is a word that was once used only by a few life scientists. Recently it has become almost a common word. Eutrophication is the name for the aging process of a body of water. In simple terms, it is the way a body of water "gets rich and dies."

How does a lake get rich and die? Every lake has shallow regions. Usually these are near an outlet where soil is always settling out of the water. Shallow areas are usually warm. Sunlight may penetrate all the way to the bottom. The warmth and

eutrophication:
the natural process by which a body of water ages

307

Challenges to Human Survival

Laboratory Activity

Why Isn't the Water Blue Any More?

PURPOSE

Eutrophication is the natural process by which a body of water gets rich in plant life, nutrients, and animal life and then dies. Large populations of decomposers and the nutrients they feed on play a vital part in eutrophication. In this investigation, you will study how the action of decomposers on nutrients changes the quality of a body of water.

MATERIALS

5 test tubes
bromthymol blue solution
masking tape
test tube holder
5 corks
plastic wrap

PROCEDURE

1. Fill five test tubes about one-half full of water. Add 5 drops of bromthymol (**brōm′thī′məl**) blue solution to the water in each test tube. The water should turn a deep blue color.
2. Use a 5-cm strip of masking tape as a label on each test tube.
3. Place the test tubes in a test tube holder or container. Secure a cork in the mouth of each test tube.
4. Label one test tube "Control." Nothing else will be added to the water in this test tube. Take the other four test tubes home with you at the end of your school day.
5. What waste or natural products in your environment would pollute the water in a lake? Garbage (potato peels, meat scraps), dead grass, decaying leaves, silt from a pond floor, mud from a puddle, and algae from a pond or lake are possible answers. To test your answers, add a small quantity, about the size of a pea, of one waste product to each test tube. Label the test tube to indicate what was added to each.
6. Bring the test tubes to school the following day. Remove the corks. Add water to each test tube so that only about 0.5 cm of air remains at the mouth of the test tube. Remember to add water to the contents of the "Control" test tube, also.
7. Wrap a 5-cm square of plastic wrap around the bottom of each cork. Insert the corks into the test tube so that the plastic-wrapped end is in the water. Some water in the test tube may overflow as you insert the corks. Be sure that there are no air spaces at the top of the test tube.
8. Mix the contents of each test tube by gently rotating the test tubes between your hands. Return the test tubes to the holder.

Pollution of the Environment

9. Observe the test tubes for 5 days. Record your observations daily. This chart will help you to record your observations. Notice that a numerical value has been assigned to each color. The color valued "0" should match the color of your control solution.

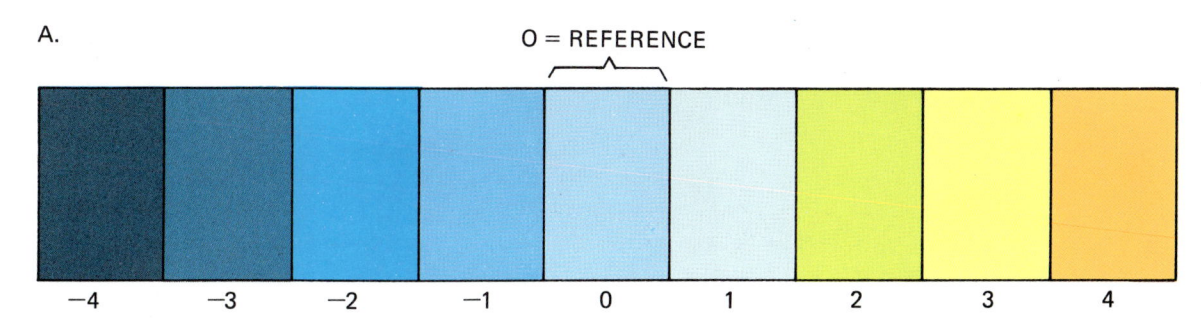

A.   0 = REFERENCE

−4   −3   −2   −1   0   1   2   3   4

10. Bromthymol blue is used to indicate the presence of CO_2. In the presence of CO_2, the bromthymol blue changes from blue to yellow. Remember, as decomposers act on nutrients in the water, they use up the O_2 dissolved in the water. They release CO_2 into the water at the same time. What color change would you expect in water samples containing decaying waste or natural products?

11. Record your results on a chart similar to this:

Contents of Test Tube	Observations				
	First Day	Second Day	Third Day	Fourth Day	Fifth Day
Control					
Meat Scraps					
Dead Grass					
Potato Peel					
Green Algae					

QUESTIONS

1. Which test tubes showed a color change toward dark blue? Which test tubes showed a color change toward a lighter or yellow color? Which waste products provided good nutrients for the decomposers?
2. Describe the dying process of a body of water.
3. Which part of the dying process was seen in this investigation?
4. If matter were added to water containing bromthymol blue and the solution immediately turned yellow, what would you conclude?

309

Challenges to Human Survival

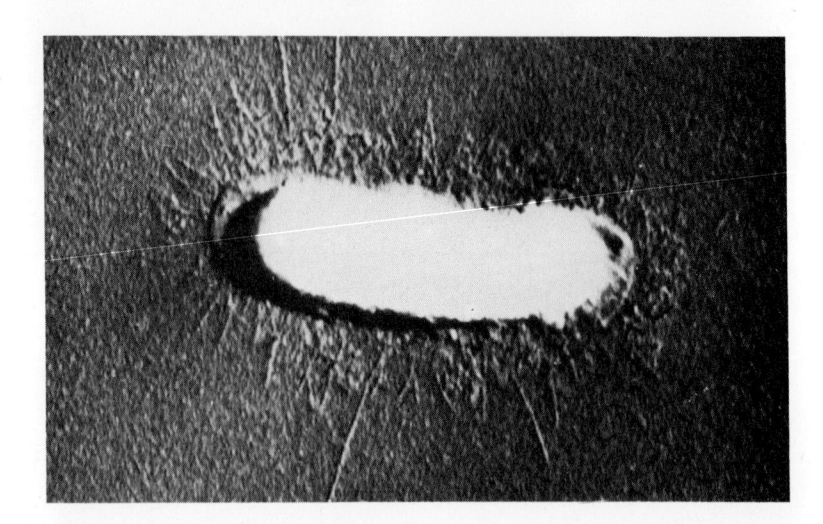

Figure 17.6: This is Salmonella typhosa, the organism responsible for typhoid fever. These bacteria grow in water polluted by human wastes.

sunlight stimulate large growths of algae. These are simple plantlike organisms. Of course, some of the algae are always dying. Their dead bodies are decomposed, and the nutrients are added to the water. The added nutrients stimulate larger growths of algae. And their dead bodies add more nutrients. In such a region, the water becomes richer and richer in nutrients. The growth of animal life, especially fish, is stimulated. Then the dying process takes over. Large quantities of dead algae demand large populations of decomposers. And decomposers use oxygen. Eventually the lack of oxygen in the more shallow areas causes a decrease in the animal life. Large mats of algae also settle to the bottom and do not get decomposed. This process, along with the settling of soil, makes the area more shallow. Eventually plants with roots, such as cattails, begin growing. A shallow lake region is now on the way to becoming a marsh.

Under natural conditions, most lakes die a very slow death. But now people are in the picture. We are greatly speeding up the eutrophication process for many lakes. Tons of waste material are being dumped into some lakes. These wastes decompose, and the by-products serve as nutrients and enrich the water. But that is not all. Fertilizers from farmlands are washed into the lakes. They also fertilize the lakes. Then there are phosphates from laundry detergents. They too are enriching the water. The results? First, there are massive growths of algae. These die, decompose, and further enrich the water. Next, the dying process. Decomposers use more and more of the lake's oxygen. Animal life declines. Mats of algae settle to the bottom, undecomposed, and decrease the depth of the water. Human activities thus speed up nature's process by hundreds or thousands of years.

Pollution of the Environment

Figure 17.7: The aging of ponds and lakes is called eutrophication. Excess nutrients cause a tremendous overgrowth of algae, which eventually die. Then the decomposers multiply and use up the available oxygen.

Career Spotlight

Although we seldom think about it, wastewater treatment and sanitation personnel are as important to our well-being as police, fire fighters, or doctors. Without clean water to drink we could have disease epidemics. This was the situation people faced before modern sewage disposal methods and wastewater treatment.

Nowadays, wastewater from homes and factories is carried in sewer pipes to special treatment plants. The people who run these plants are called *wastewater treatment plant operators.* They use special equipment to remove harmful materials from water by treating it in a series of stages. The operators need to read and interpret testing gauges. They must make adjustments that are necessary to keep the equipment working properly. In this picture, you see an operator checking a sample of water for purity at one stage of the treatment. These checks are repeated until the water is pure enough for us to use.

Wastewater treatment trainees must be high school graduates. They usually start as assistants to experienced operators and learn on the job. Because people have become concerned, water pollution standards are getting tougher. Thus, operators will have to handle more complex wastewater treatment systems in the future. Therefore, additional science education beyond high school will become important to workers in this field. However, job opportunities are expected to be very good in coming years, which can make the extra effort worthwhile.

Figure 17-8: This copper refinery uses recycled water. Pulp coming from local leach tanks is withdrawn in these circular baths. In this way, water is constantly recycled, not discharged into the environment.

Warm water can pollute. Humans can pollute a river or lake by dumping clean water into it. The water needs only to be warm. This creates a special kind of pollution called **thermal pollution.**

Several industries use water for cooling. Steel mills, nuclear power plants, and oil refineries are examples. Water for such purposes is usually pumped out of a river or lake. Then it is used for cooling, and pumped back—several degrees warmer. What is the problem?

Cold water holds more dissolved oxygen than warm water does. That is the main problem. By heating the water of a river or lake, the dissolved oxygen content is reduced. Very often it is desirable fish such as trout that suffer from the lack of oxygen. They can no longer live on the smaller quantities of oxygen. But fish such as carp and suckers, which are not desirable as game fish, often thrive in such areas. So they take over—and people who like to catch and eat game fish, such as trout, pike, largemouth bass, and so on, lose out.

thermal pollution: the upset of a freshwater ecosystem when its temperature is raised by the addition of heated waste water to it

Challenges to Human Survival

Figure 17-9: Thermal pollution is caused when water is used for cooling purposes. This causes the water temperature to increase. So when the used water is returned to the river or lake from which it was taken, it is warm, not cool. This warm water may upset the ecological balance of the waterway. Thermal pollution of this kind has been noted near nuclear power plants. It may be combated by allowing the used water to cool before it is emptied back into the waterway. This is often accomplished in large cooling towers.

soil erosion: the wearing away of soil by water or wind

Figure 17-10: A scientist tests water samples for purity. Bacterial analysis is made to safeguard drinking water.

Human carelessness can cause water pollution. In many cases, we pollute our water by accident. Many such accidents are the result of carelessness. An example is a careless farm practice that results in soil erosion. **Soil erosion** is the process by which rain and irrigation water wash soil away. Excess soil in a stream can cause severe pollution. Most of the oyster beds in our large eastern rivers have been covered by mud and destroyed. Mountain streams suffer because the soil sticks to rocks and destroys the habitat of certain insects. The insects that die are in food chains along with the fish in the stream. When the insects die, the fish populations decline or die.

Careless use of poisons can also result in severe water pollution. Usually a bad accident is noticed when people living along a river suddenly see and smell thousands of dead fish. The reason for the fish kill is often the same. Someone used a very poisonous pesticide in a nearby area. That poison had washed into the water and killed the fish.

"Oil spill" is a term that has become well known. In various types of accidents, large amounts of oil have been spilled in coastal waters. The oil floats on top of the water. Animals die because of it. The oil often washes ashore and temporarily destroys the recreational use of water and beaches. Oil spills may also cause greater damage that we do not know about. Life scientists have only recently begun to study the long-term effects of such pollution.

Radioactive wastes are a special problem. In recent years, some of our electrical energy has come from nuclear reactions. Waste products are produced during these reactions. The waste products are *radioactive*. This means that they give off harmful radiation. Many will continue to do so for fifty years or even hundreds of years. Therefore, they must be stored or disposed of in such a way that people will not be exposed to them. How can this be done?

No one has found a good answer to the problem of radioactive waste storage or disposal. Right now most of these wastes are buried in the ground. That worries a lot of people—including life scientists. There is always the possibility that the storage containers may develop a leak. Then there is the dreadful possibility that the wastes will drain into the groundwater supply. Groundwater lies underneath much of our land. The groundwater is about the last surviving body of natural unpolluted water. Our present method of radioactive waste disposal is just one of the many threats that we face.

LESSON REVIEW *(Think. There may be more than one answer.)*

1. Decomposers
    a. are more plentiful in clean water.
    b. are more plentiful in water polluted with organic wastes.
    c. use dissolved oxygen in the water.
    d. contribute dissolved oxygen to the water.

2. A fish dies in polluted water. The cause of death may be
    a. disease organisms in the water.
    b. decomposers robbing the water of its dissolved oxygen.

Challenges to Human Survival

      c. poisonous chemicals.
      d. an increase in water temperature, which decreases the water's oxygen content.

3. During the first stages of eutrophication,
    a. the water of a lake is fertilized.
    b. algae populations increase in size.
    c. decomposer populations decline in number.
    d. dissolved oxygen in the water increases.

4. During the later stages of eutrophication,
    a. dead, undecomposed algae settle to the bottom.
    b. decomposer populations increase in numbers.
    c. dissolved oxygen in the water decreases.
    d. the lake becomes more shallow.

5. The main reason that soil washing into bodies of water is damaging to all aquatic life is that
    a. the aquatic life cannot see to reproduce in dirty water.
    b. the soil plugs up the breathing organs of fish.
    c. the soil covers up rocks and destroys the habitats of insects that are in aquatic food chains.
    d. the soil uses up the dissolved oxygen and the fish and small animals die.

6. Radioactive wastes are a special problem because
    a. they are difficult to store safely.
    b. there is the possibility that buried wastes might leak into groundwater.
    c. they may start nuclear reactions if water gets to them.
    d. they may explode if exposed to polluted air.

7. Thermal pollution
    a. increases the oxygen content of an aquatic ecosystem.
    b. decreases the oxygen content of an aquatic ecosystem.
    c. may increase the number of undesirable fish in an aquatic ecosystem.
    d. will upset the balance of an aquatic ecosystem.

8. Soil erosion
    a. may be caused by careless farming.
    b. occurs naturally.
    c. can cause bottom dwellers to die in an aquatic ecosystem.
    d. may be noticed during a windstorm.

KEY WORDS

eutrophication (p. 307)   soil erosion (p. 314)
thermal pollution (p. 313)

AIR POLLUTION

Air pollution is a difficult problem. Water is scattered all over the Earth. The air is in one large body. This difference between the Earth's water and the Earth's air makes a difference in the way we must look at pollution.

We usually have a choice when we select the water we use. We can take water from several different locations. If one source is too polluted, we can take it from another. Or if all convenient sources are polluted, we can process our water and make it drinkable.

We have no such choice regarding the air that we use. We must breathe the air that surrounds us. Of course, we could live in special buildings, wear special masks, or breathe out of special tanks. But even if we did, what would we do about all the other forms of life that are exposed to polluted air?

Air pollution is truly a more difficult problem than water pollution.

How does air become polluted? One process we use causes most of our air pollution. This is the process of *combustion*, or burning. Combustion is the process by which a fuel is combined with oxygen. Energy is made available during the process of combustion. That is why we use it. But most combustion also results in by-products that can pollute the air.

Where do we use combustion? Most homes are heated by the energy from combustion. A large number of factories use combustion for their energy. Electric power plants often convert the energy from combustion into electrical energy. The major use of combustion is to power the vehicles that move us. These include cars, trucks, buses, motorcycles, ships, trains, and aircraft.

Figure 17.11: Combustion occurs when there is a source of fuel, oxygen or air, and a heat source. Combustion is the source of much of our air pollution, though it can also produce necessary energy.

Challenges to Human Survival

A temperature inversion can increase pollution. Usually the air next to the Earth's surface is the warmest. Then, as you go above the Earth, the temperature steadily decreases. When this is the case, the warmer air near the Earth's surface is always rising and being cooled. This pattern of air movement is important. It is the first step in the cleaning of polluted air.

Most air becomes polluted when it is close to the ground. When this air rises above the Earth, it mixes with unpolluted air. This decreases the pollution. Eventually, when it rains or snows, the pollutants are washed out of the air. This is nature's way of cleaning the air.

Sometimes nature's cleaning process stops working. This occurs when a warm layer of air hangs above the Earth's surface. Such a condition is called **temperature inversion.** The layer of warm air acts as a lid on the air below it. It prevents the air below it from rising. If a temperature inversion occurs over a

temperature inversion: a condition in which a layer of warm air above the Earth's surface traps air and pollutants below it. The trapped, polluted air cannot rise.

Figure 17.12: Under normal conditions, the air gets cooler as it extends up into the atmosphere. Smoke and dust particles can rise into the air and be blown away from the area. At the left, you see part of New York City under normal atmospheric conditions. During a temperature inversion, a cap of warm air prevents the air below from spreading out. The pollutants are concentrated over the city and cannot escape. At the right, you see the same part of New York City during a temperature inversion.

polluted area, the polluted air is trapped. And the longer the temperature inversion lasts, the more concentrated the pollutants become. Air pollution disasters occur during prolonged temperature inversions.

Fog can also increase air pollution. A fog is a cloud that is down on the Earth's surface. The fog (or cloud) is made up of very small droplets of water. These droplets of water float in the air. If pollutants are in the air, the droplets of water will collide with them. The pollutants may then combine with the water. If this happens, the fog is acting like a giant sponge. It soaks up the pollutants and holds them down near the ground. If a fog remains over a polluted area, the concentration of pollutants gradually increases. This often happens during a temperature inversion.

There are two types of smog. "Smog" is a term that is usually used to describe polluted air. What is smog? There are two kinds. The term "smog" was first invented to describe a combination of smoke and fog. Such combinations occur in cities that burn lots of coal. The droplets of fog combine with the smoke particles. The fog acts like a giant sponge in creating this type of smog.

The second type of smog is more common. It is properly called **photochemical smog**. *Photo* means "light." Photochemical smog is caused by sunlight reacting with certain pollutants. The chemicals in the exhausts of automobile engines are the biggest troublemakers. In the Los Angeles area, where photochemical smog is common, there are about 3 million automobiles. The bright sunlight reacts with the chemicals in the exhaust fumes. New chemicals are formed. The newly formed chemicals float in the air and give it a hazy look. If a temperature inversion exists, the photochemical smog becomes more and more concentrated. It may become so thick that it resembles a fog.

photochemical smog: a polluted air condition caused by the reaction of sunlight with automobile exhaust gases and other gaseous products of industry

There is a variety of pollutants in the air. You would have to be a chemist to understand very much about the different chemicals that pollute the air. Only a few statements about the most common pollutants will be given here.

Coal contains sulfur. When coal is burned, the sulfur combines with oxygen. The sulfur-oxygen compounds that result can cause body damage. Some kinds of oil also contain sulfur. The cheap oil that is preferred by industry contains the most sulfur. This sulfur also combines with oxygen when oil is burned. Sulfur-oxygen compounds are invisible but can sometimes be detected by their odor. It is like the odor you smell after you light a match.

Figure 17.13: The smog has almost hidden the city in the background. The automobile, because of its exhaust emissions, is most responsible for photochemical smog. To reduce the smog level, federal, state, and local governments have set up emission standards for automobiles, trucks, and buses.

There are at least three bad pollutants that may result from combustion in automobile engines. One of these is carbon monoxide. This is a colorless, odorless gas that is poisonous. Carbon monoxide is given off in the exhaust fumes of automobiles. The other two bad pollutants are produced mainly after sunlight reacts with the exhaust gases. One of these pollutants is nitrogen dioxide. This is a yellow-brown gas, which gives

Figure 17.14: Ozone formed near the Earth can be harmful. Ozone in the upper atmosphere is essential to life on Earth. Many miles above Earth's surface a layer of ozone screens out harmful radiation from the sun. In recent years, it has been learned that the propellant in many aerosol products is reacting with ozone in the upper atmosphere. Some scientists fear that unless the use of this propellant is greatly reduced, or stopped altogether, the protective ozone layer will lose much of its ability to screen out the harmful radiation. If that happens, the same scientists predict a greater incidence of skin cancers among the people of the Earth.

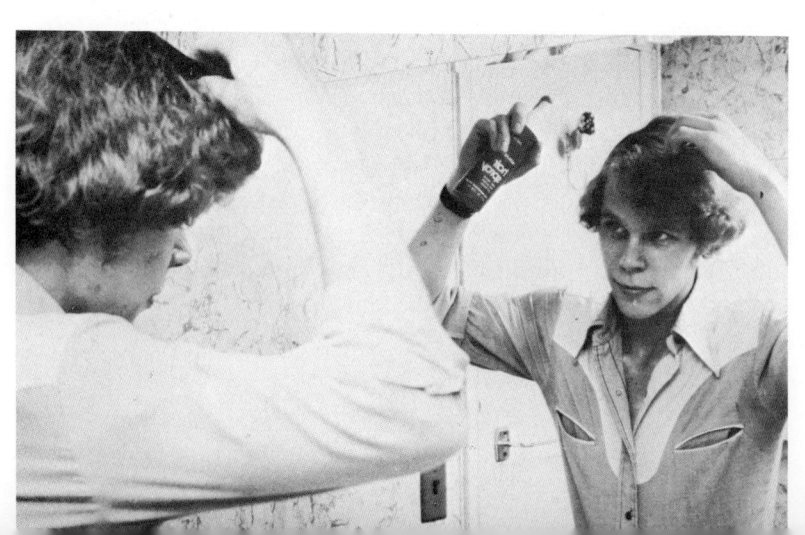

photochemical smog its color. Nitrogen dioxide has a sharp odor that is described as "sweetish." Ozone is the other pollutant that results from the photochemical process. Ozone is colorless, but it has a sharp odor. You may have smelled ozone during an electrical storm or when a piece of electrical equipment short-circuited.

What do pollutants do to the human body? As you might guess, pollutants do their greatest damage to the organs of the breathing system. What kind of damage do they do? Several things can happen. The tubes and passages of the breathing system are lined with hairlike structures called **cilia** (sĭl′ē ə). The cilia are constantly moving back and forth. They function like a broom that sweeps out foreign material that is inhaled from the air. Some pollutants can slow down these cilia—or stop them altogether. This leaves the lungs with one of their protective devices out of order.

Besides the cilia, the air passages and tubes are also lined with a sticky fluid called mucus, which you first read about in Chapter 9. (See page xxx.) The mucus traps particles that have been inhaled. Mucus production greatly increases when certain pollutants are inhaled. This is a defensive response by the body. Normally the cilia would sweep out the mucus and much of the foreign matter. But when they do not function, the mucus builds up and narrows the tubes and air passages. Coughing results. Breathing is also more difficult.

Pollutants can also cause muscle spasms in the tubes of the lungs. During the spasms, the muscle contracts and gets thicker. This narrows the passageway in the tubes and makes breathing more difficult. Along with the muscle spasms, the membranes inside the tubes may swell. This results in more narrowing of the tubes and more difficulty in breathing.

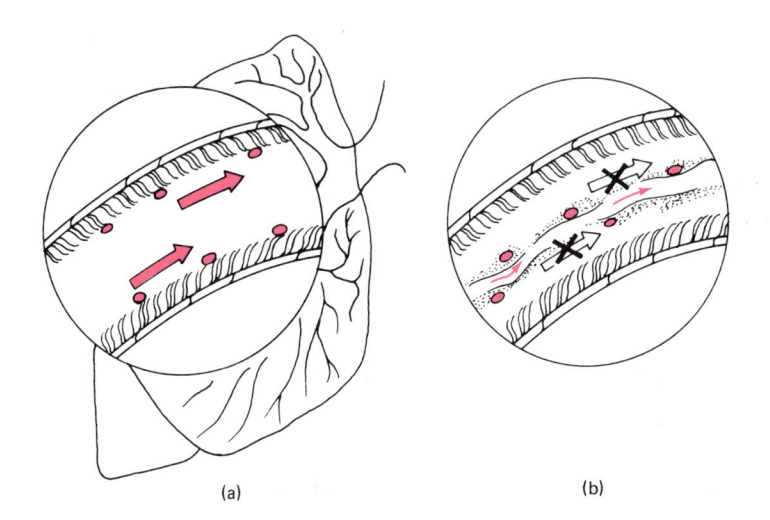

cilia: hairlike structures in organisms. In animal breathing systems, they "sweep" dust and other particles out of inhaled air.

Figure 17.15: The breathing tube on the left is normal. The ciliated cells move the drops of mucus along, and the muscle tissue is thin. Pollution causes the muscle to thicken, and the mucus layer is so thick and heavy that the cilia cannot push it along. These results you see at the right. This almost closes the tube, and gas exchange is cut off.

Challenges to Human Survival

Air pollution probably causes emphysema. Emphysema is the fastest growing cause of death in the United States. This disease destroys the elasticity and internal structure of the lungs. Smoking probably causes much of this disease. But as the following study shows, air pollution is probably another major cause. The lungs from 300 dead bodies were studied in St. Louis, Missouri. This city has typical urban pollution problems. Also studied were the lungs from 300 dead bodies from Winnipeg, Canada. This city has very little pollution. The lungs of the cigarette smokers were compared. The researchers found *four times more severe emphysema in the* lungs of the St. Louis smokers. This shows that something besides smoking is causing the emphysema. Studies of the nonsmokers from both cities also revealed some interesting facts. There were no bad cases of emphysema. But there was three times the number of light to medium cases of emphysema in the lungs of the St. Louis residents. These facts suggest that smoking may increase the damage to the lungs caused by air pollution.

Other diseases may be caused by air pollution. One big question life scientists are asking right now is this one: What other diseases may be caused by air pollution? Life scientists have not had time to come up with firm answers yet. But beginning studies are a bit frightening. For example, researchers have found that the number of lung cancer cases is twice as high in urban areas as compared with rural areas. Even when the cigarette smokers are not counted, the number of cases in urban populations is still one-third higher. Something other than cigarette smoke must be attacking the lungs of the city dweller!

How much is clean air worth? It is not going to be easy to correct the problem of air pollution. The automobile is the number one polluter. The government has passed laws requiring automobile makers to install exhaust control devices. But sometimes these do not work the way they are intended. There are also millions of cars already being used that do not have such devices. People will pay $300 extra to put an air conditioner in their cars. They will pay $125 for a stereo tape player. They will pay $90 for a radio. But will they pay $100 for a new antipollution device if it were guaranteed to reduce pollution? Find out for yourself the answer to that question. Ask car owners. Then ask a car dealer who knows people and their buying habits. Maybe you can find out for yourself how difficult the problem is going to be. Maybe you can also figure

Figure 17.16: The automobile is the nation's chief polluter of the air. Would you pay extra for a "clean car"? Would you want a plainer car than you might otherwise buy, but one that did not pollute? Do you think that technology can answer the problems of auto pollution? How?

out a better way to solve the problem. We hope so. We will all breathe a little easier if you do.

There are hard decisions ahead. You are living through a very difficult time in human history. Our supplies of oil and natural gas will slowly disappear during the next few years. Other sources of energy will have to be found to run our automobiles, trains, and planes. Other types of technology will have to be used to produce our electricity and run our factories. Because of these problems, people—you—will have many difficult decisions to make in the years ahead. For example, it is almost certain that our large reserves of coal will have to be burned to produce electricity. But how much coal should we burn? How much electricity should we produce? If we continue to use (and to waste) as much electricity as we do now, the problem will become more difficult. Increased coal burning is certain to increase air pollution that is already bad in some cities.

Now the hard questions. Will people settle for breathing more polluted air in order to obtain more electricity? Will you? Or will they settle for a smaller amount of electricity in order to breathe clean air? What about you? And what about pollution controls? Air pollution control technology is improving every year. But this technology is very expensive. Will people be willing to pay a much higher price for more electricity *and* clean air? Will you?

LESSON REVIEW *(Think. There may be more than one answer.)*

1. The reason that air pollution is more serious than water pollution is that
    a. polluted air contains more disease organisms than polluted water does.
    b. polluted air contains more dangerous chemicals than polluted water does.
    c. polluted air has less oxygen than polluted water does.
    d. polluted air cannot be avoided or processed like polluted water.

2. Air pollution from combustion is caused by
    a. factories and power plants.
    b. autos, planes, trains, and ships.
    c. schools and hospitals.
    d. homes.

Challenges to Human Survival

3. On the temperature-altitude graphs shown,
   a. graph A shows that the temperature increases with altitude.
   b. graph B shows that the temperature decreases with altitude.
   c. graph A shows the condition known as a temperature inversion.
   d. graph B shows the condition known as a temperature inversion.

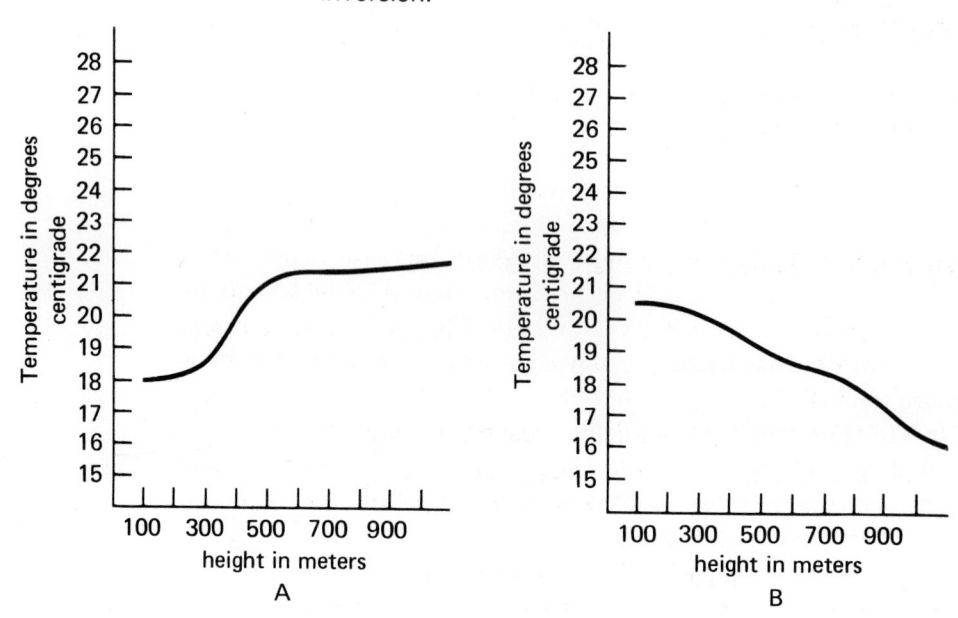

4. A temperature inversion can increase air pollution because
   a. as the temperature increases, the pollution increases.
   b. polluted air is kept near the surface and not allowed to rise.
   c. the oxygen decreases as the temperature declines.
   d. polluted air becomes more concentrated.

5. A factor that makes photochemical smog likely is
   a. high temperatures.
   b. bright sunlight.
   c. a large quantity of automobile exhaust fumes.
   d. lots of coal burning.

6. In the study of the lungs of people from St. Louis, Missouri, and Winnipeg, Canada, researchers found
   a. no emphysema in the lungs from Winnipeg.
   b. more emphysema in the lungs of St. Louis smokers than in the lungs of Winnipeg smokers.
   c. more emphysema in the lungs of St. Louis nonsmokers than in the lungs of Winnipeg nonsmokers.
   d. that severe emphysema did not exist in any nonsmokers.

7. Cilia in the breathing system
   a. function like brooms in continuous action.
   b. sweep out carbon monoxide and ozone pollution.
   c. sweep out foreign particles.
   d. are helpless against air pollution.

KEY WORDS

temperature inversion (p. 318)   cilia (p. 321)
photochemical smog (p. 319)

Applying What You Have Learned

1. Suppose you work in a factory that is polluting the environment. How could you argue to support what the factory is doing?
2. What does "balance in an ecosystem" mean?
3. A farmer sprays a field with a pesticide. Later, the balance of a stream ecosystem 5 kilometers (3 miles) away is upset. How could this happen?
4. What do you think should be done with glass bottles, aluminum and steel cans, and plastic wrappers that cannot be decomposed by bacteria or fungi?

Questions 5 and 6 refer to the following paragraph:

A certain town decides to expand its airport facilities. A natural wildlife marsh north of the town is chosen as the site of the new airport.

5. Will building an airport in this location upset the balance in the surrounding areas? How?
6. What does this decision tell you about this town's set of values?
7. What body system is most affected by air pollution?
8. Use the cost of an automobile air conditioner to support an argument for air pollution controls in an automobile.

Unit 6

Organisms

There are millions of different organisms living on the Earth. Some are green and carpet the Earth, and are called plants. Some run and swim, and graze and fly, and are called animals. Some are small and green, and wiggle and squirm in ponds, and cannot be called plants or animals. All of the Earth's organisms belong to one large community. All are interdependent. Only a few of these organisms can be brought to the pages of this unit. The ones that you will study are like representatives at a community meeting.

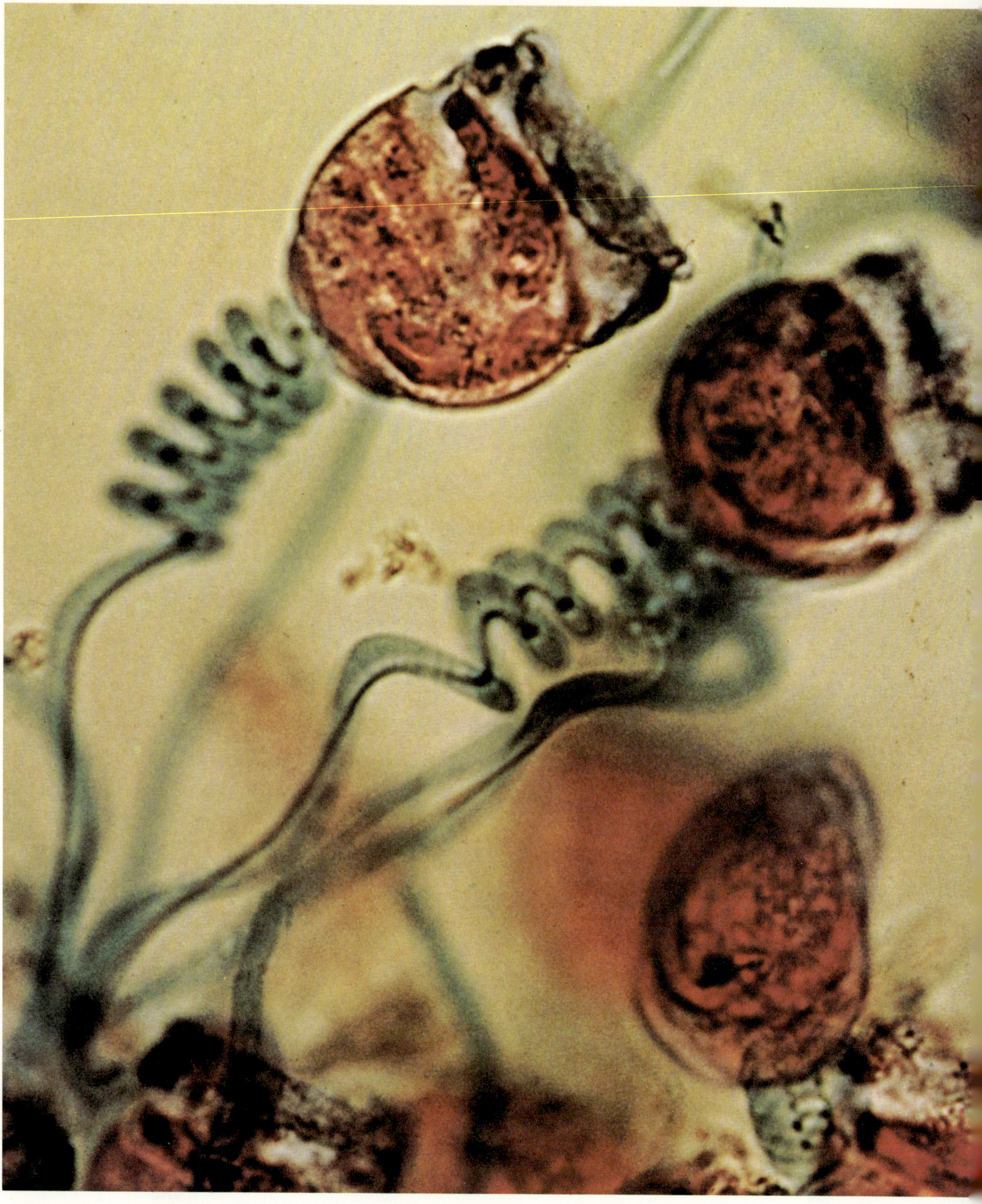

Figure 18.1: Animals or plants? Neither. *Vorticella* is a protist, one of many simple organisms that microscopes reveal.

Chapter 18

SIMPLE ORGANISMS

You can look at a tree and say: That is a plant. You can look at an elephant and say: That is an animal. But what should you say when you look through a microscope at a single-celled organism? The names "plant" and "animal" are no longer useful. Other names must be used to describe them. You will be able to name and describe the most common groups of simple organisms when you finish this chapter.

VIRUSES AND MONERANS

Viruses are not true organisms. When we use the term "organism" we mean a living individual. The **virus** is not an organism. Most life scientists would say either that the virus is "nonliving" or that it is a "borderline" form of life.

A virus is much smaller than a cell. The virus does not contain the organelles that a cell does. In fact, it is only when the

virus: a borderline form of life, much smaller than a cell. It acts like a living thing only when inside a living cell.

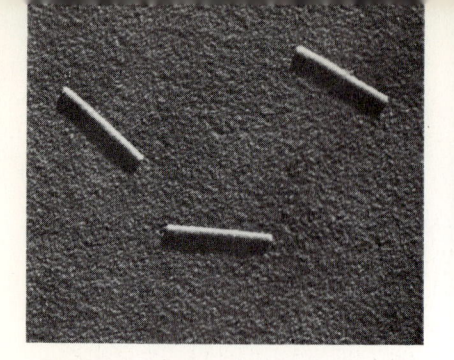

Figure 18.2: The picture on the left is an electron micrograph of tobacco mosaic viruses. The viruses on the right are the same kind, but much of their outer coat has been digested away. The barely visible strand remaining is RNA, which contains the virus genes and is similar to the DNA in ordinary cells.

virus is inside a living cell that it acts like an organism. Inside a living cell, the virus can reproduce itself. When a virus is outside a cell, it is as nonliving as a grain of sand. (See Figure 18.2.)

When a virus is inside a cell, it may reproduce itself many times. One virus in a living cell may make 200 to 500 copies of itself. The process of virus reproduction kills the cell. The cell breaks open and the viruses spill out. The viruses then enter other cells, reproduce, and destroy them. Now you know why the term "disease" is always thought of when a virus is mentioned. A disease results when viruses invade and destroy the cells of a tissue. The common cold, smallpox, chicken pox, measles, mumps, and some forms of pneumonia are all examples of virus diseases.

Bacteria are true organisms. Bacteria are very small single-celled organisms. There are about 2000 different kinds of bacteria. And they are everywhere within the biosphere. The

Figure 18.3: Most bacteria have one of these three shapes: rods, spheres, or spirals. A few bacteria cause disease, but many are important decomposers in ecosystems.

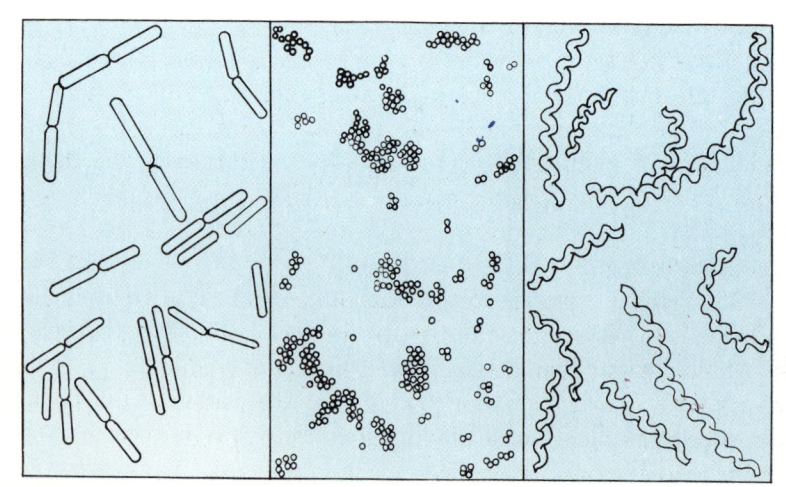

numbers of organisms within the populations are difficult to imagine. Picture in your mind a teaspoon of rich soil. Can you imagine that soil containing 2½ billion organisms? That is about the number of bacteria that such a sample of soil would contain.

Life scientists have grouped all the organisms in the biosphere into four large groups called **kingdoms.** The bacteria have been placed in the kingdom *Monera* (mŏn′ə rə). Because they are in this kingdom, we call them **monerans.** What do all the monerans have in common? The main thing is that the cells do not have a distinct nucleus. In moneran cells, the DNA (the main material of the nucleus) is scattered throughout the cell.

A few bacteria cause disease, so they are often thought of as bad organisms. It is unfortunate that bacteria are thought of in this way. You may recall from Chapter 3 the niche of the decomposers in ecosystems. A great many of the decomposers are bacteria. Through their activities, nutrient materials from dead organisms are unlocked. The nutrients are thus made available for organisms that are living. Important gases like carbon dioxide are released. Plants use the carbon dioxide in photosynthesis.

Life scientists do not like to label organisms "bad" or "good." They would rather say that all organisms "have their place." However, to emphasize the extreme importance of the bacteria, we could ask a life scientist: Are bacteria good or bad? A typical answer would likely be: Definitely good!

The blue-green algae are also monerans. Besides the bacteria, there is one other group of organisms in the kingdom of Monera. This is a group known as the blue-green algae. There are about 2500 different kinds of blue-green algae in the biosphere. A great many of them are not blue-green in color, but are red, yellow, and black.

Some of the blue-green algae exist as single cells. Others are groups of cells joined together. Often the cells are joined together in a side-by-side manner. This produces a long threadlike body called a **filament.** (See Figure 18.4.)

Blue-green algae are found almost any place there is water. An aquarium is a good place to find a specimen for examination. The kind of blue-green algae pictured in Figure 18.4 often troubles aquarium owners by growing on the sand and on the glass. Many kinds of blue-green algae can live in severe environments. For example, they are one of the few forms of life that can live in very hot springs.

Blue-green algae carry on photosynthesis. That is because their cells contain a green coloring material called **chlorophyll** (klôr′ə fil). Scientists have found that only cells that contain chlorophyll can carry on photosynthesis. Because they carry on photosynthesis, the blue-green algae are producers in the marine and freshwater ecosystems in which they live.

kingdom:
one of four large groups in which all organisms are classified. The groups are: Monera, Protista, plant, and animal.

moneran:
a simple organism of the kingdom Monera. There is no distinct nucleus in a moneran cell.

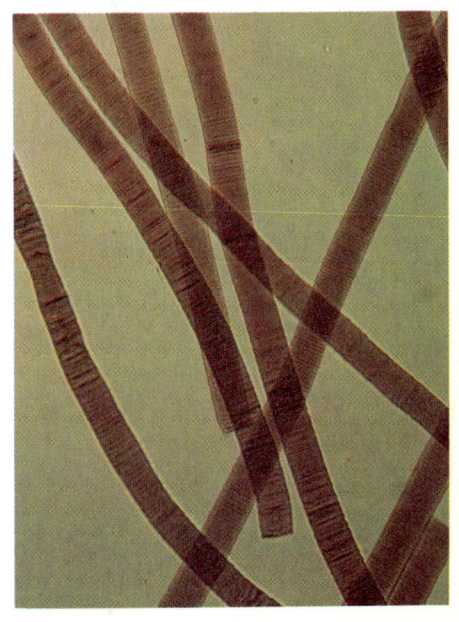

Figure 18.4: This photomicrograph shows the structure of *Oscillatoria*. These blue-green algae are a common pest in aquariums.

filament:
any threadlike structure

chlorophyll:
the green substance in cells that carry on photosynthesis. In plants, chlorophyll is found in the chloroplasts.

Figure 18.5: This is a hot spring in Yellowstone National Park. The bright colors that you see here are from the mats of algae, most of which are "blue-greens."

LESSON REVIEW *(Think. There may be more than one answer.)*

1. The cells in Figure A have a nucleus and chloroplasts. They have to be
    a. viruses.
    b. bacteria.
    c. blue-green algae.
    d. none of the above.

2. The objects in Figure B are not considered to be organisms, but they can reproduce inside cells. They have to be
    a. viruses.
    b. bacteria.
    c. blue-green algae.
    d. none of the above.

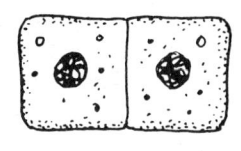

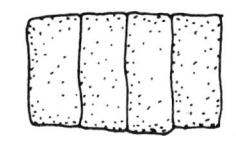

Simple Organisms

A   B   C   D

3. The cells in Figure C have no distinct nucleus but do carry on photosynthesis. They have to be
    a. viruses.
    b. bacteria.
    c. blue-green algae.
    d. none of the above.

4. The cells in Figure D are decomposers and have no distinct nucleus. They have to be
    a. viruses.
    b. bacteria.
    c. blue-green algae.
    d. none of the above.

5. Viruses
    a. are the biosphere's smallest organisms.
    b. can only reproduce inside cells.
    c. cause disease by destroying cells.
    d. are as nonliving as a grain of sand when outside cells.

6. Bacteria
    a. have no distinct nucleus.
    b. are enemies of humans in most ecosystems.
    c. usually are shaped like rods, spheres, or spirals.
    d. are abundant in the soil.

7. Which of the following terms best matches each phrase?
    a. virus   d. filament
    b. kingdom   e. chlorophyll
    c. moneran

    _____ organism without a distinct nucleus
    _____ reproduces only in cells
    _____ one of the four major groups of organisms
    _____ cells strung together in a threadlike manner
    _____ green pigment responsible for photosynthesis

KEY WORDS

virus (p. 329)
kingdom (p. 331)
moneran (p. 331)

filament (p. 331)
chlorophyll (p. 331)

THE PRODUCER PROTISTS

protist:
a simple organism of the kingdom Protista. Protist cells have a nucleus.

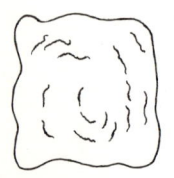

MONERAN CELL
Nuclear material (DNA) scattered throughout cell

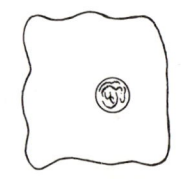

PROTIST CELL
Nuclear material confined to *nucleus*

Figure 18.6: Moneran cells do not have a distinct nucleus. Their nuclear material is scattered throughout the cell. Protist cells do have a distinct nucleus.

Most algae are protists. We have already mentioned the kingdom Monera as one of the four kingdoms to which all organisms belong. The second kingdom that we shall discuss in this chapter is the kingdom of *Protista*. The members of this kingdom are called **protists.**

The protists include a wide variety of organisms. All of the algae, except blue-green algae, are protists. What do all of the protists have in common? First of all, their cells have a nucleus in which the DNA is enclosed within a membrane. In monerans, you will recall, the DNA is scattered throughout the cell.

All of the protists are simple organisms. Many are single-celled. Those that are multicellular do not have bodies that are well organized. In some, the cells are stuck together. For these, the body forms a filament or irregular mass. And in some protists, one sees the beginnings of the tissue level of organizations. This means that such protists have groups of cells that work together as a team to perform a specific function.

Green algae are common in fresh water. There are several kinds of algae that belong to the kingdom Protista. They are placed into separate groups mainly because of different pigments (colored materials) that are in cells. Chlorophyll, as we

Figure 18.7: This is green algae floating on a pond. Such growths are sometimes called "scum" or "moss." Both names are incorrect. In the next chapter you will see that moss plants are very different from green algae.

Figure 18.8: When a bit of the green algae is removed from the pond (Figure 18.7), one can see that it is made up of threadlike filaments.

mentioned earlier, is a green pigment. But there are several kinds of chlorophyll, each a different shade of green. In addition, some kinds of algae have red, brown, or yellow pigments, or various combinations of these.

One group of algae has mostly green pigments. The members of this group are called green algae. They are very common in fresh water. In many freshwater ecosystems, they are among the key producers. Thus, green algae are at the beginning of many food chains.

In Figure 18.7, you see a typical pond in midsummer. Floating on the surface are huge masses of green algae. If you picked up a handful, it would probably feel slick or slimy. Often such growths are called "pond scum." Actually, these growths are masses of filaments—threadlike chains of cells that are attached side by side. In Figure 18.9, you can see how these filaments appear when viewed with a microscope.

Diatoms are the main producers in aquatic ecosystems. Another major group of algae is the golden-brown algae. This group contains many thousands of kinds of diatoms. The diatoms are the most important of all the algae. They are the chief producers in aquatic ecosystems, both freshwater and marine.

Organisms

Figure 18.9: A photomicrograph of *Spirogyra* shows the structure of a single filament of green algae. A blue stain marks the nuclei and cell walls. (Courtesy Carolina Biological Supply Co.)

Stop and think of the important role of the diatoms. As the chief producers in aquatic ecosystems, they are at the beginning of most of the food chains. All forms of animal life in fresh water and in the oceans depend upon them. Also remember that our oxygen supply is constantly being replaced by producers as they carry on photosynthesis. Life scientists estimate that diatoms carry on 60 to 70 percent of the photosynthesis within the biosphere. Thus, as we use up the Earth's oxygen supply, we depend mostly upon diatoms to replace it for us.

Diatoms do not look golden-brown when you see them in nature. Have you ever taken a rock from a stream or pond that is covered with a brownish slime? If so, you have probably seen diatoms as they appear in nature. It is a pleasant surprise to observe some of that brown slime with a microscope. Then you see the beautiful golden-brown color of the diatoms which make up the slime. You also see their beautiful cell walls, which are

Figure 18.10: The many shapes of diatoms are shown in these two photomicrographs. Freshwater diatoms are shown in the left photograph; marine diatoms, on the right.

Figure 18.11: This kelp was washed up on a beach near Santa Barbara, California. Many kelps are much larger than this.

made of a glasslike material formed in many patterns. Many freshwater diatoms are shaped like boats or canoes. Marine diatoms are round and triangular. (See Figure 18.10.)

Most brown and red algae are marine. The brown algae and the red algae are two other important groups of algae. Most of these algae live in marine waters.

The most common forms of brown algae are the **kelps**. The kelps are large floating seaweeds that are often found washed up on beaches of North America. Some of the kelps are very large masses of cells, many of which are organized into tissues. A few of the kelps can grow to lengths of 50 meters (55 yards) or more.

The red algae are more commonly found in warm marine waters. A large number of the so-called seaweeds are red algae. They do not grow as large as the brown algae, but some forms grow to be 3 meters (3.3 yards) or more long.

One important product is made from the cell walls of red algae. It is called **agar** (ā′gër). Agar is used to "stiffen" puddings, as capsules for vitamin pills, and as a base for cosmetics. Agar is also used in life science laboratories as a hardener for nutrient material on which bacteria are grown.

kelp:
a large, floating seaweed, the most common form of brown algae

agar:
a jellylike substance made from the cell walls of red algae

Organisms

LESSON REVIEW *(Think. There may be more than one answer.)*

1. Protists
   a. include all monerans.
   b. are all single-celled organisms.
   c. have their nuclear material scattered throughout the cell.
   d. have a distinct nucleus that is enclosed within a membrane.

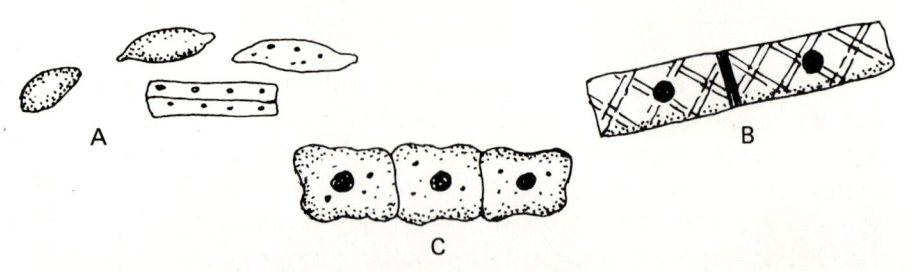

2. The protists in Figure A have cell walls of a glasslike material and are
   a. green algae.
   b. golden-brown algae.
   c. the largest algae.
   d. the chief producers in aquatic ecosystems.

3. The protist in Figure B is often called moss or pond scum. It most likely belongs to the group of
   a. red algae.
   b. brown algae.
   c. green algae.
   d. golden-brown algae.

4. The protist in Figure C is used to make agar, the hardening material in puddings. It belongs to the group of
   a. red algae.
   b. brown algae.
   c. green algae.
   d. golden-brown algae.

5. What type of algae has a species that grows up to 50 meters (55 yards) long?
   a. green algae
   b. diatoms
   c. brown algae
   d. red algae

6. Which of the following terms best matches each phrase?
   a. protist   c. agar
   b. kelp

   _____ substance used in puddings that is made from the cell walls of red algae

   _____ one of the largest algae

   _____ organism that belongs to the kingdom Protista

338

KEY WORDS

protist (p. 334) agar (p. 337)
kelp (p. 337)

THE CONSUMER PROTISTS

Consumer protists are like small animals. Would you like to see a circus? Here is how to do it. Put some pond water or water from a mud hole into a jar. Add some dead plant matter. Then wait until the jar begins to give off a bad odor. A whitish scum will usually form on the surface of the water. Use a medicine dropper to remove a drop of the water from near the surface. Make a slide of this water and observe it with a microscope. You will not see lions and tigers, or elephants and giraffes. But you will see hundreds, if not thousands, of creatures—whirling and diving, spinning and flopping. The performance of these creatures will rival that of any circus you will ever see.

What are these creatures that live in such an evil-smelling environment? The man who first saw them called them "little animalcules," meaning that they were miniature animals. Later they were called *protozoans*, which means "first animals."

But are the creatures swimming around in stagnant water really animals? Who is to say? Certainly most of them are like animals. They can move. That is an animal characteristic. Most of them are consumers. They eat bacteria, algae, and other organisms like themselves. That too is an animal characteristic. But are they animals? Today, most life scientists would probably say no. Often the name **protozoans** is still used to describe them. But from a technical standpoint, they are placed in the kingdom Protista and are called animallike protists. The term "animal" is reserved for multicellular organisms with well-organized bodies and the animal characteristics already mentioned.

protozoan: an animallike protist

The paramecium is a common protozoan. In the bad-smelling water that we described, one kind of creature will almost always be present. It is called the *paramecium* (păr′ə mē′shē əm). This is the slipper-shaped organism in Figure 18.12.

The paramecium is a single-celled organism, as are almost all the protozoans. It lives in stagnant water. It feeds mostly upon bacteria. When dead plant matter is added to stagnant water, the bacteria multiply rapidly. They decompose the plant matter, producing more nutrients and the gaseous odors which

Organisms

Laboratory Activity

Can You Find These Organisms?

PURPOSE   A jar of water collected from a mud puddle or pond may not look interesting at first glance, but take another look. The pictures on this page show just some of the organisms you may find living in that murky water. In this investigation, you will hunt with a microscope for the many fascinating organisms that make their homes in this aquatic environment.

MATERIALS

pint-size containers
stagnant water
mud
hard-boiled egg yolk

medicine dropper
slides and cover slips
microscope

PROCEDURE

1. Collect a jar of water from a pond or mud puddle. Be sure to add a handful of mud or dirt to the water in the jar.
2. Crumble a bit of hard-boiled egg yolk into the water in the jar. The egg yolk will serve as food for bacteria, and protozoans feed on bacteria.
3. Take samples of water from the top, the sides, and the bottom of the jar. Prepare slides of the water samples. Examine the slides with a microscope.
4. Look at the pictures on this page. All the organisms pictured live in still water or slow-moving water. You can find them in ponds at parks or golf courses, in mud puddles, or wherever there is stagnant water. Mud and dead plant matter in a pond or mud from a puddle often contain cysts from which protozoans develop. Which of the organisms can you identify on your slides? Make a list of them.

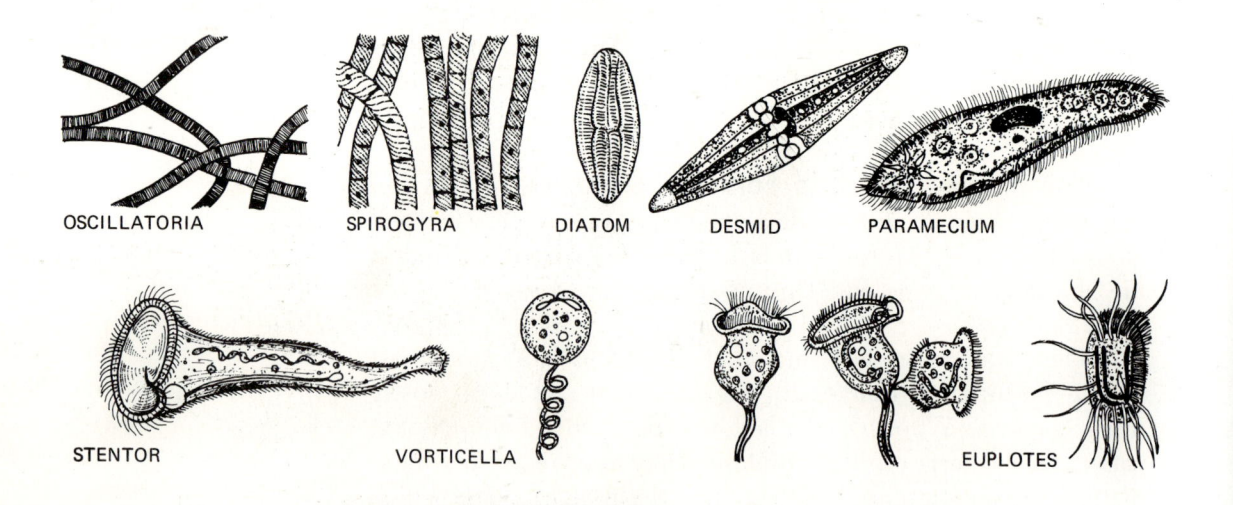

OSCILLATORIA   SPIROGYRA   DIATOM   DESMID   PARAMECIUM

STENTOR   VORTICELLA   EUPLOTES

Career Spotlight

What did it take to do the drawings on the opposite page? Artistic talent was needed. So too was training in science and technical areas. The drawings were done by a *technical illustrator*. Below, you can see one at work.

Technical illustrators have been called "the eyes of science and industry." A part of their job is to show clearly how something works or is put together. They also create drawings to show things you usually cannot see. For instance, turn to page 445, and look at Figure 24.14, which is a technical drawing of a snake's jaw. It shows you (1) a view of a snake you usually do not see, (2) what the jaw bones look like, and (3) how the jaws work. This drawing should make it clear why snakes eat in a way different from other animals.

Most of a technical illustrator's work involves making three-dimensional drawings from blueprints. Blueprints are technical plans for buildings and equipment. The technical illustrator's drawing, from the blueprint, makes it easier for untrained people to know what the final product will look like.

To be a technical illustrator demands artistic skill and proper training. Education in art techniques, drafting, and blueprint reading are essential. If, as an artist, you want to specialize in the life sciences, college courses in these areas will provide valuable background. For instance, to work in the field of medical illustration, you would study related subjects, such as anatomy.

Organisms

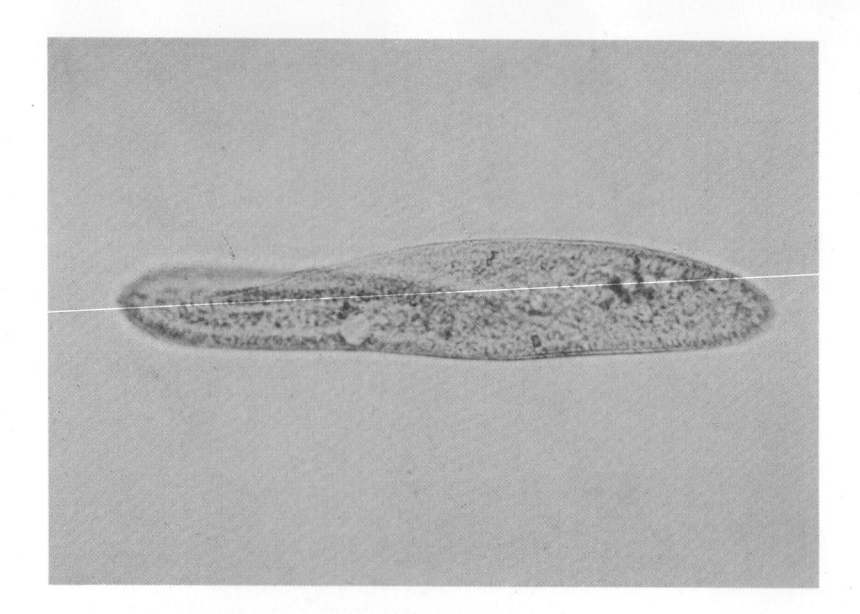

Figure 18.12: The single cell of a paramecium is covered by hairlike extensions called cilia. A paramecium swims by moving the cilia.

ciliate:
a protozoan that moves by waving the cilia that surround its cell wall

we smell. The added nutrients cause the bacteria to increase their rate of multiplication. Often there are so many that they form a scum on the water. Then the paramecia begin to multiply rapidly. After a few days, the water is teeming with millions of them.

Paramecia swim through the water by means of hairlike extensions that wave back and forth. These hairlike extensions, like those in the human breathing system, are called *cilia*. Many protozoans have cilia surrounding the cell that is their body. These protozoans are called **ciliates.**

The ameba is a very simple protozoan. Another common protozoan is the *ameba*. It is a much simpler organism than a paramecium. Its body is a cell that may take any shape. When it moves, a part of the cell will flow out and form a fingerlike extension. Then the rest of the cell will flow in that same direction.

An ameba eats by flowing around and surrounding its food. The food becomes enclosed in its body. (See Figure 18.13.)

An ameba reproduces by splitting in two. First the nucleus divides into two parts. Then the cell pinches in until two small amebas result. This method of reproduction is typical of all the protozoans.

The cyst enables protozoans to survive drought.
What happens when a pool of stagnant water evaporates? Do all of the protozoans die? Not all of them. Many survive.

Under difficult living conditions, such as lack of water, many protozoans make a tough outer shell-like covering that encloses

Simple Organisms

Figure 18.13: The shape of an ameba is always changing as the cytoplasm flows into the fingerlike projections. The nucleus is the circular saucer-looking body near the middle. (Courtesy Carolina Biological Supply Co.)

their body. This is called a **cyst** (sĭst). In the cyst, they can live for many months. The cysts are light and can be blown great distances by the wind. When favorable conditions exist, the wall of the cyst will break. The protozoan then resumes an active life.

One can often take grass or hay, put it in water, and have protozoans within a few days. The reason: Cysts containing live protozoans are on the plants. In the water, they hatch out of the cyst, reproduce, and begin again to lead their busy way of life.

cyst:
the hard, shell-like covering protozoans form in time of drought

Figure 18.14: An ameba feeds on a paramecium by first surrounding it and then taking it into its body.

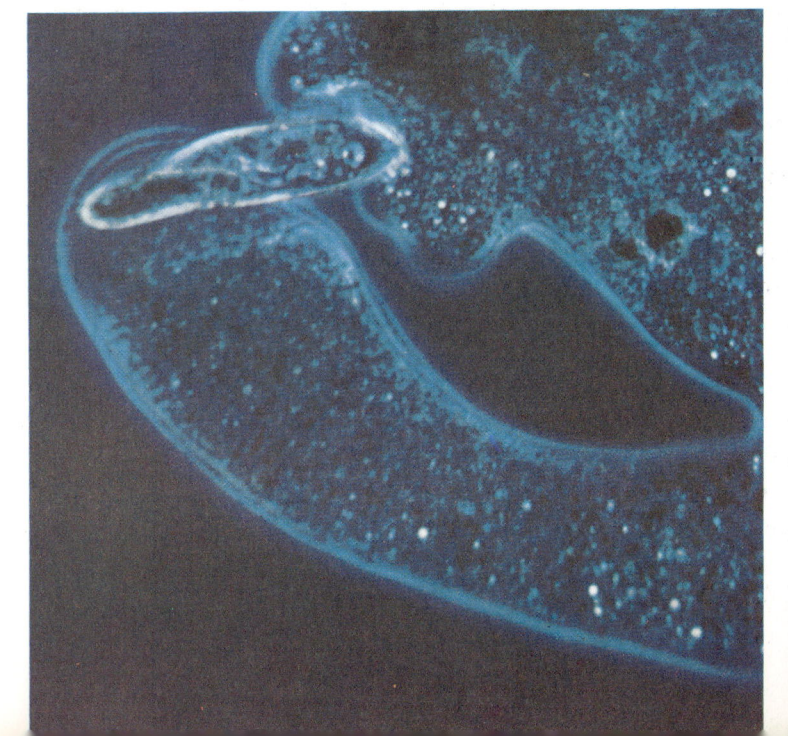

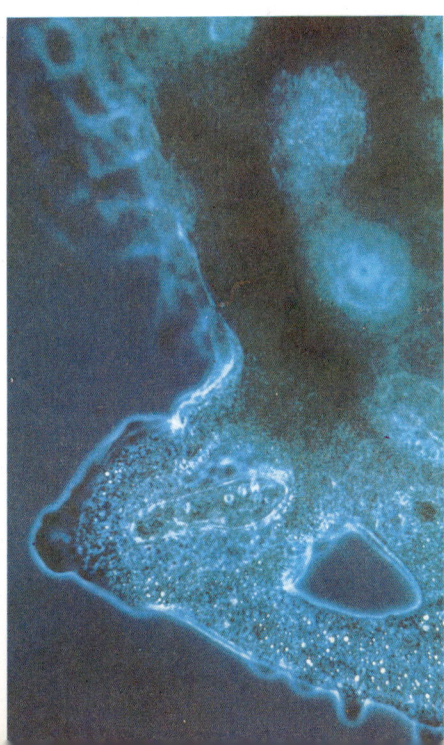

Organisms

LESSON REVIEW *(Think. There may be more than one answer.)*

1. Protozoans
   a. move like animals.
   b. are producers.
   c. are correctly called animallike protists.
   d. are correctly called animals.

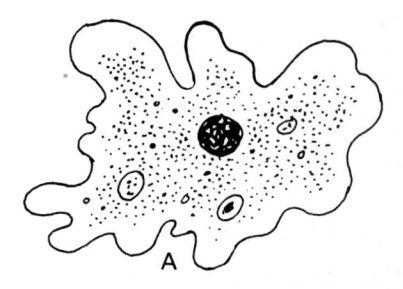

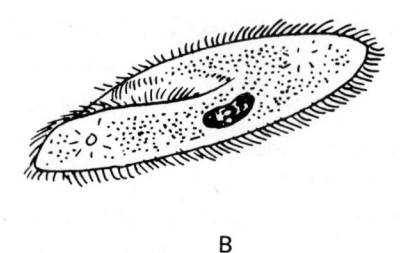

A   B

2. The protist in Figure A
   a. is an ameba.
   b. is a paramecium.
   c. moves slowly.
   d. eats by flowing around its food.

3. The protist in Figure B
   a. is an ameba.
   b. is a paramecium.
   c. feeds on decomposers.
   d. is a ciliate.

4. Protozoans
   a. can sometimes survive out of water.
   b. can sometimes be transported by the wind.
   c. reproduce in structures called cysts.
   d. undergo development from single to multicellular organisms in cysts.

5. Which of the following terms best matches each phrase?
   a. protozoan   c. cyst
   b. ciliate

   _____ tough outer covering that enables many protozoans to survive out of water

   _____ an animallike protist

   _____ a protist that moves by the beating motion of hairlike extensions

---

KEY WORDS

protozoan (p. 339)   cyst (p. 343)
ciliate (p. 342)

344

Applying What You Have Learned

1. What characteristic do members of the kingdom Monera have in common?
2. Name the members of the kingdom Monera and describe their roles in nature.
3. How are algae separated into different groups?
4. List two important niches of diatoms.
5. In what habitat would you find diatoms shaped like triangles?
6. List some diseases caused by viruses.
7. Are viruses living organisms? Give reasons for your answer.
8. What do people usually mean when they say that a pond is covered with moss or scum?
9. How do some protists show the tissue level of organization?
10. Describe how an ameba moves from place to place.
11. Why are the blue-green algae placed in the kingdom Monera and not in the kingdom Protista with the rest of the algae?
12. What are cilia in protozoans used for? What are they used for in land animals?
13. When all the water in a pond evaporates, what happens to the protozoans?
14. Give an example of simple organisms that will fill: (a) a decomposer niche, (b) a producer niche, (c) a consumer niche.
15. Are all producers green in color? Explain your answer.

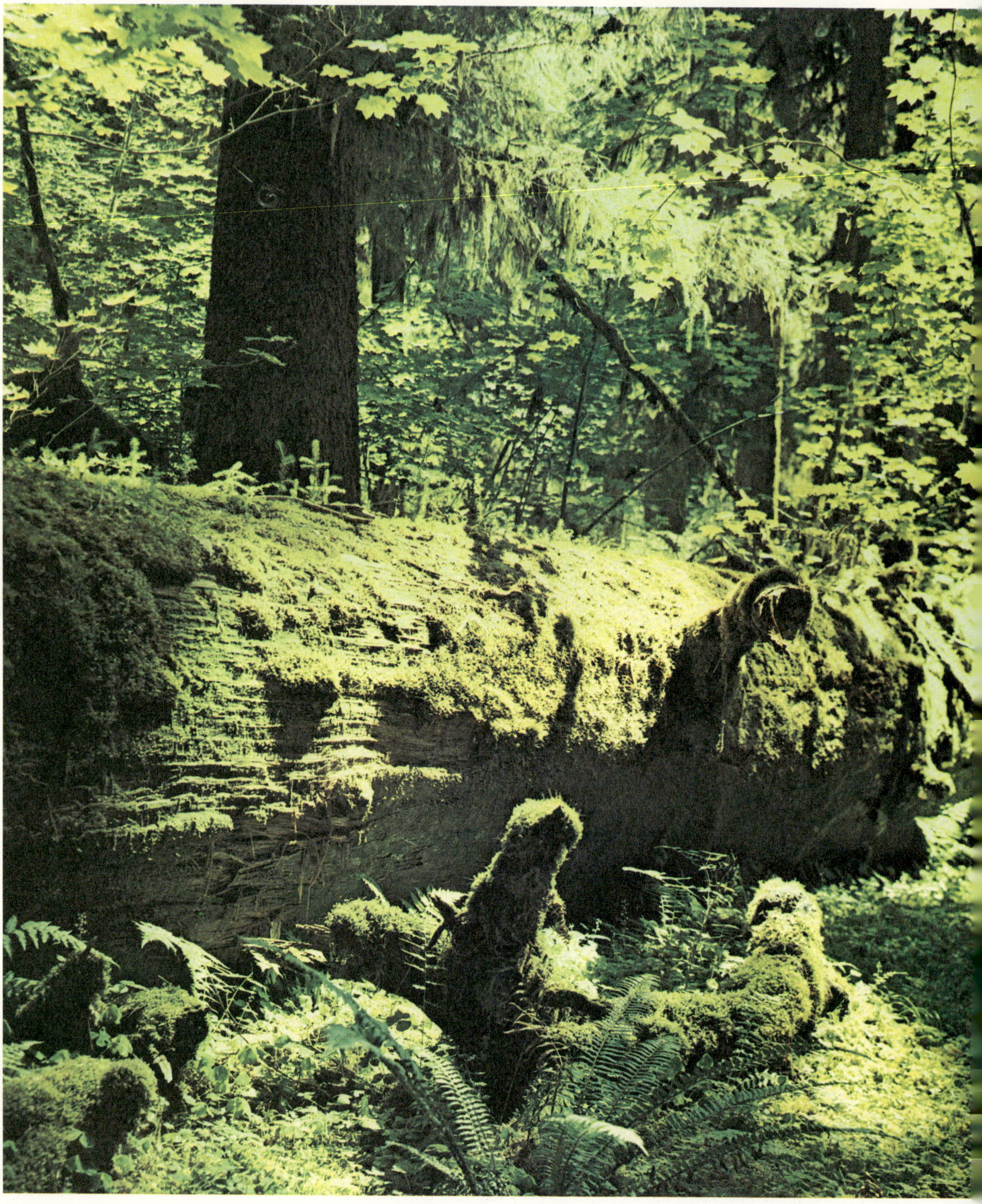

Figure 19.1: Fungi, mosses, and ferns are a major part of the lush vegetation of Olympic National Park in the northwestern part of the state of Washington.

Chapter 19

FUNGI, MOSSES, AND FERNS

This chapter is about plants and some organisms that are almost plants. But it is not about the kind of plants that you see everywhere in lawns, parks, fields, and forests. Those familiar organisms are seed plants, and we shall deal with them in the next chapter. The plants and "almost" plants that you will study in this chapter are less visible in our landscapes. As you will discover, this does not make them less important.

FUNGI

What are fungi? The *fungi* include many thousands of kinds of organisms. Some are single-celled; others are multicellular. Some of the multicellular fungi have tissues and organs. Then how are they all alike? It is what they do not have that counts: All of them lack chlorophyll. They do not have machinery for carrying on photosynthesis. All of the fungi obtain their nu-

Organisms

saprophyte:
an organism, such as a fungus, that gets nourishment from dead organisms and in the process helps decay them

parasite:
any plant or animal that does not make its own food, but lives on or in living plants or animals

mold:
a type of fungus

trients from other organisms. Some live on dead organisms or on nutrients in the soil that were once a part of organisms. These are **saprophytes** (săp′rə fīt′). Others live in or on the bodies of living organisms. These are **parasites.**

The fungi are usually placed in the kingdom Protista, along with protozoans and algae such as those you studied in the last chapter. But some life scientists prefer to call them plants. Others consider them "almost" plants. Still others consider fungi less plantlike than algae.

Molds are common fungi. There are a great number of different fungi that are called **molds.** A typical mold starts its life as a single cell enclosed within a protective covering. This cell with its protective covering is called a *spore*. In the photo, the round objects are spore cases filled with spores. Under favorable conditions, the spore covering breaks. The cell enlarges and the nucleus divides many times. The cell continues to grow, but for a long time it does not divide. (This is unusual. In other organisms, the cell from which they develop grows and divides.) Eventually the mold growth becomes a large mass of threadlike filaments.

Chemicals produced by the mold filaments flow out on the material that the mold is growing upon. The chemicals break down or digest the material. Some of the digested material is taken into the mold and used as food. Eventually parts of the filaments will begin to divide and produce a mass of cells. These develop into a reproductive organ, which in turn produces more spores. (See Figure 19.2.)

Mold spores are very light. They are easily carried by movements of air. Thus, they can be found in many samples of air. Your own classroom probably has at least a hundred mold spores in the air. Many hundreds (or thousands) more are present in dust on the window ledges, furniture, and floor.

Figure 19.2: A mold spore germinates (a) and develops into a mass of filaments (b). Chemicals from the filaments digest the bread (c), releasing nutrients that the mold uses as food.

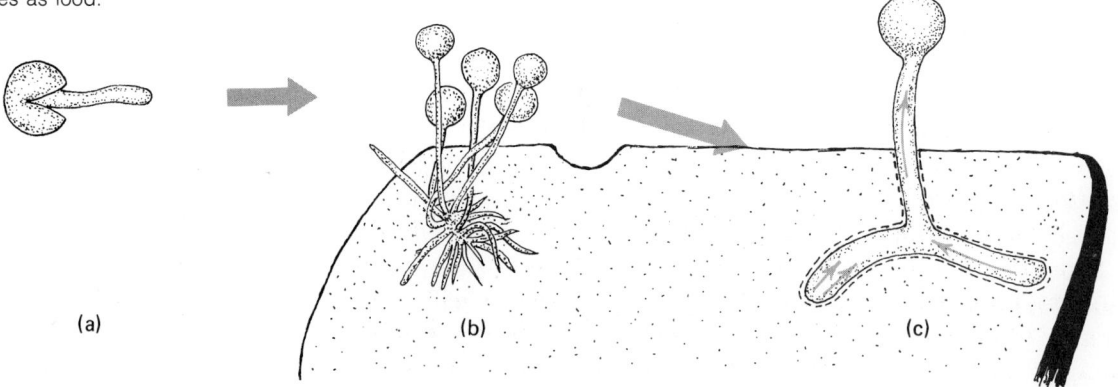

(a) (b) (c)

Fungi, Mosses, and Ferns

Figure 19.3: Molds may function as decomposers in aquatic ecosystems also. The mold covering the dead fish above will help eliminate it from the stream in which it lived. The mold at the right is the mold that produces the antibiotic called penicillin. It is growing in a laboratory culture dish.

Molds are important fungi. Why are molds important? There are at least three good reasons. First, they are a key group of decomposers in terrestrial ecosystems. They are very important in forest ecosystems, where they break down dead trees and leaves.

A second reason for their importance has to do with antibiotics—chemicals that kill bacteria. Penicillin is a well-known antibiotic. It is used to kill certain kinds of disease-causing bacteria. Penicillin was first discovered as a product of a mold. When its importance became known, whole factories were developed for growing the mold that produced penicillin. Now penicillin, like most other mold antibiotics, is made in chemical laboratories. Life scientists duplicated the mold's manufacturing process.

Do you like cheese? Even if you do not, you will have to admit that cheese is an important food. Cheeses are made when certain kinds of mold grow in milk. Each of the different cheeses that you will find in a supermarket was made by a different kind of mold. This is a third reason why molds are important.

There are many other kinds of fungi. There are many other kinds of fungi besides molds. Have you ever seen or heard of yeasts, rusts, smuts, toadstools, puffballs, and mushrooms? They are all fungi.

Yeasts are single-celled fungi. They are found on most fruits. Chemicals from the yeast break down the sugar in fruits. In the process, they make an alcohol. This is the way wine is made from grapes. When yeasts break down sugar, carbon dioxide is a by-product. That is why yeasts are used in making bread. The carbon dioxide gas produced in bread dough causes it to expand, or rise.

Organisms

Laboratory Activity

How Many Mold Spores Can You Collect in Five Minutes?

PURPOSE

Mold spores are light in weight and easily carried in the air. If they fall on a favorable location, they will grow to produce large masses of threadlike filaments. In this investigation, you will prepare a mixture of materials to support the growth of mold. You will also determine how many mold spores are present in different air samples. You should find out if the number of mold spores in the air varies in different locations. You should also find who can collect the most mold spores in 5 minutes. You will work in a team of five.

MATERIALS

1 medium size potato
paring knife
1-cup measuring cup
2 one-quart saucepans and covers
hot plate
potholder
3 envelopes of unflavored gelatin
plastic food wrap
6 bowls
4 partners

PROCEDURE

A. Prepare the *medium,* or food source, that will support the growth of mold.

1. Wash and peel one potato. Slice the potato into 1-cm cubes. Add the potato cubes to 1½ cups of water in a saucepan. Cover the pan. Boil the potatoes for 15 minutes.

2. Use a potholder and pour 1 cup of the water from the boiled potatoes into a measuring cup. CAUTION: Do not burn yourself. Transfer that water to another saucepan. Pour the remaining water from the boiled potatoes into the measuring cup. Add water to the measuring cup to bring the volume to 1 full cup. Add this cup of water to the second saucepan. Discard the potatoes.

3. Heat the 2 cups of potato water to boiling. Add 3 envelopes of gelatin. Stir and cover. Boil the mixture for 3 minutes.

Fungi, Mosses, and Ferns

4. Remove the saucepan from the heat. Add 4 teaspoons of sugar to the water mixture. Cover the saucepan and cool for about 5 minutes.

5. Prepare six pieces of plastic wrap. Each piece of plastic wrap should be large enough to cover one of the six bowls.

6. Pour some of the cooled solution into each of the six bowls. The amount of solution in each bowl should be about 1 cm in depth. Cover the bowls with plastic wrap.

7. Let the medium harden. This should take about 3 hours at room temperature.

B. Collect the mold spores.
1. Set aside one covered bowl to be used as a control. Label this bowl "Control."

2. Collect as many mold spores in each of the five remaining bowls as you can in 5 minutes. To do this, remove the plastic wrap from the bowl. Place the bowl in an area you think will be rich in mold spores. Or walk in several different locations with the bowl. The methods of collecting are up to you. Remember, the mold spores must be collected from the air. You must not touch the medium with dust or debris. Plan your strategy before you begin collecting. Each of your four partners will collect mold spores by the technique he or she develops.

3. Cover the bowl with plastic wrap at the end of the 5-minute collecting period. Label it with the location or locations where you did your collecting.

4. Mold grows best at room temperature in moderate light. Find a location where the bowl will be undisturbed for several days.

C. Count the mold colonies in your bowl.
1. The spores that you collected will grow into fuzzy, circular colonies. Each circle represents the growth from one spore. Count the number of colonies in each bowl. Record this number. Count the number of colonies in the control medium. Subtract the number of colonies in the control medium from the number of colonies in each bowl.

2. Compare the number of mold spores collected by different members of the class. Who collected the most mold spores? Where did they collect them? What techniques did they use? What was the total number of spores collected by the class?

Figure 19.4: These toadstool mushrooms are growing in a forest floor of dead leaves. Each has a network of filaments underground.

The holes or spaces in bread were formed by little pockets of carbon dioxide that were in the bread dough.

Rusts and *smuts* are parasitic fungi. They grow on agricultural crops such as wheat and corn. They are pests that farmers would like to rid their crops of.

Toadstools and *puffballs* are large fungi that you may find growing in meadows and forests. For every toadstool or puffball that you see, there is a mass of filaments hidden from view. The part that you see is the reproductive structure. The unseen filaments feed the organism with products of their decomposition activities.

Mushrooms are fungi that can be used for food. They can be collected from nature. However, this is very dangerous unless you are an expert on this subject. Some fungi that look like edible mushrooms are poisonous when eaten. Therefore, you should be trained by an expert before you collect and eat wild mushrooms. Otherwise, it is safer to buy mushrooms that have been grown and packaged for you.

LESSON REVIEW *(Think. There may be more than one answer.)*

1. The protist in Figure A often grows on bread. It
    a. has single cells packed inside the round, black objects.
    b. has chlorophyll in its filaments.
    c. is a saprophyte.
    d. has round, black objects that are called spores.

Fungi, Mosses, and Ferns

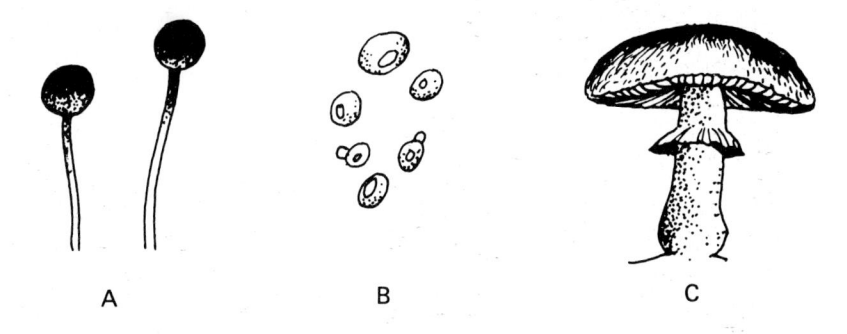

A   B   C

2. The protist in Figure B is single-celled and produces alcohol from sugar. It
   a. also produces penicillin.
   b. also produces carbon dioxide from sugar.
   c. is a parasite.
   d. is sold in supermarkets.

3. The protist in Figure C grows amid dead leaves on the forest floor. It
   a. could be a toadstool.
   b. could be a mushroom.
   c. is a saprophyte.
   d. could be edible or poisonous.

4. Which of the following terms best matches each phrase?
   a. saprophyte
   b. parasite
   c. mold

   _____ an organism that lives on dead organisms

   _____ a fungus that may produce an antibiotic

   _____ an organism that lives on or in another live organism

KEY WORDS

saprophyte (p. 348)   mold (p. 348)
parasite (p. 348)

MOSSES AND FERNS

What are mosses? Life scientists consider the **mosses** to be true plants; hence, they are placed in the plant kingdom. But there are about 25,000 different kinds of plants that are mosses, or are closely related to the mosses. What do they all have in common?

moss:
a small plant that grows in mats or clumps and is most abundant in moist habitats

353

Organisms

Figure 19.5: This is what a typical clump of moss plants looks like. The individual plants are quite small. Note how they grow very close to each other.

All of the mosses are small plants. None of them have a *vascular system*. That is, the moss plant lacks an inner transport system. There are no tubes in a moss plant to carry water and food throughout the plant. Individual moss plants grow very close together, forming mats or clumps. (See Figure 19.5.) Mosses are typically found growing on rocks and at the bottom of trees. They are most plentiful in habitats that are moist.

Why do moss plants grow close together? There is a good reason why moss plants grow close together. During part of their life, they reproduce sexually. This means that two different cells, one a sperm cell and the other an egg cell, must come together. This is called *fertilization*. The single cell that results from fertilization is called a fertilized egg cell. The fertilized egg cell divides and the process continues until a new moss plant results. Now you can understand why moss plants must be close together. In order for fertilization to occur, the sperm must travel from one plant to the egg in another plant. In nature, a single drop of water may be enough to serve two small moss plants. The sperm can swim to the egg in the water. In larger plants, the sperm may get to the egg by the splashing of raindrops.

Mosses have an important niche in terrestrial ecosystems. Mosses grow over the ground like a carpet. Many mosses have leaflike parts that soak up and hold water. The way they grow and their water-holding ability cause mosses to have an important niche in terrestrial ecosystems.

Mosses often grow in shady places, where seed plants will not grow. In such places, they spread out and cover bare ground. When it rains, the soil underneath the mosses is protected. Otherwise it would be lost by erosion, the process by which soil is washed away by moving water.

Fungi, Mosses, and Ferns

Mosses are often the first plants to invade an area where the plant life has been destroyed. They help protect the soil in such an area from erosion. By holding moisture, they also act as an ideal seedbed. Seeds from larger plants may land in a moist clump of mosses. There they have an ideal location in which to grow and develop.

Ferns are vascular plants. A **vascular plant** is one that has a system of tubes through the roots, stem, and leaves. The tubes transport water and food throughout the plant. Mosses do not have a vascular system. That is one reason you never see large moss plants. **Ferns** do have a vascular system. In ferns, water can be taken in by the roots, then transported up to the leaves. Food manufactured in the leaves can be transported down to the roots. Ferns are generally much larger than mosses. Their vascular system helps make the greater size of ferns possible.

Rows of long green leaves give the fern a shape that is easy to recognize. (See Figure 19.6.) The fern "plant" that you see is generally just its leaves. The stem of the fern grows under the ground, and it grows horizontally. New ferns usually grow up from several different points along the horizontal stem.

If you look at a fern closely on an autumn day, you will probably see lines of brown spots on the lower sides of the leaves. These spots are spore cases filled with spores—reproductive cells. Ferns reproduce by asexual reproduction during part of

vascular plant:
a plant that has a tubular transport system

fern:
a nonflowering vascular plant whose stem grows under the ground

Figure 19.6: The dark splotches on the underside of this fern are structures that contain spores. Ferns use sexual and asexual reproduction at different stages of their life cycle. The spores are a part of the asexual stage in fern reproduction.

Figure 19.7: In the Pyrenees Mountains between Spain and France, Basque farmers cut and stack ferns in the fall. They will be used as animal feed during the winter months.

their lives. Under favorable conditions, a fern spore grows and divides. Eventually, a new fern is produced.

Ferns once dominated the landscape. Figure 19.6 shows a typical kind of fern. Ferns are fairly common in many regions of the United States. They are more plentiful in moist areas. In many sections of the country, they grow as weeds along roadsides.

Ferns are much more common in some other parts of the world. In certain tropical regions, ferns grow to be as tall as trees. There are parts of Europe where solid growths of ferns cover whole hillsides. In these regions, the ferns are a valuable crop. They are cut and stored to be used as hay—food that will be fed to animals during the winter. (See Figure 19.7.)

If you had lived about 350 million years ago, you would have seen ferns everywhere. Much of the Earth was wet and swampy at that time. Many ferns were large and treelike. They dominated the landscape the way our large seed plants—the trees—do now. Our coal deposits date back to that period in our Earth's history. Dead ferns made up much of the plant matter that was formed into coal.

LESSON REVIEW (Think. There may be more than one answer.)

1. Mosses
    a. are most plentiful in dry habitats.
    b. only reproduce asexually.
    c. sometimes grow more than a meter tall.
    d. often form a bed where seeds can develop.

2. Ferns
   a. can be found growing as trees.
   b. have to grow close together in order to reproduce.
   c. are used for animal feed.
   d. were more abundant in past years than they are now.
3. Ferns
   a. have a water transport system in their stems and leaves.
   b. have a stem that grows horizontally above the ground.
   c. reproduce asexually by spores that grow on the bottom of the leaves.
   d. are usually taller than mosses.
4. A vascular plant
   a. may be a moss.
   b. may be a fern.
   c. has a tubelike transport system.
   d. carries on photosynthesis.

KEY WORDS

moss (p. 353)   vascular plant (p. 355)   fern (p. 355)

Applying What You Have Learned

1. Give at least one reason why vascular plants such as the ferns are usually much larger than nonvascular plants such as the mosses.
2. Which of the three basic niches do mushrooms and other fungi belong in?
3. What part of the physical environment helps the moss plant to reproduce? How?
4. Wheat rust is a type of fungus that invades the live wheat plant. It reduces the amount of wheat that can be harvested. Is the wheat rust a parasite or a saprophyte? Why? Why do farmers consider wheat rust to be a pest?
5. How are producers dependent upon saprophytes?
6. Which of the ten organ systems in humans could the system of tubes found in vascular plants be compared to?
7. What part of the cell forms the system of tubes in vascular plants?
8. Would it be possible to have a food chain that consisted of only a producer and a decomposer? Why?
9. List three ways in which people depend upon fungi.
10. How do mosses form an ideal seedbed for larger plants?
11. The burning of coal releases energy in the form of heat. Where did this energy come from and how did it get stored in the coal?
12. What part of the physical environment helps spread the spores produced by puffballs and mushrooms?

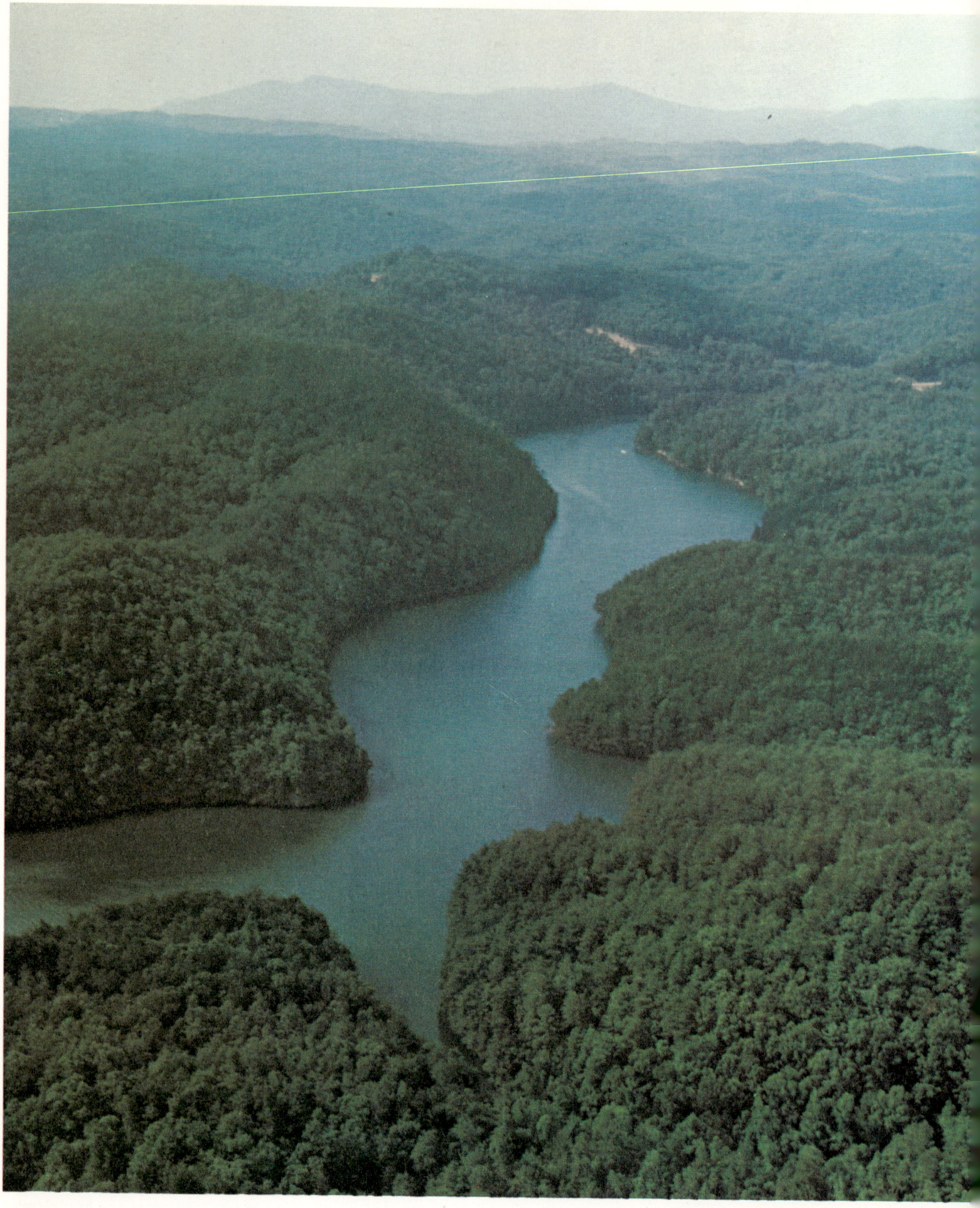

Figure 20.1: Green plants give much of the Earth's surface a "blanket" of color, and seed plants dominate this greenery.

Chapter 20

SEED PLANTS

This chapter is about the two most important groups of seed plants—the conifers and the flowering plants. The seed plants are by far the most important plants. They are the most plentiful and are at the beginning of most animal food chains. Most animals depend upon seed plants for their habitat, or shelter. Finally, they are important for a reason that few people appreciate. You will recall that during photosynthesis they release oxygen as a by-product. This oxygen enters the atmosphere and helps replace the oxygen that is being used up.

CONIFERS

Conifers are cone-bearing plants. *Conifers* do not have flowers. Their seeds are produced in a woodlike structure called a cone. Actually, each conifer produces two kinds of cones. One is small and produces **pollen**. Pollen contains the male re-

pollen:
small particles that contain a plant's male reproductive cells

Figure 20.2: In this closeup of a lodgepole pine tree, you can see a large seed cone on the left and smaller pollen cones on the right.

Figure 20.3: These spruce trees in the Rocky Mountains of Colorado are typical of the conifers that cover large areas of the Earth.

productive cells. The larger cones contain egg cells, the female reproductive cells. Pollen is transported to the egg cells by currents of air. When the egg is fertilized, it develops into a seed. Hence, the large cones produced by conifers are usually called seed cones. In Figure 20.2, you can see the two types of cones from a lodgepole pine tree, a conifer that grows in the western United States.

Seed Plants

Most conifers are trees. Life scientists usually place a plant in one of three general categories: tree, shrub, or herb. You know what a **tree** is—it is an upright plant, usually with one strong, woody main stem, or trunk. A **shrub** is also a woody plant, usually smaller than a tree, with many stems growing out of the ground. "Bush" is another name for a shrub. An **herb** is a plant without a woody stem. Grasses and most ornamental flowering plants are examples of herbs.

Most conifers are trees. A few are shrubs. No conifers are herbs. The trees and shrubs that are conifers are often called "evergreens." They are called by this name because they usually do not shed their leaves in the autumn. Their leaves, which are ordinarily needlelike or flat and scalelike, usually last for 2 or more years.

tree:
an upright plant, usually with one strong, tall, woody main stem, or trunk
shrub:
a woody plant, usually shorter than a tree, with several stems growing out of the ground. Also called a bush.
herb:
a plant without a woody stem

The oldest and the largest organisms are conifers. The oldest known organism on the Earth is a conifer. It is one of a group of very old bristlecone pine trees growing on a dry mountain in east central California. The oldest tree in this group was discovered in 1957. It is estimated to be over 4600 years old.

The largest organisms on Earth are also the conifers. One of them, the famous "General Sherman" redwood, is described in detail on page 97 in Chapter 6. (See Figure 20.5.)

Figure 20.4: Most of these bristlecone pine trees are over 3000 years old. One bristlecone pine is the oldest known organism.

Organisms

The conifers are important plants. More than 75 percent of all the lumber that we use in the United States is produced by conifer trees. The western white pine, the Douglas fir, and the ponderosa pine (western yellow pine) are three of the most important lumber producers. Other important lumber-producing trees include certain cedars and spruces and the redwood and cypress.

About 90 percent of wood pulp, which is used to make paper, is furnished by conifer trees. Various kinds of pines, hemlocks, and spruces are widely used for making pulp.

There are two main reasons why conifers are valuable as lumber and pulp producers. First, most conifers grow faster than other kinds of trees. As an example, in certain regions of the South, pine trees are planted as a "tree crop." They are harvested just 20 years later. The second reason conifers are valuable is that the wood they produce is soft and easily worked. For this reason the wood of conifers is called "softwood." (Flowering trees produce "hardwood" and the trees themselves are called "hardwoods.")

Figure 20.5: This giant sequoia has a name of its own: the General Sherman. Notice how the people are dwarfed by this giant plant.

Figure 20.6: Ornamental junipers like these shrubs are typical of conifers that are used for landscaping buildings and parks. They are used largely because they remain green all year around.

Conifers are also valuable as ornamental trees and shrubs. There are few parks or landscaped yards in America without at least one evergreen tree or shrub. They are admired for both their beauty and for the fragrant odor that they produce.

LESSON REVIEW *(Think. There may be more than one answer.)*

1. The larger cone of a conifer
    a. produces male reproductive cells.
    b. produces pollen.
    c. produces female reproductive cells.
    d. produces egg cells.
2. The smaller cone of a conifer
    a. produces male reproductive cells.
    b. produces pollen.
    c. produces female reproductive cells.
    d. produces egg cells.
3. The conifers have the distinction of having species that are
    a. the oldest organisms on the Earth.
    b. the most plentiful organisms on the Earth.
    c. the most useful organisms on the Earth.
    d. the largest organisms on the Earth.
4. Conifers are valuable because
    a. they grow slowly and produce hard wood.
    b. 90 percent of our wood pulp for paper comes from them.
    c. their wood is hard and can be used in good furniture.
    d. people like them for use as ornamentals.

Organisms

5. Which of the following terms best matches each phrase?
   a. pollen
   b. tree
   c. shrub
   d. herb

   ____ a plant without a woody stem

   ____ small grains that contain male reproductive cells

   ____ a tall plant, usually with one woody stem, or trunk

   ____ a woody plant with several stems

KEY WORDS

pollen (p. 359)   shrub (p. 361)
tree (p. 361)   herb (p. 361)

FLOWERING PLANTS

What is a flower? Though important, the conifers are just a small fraction of the whole group of seed plants. Most seed plants have flowers. What is a **flower**? It is a reproductive system. Most flowers contain both male and female reproductive organs. The male reproductive cells are produced in **anthers**, little sacs

flower:
the reproductive system of all seed plants except conifers

anther:
a sac in a flowering plant that produces male reproductive cells. The anthers are at the ends of stamens.

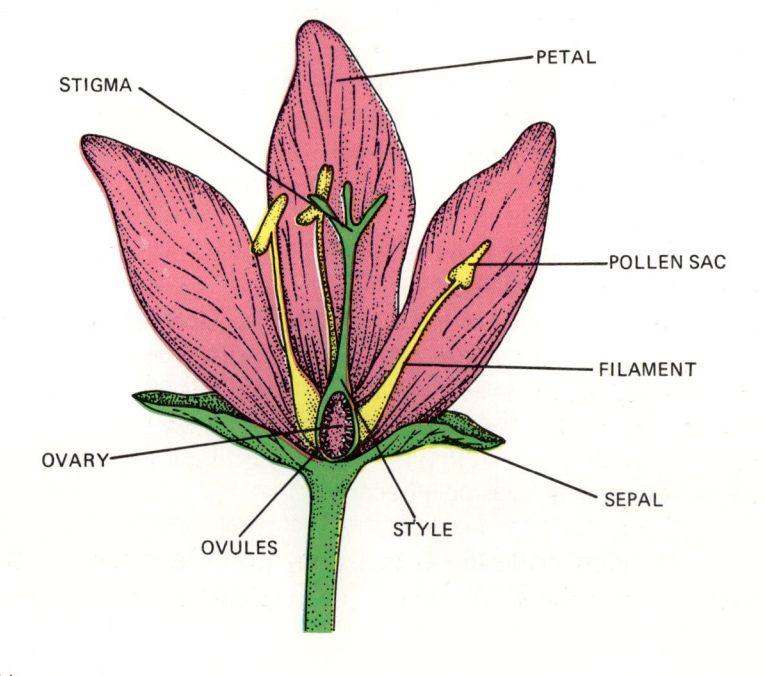

Figure 20.7: A flower has been pollinated when pollen from the stamens gets to the stigma. There the pollen grain breaks open and male reproductive cells bore down through the style to the ovary. When they reach the egg cells, fertilization occurs.

at the ends of the **stamens.** (See Figure 20.7.) These male cells are enclosed in small grains of pollen. Note in the flower diagram that the female egg cell is enclosed in a structure called the *ovary*.

Many flowers also have organs such as *sepals* (sē′pəl) and *petals*. (See Figure 20.7.) Often the petals, and sometimes the sepals, are brightly colored. If so, they may help attract insects such as bees, which help pollinate the flower.

What does it mean "to pollinate" a flower? It means that the pollen containing the male reproductive cells gets to the **stigma.** When this occurs, the flower is said to be pollinated.

The pollen grain breaks open after it lands on the stigma. Then it bores its way down through the **style,** which is usually a thin stalklike projection above the ovary. Eventually, the male reproductive cells reach the egg cells, which are inside the **ovules** (ō′vyool). Fertilization occurs when the nucleus of a male cell joins with the nucleus of a female egg cell.

Many flowers are cross pollinated. This means that pollen from one flower lands on the stigma of another flower. How does the pollen get from one flower to another? Wind carries some pollen. (Hayfever sufferers will agree with that fact.) Also, insects, such as bees, carry pollen from one flower to another on the hairs of their bodies. Some plants are completely dependent on bees for their pollination. Thus, any pesticide that kills bees may do more harm than good.

The ripened ovary is a fruit. After the egg cell is fertilized, it develops into a tiny plant called an embryo. While this is happening, the petals, sepals, and stamens usually dry up and fall off. The ovary which is left usually begins to grow. Sometimes, other parts of the flower—sepals, for example—will remain attached to the ovary and grow with it.

Eventually, the ovary stops growing. At about the same time, the ovules inside the ovary have completed their development and have become seeds. Then the ovary is said to be "ripe." Now it has a new name. It is called a **fruit**.

You usually think of apples and bananas when you hear the term "fruit." Apples and bananas are fruits. But so are cockleburs, walnuts, acorns, corn ears, and maple seeds. (See Figure 20.8.)

Every fruit contains one or more seeds. Sometimes the fruit breaks open and the seeds are released and scattered. The bean and pea pod are such fruits. In other cases, it is the whole fruit that is scattered. The seed or seeds stay inside the fruit until they begin to grow. The dandelion fruit is an example. (See Figure 20.9.)

The seed contains a dormant plant embryo. We have said that the seed contains a tiny plant called an embryo.

---

stamen:
the stalk in a flower that has a sac (anther) at its end in which male reproductive cells are produced

stigma:
the uppermost part of a style on which pollen first lands

style:
the region between the ovary and the stigma in a flower

ovule:
a rounded body inside the ovary of a flowering plant that holds the female reproductive cell

fruit:
the ripened ovary of a seed plant, in which the ovules have become seeds.

Organisms

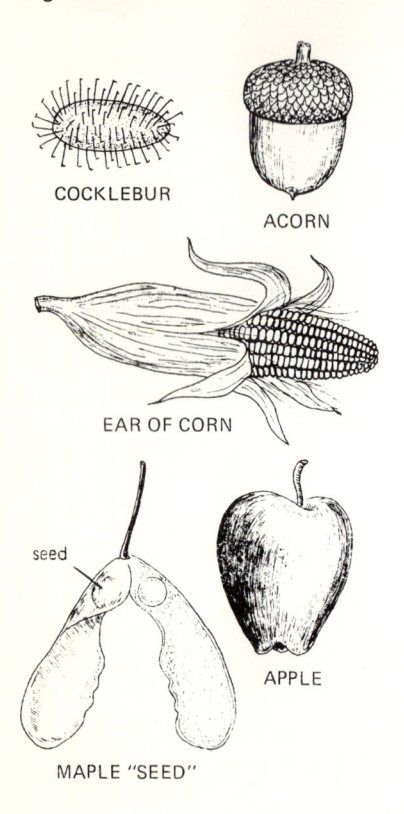

Figure 20.8: The maple "seed" is really the whole fruit, with two seeds inside it.

Figure 20.9: The many individual fruits of the dandelion are in a cluster, ready to be picked up and spread by the wind. Each of these fruits has a single seed.

Now we will add one more term. The seed contains a *dormant* plant embryo. What does dormant mean? It means that the embryo is not active. You could appreciate what this means if you could somehow look inside the cells of the embryo. In another chapter, we said that the inside of a cell is a very busy place. Thousands of chemical reactions and processes are going on. This is not the case inside the cells of the dormant plant embryo. One could say that things are barely moving. Just enough activities take place to keep the cells alive.

The plant embryo can remain dormant for long periods of time—often for many years. What causes it to start growing again? Several things may have to happen to the seed. Sometimes the seed must go through periods of freezing and thawing temperatures. Sometimes it must be exposed to light for certain periods. Water is *always* required to break the dormant stage of a plant embryo.

Figure 20.10: Water was used to end the dormancy of this lima bean. The plant embryo inside started to grow, breaking out of the seed.

After the dormant stage is broken, the embryo starts to grow again. It breaks out of the coating that surrounds the seed. If it is in a favorable environment, it will continue to grow until it becomes an adult plant.

LESSON REVIEW *(Think. There may be more than one answer.)*

1. A flower
    a. is a reproductive tissue.
    b. is a reproductive organ.
    c. is a reproductive system.
    d. always has brightly colored petals and sepals.
2. The female reproductive organ of a flower is a(n)
    a. stamen.
    b. anther.
    c. petal.
    d. style.
3. The male reproductive organs of a flower are
    a. stamens.
    b. anthers.
    c. petals.
    d. styles.
4. If the sepals and petals of a flower are small and not colorful, it will
    a. not be able to reproduce new plants.
    b. probably be wind pollinated.
    c. probably be insect pollinated.
    d. wither and die.
5. At the time of pollination, pollen grains land on the structure labeled
    a. ovules.
    b. stigma.
    c. anther.
    d. petals.
6. Which of the following terms best matches each phrase?
    a. flower   c. ovary   e. style   g. ovule
    b. stamen   d. stigma   f. anther   h. fruit

    _____ contains the ovules

    _____ the ovary is at its base

    _____ supports the anther

    _____ system responsible for reproducing seed plants

    _____ a ripened ovary

    _____ pollen has to land on it for fertilization to be possible

    _____ egg cell is inside one

    _____ pollen is produced in it

367

Organisms

KEY WORDS

flower (p. 364)
anther (p. 364)
stamen (p. 365)
stigma (p. 365)

style (p. 365)
ovule (p. 365)
fruit (p. 365)

THE KINDS OF FLOWERING PLANTS

There are two groups of flowering plants. Life scientists have classified the flowering plants into two groups. (We will use the common names rather than the scientific names to describe them.) The members of one group are called **monocots.** Plants in this group have several things in common. Most of them have leaves in which the veins are parallel to each other. (See Figure 20.11.) Also, the parts of the flowers are usually in threes or multiples of threes. As an example, many will have three petals, three sepals, and three or six stamens. (Remember, the stamens are the male reproductive organs in the flower.)

Plants in the second group of flowering plants are called **dicots.** The leaves of dicots are net-veined. This means that there is one main vein in the leaf and that all others branch off from it.

monocot: a member of one of two groups of flowering plants. Most monocots have leaves with parallel veins and flower parts that occur in threes or multiples of three.

dicot: a member of one of two groups of flowering plants. Most dicots have leaves with netted veins, and flower parts usually occur in fours or fives.

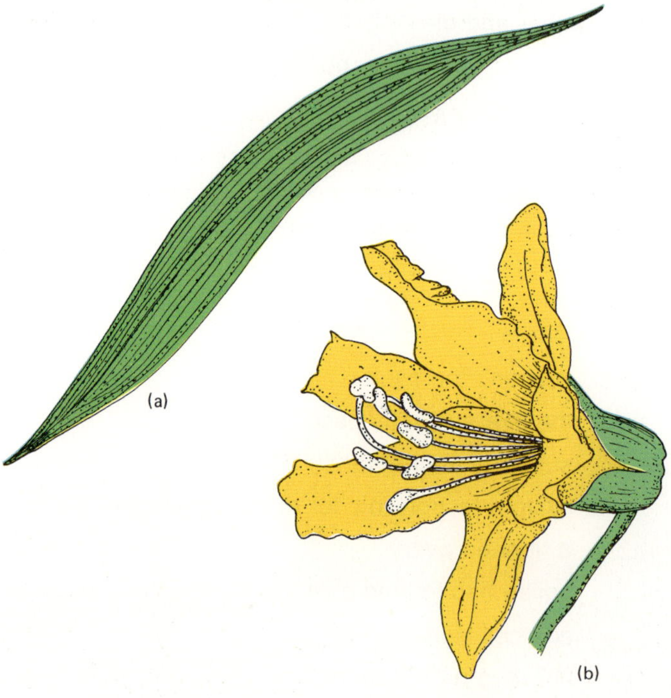

Figure 20.11: A typical monocot has leaves with parallel veins (a) and three petals, three sepals, and three or six stamens (b).

Seed Plants

Figure 20.12: The leaf of a dicot is net-veined (a). The flower parts of a typical dicot are in multiples of four or five (b).

(See Figure 20.12.) Usually, the flower parts of dicot flowers are in fours or fives. Take a look at a leaf on a tree, or at a fallen leaf. What kind of pattern do you see in the leaf veins?

Maple trees are dicots. So are oak trees. In fact, almost all flowering trees are dicots. There is one main exception: Palm trees are monocots. Not all flowering plants are trees, of course. Some flowering plants are shrubs. And most kinds of flowering plants are herbs.

The grass family is an important group of monocots. The "family" is one type of group that the life scientist uses in classifying related plants. There are several families of monocot plants. We will introduce you to two of them.

Figure 20.13: This large combine is harvesting a field of wheat. Wheat is one of many important "cereal grasses" grown by people.

Organisms

The grass family is a very important family of monocots. There are about 7000 different kinds of grasses in this family. All of the cereal grains—for example, wheat, oats, corn, rye, barley, and rice—are members of the grass family. People depend directly upon these grains for food. Many other grasses are at the beginning of food chains for animals that people eat. Thus, people also depend indirectly upon those other grasses for their food.

The flowers on grass plants do not have bright colored petals and sepals. A head of wheat hardly looks like a collection of flowers to most people. Yet within each flower are the male and female reproductive organs: stamens and ovary. As you might guess, grass flowers do not rely on insects for pollination.

The lily family has attractive flowers. The lily family is another common family of monocots. The flowers of this family usually have petals, and sometimes sepals, that are attractive and showy. Many, such as the Easter lily, tiger lily, tulip, hyacinth, and crocus, are grown for their beauty. Many of our most attractive spring wild flowers also belong to the lily family. Lily of the valley, trillium, adder's tongue, and clintonia are just a few examples.

Some of our best-known food plants also belong to the lily family. These include asparagus, onions, leeks, and chives.

Figure 20.14: These are flower clusters of timothy, a member of the grass family that is grown as a hay crop.

Figure 20.15: These spring wild flowers belong to the lily family. This kind of dogtooth violet is usually called the glacier lily.

The pea family is a large family of dicots. The members of the pea family are called **legumes**. There are over 15,000 different members in this group. Besides peas and beans, which are well-known garden legumes, people also cultivate soybeans, peanuts, alfalfa, and vetch. Legumes grown for their beauty, not for their food value, include the sweet pea, wisteria, lupines, and Scotch broom. Locust trees are legumes that are commonly planted as shade trees in some parts of the United States. Mesquite (mĕ skēt') is a legume that grows as a wild tree in Texas.

The rose family has many important fruit producers. The rose family is an important dicot family. Many plants well known for their beauty belong to this family. The roses, spirea, hawthorns, and flowering quinces are examples. However, the family is more important because of the fruits produced by certain of its members. These include the apple, plum, cherry, peach, pear, apricot, blackberry, raspberry, and strawberry.

The composite family is very large. The composite family is a dicot family. It is the second largest family in the world, and the largest in the United States. (The largest family of flowering plants is the orchid family. It is a monocot family, and most of its members live in tropical and subtropical habitats.) Dandelions, asters, goldenrods, sunflowers, daisies, chrysanthemums, zinnias, and marigolds are a few of the best-known composites. In middle and late summer, the composites are the dominant wild flowers in most fields and roadsides. Many troublesome weeds, such as ragweed, sagebrush, thistles, and cocklebur, also belong to this family.

legume:
a plant of the pea family. All legumes have their seeds in pods.

Figure 20.16: The Texas state flower, the bluebonnet, provides a sea of color over the landscape. The bluebonnet belongs to the legume family.

Figure 20.17: These flowers are wild strawberries, which belong to the rose family.

Organisms

Laboratory Activity

How Are Flowers Alike? How Are They Different?

PURPOSE Most people take flowering plants for granted. There is no better way to appreciate the diversity among flowers than to examine the many kinds of flowers there are. You can do this in fields, in a greenhouse, or at a florist shop. In this investigation, you will probe into several flowers to discover the parts of a flower.

MATERIALS
several flowers
black construction paper
knife
magnifying lens
centimeter ruler
tweezers
toothpicks

PROCEDURE
A. Examine a typical monocot flower, the daffodil. The daffodil belongs to the amaryllis family.
1. Locate the sepals and petals. They are the same color. Therefore, they may all appear to be petals on the flower. The sepals are located on the outside of the petals. How many sepals and petals are there?
2. Remove the sepals and the petals. Study the reproductive organs of this flower. Locate the style and the stamens. How many of each are there?
3. Examine the style. You may want to place the parts of the flower on black paper. The flower parts will show up better against the dark background. Locate the stigma and the ovary. Feel the stigma. How does it feel? What advantage might this be?
4. Cut open the ovary with a knife. Locate the ovules, which contain the egg cells. Use a magnifying lens.
5. Measure how far the sperm cells must travel in order to reach the ovary.
6. Examine the stamens. Locate the anther at the end of each. Break open an anther. Examine the pollen grains with a magnifying lens. Describe what you see.

B. Examine a typical dicot flower, the rose.
1. Locate the sepals and the petals. How many of each are there?
2. Study the reproductive organs of the rose. Remove the sepals. Grasp the flower around its base and gently tear the flower apart. Locate the many stamens and styles. How many stamens and styles do you count?

Seed Plants

3. With tweezers, remove the styles and stamens. The styles are very small. Locate the stigma and the ovary at each end of a style. The ovary contains many ovules, and each ovule contains an egg. Cut open an ovule, and examine it with a magnifying lens. Describe what you see.

C. Examine a typical composite flower, the chrysanthemum.
1. Examine the flower carefully. Use a magnifying lens. Describe what you see.
2. Locate the ray flowers on the outside of the composite flower.
3. Locate the disc flowers in the middle of the flower.
4. Locate the sepals of the chrysanthemum. They are small and joined together.
5. Look carefully at the petals. They are also joined together. Many petals fused together give the appearance of one long petal.
6. Locate the stamens. They are united in a ring around the single style. Examine the stamens to find the anthers.
7. Look carefully at the style. What is different about it?
8. Cut open the ovary. How many ovules do you find?

QUESTIONS

1. How are the flowers that you studied alike? How are they different?

MONOCOT

daffodil

DICOT

rose

COMPOSITE

chrysanthemum (a dicot)

373

Figure 20.18: Most people would say that this is one sunflower. But this sunflower "head" is actually made up of hundreds of tiny flowers. You can see these individual flowers if you look closely at the picture.

Members of the composite family are unusual flowers. Most people will look at a dandelion and think that it is one flower. Actually, one dandelion may be one hundred flowers. Composite flowers are small and packed tightly together in a "head." There are often two types of flowers in the head of a composite. The sunflower is an example. (See Figure 20.19.)

Figure 20.19: A sunflower head is made up of disc flowers (which contain reproductive organs) and ray flowers (which are sterile).

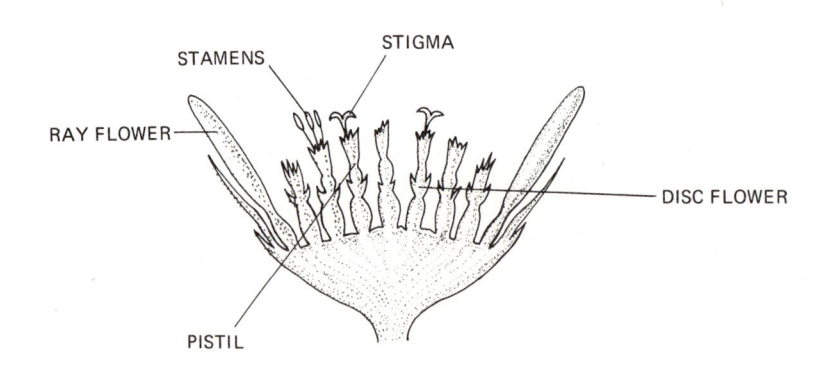

374

Seed Plants

LESSON REVIEW *(Think. There may be more than one answer.)*

1. A plant with parallel-veined leaves would probably have a flower with
    a. four petals.
    b. three sepals.
    c. five stamens.
    d. four stigmas.

2. Of the fruits shown below,
    a. D is produced by a member of the grass family.
    b. E is produced by a member of the lily family.
    c. C is produced by a member of the pea family.
    d. B is produced by a member of the lily family.

3. Of the fruits shown below,
    a. A and E were produced by a dicot plant.
    b. B and D were produced by a monocot plant.
    c. B is a legume.
    d. C is a legume.

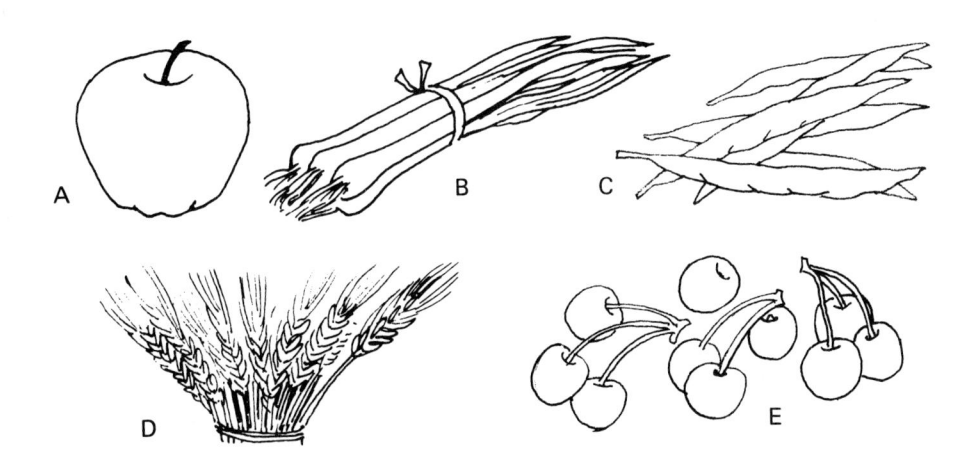

4. The family with flowers that are unusual and confusing for most people is
    a. the rose family, because most of the flowers look like fruits.
    b. the composite family, because their flowers are small and clustered together in a head, which is called "a flower."
    c. the pea family, because the flowers are actually pods, such as the pea pod and the bean pod.
    d. the lily family, because their flowers are seldom seen.

5. Which of the following statements is true?
   a. Most monocots have leaves with parallel veins.
   b. Most dicots have leaves with netted veins.
   c. Most monocots have flower parts that occur in threes or multiples of three.
   d. Most dicots have flower parts that usually occur in fours or fives.

KEY WORDS

monocot (p. 368)
dicot (p. 368)

legume (p. 371)

Applying What You Have Learned

1. Briefly describe the function of each of the following plant systems: root, stem, leaf, and flower.
2. In what part of the plant would you find plant cells with the greatest amounts of chlorophyll?
3. How do flowers rely on populations of insects such as the bee?
4. What do pine needles do? What makes them green?
5. When the air temperature drops below freezing, the leaves of flowering plants turn different colors. What prominent leaf pigment has disappeared? What other leaf pigments become prominent?
6. Where could you find cell walls functioning as vascular tissue in the flowering plants?
7. How could you tell if a plant is a monocot if it does not have any flowers?
8. Would you expect to find large, brightly colored petals and sepals on flowers that are pollinated by wind-carried pollen? Why?
9. Do you think cone-bearing and flowering plants have complex vascular systems? Why?
10. You have found a flower with four sepals, four petals, and five stamens. What major group of flowering plants would you put it in?
11. Certain flowering plants, such as the potato, onion, carrot, and turnip, store food energy in the root or underground stem. How do you think this stored energy may be used by the plant? Where did this food energy come from and how did it get to the root?

Seed Plants

12. The fruits and seeds of plants such as tomatoes, cherries, peas, and beans contain stored food energy. How do you think this stored food energy will be used by the seed embryo? Where is this stored food energy produced and how does it get to the fruit?

13. Describe some ways in which plant seeds are transported to new environments.

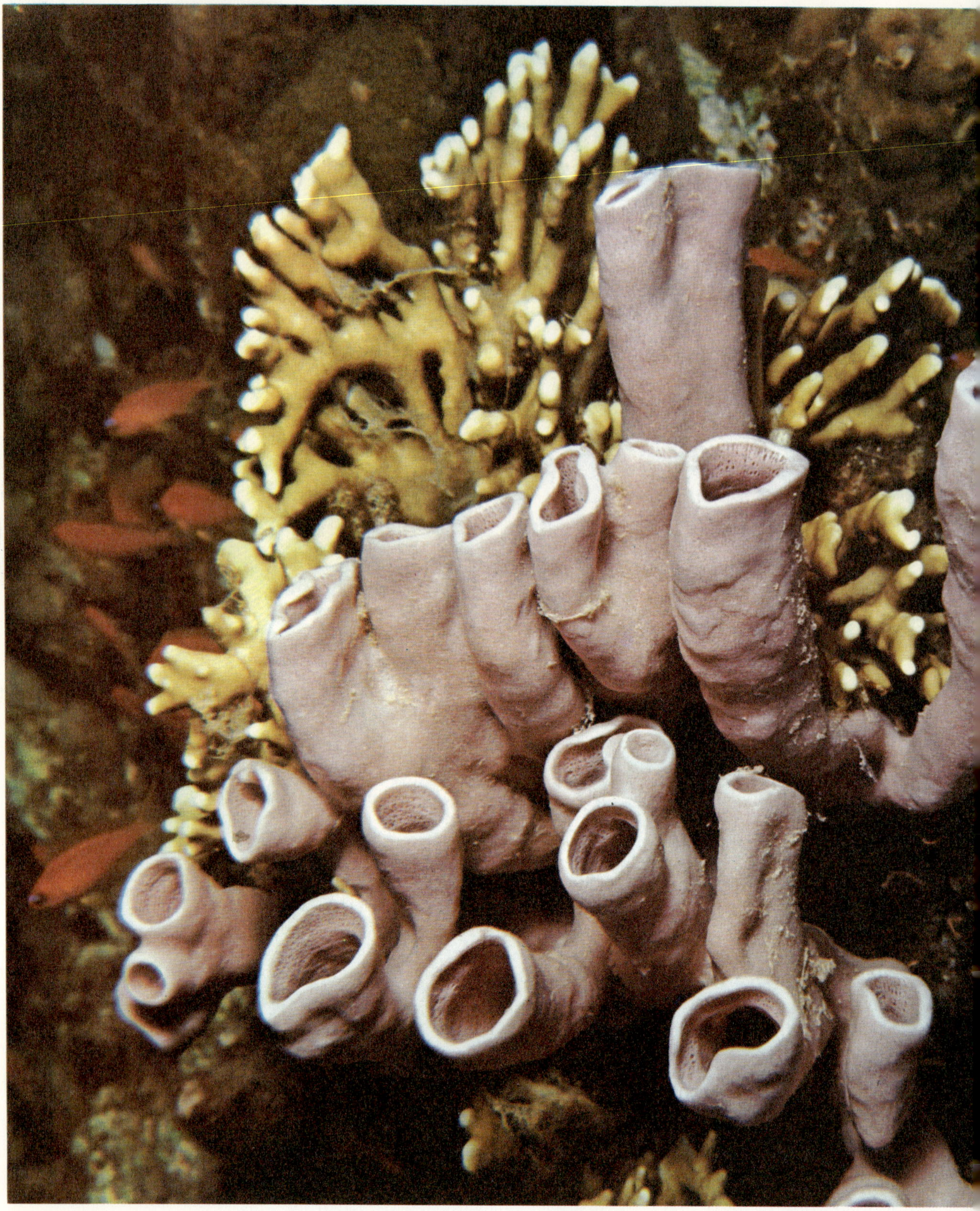

Figure 21.1: For years, it was thought that sponges were plants. These are sponges. Do they look more like animals or more like plants?

Chapter 21

SPONGES AND COELENTERATES

This is the first of four chapters about the animal kingdom. Your study of these chapters will be like watching a parade of animals march by you. Who shall lead the parade? In what order should the animals follow each other? Will all of the Earth's animals be in the parade?

The sponges will lead the parade. They are the simplest of all the animals. The sponges will be followed by the coelenterates (sǐ lĕn'tə rāt'). They are the next most complex animals. The three chapters after this one will continue presenting the Earth's animals in order of their complexity—from the simplest to the most complex.

About 1,250,000 different kinds of animals have been discovered on the Earth. Life scientists think that there are thousands of more kinds of animals yet to be discovered. Think how long it would take you to count to 1,250,000. This will help you understand why it is impossible to study all the animals. No one has ever done it. There are too many of them. In these four chapters, you will study only the most common groups of animals. But the animals that you will study are in many ways like all the others. You will therefore know many things about most of the animals you may ever see.

Organisms

SPONGES

sponge: an aquatic creature that has the simplest structure of all the animals. A sponge has specialized cells, but no tissues, organs, or systems.

Sponges are the simplest animals. If you ever have to memorize all the parts of an animal, pick a **sponge**. There is not much to memorize. The sponge has no tissues, organs, or systems. Sponges do have specialized cells, but these are of only a few different kinds. There are two parts to the body of a sponge. First, there are the cells. They make up the living portion of the sponge's body. Most sponges have no definite shape. Their cells are massed together in a shape that could be described as a blob. The second part of the sponge's body is the skeleton. (A few sponges do not have a skeleton.) The skeleton of a sponge is not like that of more complex animals. It is nonliving material that is produced by the sponge cells. The fiberlike material of the bath sponge is one type of sponge skeleton. (See Figure 21.2.) Other sponges have a harder skeleton made of small pieces of limestone. (See Figure 21.3.)

Most sponges live in marine ecosystems. There are about 5000 different kinds of sponges. Most of them live in marine ecosystems. A few live in freshwater ecosystems. None of them live on the land.

The marine habitats of sponges vary greatly. Some sponges live on rocks along the shores of seas and oceans. Others can be found at the bottom of the deepest oceans. Freshwater sponges

Figure 21.2: This "bath sponge" used to be a living animal, and its home was the sea. (Courtesy Carolina Biological Supply Co.)

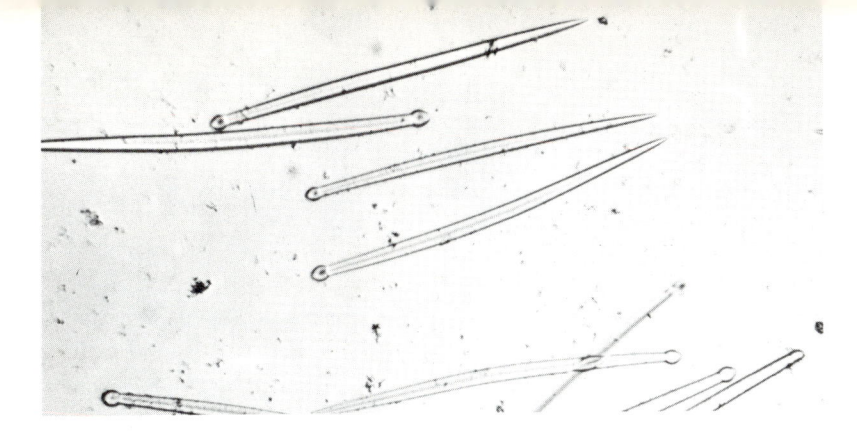

Figure 21.3: These needlelike bodies were removed from a sponge. They, and hundreds of others like them, made up the skeleton in the animal's body. (Courtesy Carolina Biological Supply Co.)

are found in ponds, lakes, streams, and rivers. They are often attached to rocks and old logs that are underwater.

The marine sponges also vary greatly in size. Some are only about 0.1 centimeter (0.04 inch) tall when fully grown. Others are as large as a barrel. Few people would recognize a freshwater sponge even if they were looking right at it. Freshwater sponges are usually green, and they often look like clumps of moss. In fact, life scientists used to think that sponges were simple plants.

Figure 21.4: This colorful sponge was photographed in the warm waters of the Bahama Islands, about 250 kilometers (155 miles) east of Miami, Florida.

Organisms

Figure 21.5: Parts of this freshwater sponge show the brown color it has in winter. Why might freshwater sponges turn brown in winter?

collar cell:
a specialized cell in a sponge that moves water through its body, which helps feed the sponge

flagellum:
a small, whiplike structure. In the collar cells, it helps to move water through the body of a sponge.

wandering cell:
a specialized cell in a sponge that moves through the body of the animal, distributing food to other cells

Sponges are full of holes. What is the most interesting thing about the sponges? It is probably the fact that they are the only animals that are full of holes.

If you are thinking like a life scientist, you know that the holes in the sponge probably serve some purpose. They do. Water is always flowing into the body of the sponge—through the holes and then through tiny canals on the inside of the sponge. The water flows out of the sponge's body through special openings.

Why does water flow through the sponge's body? Inside the body of the sponge there are specialized cells called **collar cells.** (See Fig. 21.6.) Each cell has a little whiplike **flagellum** (flə jĕl′ əm) that moves back and forth rapidly. The movement of the flagellum creates a tiny whirlpool. The collar cells line the canals that are inside the sponge's body. Thousands of little whirlpools are enough to cause water to circulate through the sponge.

Why does the sponge need water flowing through its body? The moving water feeds the sponge. Each of the collar cells has its own little whirlpool. You probably know that when an object gets trapped in a whirlpool it is usually pulled down to the bottom. At the bottom of these tiny whirlpools is the main cell body. Tiny bits of food material, including such things as bacteria, are taken into the cell body. Much of the food flows on through the cell body and is passed on to **wandering cells.** The wandering cells move around like amebas. As their name suggests, they wander through the body of the sponge and distribute food to other cells.

What is the sponge's niche in aquatic ecosystems? What good is an old dead tree that is full of holes? You will find that it may serve as a home for numerous insects, owls, squirrels, and raccoons—just to mention a few animals. The sponge likewise often serves as a home for numerous other aquatic organisms. There are usually two or more kinds of algae living in a sponge—that is what gives freshwater sponges their green color. In marine ecosystems, very large sponges may house as many as 16,000 small animals—mostly shrimp and crabs. Just imagine the number of animals that would be without a habitat if there were no sponges!

Sponges still have commercial value. Years ago, someone who bought a "sponge" bought the soft skeleton of a sponge after the cells had been destroyed. But someone buying a sponge today usually gets a piece of cellulose or rubber that is full of holes. This artificial sponge functions like the skeleton of a true sponge, and has thus taken its name.

Because of their commercial value, bath sponges have been collected by people for many years. And even though artificial sponges are widely used today, there is still enough demand for real sponges to make diving for them profitable. Some of the largest sponge "farms" are found in the West Indies and off the coast of Greece.

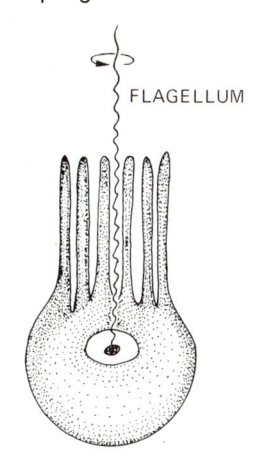

Figure 21.6: Collar cells line the canals inside a sponge. Movements of the whip-like flagellum create a tiny whirlpool.

LESSON REVIEW *(Think. There may be more than one answer.)*

1. Sponges
    a. have specialized tissues but no organs or systems.
    b. are made up of cells and cell products.
    c. usually have a skeleton.
    d. may have limestone in their body.

2. Sponges are
    a. found only in marine waters.
    b. found only in fresh water.
    c. all less than 1 centimeter tall.
    d. sometimes thought of, incorrectly, as clumps of moss.

3. Sponges
    a. feed by surrounding their food like an ameba.
    b. feed by stinging their food and capturing it with their tentacles.
    c. feed by taking food out of the water that flows through them.
    d. do not feed on anything. They are producers.

4. The sponge's niche is
    a. to grow on the bodies of aquatic animals and hide them from their enemies.
    b. to produce oxygen for all other forms of aquatic life.
    c. to decompose all forms of dead aquatic life.
    d. to provide a habitat for many forms of aquatic life.

Organisms

5. Which of the following terms best matches each phrase?
   a. sponge   c. flagellum
   b. collar cell   d. wandering cell

   _____ whiplike structure

   _____ animal that is full of holes and is the simplest of all animals

   _____ cell in a sponge that captures food

   _____ cell in a sponge that distributes food

KEY WORDS

sponge (p. 380)   flagellum (p. 382)
collar cell (p. 382)   wandering cell (p. 382)

COELENTERATES

Coelenterates are aquatic animals. There are at least 10,000 kinds of animals called **coelenterates**. All of them live in aquatic ecosystems. Most of them are marine animals. Only a few live in fresh water. How are the coelenterates different from other animals?

The body plan of a coelenterate is shown in Figure 21.7. If you were to cut a coelenterate down the middle—from top to bottom in the picture—you would end up with two equal halves. This is like cutting a round cupcake in half vertically. It would

coelenterate: an aquatic animal that has three distinct features—radial symmetry, stinging cells, and a large hollow cavity into which food enters and wastes leave

Figure 21.7: A coelenterate has radial symmetry. It has no "right" or "left"—any way the body is cut, the halves will be identical.

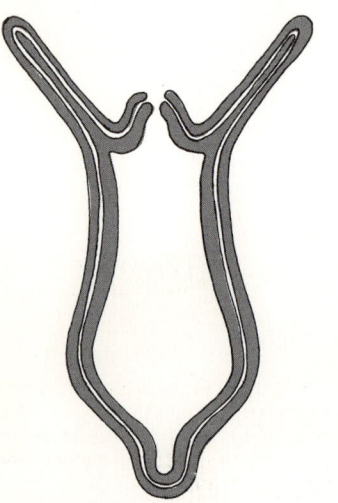

Figure 21.8: In a nematocyst that is ready to shoot out (a), the sharp threadlike part is coiled inside. When the nematocyst discharges (b), the thread may coil around whatever is being stung.

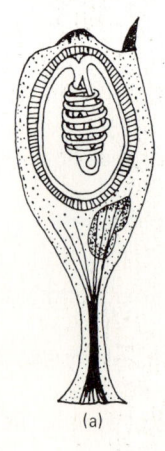

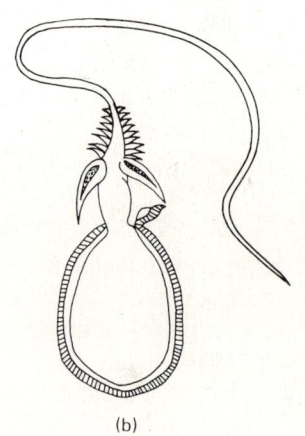

(a)   (b)

384

not matter which way you cut, as long as you cut down the middle. Both halves would always be the same. Like a cupcake, a coelenterate has no right or left side. An animal that has this type of structure is said to have **radial symmetry.**

The coelenterates have several other distinct features. They all have stinging cells called **nematocysts** (nēm′ə tə sĭst′). (See Figure 21.8.) The nematocysts are usually in the tentacles, though sometimes they are found throughout the body of the coelenterate. The nematocyst is a coiled thread inside a small capsule. The thread has a sharp point which may be barbed like a fishhook. When the capsule comes in contact with a foreign object, it may break open. If it does, the nematocyst shoots out like a harpoon and stabs whatever is in its way. Many nematocysts contain a poison that paralyzes animal tissue. A creature such as the Portuguese man-of-war, which is a large colony of coelenterates, can cause serious injury or death to a human being. (See Figure 21.9.)

radial symmetry:
a pattern in which parts are arranged around a central point or line so that any vertical cut through the center divides the whole into two identical halves

nematocyst:
the stinging cell of a coelenterate. The nematocyst shoots out a barbed thread coiled inside a capsule to sting prey.

Figure 21.9: A Portuguese man-of-war is not one animal but a colony of coelenterates. Their tentacles, which may hang down 2 or more meters, can be very dangerous if touched.

Organisms

gastrovascular cavity:
a large hollow space in the body of a coelenterate with one opening through which food enters and wastes leave

polyp:
a coelenterate with upright tentacles and the opening to its gastrovascular cavity on its "top" side

medusa:
a coelenterate with tentacles that hang from its body and the opening to its gastrovascular cavity on its "bottom" side

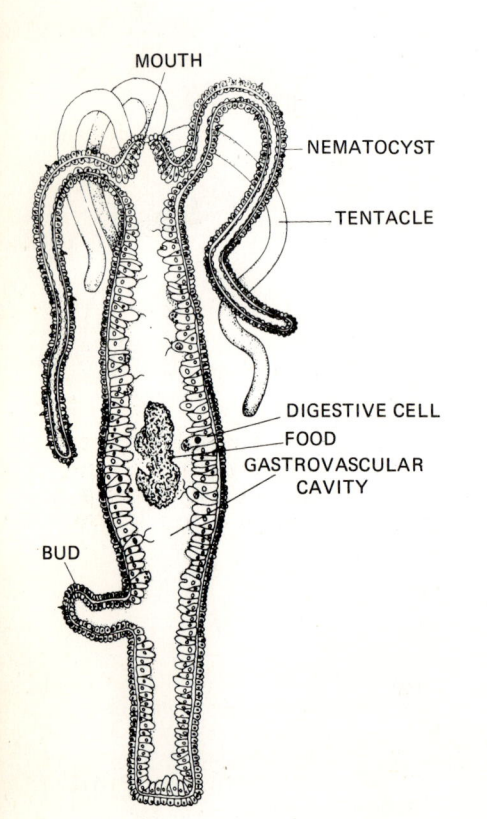

Coelenterates have a gastrovascular cavity. The coelenterates have one more distinct feature. It is a single large hollow space inside their bodies called the **gastrovascular cavity**. In Figure 21.10, you can see the basic structure of this baglike cavity. It has one opening. All food enters that opening—usually after being stung by nematocysts and captured by the animal's tentacles. Once the food is inside, certain of the cells that line the cavity release digestive enzymes. These enzymes break down the food so that it can be absorbed by other cells. Waste materials then leave the body cavity by the same opening that food entered.

Coelenterates have two body forms. In Figure 21.11, you can see the two basic body forms of coelenterates. One form, called a **polyp** (pŏl'ĭp), has upright tentacles. In a polyp, the opening of the gastrovascular cavity is on the "top." The other form of coelenterate, called a **medusa** (mə doo'sə), has tentacles that hang down below the main body. Its opening is on the "bottom." The hydra, a freshwater coelenterate, has a polyplike body. (See Figure 21.12.) One polyplike marine animal is the sea anemone (ə nĕm'ə nē). The jellyfish, which is a common name for many coelenterates, has tentacles that hang down from its body. It is a typical medusa.

Figure 21.10: The diagram at the left shows a vertical cross section of a hydra. The bulge on the side of the animal is a "bud," the beginning of another hydra. The bud will continue to develop and finally break off from the parent. This is a method of asexual reproduction.

Figure 21.11: These are the two general body forms of coelenterates. A medusa is basically an upside down polyp.

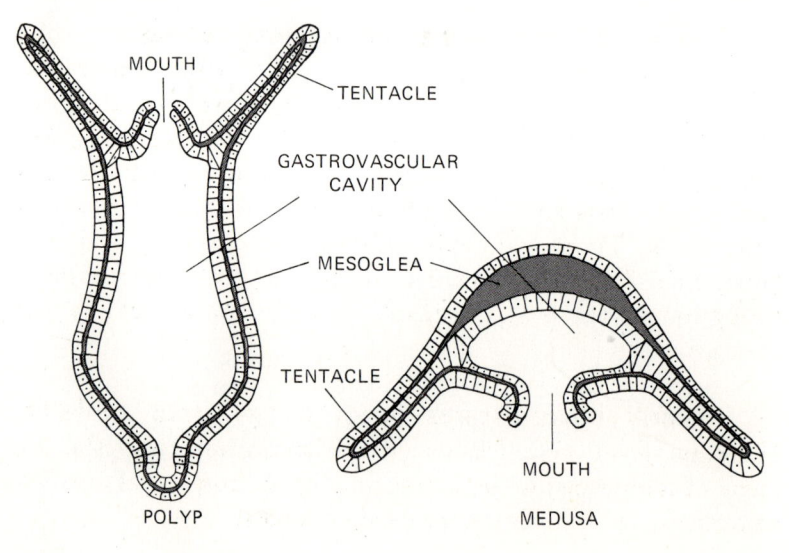

386

Figure 21.12: A hydra waves its tentacles (above) at a water flea, which it is likely to capture and eat. If caught, the water flea will be paralyzed by the stinging cells in the hydra's tentacles. This jellyfish (left) is a typical medusa-form coelenterate. It is a marine animal.

The mesoglea serves as an internal skeleton. The bodies of coelenterates have two layers of cells. Between the two layers is a protein material that strengthens the body structure. It is called the **mesoglea** (mĕ zō glē′ə), which means "middle jelly." In polyp-form coelenterates, the mesoglea is usually thin. In medusa-form coelenterates, the mesoglea is usually quite thick. There are some jellyfish that measure 4 meters (more than 12 feet) across. The mesoglea is very important in strengthening these large animals. It can be thought of as a type of internal skeleton.

Several coelenterates have an external skeleton. The polyp-form coelenterates usually have a thin mesoglea. But many of them must withstand strong forces from water currents and waves. How do they survive such forces?

mesoglea: a layer of protein material between the two layers of cells in a coelenterate. In medusa-form coelenterates, the mesoglea acts like an internal skeleton.

387

Organisms

Laboratory Activity

Observe a Freshwater Coelenterate.

PURPOSE Hydras are freshwater coelenterates found in lakes and ponds. Hydras attach themselves to the underside of floating leaves of water lilies or to the submerged stems of water plants. In this investigation, you will observe a hydra and study its reactions to various stimuli.

MATERIALS

| small dish | live hydra | toothpicks | meat juice |
| medicine dropper | hand lens | flashlight | |

PROCEDURE

1. Place 3 drops of water in a small dish.
2. Examine carefully the hydra culture your teacher will provide. Locate one or more hydra. Use a medicine dropper to pick up one hydra. Transfer it to the water in the dish.
3. Use a hand lens to observe the hydra immediately after transferring it to the dish. Why do you think the hydra changed its form?
4. Continue to observe the hydra. The animal will gradually relax into its natural threadlike form. Notice the tentacles that surround the mouth. What purpose do you think the tentacles serve?
5. Study the hydra's body. The hydra has radial symmetry. Its left side is identical to its right side. Describe what you see.
6. Do hydras swim around in the water? Does the animal attach itself to the bottom of the dish?
7. How does the hydra respond to touch? Gently touch one of the tentacles with a toothpick. Move the dish suddenly so that the water is disturbed. How does the hydra respond?
8. Observe the hydra undisturbed for several minutes. A hungry hydra is more active than a well-fed hydra.
9. Shine a flashlight into the water. How does the hydra react to light? How much light is present in the hydra's natural environment?
10. With a medicine dropper, add 1 drop of meat juice to the water. How does the hydra react to the meat juice?
11. Does the hydra have any buds? Buds are the beginnings of new hydras. Buds form about midway on the hydra's body. The hydra may also reproduce sexually.

QUESTIONS

1. Consider the simple structure of the hydra. How do you think the animal breathes?
2. What kind of stimuli can the hydra detect?
3. How do hydras reproduce?
4. How is the hydra a more complex animal than the sponge?

Many polyp-form coelenterates produce hard materials that surround the main part of their bodies. The **coral** is such an animal. It produces a limestone case that surrounds its body. The coral can extend its body and tentacles above the limestone case to capture food. Or it can contract and pull itself down inside the limestone case to be protected. (See Figure 21.13.)

Coral grows in large colonies in warm tropical waters. When the animal dies, the limestone case remains. After many years, the limestone from coral colonies builds up and forms large reefs. Many islands in the South Pacific Ocean are completely surrounded by large coral reefs.

Coelenterates have varied niches. What is the niche of coelenterates in their ecosystems? There is no single answer to that question. They have a variety of niches. Coral illustrates this type of variety.

Because coral is an animal, it acts as a consumer in the marine ecosystems in which it lives. But that is not the whole story. Many kinds of coral have single-celled algae living in the tissues of their bodies. These algae carry on photosynthesis, producing food and giving off oxygen. Because of these algae, the coral has, in a sense, the role of producer as well as consumer.

Anyone who has ever viewed the underwater scenes near a coral reef knows the coral's major role in marine ecosystems. Coral reefs are teeming with animal life. Thousands of animals, of great variety, swim in and out of the nooks and crannies of the reef. Thousands of others crawl on the coral or rest near it. The coral serves as a habitat for all of these thousands of animals. It also serves as a food source for many other animals.

Why is coral a good habitat? First of all, it has nooks and crannies where animals can hide and be protected from larger predators. Most coral reefs are brightly colored. Fishes and other small animals with the same bright coloration can live near the coral and escape being seen by predators. Because of the variety of habitats in coral reefs, the reefs themselves can easily be compared to the trees of a jungle.

coral:
a polyp-form coelenterate that produces a hard limestone case around its body

Figure 21.13: The living coral animals are polyps that secrete a protective coating of limestone.

LESSON REVIEW *(Think. There may be more than one answer.)*

1. Coelenterates
    a. are all marine animals.
    b. all have tentacles with stinging cells.
    c. are the biosphere's smallest animals.
    d. have one body opening.

Organisms

A   B

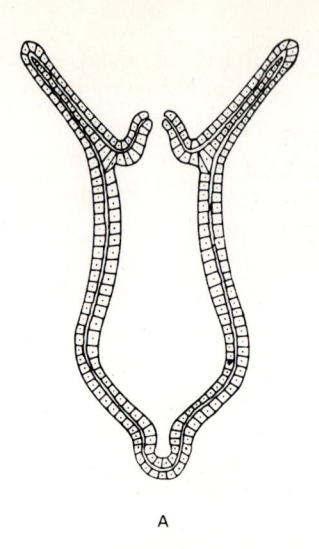

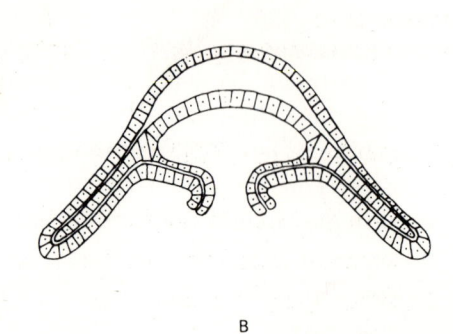

2. Figure A above is the coelenterate body form
    a. called a polyp.
    b. called a medusa.
    c. of a coral.
    d. of a hydra.

3. Figure B above is the coelenterate body form
    a. called a polyp.
    b. called a medusa.
    c. of a jellyfish.
    d. of a hydra.

4. Coral
    a. is not an animal.
    b. is found in colonies.
    c. is a consumer but can also be thought of as a producer.
    d. has the niche of providing a habitat for a variety of populations.

5. Which of the following terms best matches each phrase?
    a. coelenterate   e. polyp
    b. radial symmetry   f. medusa
    c. nematocyst   g. mesoglea
    d. gastrovascular cavity h. coral

    __e__ coelenterate form with tentacles upright

    __c__ a stinging cell in a coelenterate

    __b__ characteristic shape of all coelenterates

    __g__ protein material between the inner and outer layer of coelenterate cells

    __d__ the space inside coelenterates

    __a__ has stinging cells

    __f__ coelenterate form with tentacles hanging down

    __h__ coelenterate that secretes limestone around its body

KEY WORDS

coelenterate (p. 384)
radial symmetry (p. 385)
nematocyst (p. 385)
gastrovascular cavity (p. 386)

polyp (p. 386)
medusa (p. 386)
mesoglea (p. 387)
coral (p. 389)

Applying What You Have Learned

1. How are coelenterates more complex than sponges?
2. Name an object that has radial symmetry.
3. What levels of organization are absent in sponges and coelenterates?
4. How do the wandering cells in the sponge and the gastrovascular cavity in coelenterates perform similar functions?
5. Describe the main difference between an internal skeleton and an external skeleton.

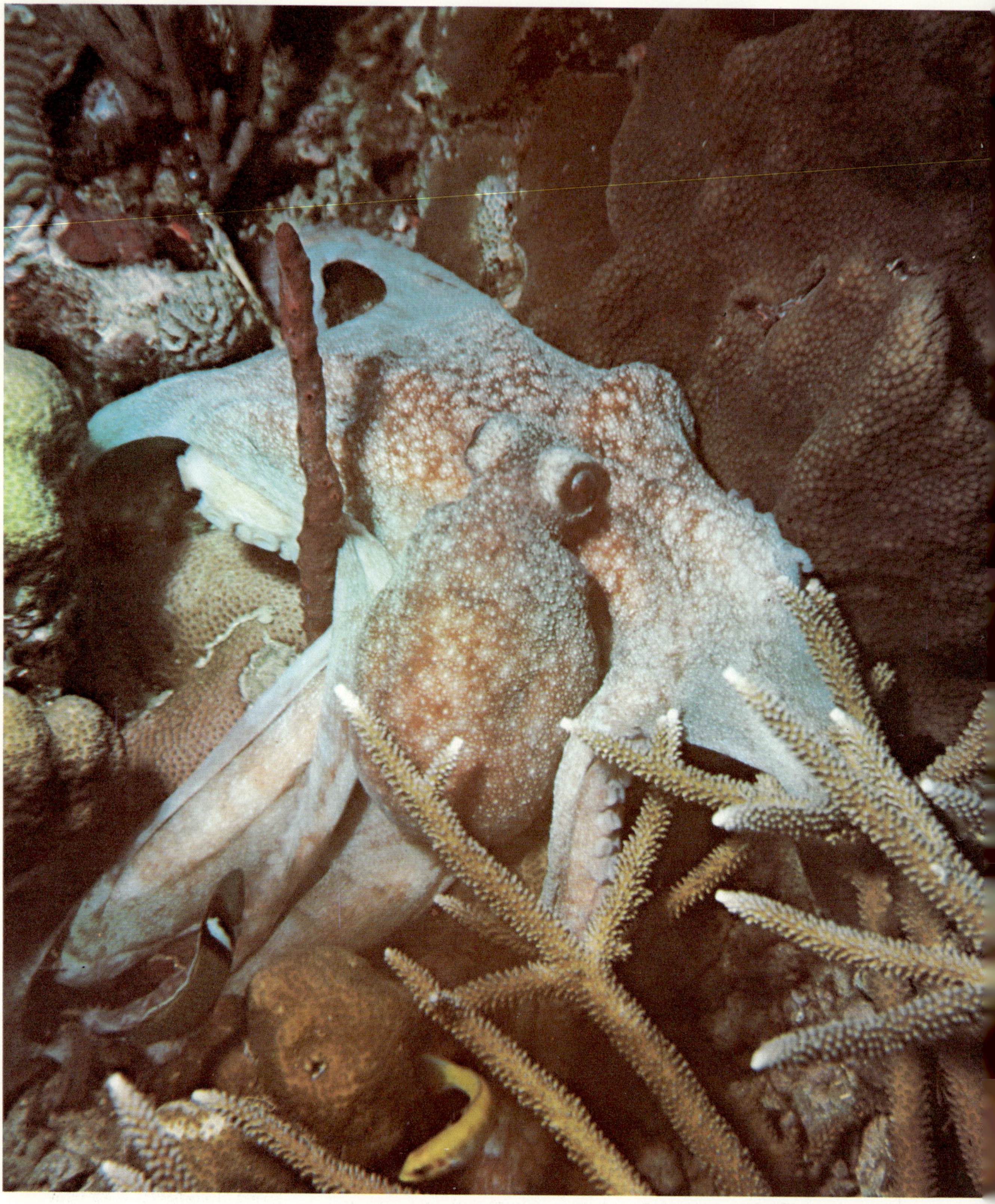

Figure 22.1: The octopus uses a jet of water for rapid movement. This mollusk can hide by releasing its own inky "smoke screen."

Chapter 22

WORMS AND MOLLUSKS

What are "worms"? Actually, that name has been given to thousands of different animals. To be called a "worm," an animal must usually be smaller than a snake, longer than it is wide, and without arms and legs. In this chapter, you will study three specific groups of animals called worms. These are the flatworms, the roundworms, and the segmented worms.

The mollusks make up another specific group of animals. The animals in this group have some things in common, but there are also many differences among them. Some of the common mollusks you may have seen are clams, snails, slugs, and squids.

FLATWORMS

Most flatworms have a flat body. As you might guess, flatworms get their name from the flat shape of their bodies. They also have the same type of body symmetry that you and

flatworm:
a kind of worm with a flat body

Organisms

bilateral symmetry: a pattern in which parts are arranged around a line, so that each half of the pattern is a mirror image of the other

most other animals have. It is called **bilateral symmetry.** The cutting of a cupcake was used to help you picture radial symmetry. Now substitute a gingerbread "man" for the cupcake. There is one place where you can cut a gingerbread man and get two equal halves. Where?

There are more than 7000 kinds of flatworms. About two-thirds of them are parasitic, meaning that they live in or on the bodies of other animals.

The flatworms are much more complex than the coelenterates. Flatworms have a nervous system and a small brain. They also have well-developed reproductive, excretory, and digestive systems. They do not have a breathing system or a skeleton. There are three main groups of flatworms.

Turbellarians are mostly free-living.

turbellarian: a member of one of the three main groups of flatworms. Turbellarians live in aquatic or moist terrestrial ecosystems.

The members of one of the three main groups of flatworms are called **turbellarians** (tûr′bə lâr′ē ən). Few of the turbellarians are parasites. Most live independently and are thus called free-living. These worms range in length from less than 0.1 centimeter to over 50 centimeters (0.25 inch to over 20 inches). Most of them live in aquatic ecosystems. Some live in moist terrestrial environments.

The planarian is a typical turbellarian.

planarian: a nonparasitic freshwater flatworm

The **planarian** (plə nâr′ē ən) pictured in Figure 22.2 is a very common turbellarian. Planarians are usually found in freshwater habitats. They are almost always found on rocks in clear streams. In this habitat, they feed on a variety of diatoms and small animals.

regeneration: the process by which some organisms grow back a lost part of the body

The planarian is one of the life scientist's favorite experimental animals. Cut a planarian into five pieces and you will usually end up with five planarians. This **regeneration** ability—the ability to grow new body parts—is the basis of many experiments. Other experiments are designed to study the planarian's mental ability. The planarian has a small cluster of nerve cells in the "head" end of its body. This is the simplest kind of brain. Life scientists are trying to learn how this brain functions. They can use this information in learning about the function of other animal brains.

Figure 22.2: Planarians have a pair of light-sensitive spots on the "head end." (Courtesy Carolina Biological Supply Co.)

Worms and Mollusks

Figure 22.3: This is the adult stage of a liver fluke, stained to show its internal structure. It requires three hosts. Humans are the host animals for the adults. This fluke spends another part of its life cycle inside a snail, and then it must live in a fish. (Courtesy Carolina Biological Supply Co.)

Flukes are parasitic. The animals belonging to the second main group of flatworms are called **flukes**. All the flukes are parasitic. Some attach themselves to the outside of animal bodies. Some use a sucker as a mouth for feeding upon the animal. Other flukes live as parasites on the inside of animal bodies. Many of their names—such as "liver fluke," "lung fluke," and "blood fluke"—give clues to some of their habitats.

fluke: a member of one of the three main groups of flatworms. All flukes are parasitic.

Some flukes lead complex lives. People usually think of themselves as leading complex lives. Yet we cannot begin to compare the complexity of our lives with that of some flukes. For example, some flukes undergo four or five separate stages of development. During each of these stages, the body structure is very different. In one stage, the creature may swim in a freshwater pond and look like a protozoan. At another stage, the creature might have a worm shape and live coiled up in a cocoon-like protective case.

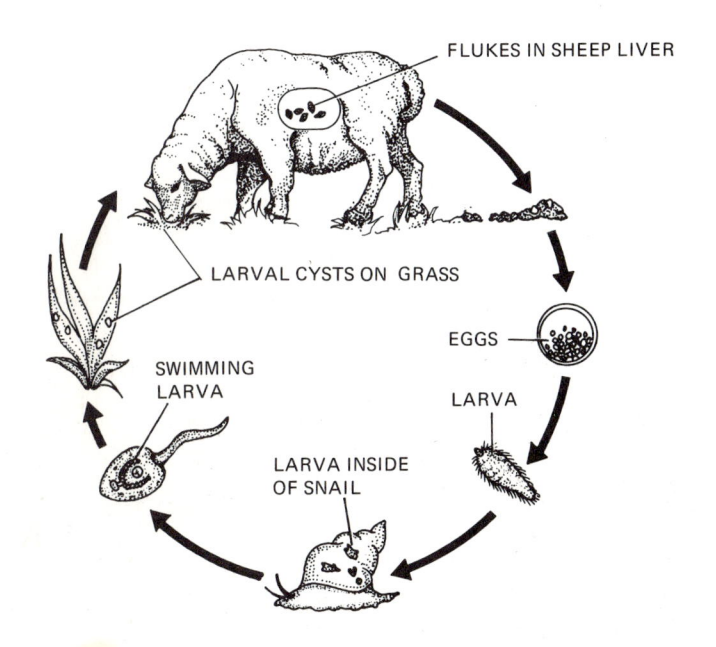

Figure 22.4: The sheep liver fluke has two hosts. Its life cycle can be completed when sheep graze near a pond or in a marshy area.

395

Organisms

host:
the organism on or in which a parasite lives

A variety of hosts may also make the fluke's life more complex. The **host** is the animal that a parasite lives on or in. Before some flukes can mature they must live first in the body of a snail and then in the body of a vertebrate animal. (Vertebrates—for example, mammals, birds, and fish—are animals that have backbones. You will learn about them in Chapter 24.) Such flukes require two hosts. Others must live first in a snail, then in a crayfish or fish, and then in some vertebrate. That is three hosts, but still not the end of the story. There are flukes that must live in four hosts before they mature! One kind of fluke must live in a snail, an amphibian (such as a frog), a rat or mouse, and a cat, dog, mink, or weasel. As you can imagine, life scientists have had lots of fun—and headaches—trying to study the lives of such flukes.

Some flukes cause human diseases. In Chapter 15, we discussed the agent causing swimmer's itch. This is a mild disease causing redness and irritation of the skin. The agent that causes the disease is a fluke that has left the body of a snail and is looking for its second host—a bird. It bores into the human skin and then dies. For this fluke, we are not a suitable substitute for a bird.

Humans are a suitable host for certain lung flukes, liver flukes, and blood flukes. These flukes cause serious diseases in people. They are most common in countries of the tropical Far East and Africa. For example, blood flukes are common in people who work in rice paddies and other irrigated places. These flukes have left the body of a snail. They swim in the water of the rice paddy and bore through the skin of the victim. Once inside the victim they live in the bloodstream and cause all sorts of problems.

One unexpected side effect of the construction of the Aswan High Dam in Egypt was an increase in blood fluke disease. As water covered a greater area behind the dam, snails carrying blood flukes were able to travel to new areas where they had not previously been a problem.

tapeworm:
a member of one of the three main groups of flatworms. All tapeworms live as parasites in animal intestines.

Tapeworms are internal parasites. The third main group of flatworms includes all the worms called **tapeworms**. All the tapeworms live within the intestines of animals.

Figure 22.5: These blood flukes can cause human disease after leaving their snail host. The flukes infect a person by boring through the skin and then living in the human host's bloodstream

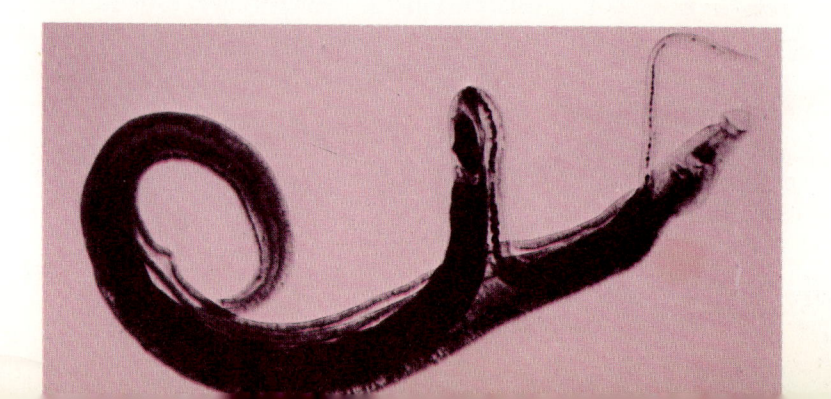

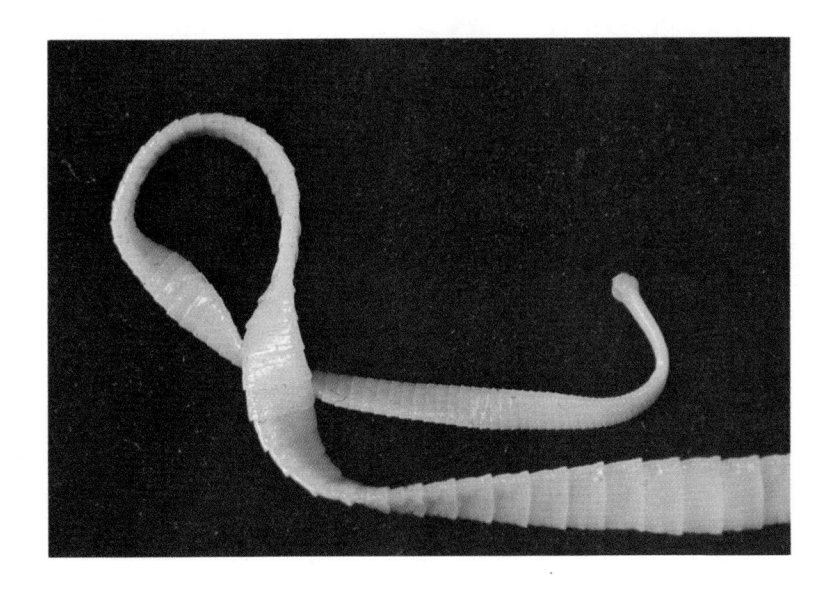

Figure 22.6: This is just a part of a tapeworm. (Tapeworms may be several meters long.) Note the body segments and the small round head at the end. The head has tiny hooks that the tapeworm uses to anchor itself on the wall of its host's intestine.

The tapeworm is a much simpler animal than the other types of flatworms. For example, the tapeworm does not have a mouth or a digestive system. It does not need one! Most tapeworms live in the intestines of vertebrate animals. In that environment, the worm just absorbs food through the layers of its body. The food has already been digested by the worm's host.

Humans are the host for some tapeworms. The beef, pork, and fish tapeworms are all tapeworms for which people are the main host. These tapeworms take their names from the other host in which they may live. When in the human body —in the intestine—the tapeworms may grow to be several meters long. The fish tapeworm, which is the most harmful, may grow to be 15 or more meters (50 feet) long.

Tapeworms cause problems in several ways. They rob the host of food. This can be a very serious problem in countries where food is in short supply. Tapeworms also irritate the walls of the intestine, which results in pain and physical damage.

How do people become infected with tapeworm? Usually by eating meat that is not thoroughly cooked. When beef, pork, or fish tapeworms live in the animal for which they are named, they live in the muscles. The muscles, of course, are what we eat as meat. Does this mean that we should always eat well-cooked beef, pork, or fish? Not necessarily. All three of these tapeworms are very rare in the United States. Beef raised in the United States is rarely infected with any human parasite. Pork should be well cooked because there are other parasites (the trichina worm) that may be in the muscles. Fish may or may not be infected with human parasites. There is always some risk in eating uncooked or incompletely cooked fish.

Organisms

LESSON REVIEW *(Think. There may be more than one answer.)*

1. Of the three flatworms pictured,
    a. A would be found in most uncooked beef.
    b. B could be found in some humans.
    c. C would be the easiest to find in an outdoor habitat.
    d. A can be many times larger than B or C.

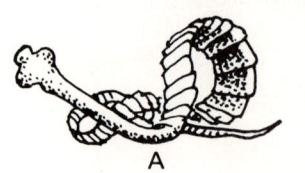

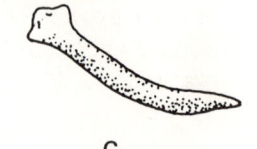

A   B   C

2. Flatworms
    a. are mostly parasitic.
    b. are mostly free-living.
    c. have tissues and organs but no systems.
    d. have the same body symmetry as you have.

3. The planarian
    a. is a parasitic turbellarian.
    b. has a simple brain.
    c. can regenerate body parts if cut into pieces.
    d. can be found on rocks in clear streams.

4. Flukes
    a. are all parasitic.
    b. live either outside or inside other animals.
    c. may live in as many as four different hosts during their life.
    d. may bore into or through the skin.

5. Human tapeworms
    a. can be as long as your arm spread.
    b. absorb food from the human intestine where they live.
    c. will be contracted any time one eats poorly cooked beef or pork.
    d. never cause pain or physical damage.

6. Which of the following terms best matches each phrase?
    a. flatworm   e. regeneration
    b. bilateral symmetry f. fluke
    c. turbellarian   g. host
    d. planarian   h. tapeworm

    ____ organism that a parasite lives in or on

    ____ mostly free-living flatworms

    ____ long, with bodies made up of segments

    ____ all are parasitic; some have complex lives

    ____ process of growing missing parts

398

_____ a kind of worm with a flat body

_____ a common, free-living flatworm

_____ characteristic of the body form of all flatworms

KEY WORDS

flatworm (p. 393)
bilateral symmetry (p. 394)
turbellarian (p. 394)
planarian (p. 394)

regeneration (p. 394)
fluke (p. 395)
host (p. 396)
tapeworm (p. 396)

ROUNDWORMS

Roundworms are almost everywhere. There are many different kinds of **roundworms** and closely related worms. About 14,000 kinds have been identified. Some life scientists estimate that there may be 500,000 different kinds on the Earth. There are a lot more to be studied and identified! Large numbers must also be used to describe many of the individual populations of roundworms. Some populations have a fantastic reproductive ability. For example, someone went to the trouble of counting (or estimating) the number of roundworms in one decaying apple. The number: 90,000.

Roundworms may live in the soil, in fresh water, and in marine ecosystems. Many of them are parasites in both plants and animals. The roundworms vary greatly in size. Some are microscopic. And there are some roundworms that are over a meter (more than a yard) long.

Nematodes are the most abundant group of roundworms. By far the largest number of roundworms belong to

roundworm:
any of a major group of worms that has a round body

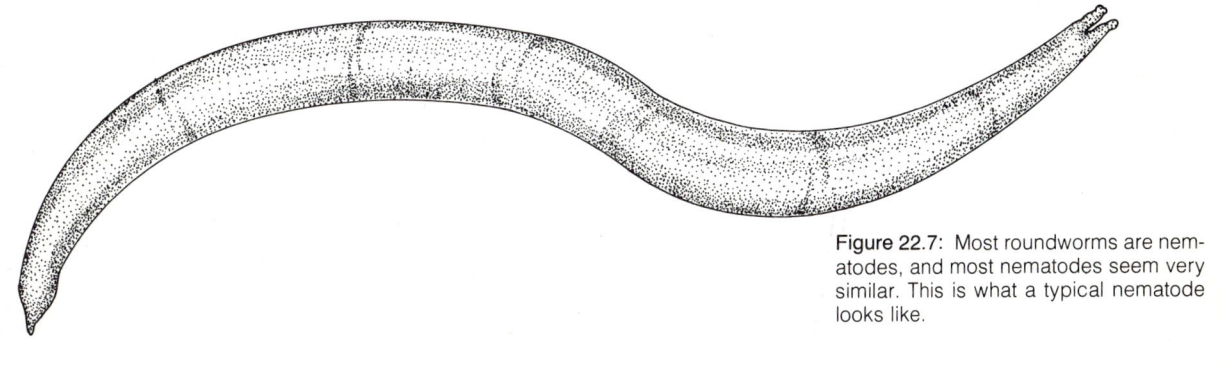

Figure 22.7: Most roundworms are nematodes, and most nematodes seem very similar. This is what a typical nematode looks like.

Organisms

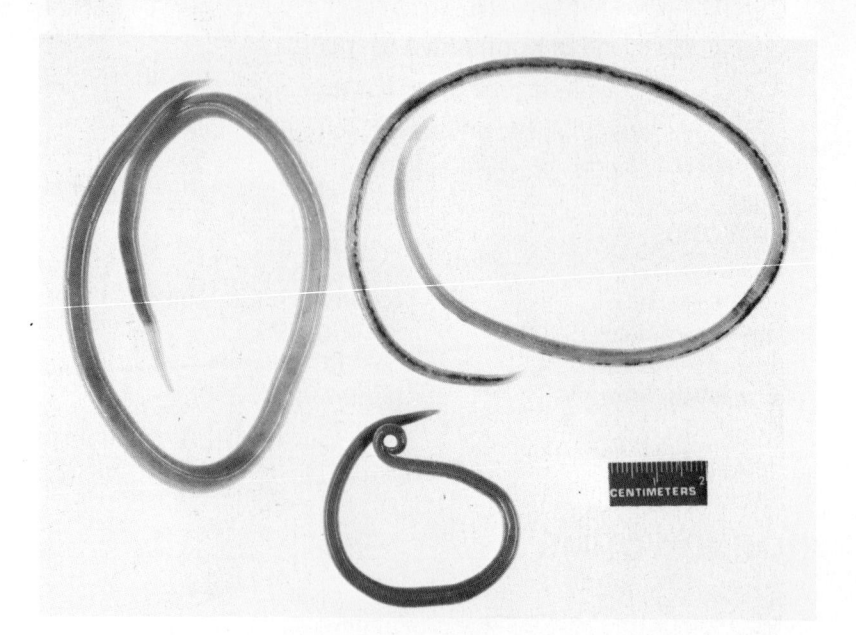

Figure 22.8: Nematodes are important decomposers in terrestrial ecosystems.

nematode:
a member of the largest group of roundworms

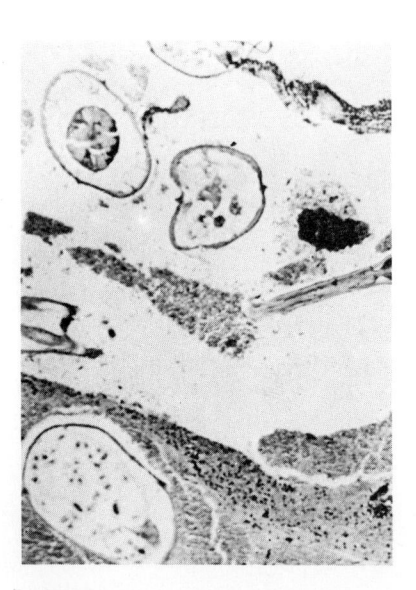

Figure 22.9: These roundworms are pinworms—nematodes that live as parasites in the intestines of the human body.

a subgroup called **nematodes** (nĕm'ə tōd). Over 12,000 kinds of nematodes have been identified. There is one strange fact about them. Of all the many kinds, about 10,000 different kinds of these worms look alike. A diagram showing the general structure of a nematode body is shown in Figure 22.7.

If most of the nematodes look alike, how do they differ from each other? In many ways. They differ in where they can live. They differ in what they can eat. They differ in size. They differ in the way they reproduce. They differ in internal structure.

Nematodes have an important niche. Nematodes are most abundant in good fertile soil. What do they do there? Their main activity is feeding on dead plant and animal matter. In doing this, they play a very important role in recycling nutrient materials in the ecosystems to which they belong. Their role is very similar to the role of the decomposer bacteria and fungi.

Parasitic nematodes are a problem. Almost every vertebrate animal is a host for one or more kinds of parasitic nematodes. The human alone may be a host for over fifty different kinds. These figures may give you some idea of the importance of nematodes as parasites.

How do people become infected by nematode parasites? There are several ways. One way is to swallow their tiny eggs. Ascaris, a common parasitic nematode in both pigs and human beings, is gotten in this way. The eggs hatch in the intestine, and the young worms bore through the intestine wall and enter the bloodstream. They leave the bloodstream in the lungs, where they continue their development. Eventually, they are

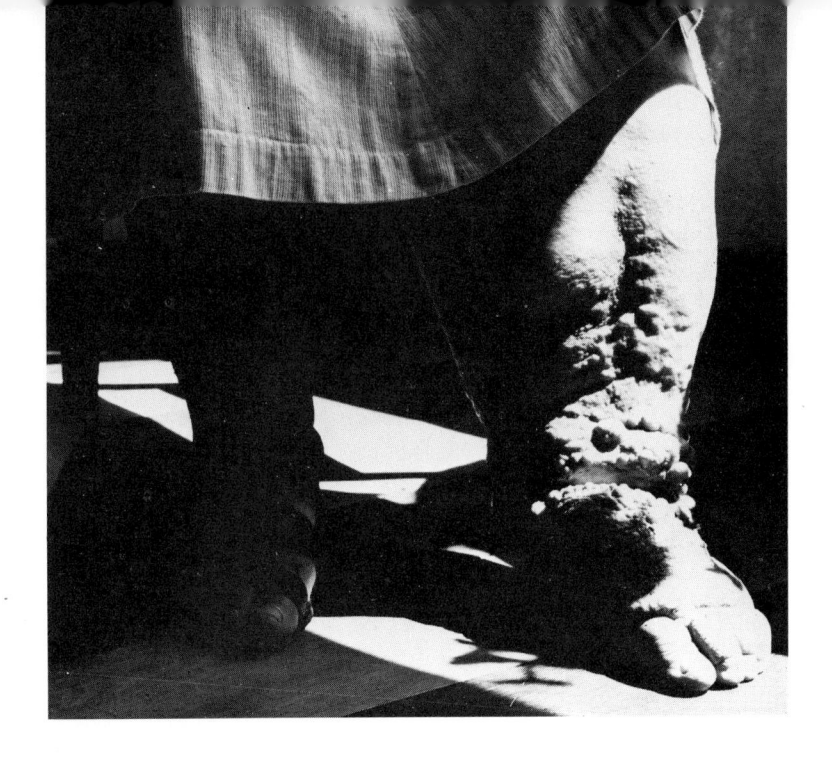

Figure 22.10: This woman is a victim of elephantiasis, which is caused by a mosquito-transmitted roundworm. No treatment has yet been found for this tropical disease.

coughed up into the throat, then swallowed. They live their adult lives in the intestine. This is the same way that a pet dog or cat is likely to be infected with various kinds of worm parasites. The trichina worm, the parasite that is a disease hazard in undercooked pork, is also a nematode worm.

Some nematode worms have two hosts. The filaria worm is an example. This parasite is common in some tropical areas of the world. A mosquito carries the microscopic immature worms in its body. When it bites a human being, some of these worms invade the bloodstream. They usually migrate to a lymph node (a reservoir for lymph) in the arm or groin. As the worms grow and multiply they plug up the lymph circulation. This causes a swelling of the leg or arm. Eventually, an ugly-looking elephant-like arm or leg is produced. One name for this disease is elephantiasis (ĕl′ə fân tī′ə sĭs).

Rotifers are the "wheel animals." Sometime every person should have the opportunity to watch **rotifers** (rō′tə fər). They are fascinating creatures. All of them are microscopic. Most of them live in fresh water. Almost any sample of pond water (in which there is some algae) will have several rotifers. The rotifers make up a small subgroup of the roundworms. About 1500 kinds of rotifers have been described.

How would you know a rotifer if you saw one? If you looked at the front end of the animal, you would see what appears to be one or two wheels spinning rapidly. When rotifers were first discovered, they were called "wheel animals." What look like wheels are really bands of hairlike cilia waving in one direction. They create a whirlpool, which draws food down into the rotifer's

rotifer:
a microscopic freshwater roundworm. Rotifers have bands of cilia on their head ends that look like spinning wheels.

Organisms

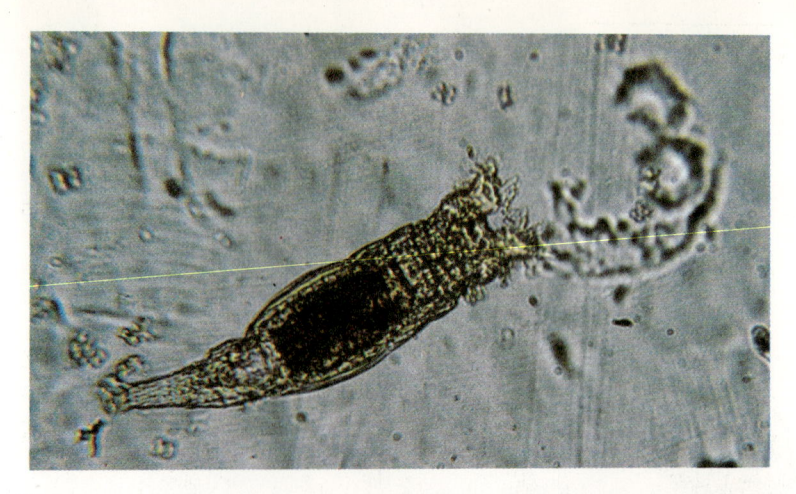

Figure 22.11: This is a typical rotifer, such as you might find in a sample of pond water. If you look closely, you can see the two whirling bands of cilia at the front end of the animal.

mouth and jaws. (What other animal gets food this way?) The jaws are usually busy moving back and forth. (See Figure 22.11.) Many who observe the jaws in action for the first time think that they see a heart beating.

Gastrotrichs are also interesting creatures. Look at the fuzzy looking creature in Figure 22.12. It is a **gastrotrich** (găs′trō **trĭk**′). Only about 200 different kinds have been discovered. But they are fairly common in freshwater ecosystems. If you ever study a sample of pond water looking for a rotifer, you are also likely to see a gastrotrich. If you see such an animal and wonder how it lives, you might as well begin your own study. Life scientists know little about them.

gastrotrich: a microscopic, freshwater animal related to roundworms. Gastrotrichs have cilia on their bodies that give them a fuzzy appearance and help propel them through water.

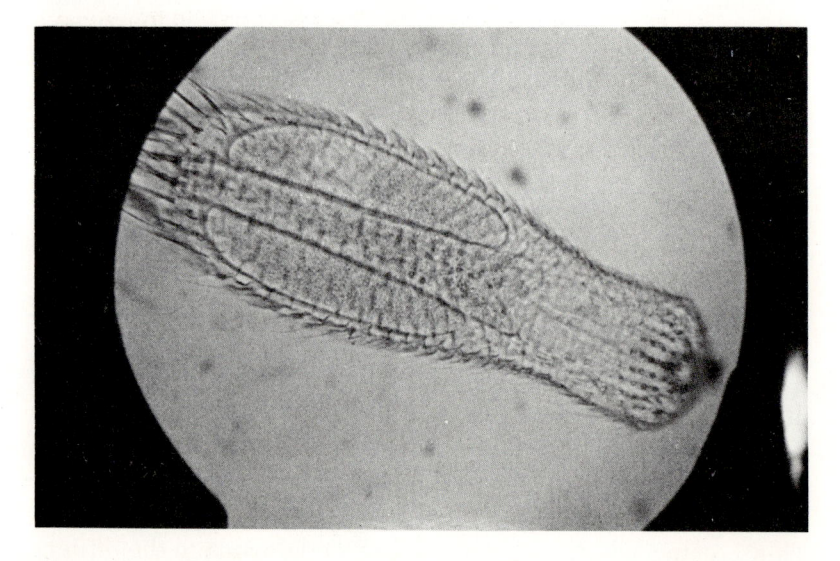

Figure 22.12: If you look for rotifers in pond water, you may also see an animal like this. It is a gastrotrich, and not much is known about it.

Worms and Mollusks

LESSON REVIEW *(Think. There may be more than one answer.)*

1. Of the three animals pictured,
    a. A is more likely to be found in the soil than B or C.
    b. B has two wheels that rotate at the end of its body.
    c. C is the roundworm that life scientists know the most about.
    d. A could be parasitic or free-living.

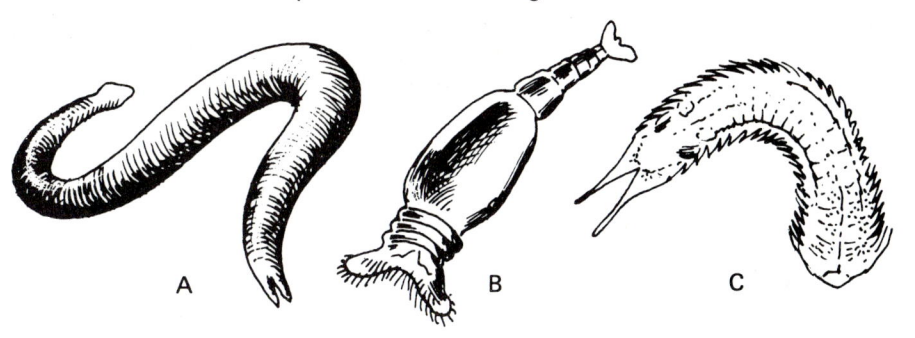

2. Roundworms
    a. are all smaller than 1 mm (0.04 inch).
    b. live only in the soil and fresh water.
    c. are all parasites.
    d. have a fantastic reproductive ability.

3. Nematodes
    a. make up the largest group of roundworms.
    b. mostly look alike.
    c. differ in many ways.
    d. have a main niche similar to that of decomposer bacteria and of fungi.

4. Parasitic nematodes
    a. can infect humans.
    b. include ascaris which may be found only in the human intestine.
    c. may have two hosts.
    d. include the trichina worm which might be contracted from uncooked beef.

5. Rotifers
    a. are all microscopic.
    b. are common in pond water.
    c. have jaws that whirl around their head.
    d. feed by creating a whirlpool.

6. Which of the following terms best matches each phrase?
    a. roundworm   c. rotifer
    b. nematode   d. gastrotrich

    _____ one of the most common types of roundworm

    _____ any of a major group of worms that has a round body

403

Organisms

_____ a microscopic roundworm that may be called a "wheel animal"

_____ a freshwater animal related to roundworms which has a fuzzy body

KEY WORDS

roundworm (p. 399)
nematode (p. 400)

rotifer (p. 401)
gastrotrich (p. 402)

MOLLUSKS

mollusk: a small animal that has a soft body. Some, but not all, mollusks have a shell, tentacles and eyes, and a muscular foot.

Mollusks are soft-bodied animals. The term **mollusk** means "soft-bodied." A snail is a mollusk. If you could pull one out of its shell and pinch it, you would see that it fits that description.

There are about 50,000 different kinds of mollusks. Here are four characteristics that most of them have in common: (1) They have a well-developed head region, with tentacles and eyes. (2) They have a muscular foot that occupies most of the underside of the body. (3) They have a region in the top of the body, where most of the internal organs are massed together. (4) Finally, they have a sheet of tissue over the top of the body called a **mantle**. The mantle usually produces a limestone type of shell.

mantle: a tissue sheet over the top of a mollusk's body that usually produces a hard shell

Figure 22.13: This land snail is creeping along on its muscular "stomach foot." Its eyes are at the ends of the two tentacles. The snail's twisted limestone shell was produced by the mantle, which is hidden from view inside the shell.

Figure 22.14: Looking down at this chiton, you can clearly see the eight plates produced by its mantle.

Figure 22.15: The largest clams known live near coral reefs in the Indian and Pacific Oceans. The shell of the giant clam may grow to a width of 1 meter.

Chitons have shells of eight plates. Chitons (kɪt′ən) are flat mollusks that have eight separate plates over their backs. These plates, which are produced by the mantle, overlap each other and function like the armor plating of ancient warriors.

Chitons are commonly found on rocks and in pools along ocean shores. They feed on algae which they obtain with a rough-edged tongue that moves in and out of the mouth.

Bivalves have no head. Clams, oysters, and scallops are some of the most common **bivalve mollusks.** The name "bivalve" refers to the shell of these animals, which is divided into two halves. The bivalves do not have a well-developed head like the other two mollusks. The body is flat and sandwiched between the two parts of the shell that encloses them. Bivalves have a very strong foot that they can extend beyond the shell. Besides its function during movement, the foot can also anchor the animal in moving water. Clams move very little and often appear lifeless when seen in shallow water. But some might surprise you. When alarmed, some clams use the foot like a shovel and rapidly dig themselves out of sight beneath the sand.

Gastropods are the largest group of mollusks. About 40,000 different **gastropods** have been identified. Many more remain to be discovered. This makes the gastropods the largest group of mollusks. What are gastropods? The name means "stomach foot." The common members of this group are

chiton:
a mollusk with a shell made of eight overlapping plates

bivalve mollusk:
a mollusk that has no head inside the two halves of its shell

gastropod:
a member of the largest group of mollusks. A large muscular foot is the common characteristic of all gastropods.

Organisms

Figure 22.16: This gastropod is a marine slug. Like most marine slugs, it is brightly colored. The terrestrial slugs that you are likely to see are almost colorless.

cephalopod: the largest of the mollusks. The foot of a cephalopod is wrapped around its head and divided into tentacles.

the snails and slugs. They have a large foot beneath the length of their bodies. They glide along on this foot as though they were sliding on their stomachs.

Snails are easily recognized because of their twisted shells. Most of the internal systems of a snail are packed within the cramped space of the shell. The snail's foot and head can be extended or drawn up into the shell. Slugs are usually described as "snails without shells." The internal organs of the slug are arranged differently from the snail's. Except for that fact, the description just about fits the slug.

There is a great variety of gastropods in marine ecosystems. These animals are the most spectacular gastropods in size, shape, and color. Anyone who has seen a seashell collection will agree. Most seashells once covered the backs of marine gastropods.

Cephalopods are the largest mollusks. The term **cephalopod** (sĕf′ə lə pŏd′) means "head foot." The foot of these mollusks is wrapped around the head and divided into armlike tentacles. The squid, octopus, and nautilus are all cephalopods. The squid has ten tentacles, the octopus has eight, and the nautilus has seventy or eighty.

The nautilus is the only cephalopod that has a real shell. The mantle of the squid produces an internal structure that functions like a small skeleton. It gives the animal a rigid structure. This enables some squids to grow very large and survive the dangers of water currents. Some giant squids are real sea monsters. About 100 years ago, a squid was caught that was 15 meters (50 feet) long—not counting the tentacles.

The octopus does not have any type of shell. It is much more flexible than the squid. But seldom is the body of an octopus more than 35 centimeters (14 inches) wide or long.

Figure 22.17: Squids—especially small ones—are highly prized as food. But how are they caught? This Basque "fisherman" is using a technique that is common in Spain. A heavy, brightly colored, multihooked lure is tied to a long cord and dropped into the water. The "fisherman" bounces the lure up and down. This apparently angers or attracts the squid; it grabs the lure and gets itself hooked. Then it is slowly pulled to the surface and held there until it has "shot its ink" (above). The remaining ink and the ink sac are removed when the squid is cleaned. Later the squid may be cooked in a sauce made from this ink.

Cephalopods have two interesting adaptations—characteristics that suit them for survival in their environment. They move through the water by a kind of jet propulsion. They take water into their bodies, then send it out through a tube. They can send the water out slowly, which causes them to move slowly. Or they can send the water out rapidly. This enables them to make quick movements. The squid and the octopus can turn the tube in just about any direction, and so they can shoot off in any direction.

The second interesting adaptation of cephalopods is their special method of protecting themselves. They have ink sacs in their body. When alarmed, they release the ink, clouding the water. This "smoke screen" enables them to hide or escape from their enemies.

The cephalopods have one other unique characteristic. They have the best developed eyes and brains of any one of the invertebrate animals. (Invertebrates are animals without backbones. All of the animals presented in Chapters 21, 22, and 23

Organisms

are invertebrates.) The eyes of cephalopods are much like your own in structure. Life scientists are doing a great deal of research on the nervous system and brain of the squid. Research on these animals has helped us to understand our own nervous system better.

LESSON REVIEW *(Think. There may be more than one answer.)*

1. Of the animals pictured,
    a. A is the only one found in all three major habitats (terrestrial, freshwater, marine).
    b. B is a parasite.
    c. C is incapable of movement.
    d. one related to D was found that was longer than your classroom.

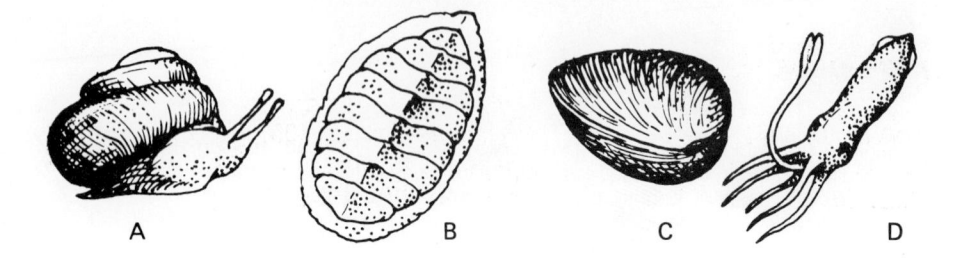

2. Mollusks
    a. are hard-bodied animals.
    b. include snails, coral, and chitons.
    c. usually have a muscular foot.
    d. have a mantle that may produce a limestone shell.

3. Chitons
    a. are flat mollusks.
    b. have a one-piece shell.
    c. feed on a variety of small animals.
    d. are never longer than the width of your finger span.

4. Bivalves
    a. are snails.
    b. have a shell with two parts.
    c. are never longer than one of your arms.
    d. are almost lifeless and incapable of rapid movement.

5. Gastropods
    a. all have shells.
    b. are the largest group of mollusks.
    c. move by gliding along on their foot.
    d. are found only on land and in fresh water.

6. Cephalopods
    a. may have a shell or an internal skeleton.

b. are sometimes longer than your arm spread.
c. move by a method similar to jet propulsion.
d. have the best developed eyes of any invertebrate animal.

7. Which of the following terms best fits each phrase?
   a. mollusk   d. bivalve mollusk
   b. mantle   e. gastropod
   c. chiton   f. cephalopod

   _____ flat mollusk whose shell consists of eight plates
   _____ produces ink to defend itself
   _____ member of the largest group of mollusks
   _____ member of a group of mollusks that have no head
   _____ soft-bodied animal that sometimes has a shell, tentacles and eyes, and a muscular foot
   _____ top layer of tissue that secretes the shell of mollusks

KEY WORDS

mollusk (p. 404)
mantle (p. 404)
chiton (p. 405)

bivalve mollusk (p. 405)
gastropod (p. 405)
cephalopod (p. 406)

ANNELIDS

Annelids are segmented worms. The term **annelid** (ăn′ə lĭd) means "ringed." The body of an annelid worm consists of many segments, which gives it a ringed look.

The annelids are the most complex of all the animals that could be called worms. They have well-developed transport, digestive, nervous, reproductive, and excretory systems. They are much more complex than the roundworms, which they sometimes resemble.

There are three main groups of annelids. Two of the groups are free-living; one is parasitic.

Polychaete worms are mostly marine. One main group of annelids is made up of **polychaete** (pŏl′e kāt′) **worms**. *Polychaete* means "many bristles." That is a good descriptive name for these annelids. This group does have many bristles, which extend out from the body segments. Some of the bristles are thickened and function much like paddles for moving the worm through its watery environment.

Most of the polychaete worms live in the intertidal zone of

annelid:
any of a major group of worms with a body divided into ringlike segments

polychaete worm:
an annelid that has many bristles and that usually lives in an intertidal marine habitat

Figure 22.18: The clamworm (left) is a polychaete that moves about in its environment. (Courtesy Carolina Biological Supply Co.) The featherduster worm (right) is a "tube worm," anchored in one spot all its life. What function does the feathery crown serve?

marine waters. This is the area that is constantly changing in depth because of the movement of the tides. Most of these worms feed on the top layer of mud that is rich in microorganisms and in dead plant and animal matter. The worms actually eat the mud. It then passes through the digestive canal, where nutrient materials are digested and absorbed. The sand and other material that cannot be used for food are released as waste material.

There are two basic types of polychaete worms. One type crawls and swims about its environment. An example is the "clamworm" pictured in Figure 22.18. The other type buries itself in a tubelike hole in the mud and lives its life in one area. This second type is commonly called a "tube worm." A type of tube worm called the "feather duster" worm is also shown in Figure 22.18.

Oligochaetes are freshwater and terrestrial. Another major group of annelids is made up of **oligochaete**

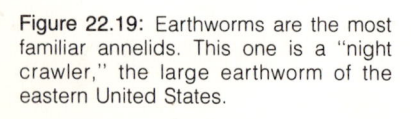

Figure 22.19: Earthworms are the most familiar annelids. This one is a "night crawler," the large earthworm of the eastern United States.

(ŏl'ĭ gō kēt') **worms.** *Oligochaete* means "few bristles." The earthworm is the best known oligochaete. It has very short bristles on its body that can be felt but cannot be seen without magnification.

The habitat of the earthworm is well known by anyone who has lived in a rural area. They are common inhabitants of soil that is rich in decaying plant and animal matter. They burrow through the soil, eating the soil as they go. The soil passes through the worm's digestive canal, where the nutrients are digested and absorbed. The earthworm helps the soil by loosening it. This enables water to soak in and also allows the soil to contain more air, which is needed by the cells of plant roots.

Leeches are parasites. The last major group of annelids is made up of the **leeches**. Leeches are usually flat and are dark red or brown in color. They have one or two suckers that they use to attach themselves to other animals. They are usually found in freshwater ecosystems, and they are parasites on fish and other animals of this environment. They feed on blood, which they suck from the host's body. Leeches do not have any bristles on their bodies.

Leeches used to be grown and carried by doctors as they visited their patients. This was back in the days when bloodletting, or the draining of blood, was thought to be good for curing diseases. What better way to drain blood than to put leeches on the patient's body! Actually, this is not as painful as you might think. If you have ever had a leech on you, you know that you seldom feel its bite or the removal of the blood.

oligochaete worm:
annelid that has few bristles, which are too small to be seen, but can be felt

leech:
a parasitic annelid that attaches itself to its animal host by suckers

Figure 22.20: Leeches are parasitic annelids in freshwater ecosystems. This leech is taking a meal from the underside of a turtle.

Organisms

Laboratory Activity

Find the Body Systems in an Earthworm.

PURPOSE — The earth worm is probably the most important worm to people because it is the primary builder of topsoil. Earthworms burrow through the soil by eating their way through the leaves, seeds, plants, and tiny animals in the ground. In this investigation, you will dissect an earthworm and study its parts.

MATERIALS

earthworm straight pins
hand lens pointed scissors
dissecting pan tweezers

PROCEDURE

1. Observe the outside appearance of the earthworm. Notice the many segments which make up its body. Are both ends of the earthworm alike? Does the earthworm have a head and a tail?
2. Feel the skin of the earthworm. Are the top and bottom sides of the earthworm alike? Do you feel the hairlike bristles, or *setae* (**sē′tē**), on the bottom of the earthworm? The setae are arranged in four rows along the length of the underside of the worm. The bristles help the worm in moving.
3. Look at the first segment of the earthworm. Find the lip. The lip is used by the earthworm in digging through the soil.
4. Notice the light, saddle-shaped band around the outside of the earthworm (segments 31 to 37). This is the *clitellum* (klĭ **tĕl′**əm). It secretes a liquid that forms a cocoon from which the tiny worms hatch.
5. Place the earthworm in the dissecting pan. The top surface of the earthworm should be facing you.
6. Pin the first and last segments of the earthworm to the pan. With a sharp scissor, cut through the top layer of skin, starting at the head end. Cut through only five or six segments. Be careful not to cut into the internal organs.
7. Using tweezers, force the skin open. Pin back the skin. Continue cutting and pinning for about one-third the length of the worm.
8. Note how the inner cavity is divided by thin walls. These walls appear at the same intervals as the external segments.
9. The *digestive system* is a long tube that runs through the entire length of the body. Locate the *pharynx*. It is below the mouth. The pharynx is muscular and sucks food and soil into the worm.

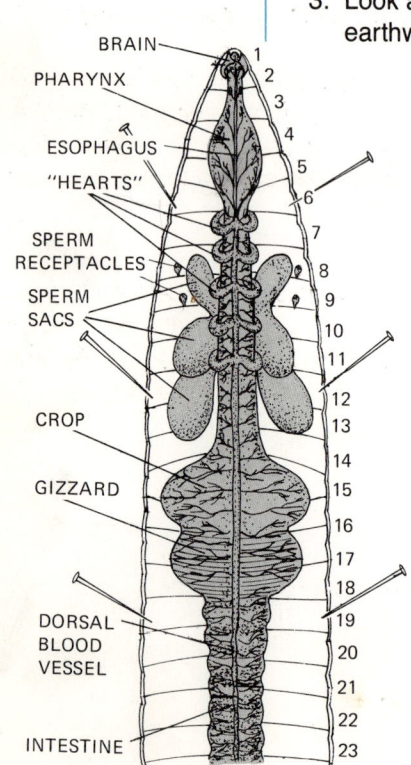

10. From the pharynx, the food passes to a long narrow tube, the *esophagus*. Locate the esophagus.
11. The food is then stored in the thin-walled *crop*. Locate the crop.
12. The food is slowly released from the crop into the *gizzard*. As its muscular walls move back and forth, sand in the gizzard grinds the food. Locate the gizzard.
13. From the gizzard, the food passes to the long *intestine*. Here enzymes are secreted that aid in digestion. Locate the intestine.
14. The *anus* is found in that last segment of the worm. Undigested food and soil leave the body through the anus. Locate the anus.
15. Identify parts of the *circulatory system*. The five pairs of hearts which surround the esophagus pump blood through the large *ventral blood vessel*. Locate the hearts. Locate the ventral blood vessel by looking beneath the intestine.
16. Blood flows from the ventral blood vessel into smaller vessels in each segment of the worm's body. These vessels carry blood to the capillaries which supply the cells. The blood returns to the hearts through small vessels in each segment and then to the dorsal blood vessel. Locate the *dorsal blood vessel* on top of the intestines.
17. Identify parts of the *nervous system*. The two tiny, round, white bodies above the pharynx make up the brain. Locate the brain.
18. Identify parts of the *reproductive system*. Earthworms are *hermaphrodites* (hər **măf′**rə dīt′)—that is, each earthworm contains both male and female reproductive organs. Locate the three pairs of white bodies surrounding the esophagus. These are the sperm sacs in which the sperm mature and are stored. The testes are located in the sperm sacs.
19. Locate the *sperm receptacles* (segments 9 and 10). These store sperm until eggs from the ovaries mature.
20. The tiny ovaries lie below the third pair of sperm sacs. Locate the ovaries with a hand lens.
21. Identify the parts of the *respiratory system*. The earthworm has no lungs. It is a "skin breather." The thin skin is kept moist by mucus secreted from the glands below the skin's surface. Oxygen in air diffuses through the skin into the blood in the many capillaries lying close to the skin's surface. Use a hand lens to locate some of these tiny blood vessels.

QUESTIONS

1. What is the importance of topsoil to people? How do earthworms help to prepare topsoil?
2. Briefly compare the following organ systems in the earthworm and in the human:
    digestive system
    circulatory system
    nervous system
    reproductive system
    respiratory (breathing) system
3. Why do you think the earthworm has neither eyes nor ears?

Organisms

LESSON REVIEW *(Think. There may be more than one answer.)*

1. Annelids
   a. have bodies made of segments.
   b. are the most complex of all worms.
   c. have well-developed systems.
   d. are all free-living.

2. The earthworm
   a. is the only annelid without bristles on its body.
   b. lives in ponds and streams and is the main food for several kinds of fish.
   c. is incapable of moving from the tube that it lives in.
   d. takes food from soil that moves through its body.

3. Polychaetes
   a. have many bristles on their bodies.
   b. live mostly in fresh water.
   c. feed on small animals, which they sting with their bristles.
   d. all are attached and remain in one place throughout their lives.

4. Oligochaetes
   a. live only in terrestrial habitats.
   b. have few bristles on their bodies.
   c. include the earthworm, which is the best-known member of the group.
   d. all feed on nematodes.

5. Leeches
   a. are parasites.
   b. suck blood from their hosts.
   c. were once used by doctors.
   d. are poisonous and very painful when attached to humans.

6. Which of the following terms best matches each phrase?
   a. annelid   c. oligochaete worm
   b. polychaete worm   d. leech

   _____ a many-bristled annelid

   _____ an annelid with few bristles

   _____ a member of a major group of worms all of which are segmented

   _____ a parasitic annelid that feeds on the body fluids of a host

KEY WORDS

annelid (p. 409)   oligochaete worm (p. 411)
polychaete worm (p. 409)   leech (p. 411)

Applying What You Have Learned

1. What systems have the mollusks and worms developed that are not present in sponges and coelenterates?
2. What adaptation enables the annelid to move on land?
   Questions 3 through 6 are based on this diagram of the life cycle of a fluke:

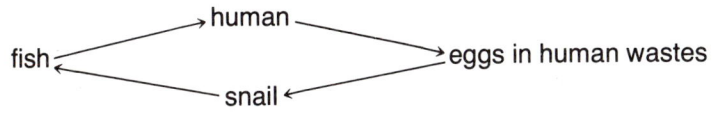

   fish → human → eggs in human wastes → snail → fish

3. How do you think people get infected with this fluke?
4. Explain three ways that people might prevent this life cycle from being completed.
5. Describe the kind of sanitary facilities that may be common in an area where this parasite infects a lot of people.
6. How do modern methods of waste disposal decrease the chances of a parasite's developing to maturity?

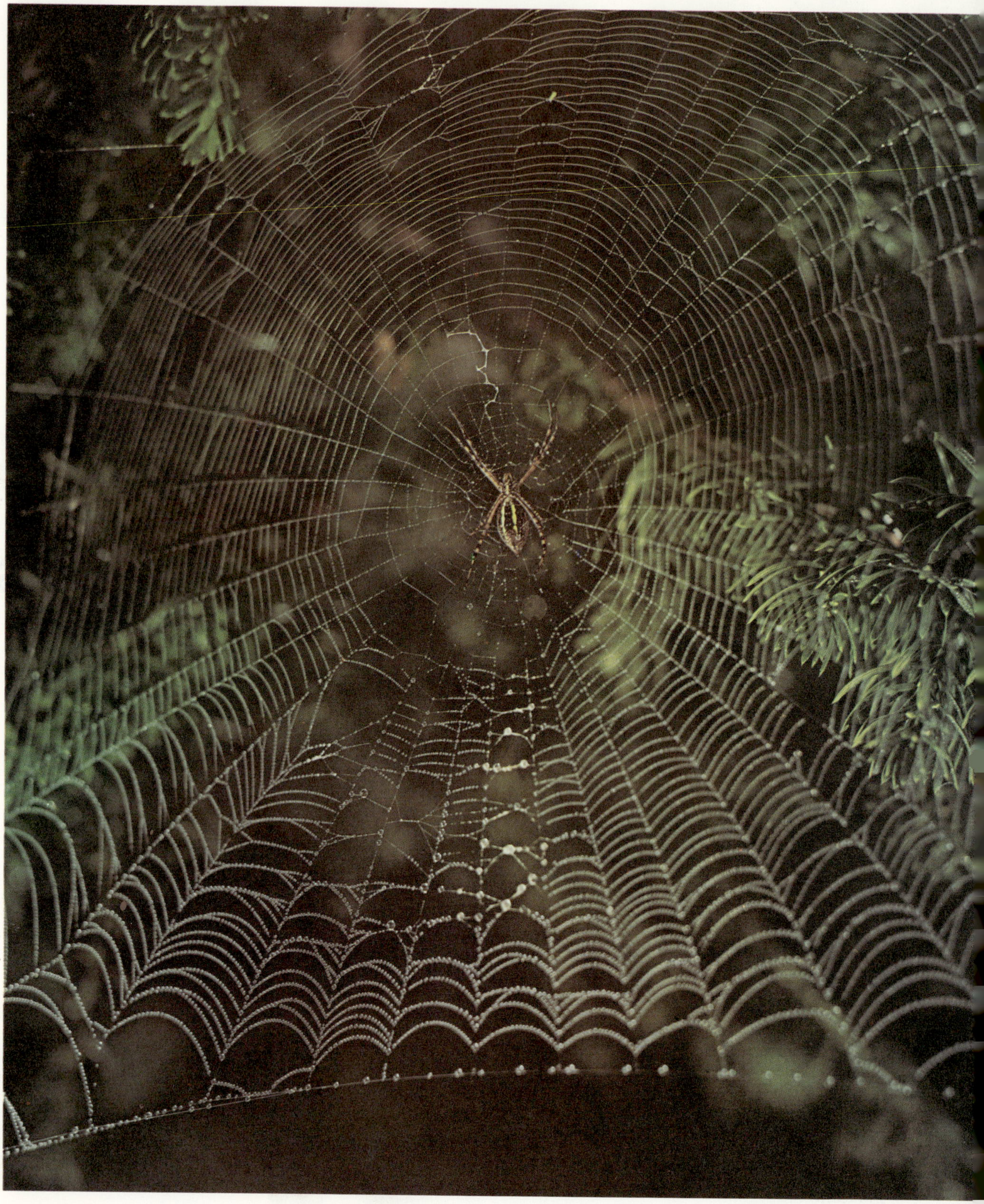

Figure 23.1: This golden garden spider is waiting for its prey. Spiders belong to the largest group of animals, the arthropods.

Chapter 23

ARTHROPODS AND ECHINODERMS

The arthropods make up the largest group of animals. There are more kinds of arthropods than of all the other animals put together. The arthropods include insects, spiders, centipedes, and lobsters.

The echinoderms (ĭ kī′nə dûrm′) are marine animals. They are found only in saltwater environments. Echinoderms include such animals as the starfish, sea cucumber, and sand dollar.

ARTHROPODS

What are arthropods? Arthropod means "jointed feet." All of the **arthropods** have legs and feet that are jointed.

Another thing all arthropods have in common is an **exoskeleton**. This is a skeleton that is on the outside of the body—like the suit of armor on a knight. The exoskeleton is made up of many separate plates. This allows the animal freedom of movement.

arthropod:
one of a major group of animals with jointed legs and feet, and an exoskeleton

exoskeleton:
a kind of skeleton on the outside of an animal's body

Figure 23.2: The name millipede means "one with a thousand feet." This is an exaggeration, but millipedes may have 150 or more legs.

Over a million different arthropods have been identified so far. Life scientists estimate that there are many thousands (perhaps millions) yet to be discovered.

Millipedes have many legs. The **millipedes** make up one of the smaller groups of arthropods. These arthropods are somewhat like worms in shape, but they have many legs. How many legs do they have? Sometimes they are called "thousand-legged worms." They do not have a thousand legs. Seldom do they have as many as 200. But even 150 legs on a 3-centimeter (about 1 inch) worm is a lot of legs!

Millipedes are vegetarians, meaning that they eat only plant matter. Some of the millipedes have a bad reputation because of the stink glands that they have along their sides. These glands give off a chemical that can kill or drive away insect predators.

millipede: a wormlike arthropod with two pairs of legs on each body segment

centipede: a wormlike arthropod with one pair of legs on each body segment

Centipedes are predators. The **centipedes** are long, wormlike arthropods. But you will not confuse them with millipedes if you remember this: Centipedes have only one pair of legs on each body segment. (See Figure 23.3.) Millipedes have two pairs of legs on each segment. Another difference is that

Figure 23.3: A centipede has one pair of legs for each body segment and looks less wormlike than a millipede. Centipedes are predators that eat other animals, but millipedes are vegetarians.

centipedes are predators. This means that they hunt and kill other animals for food. Most centipedes have poison claws near their heads. These claws are used for killing their prey.

The largest centipedes are found in the tropics. One kind that lives in Central America is over 25 centimeters long and 2.5 centimeters wide (10 inches by 1 inch). It feeds on lizards, mice, and large insects.

Centipedes are night creatures. During the day, they hide under logs and stones. Some can give a painful sting with their poison claws, but even the large tropical centipedes are not considered deadly to humans.

The insects make up the largest group of animals.

There are more than twice as many kinds of insects known as all the other animals put together. For example, there are about 250,000 different kinds of beetles! And the beetles make up only one of twenty-six major groups of insects.

How can you know whether or not an animal is an insect? There is one simple way—if the insect is an adult. Count the legs on the animal. If the total is six, you are looking at an insect.

Why are there so many insects? A life scientist might answer such a question by saying that insects are very successful. What is success? Success is finding a niche in an ecosystem that will allow a population to survive. Insects have had great success in finding available niches in ecosystems. Study any ecosystem carefully and you can usually find a number of different insects living close to each other. But each has its own niche and seldom competes with another.

Insects develop in stages.

Most insects go through different stages during their development. Many go through a four-stage development. Many others go through a three-stage development. The four-stage development begins with the egg. The egg hatches into a wormlike creature called a

Figure 23.4: Both the larva and the adult insect may compete with people for food. The adult insect is formed during the pupa stage.

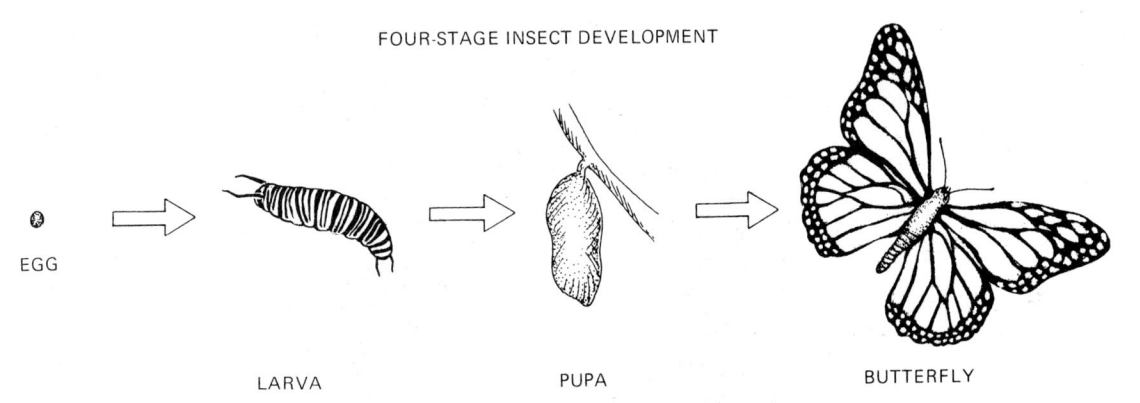

FOUR-STAGE INSECT DEVELOPMENT

EGG → LARVA → PUPA → BUTTERFLY

Organisms

THREE-STAGE INSECT DEVELOPMENT

EGG → NYMPH → GRASSHOPPER

Figure 23.5: Some insects go through more than one nymph stage before becoming adults. Both nymphs and adults have big appetites.

larva: the second of four developmental stages in certain insects. It is worm-like in nature.

pupa: the third of four developmental stages in certain insects. It is the stage in which a larva changes into an adult.

nymph: the second of three developmental stages in certain insects. It is the wingless insect that hatches from an egg.

social insect: an insect that lives in organized groups. Within the groups, specific jobs are done by specific members.

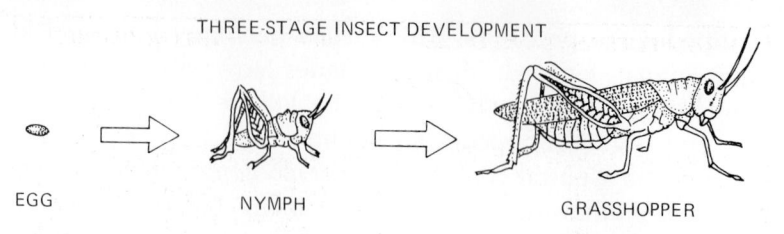

larva. The larva stage may last for days—or months. Eventually, the larva will make some kind of a protective case for itself. In this protective case, it is sealed off and protected from its environment. Then it enters the **pupa** (pyoo′pə) stage of development. Drastic changes occur during the pupa stage. The larva changes from a wormlike creature into an adult. The fourth stage of development is complete when the adult emerges from the pupa case.

The three-stage pattern of development also begins with the egg. But the creature that hatches from the egg looks less like a worm and more like a small adult insect. It is usually called a **nymph**. The nymph grows, and after some time it develops wings and a mature reproductive system. Then it is considered an adult.

The development of insects helps explain why they are such serious competitors of humans. During their development, they may occupy two niches. For example, the niche of a larva may be that of a leaf-eater on a tree. Later, that larva may spin a cocoon (a type of pupa case) and change into a butterfly. The butterfly, by feeding on flower nectar and by having the ability to fly, is able to live an entirely different way of life.

Some insects live together in groups. Some insects are called **social insects**. They are given this name because they live together in groups that seem a little like our own society. Within our society, individuals have different roles that help the society survive. We have nurses, soldiers, and garbage collectors. They all have important roles. The same thing is true of insect societies. Some of them even have specialized individuals that could be called "nurses," "soldiers," and "garbage collectors."

Bees, ants, and termites are the insects with the most highly organized societies. The honeybees are the best known. Their society is the beehive. Within the beehive, various individuals have their specific niches. The "workers" gather nectar and pollen and tend to the important chores within the hive. The "drones" are males that function only during reproduction. Their only role is to provide male sperm. The "queen" has the special task of laying all the eggs for the group.

Insects have many special talents. If you want to find the

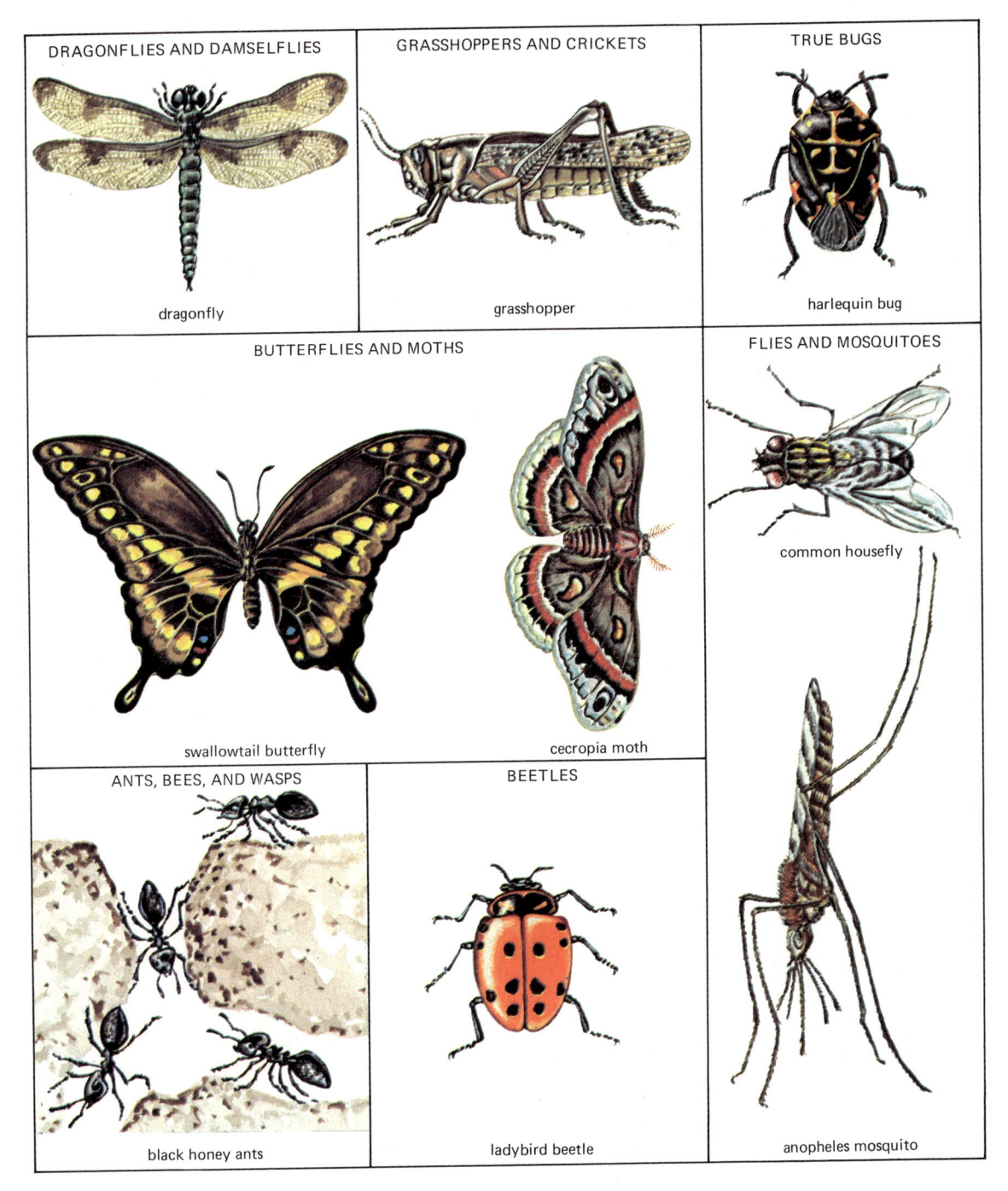

Figure 23.6: Extremely varied in shape and way of life, insects fill many niches in different ecosystems. They are our toughest competitors. Not all are pests. Honeybees benefit people, and some crops fail if insects do not cause pollination by collecting pollen.

fastest digger in the animal kingdom, you can almost bet that it will be an insect. You will also find champion fliers, biters, smellers, jumpers, hunters, fighters, and weight-lifters among

421

Figure 23.7: Ostracods (left) are tiny freshwater crustaceans that look like a mixture of shrimp and clam. The ostracod has a bivalve shell, but the jointed legs sticking out mark it as an arthropod. The crayfish (right) is a freshwater version of the sea lobster.

the insects. Insects have a great variety of outstanding adaptations that enable them to survive competition for their niches. Just try taking over the territory of a group of bees or ants and you will understand their ability to compete! Insects are the strongest competitors of humans. As you learned in Chapters 3 and 17, some difficult problems can be caused by the ways that we fight insect pests.

Crustaceans are used for food. The **crustaceans** (krŭ stā'shən) include a wide variety of animals. Most of them are aquatic. This group includes several members that are highly prized as human food. Among these are the lobster, crayfish, shrimp, and prawn.

crustacean: usually, an aquatic arthropod with an exoskeleton

Because of their aquatic way of life, crustaceans seldom compete with people. So, unlike insects, they are seldom considered troublesome. There is one main exception—the barnacle. The barnacle lives inside a part of its exoskeleton that has grown around it. In other words, it has its own protective case. It attaches itself to objects in the water and then spends the rest of its life in one place. Barnacles create problems by attaching themselves to ship bottoms. People spend many thousands of dollars every year for special repellent paints or for the scraping of ship bottoms to remove the barnacles.

The arachnids have four pairs of legs. The **arachnids** (ə răk'nĭd) have a bad reputation. Not many people care to make friends with spiders, ticks, mites, and scorpions. There are some good reasons for disliking them. Some have poisonous

arachnid: a terrestrial arthropod that has four pairs of legs

Arthropods and Echinoderms

Laboratory Activity

Can You Catch an Insect?

PURPOSE In this investigation, you will collect several different insects and study their structures.

MATERIALS plastic bags with ties magnifying glass
several insects

PROCEDURE
1. Capture several different insects. (Devise your own method.) Transfer each insect to a plastic bag, and close the bag with a tie strip. Observe your insect.
2. How many pairs of wings does each insect have? Most flying insects have two pairs of wings. How many wings does a housefly or mosquito have?
3. Which of the insects flies? Most adult insects can fly. Beetles fly poorly. If your insect cannot fly, it is probably an immature form.
4. How does each insect eat its food? Most insects chew their food or suck the food up through some kind of a tubelike structure.
5. Look for two kinds of eyes in each insect's head. Most insects have one pair of large eyes called compound eyes on the sides of the head. Usually, insects have other eyes, called simple eyes, located in the front of the head.
6. Look for three parts of each insect's body. All insects have three body regions: head, thorax, and abdomen. The thorax is behind the head and is made up of three segments. One pair of legs is attached to each segment of the thorax. The region behind the thorax is the abdomen.
7. How would you describe each insect's antennae? Antennae are sensitive to touch and odor. Some look like a string of beads; some are swollen or club-like on the end; some are featherlike; and some are like bristle hairs.
8. Look for special adaptations that help each insect survive in its environment. Describe the adaptations. Many insects have special adaptations such as wings, long legs, or stingers. Some give off bad odors. Others are bad tasting (do not taste yours)—such insects are often brightly colored. What adaptations did you find?
9. Place two or more insects together in one bag and slowly crowd them together. How does one insect react to the presence of other insects?
10. Try to identify the insects that you collected. Some insects are difficult to identify. One identification book will not include all the different kinds of insects.
11. Try to discover the niche of each insect you captured. All insects are consumers. Which of your insects is a first-, second-, or third-order consumer?

QUESTIONS
1. Consider the methods you used to capture different insects. How could each method be used to get rid of insect pests?
2. How can adaptative coloration help an insect?

Figure 23.8: The red "hourglass" on the underside of a large black abdomen is the key identification mark of the female black widow spider. The female black widow is one of the few poisonous spiders; most spiders are harmless to human beings. The harmful effects of the venom can be counteracted by proper medical care.

venom. Others bite, sting, or carry diseases. All have four pairs of legs. Studying these arthropods obviously calls for special care. But it is worth it to the life scientists who study arachnids. They are fascinating animals.

How would you like to drop from a tree, making your own rope as you go? That is easy for some spiders. Do you know of any other animal that can do that? Spiders and their relatives are very successful hunters. They use all sorts of devices like trapdoors and webs to catch their prey. Their mating and courtship behavior is interesting. They perform a great variety of dances and other mating activities.

LESSON REVIEW *(Think. There may be more than one answer.)*

1. Arthropods
    a. are animals with the greatest number of members.
    b. include insects, spiders, centipedes, and lobsters.
    c. all have an internal and an external skeleton.
    d. all have jointed feet.

2. Of the animals pictured,
    a. A is an insect and has more different relatives than any other kind of animal.
    b. B is a crustacean and lives in aquatic habitats.
    c. C is an arachnid and may be poisonous.
    d. D is a centipede and may have poison claws near its head.

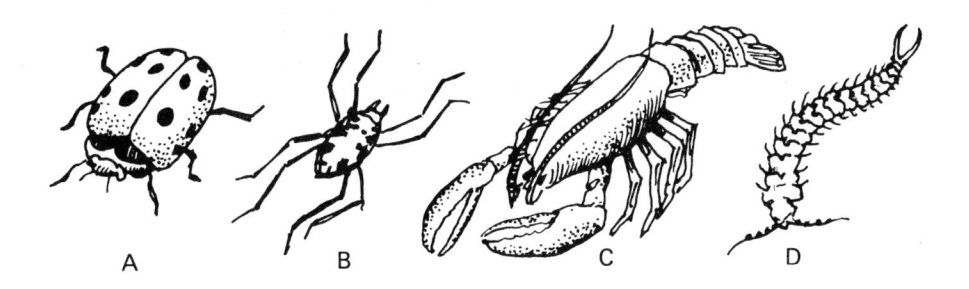

A B C D

3. A wormlike arthropod
    a. with one pair of legs per body segment would be a millipede.
    b. that is a predator could be a centipede.
    c. that has poison claws would be a millipede.
    d. that is as long as your finger span would be a centipede.

4. Insects
    a. are more than twice as large a group as all other animals put together.
    b. all have six legs as adults.
    c. occupy a great many different niches.
    d. all go through four stages of development.

5. An insect
    a. may be a wormlike larva.
    b. may spend part of its lifetime sealed off in a small case.
    c. may resemble its parents when it is born.
    d. may occupy two completely different niches during its lifetime.

6. Crustaceans
    a. are mostly terrestrial.
    b. are mostly serious competitors of people.
    c. have many members that are prized as food.
    d. are all poisonous if eaten.

7. Arachnids
    a. have the same habitat and niche as crustaceans.
    b. have some members that are poisonous.
    c. have a bad reputation.
    d. include spiders and ticks.

Organisms

8. Which of the following terms best matches each phrase?
   a. arthropod   f. pupa
   b. exoskeleton   g. nymph
   c. millipede   h. social insect
   d. centipede   i. crustacean
   e. larva   j. arachnid

   _____ all arthropods have one on the outside of their body

   _____ worm stage of insect development

   _____ a member of a major group of animals that all have jointed feet

   _____ sometimes called "thousand-legged worm"

   _____ wormlike arthropod with one pair of legs per body segment

   _____ a honeybee is an example

   _____ a spider is one

   _____ immature insect stage that often looks like the parent

   _____ a lobster is one

   _____ an insect stage, often enclosed in a case

KEY WORDS

arthropod (p. 417)
exoskeleton (p. 417)
millipede (p. 418)
centipede (p. 418)
larva (p. 420)

pupa (p. 420)
nymph (p. 420)
social insect (p. 420)
crustacean (p. 422)
arachnid (p. 422)

ECHINODERMS

echinoderm: any of a major group of marine animals with a limestonelike skeleton just beneath a bumpy or spiny skin. Echinoderms have tube feet.

The echinoderms are marine animals. The echinoderms are the "spiny-skinned" animals. If you want to see one of these animals in nature, you will have to visit an ocean. They are strictly marine animals.

Five is a key number to remember when thinking of echinoderms. Most of them have their bodies divided into five equal parts. In the starfish, which is the most common example, there are usually five arms. These arms extend from a central area like spokes on a wheel. Do echinoderms have radial symmetry?

All of the echinoderms have a limestonelike skeleton just beneath the skin. Most of them are covered with bumps or spines —hence their name.

Figure 23.9: This starfish is lying on the ocean bottom. You can see the spines of the limestonelike skeleton on top of the animal.

Two other outstanding echinoderm features are their system of water canals and their **tube feet.** The water canals aid in digestion and movement. The tube feet, pictured in Figure 23.10, have little suction cups on the end. They grasp objects and enable the echinoderm to crawl along. The tube feet also help it to obtain food.

The starfish is a common echinoderm. People did not invent the tug-of-war. This kind of test of strength has been going on for millions of years. Typical contestants are a starfish and a clam. The starfish uses its many tube feet and attaches its

tube feet:
feetlike organs on echinoderms that have suction cups at their ends. Tube feet help echinoderms to move and to get food.

Figure 23.10: This closeup of a starfish arm (left) shows the tube feet in action, walking the starfish over the sand. Notice the little suction cup at the end of each foot.

Figure 23.11: This starfish (right) has begun to attach its tube feet to a clam. Soon there will be a classic tug-of-war.

Organisms

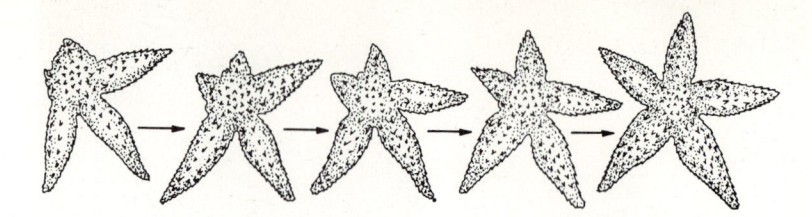

Figure 23.12: Loss of an arm is not a serious problem for the starfish. The missing arm will grow back. And if the "lost" arm has a bit of the starfish's center attached, it will usually regrow the missing body as well as the other four arms.

arms to the two "valves" of the clam. It pulls and pulls, trying to open the valves. The clam, using muscles on the inside, pulls and pulls to keep the valves closed. The strength of the two animals is usually about equal. But the starfish usually wins the contest because it can pull for a longer time.

All the starfish needs is a tiny opening between the two valves. Then it pushes its stomach outside of its body. (Actually, it turns inside out, the way a sock is turned "wrong-side out.") The stomach is pushed through the tiny opening between the valves. The stomach releases enzymes that kill the clam and begin the digestive process. Since the starfish has no teeth or jaws to chew with (it does not even have a head), it must wait until the clam is soft like jelly. Then it brings stomach and food back into its body.

The five arms of a starfish are almost completely independent. If one of them breaks off, the arm usually continues living

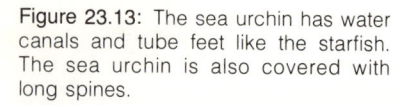

Figure 23.13: The sea urchin has water canals and tube feet like the starfish. The sea urchin is also covered with long spines.

Figure 23.14: The sand dollar has no protective spines. It buries itself in the sand if a predator such as a starfish comes near.

and grows a complete body, including four more arms. (See Figure 23.12.) Meanwhile, the rest of the original starfish grows back the missing arm. What is the ability to grow new body parts called?

Starfish vary greatly in size. Some kinds are less than 2 centimeters (less than 1 inch) wide when fully grown. There are others that are 1 meter (more than 1 yard) across at maturity. Although five is the usual number of arms, there are some starfish with as many as forty or fifty.

The sea urchin looks like a ball of spines. The starfish make up one major group of echinoderms. Another major group of echinoderms includes the sea urchin and its relatives. Most of them are easy to identify. A sea urchin usually looks like a ball covered with long spines. It uses the spines for defense (some are poisonous) and for turning itself right-side up if it gets rolled on its back.

Most sea urchins live in shallow water. They are a common sight on rocky beaches. Many bore their way into spaces in rocks and use food that is washed into them. They have a larger mouth than starfish. They also have teeth surrounding the mouth, so that they do not have to turn their stomachs inside out the way that starfish do.

The sand dollar is a relative of the sea urchin. It is much more flattened, and it does not have long spines. The sand dollar is a common sight on the beaches of the Pacific Ocean. When the tide is in, the sand dollar turns on its side and feeds on the matter that is brought to it. When the tide goes out, it lies flat in the sand and buries itself. If a starfish comes by while it is feeding, the sand dollar buries itself and stays in the sand until long after the starfish is gone. The starfish is a common predator of the sand dollar.

429

Figure 23.15: They may look like plants, but feather stars are animals. These graceful echinoderms are common around coral reefs.

There are three other groups of echinoderms.
There are three other groups of echinoderms, but they are not as commonly seen as starfish and sea urchins. One group contains the brittle stars. These are mostly small delicate starfish that live at great depths in the oceans. They feed on small animals. They do not have the strength for the tug-of-war that their larger relatives have.

The sea cucumbers do somewhat resemble cucumbers lying on their sides. They lie in sand or mud and push decaying material into their mouths. When threatened by a predator, a sea cucumber can shoot part of its insides out the rear opening of its body. This often causes the predator to leave it alone. Sea cucumbers have a great ability to grow back missing body parts.

The sea lilies and feather stars are delicate echinoderms that look much like plants. The sea lilies attach themselves to the bottom of the ocean and capture food with their waving arms. The feather stars live in shallower water but also attach themselves to the bottom. Not much is known about the life of either of these two types of animals.

LESSON REVIEW *(Think. There may be more than one answer.)*

1. Echinoderms
    a. are all marine animals.
    b. have radial symmetry.
    c. are mostly covered with bumps or spines.
    d. have an exoskeleton like arthropods.

2. Of the animals pictured at the top of page 431,
    a. A is a sea urchin and may have poisonous spines.
    b. B is a sand dollar and can bury itself in the sand along a beach.
    c. C is a starfish and may prey upon sand dollars.
    d. A, B, and C are the only three kinds of echinoderms.

A   B   C

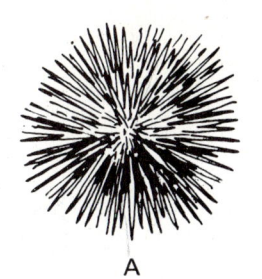

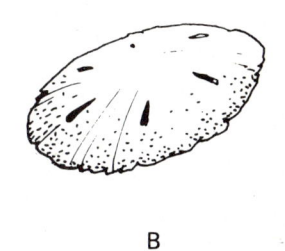

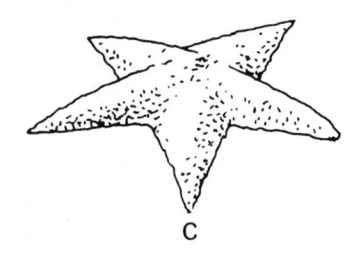

3. Starfish
    a. have no tube feet.
    b. eat clams by taking the whole clam into their bodies.
    c. have the ability to regenerate missing parts.
    d. may be wider than the length of your arm.

4. The sea urchin
    a. is covered with long spines.
    b. lives in deep water.
    c. feeds differently than the starfish.
    d. has a flat relative called a sand dollar.

5. An echinoderm
    a. that looks like a plant is a brittle star.
    b. that resembles a garden vegetable is a picklefish.
    c. that is named after a flower is a sea lily.
    d. that shoots out part of its insides to scare predators is a feather star.

6. Tube feet
    a. help an echinoderm to move.
    b. help an echinoderm to get food.
    c. have suction cups on their end.
    d. are an unnecessary adaptation.

KEY WORDS

echinoderm (p. 426)
tube feet (p. 427)

Applying What You Have Learned

1. Why are insects people's strongest competitors?
2. What term describes the echinoderm's ability to regrow missing body parts?
3. Describe the differences between three-stage and four-stage development of insects.
4. List some of the habitats that insects have adapted to.
5. Echinoderms have radial symmetry. What does that mean?
6. How are the spider's food habits beneficial to people?
7. How does the starfish break down and digest its food?
8. How are centipedes different from millipedes?

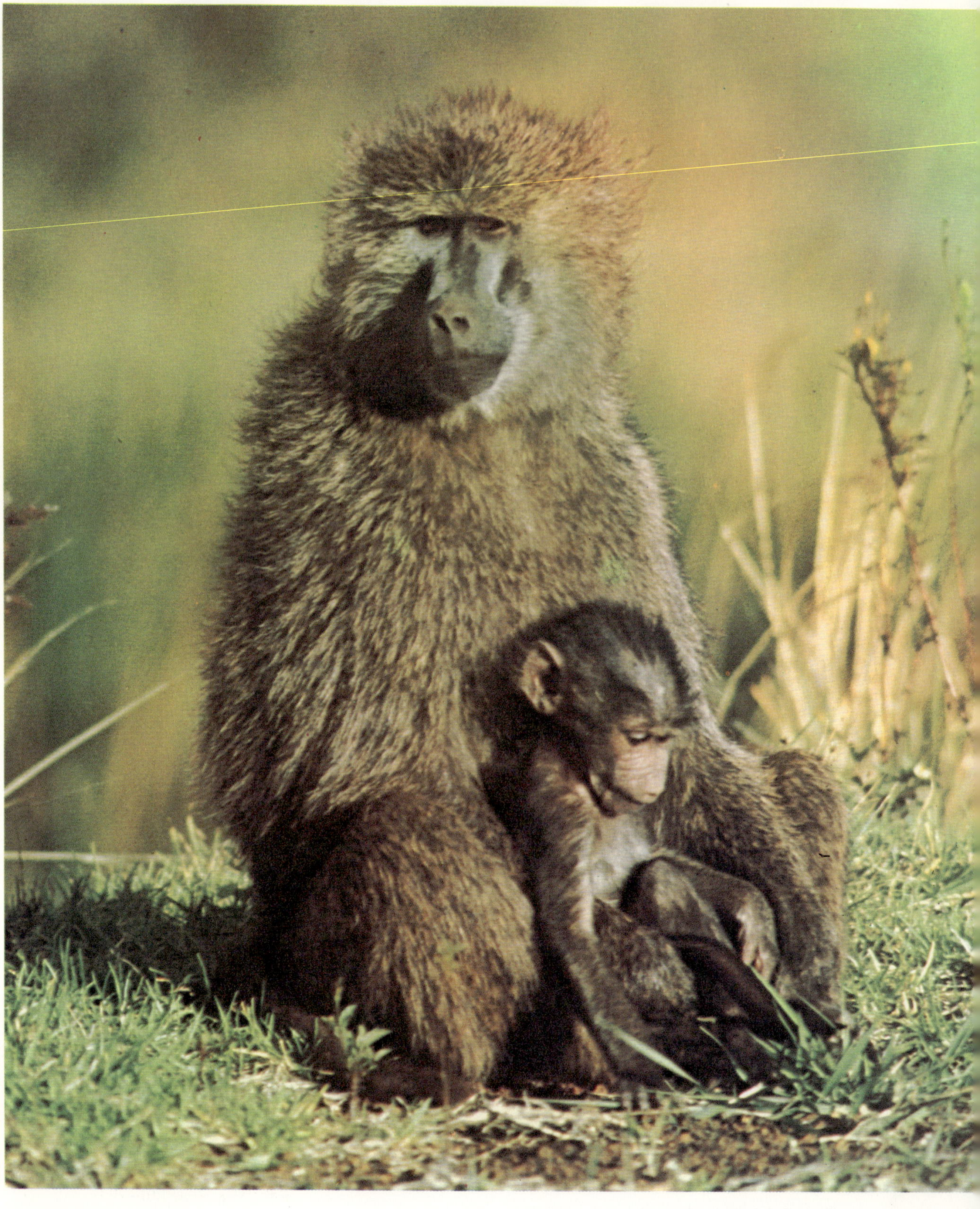

Figure 24.1: This baboon mother and child are mammals. Mammals provide greater parental care than any other vertebrate group.

Chapter 24
VERTEBRATES

Vertebrates are animals with a spinal column. The column is made up of individual units called vertebrae (vêr′tə brē). The vertebrae are usually made of bone. However, in fish such as the shark, they are made of cartilage. (Remember, cartilage is the semisolid material in the tip of your nose.)

Vertebrates are the most highly organized animals. There are five major groups of vertebrates: fish, amphibians, reptiles, birds, and mammals.

vertebrate:
an animal with a spinal column that is usually made of bony units called vertebrae. In some, the vertebrae are made of cartilage

FISH

Fish are the oldest vertebrates. When organisms die, their skeletons or some other parts of their bodies may be preserved. Such remains are called **fossils**. People have been studying vertebrate fossils for many years. Fossils can be dated by various techniques. This means that life scientists can find out

fossil:
an impression, found in rock or other material, of the skeleton or other parts of a dead organism

Organisms

Figure 24.2: Fish come in all shapes and sizes, but most fish have a long flattened body ending in a rayed tail. As you can see from this illustration, there are many exceptions to this rule. Fish have a great many adaptations suited to particular niches.

when the animal was alive. The bones of dinosaurs are probably the best-known vertebrate fossils. But dinosaurs are not the oldest vertebrates. Fish were on the Earth millions of years before the dinosaurs. Fish are the oldest of all the vertebrates. They are also the most plentiful. Almost half of all the vertebrates are fish.

In the last chapter, you learned about the great variety of insects, and how they have found and filled so many niches. But the insects are mostly terrestrial animals. About three-fourths of the Earth is covered by water. This is the home of the fish. And they have filled every possible niche in it. In doing this, fish have developed a variety of adaptations that is almost unbelievable. Just suppose that you were to take paper and pencil and draw the wildest and weirdest looking fish that you can imagine. A life scientist who is an expert on fish could probably find you a picture of a real fish to match your creation.

Hagfish and lampreys are the most primitive fish. In every animal group, there are members that people like, and members that people do not like. If a popularity contest were to be used in rating fish, the *hagfish* and *lampreys* would probably come out near the bottom. They are the oldest of all living fish. But their age is not the reason for their unpopularity. They are long and slender, almost snakelike in appearance. Their bodies are soft and slimy. They have no jaws. They have a round mouth, and most of them have a sucker surrounding the mouth. They feed by attaching themselves to the bodies of other fish. The hagfish rasps its way into the muscles of its victim, burying its head as it eats its way inside. The lamprey bites a hole and sucks blood and juices from its victim. Many of its victims are fish that

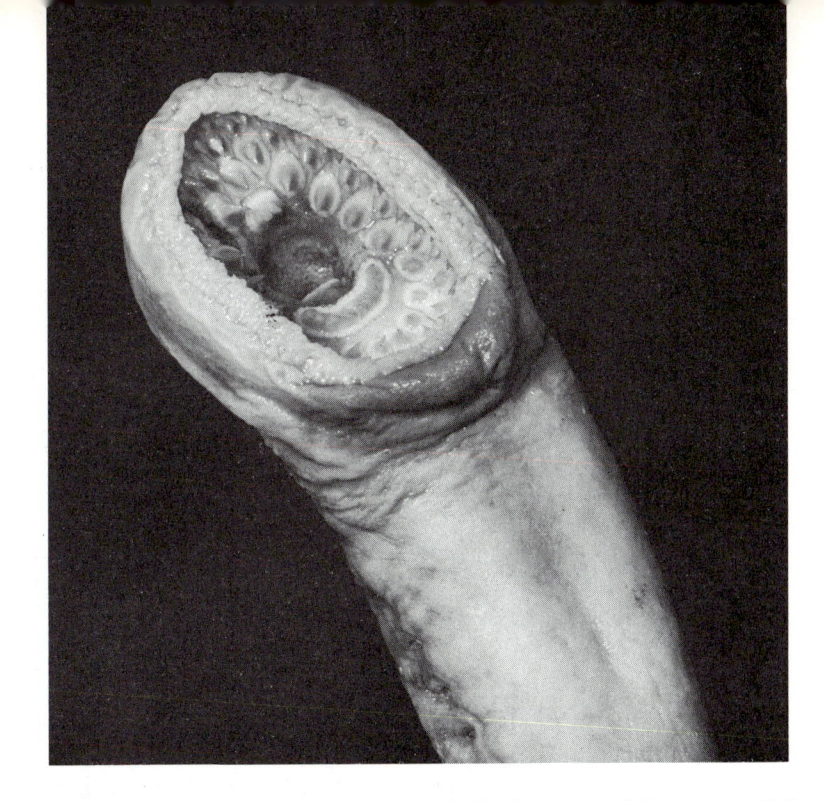

Figure 24.3: This is the round suction type of mouth that a lamprey eel uses to attach itself to a fish. After the teeth have opened a hole, the eel sucks the fish's body fluids.

we prize for food. Lampreys have killed many of the game fish in the Great Lakes. Are these facts enough to explain the unpopularity of the hagfishes and the lampreys?

Sharks and rays would win another contest. If the hagfish and lampreys are the most unpopular, the *sharks* and *rays* would certainly be voted the "most feared." The thought of a shark is enough to panic many people who swim in ocean waters.

Figure 24.4: Like all sharks, this gray shark has a skeleton of cartilage. Sharks must swim constantly in order to stay at a given depth. A stationary shark would sink to the bottom.

Figure 24.5: Rays also have cartilage skeletons. The manta ray feeds on small animals as it "flies" along over the ocean floor.

Just a look at a shark is enough to cause fear. Sharks have powerful torpedolike bodies. Their skin is covered with rasplike scales. They have powerful jaws. Their mouths have hundreds of sharp daggerlike teeth. The behavior of predator sharks matches their fierce looks. They are vicious killers and they will eat just about anything, including people. The *tiger shark* probably has the champion reputation for eating "anything." In the stomach of one tiger shark, there were found fish bones, grass, feathers, bones of marine birds, pieces of turtle shell, some old cans, pieces of a dog's backbone, and a cow's skull. Others have been found with such things as a large oil can, a horse's head, and pieces of a bicycle in their stomachs.

Rays, and their smaller relatives called *skates*, are batlike fish. Their side fins extend like wings on either side of the body. Life scientists often describe their movement as "flying" through water.

Rays are also frightening creatures, but few of them deserve their reputation. They look frightening, especially when they are as large as the manta ray. (See Figure 24.5.) Actually, most rays are peaceful creatures that swim along the bottom of the sea feeding on small animals. One small group, the sting rays, have a sharp spine and poison glands near the tail. These rays are dangerous. Their stings are very painful, and the effects of the sting can cause death by heart failure. Electric rays also give the group a bad reputation. These are rays that carry a charge of electricity on their bodies. Life scientists are not sure how these rays use their electricity.

The bony fish are the most plentiful. Hagfish, lampreys, sharks, and rays have skeletons of cartilage. All other fish have normal "hard" bones, and this group as a whole is called the *bony fish*. Most fish belong to this group.

All fish need oxygen. They get it from the oxygen dissolved in water. You have lungs to take oxygen from the air. Most fish have **gills,** which are organs that can take oxygen from the water. A fish's gills are located on each side of its head. The water goes in the mouth, over the gills, and out the sides of the fish's head. In passing over the gills, the water is separated from tiny capillaries by only a thin wall. Oxygen passes through that wall and the capillary wall, and thus reaches the blood.

It was mentioned earlier that fish have filled every possible niche in water. You may think: How can there be so many different niches in a uniform substance like water? But water is by no means uniform. It differs in the chemicals that are in it. There are differences in water temperature. There are great pressure differences in bodies of water. The quantity of light decreases with the depth of the water. There are differences in terrain—for example, there are underwater mountains, valleys, and reefs. The effects of these physical factors on life in aquatic ecosystems were discussed in Chapters 4 and 5.

gill: a breathing organ by which fish and certain other aquatic animals take oxygen from water

Figure 24.6: This is a largemouth bass, a popular freshwater game fish found in many parts of the United States. Freshwater and marine bass are among the many bony fish found in the waters of the Earth.

Organisms

LESSON REVIEW *(Think. There may be more than one answer.)*

1. Of the animals pictured,
    a. A is a bony fish and belongs to the largest group of fish.
    b. B is a shark and has a skeleton of cartilage.
    c. C is a ray and most of its group are harmless.
    d. D is a lamprey and has a bony skeleton.

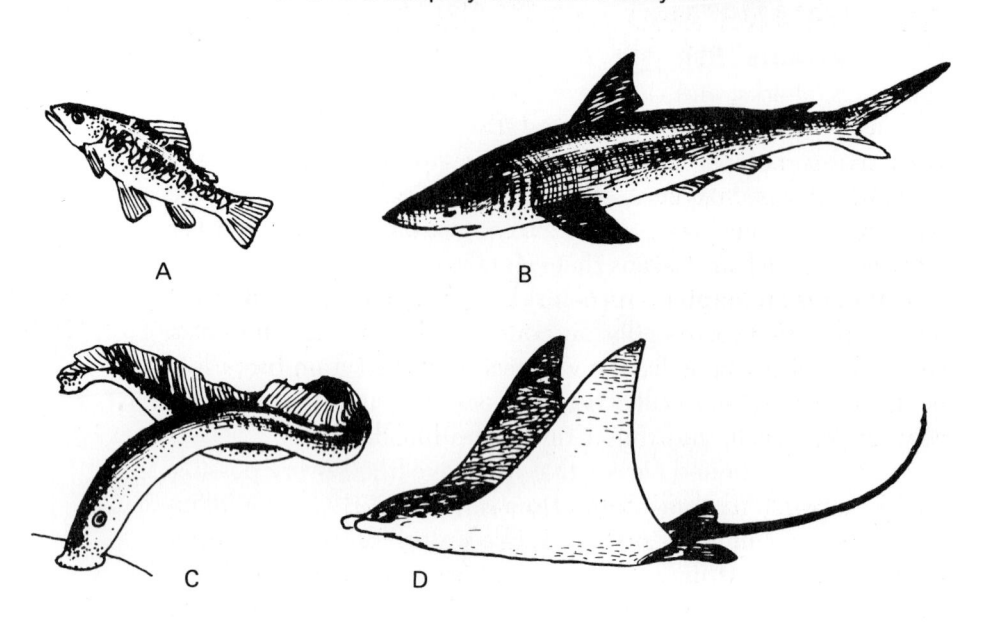

A   B

C   D

2. Fish
    a. are the oldest vertebrates.
    b. make up about one-half of all vertebrates in number.
    c. occupy one niche in both the freshwater and marine habitats.
    d. all have the same general body structure.

3. If a fish
    a. had a cartilage skeleton, it could be a shark.
    b. sucks blood and juices from its victim, it could be a ray.
    c. has a bony skeleton, it is in the minority of all fish.
    d. is very dangerous, it could be a skate.

4. Fish breathe
    a. with the same organs that you do.
    b. through lungs that are attached to gills by tubes.
    c. oxygen from the water.
    d. only when they come to the surface.

5. Which of the following would be least likely to leave a fossil?
    a. a dinosaur
    b. a bony fish
    c. a shark
    d. a human

KEY WORDS

vertebrate (p. 433)   gill (p. 437)
fossil (p. 433)

AMPHIBIANS AND REPTILES

Most amphibians are extinct. Fish are the oldest vertebrates. The amphibians are the next oldest. But the study of fossils shows that most amphibians are now extinct. There were once eleven major groups of amphibians. Only three are still represented on Earth.

What is an **amphibian**? Generally, an amphibian is defined as an animal that can live both in water and on land. But not all of the living amphibians fit this definition. Some amphibians live in the water. Some live on the land and return to the water to reproduce. And some live on the land all of the time.

amphibian: a vertebrate that lives part of its life in water and part on land

The wormlike amphibians are terrestrial. Of the three groups of amphibians, the smallest and rarest is the group of *wormlike amphibians*. Unlike all other amphibians, both extinct and living, the wormlike amphibians do not have legs. As you can see in Figure 24.7, they very much resemble the earthworm. The obvious differences are that the amphibian's body is not segmented, and that it has a mouth and eyes.

The wormlike amphibians are found mostly in tropical countries. Even there, they are a rare sight, for they spend most of their time in the moist soil. They probably feed on worms and termites. A few of them reproduce in water, but they are otherwise considered terrestrial animals.

Newts and salamanders are amphibians with tails. The *newts* and *salamanders* have two pairs of legs. They also have a long tail, which is useful for swimming in the aquatic environment in which most of them live. Some salamanders live in terrestrial habitats, but they return to the water to reproduce.

How does an amphibian like a salamander manage to breathe both in the water and on the land? There are three ways

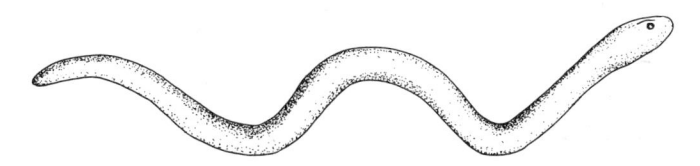

Figure 24.7: This drawing shows one of the wormlike amphibians. Not much is known of their life habits because of their rarity.

Figure 24.8: This is a red-spotted newt. It has a complex life pattern that alternates between aquatic and terrestrial habitats.

that amphibians may breathe. When they are in the young *tadpole* stage and living in water, they have gills like a fish. The gills absorb oxygen right out of the water. Some adults, like the newt in Figure 24.8, also have gills. They look different from a fish's gills, but they function the same way.

The second way amphibians can breathe is through the skin. Usually their skin is moist, and there are many capillaries just under the surface. Oxygen dissolves in the moisture and then passes through the skin to the capillaries.

The third way some amphibians breathe is with lungs. With lungs, the amphibian is able to lead an active life on the land.

Frogs and toads are amphibians without tails. *Frogs* and *toads* lay their eggs in the water. The eggs hatch into tadpoles. The tadpole may live in the water for a few weeks, or a few years. But eventually it loses its tail, develops legs, and usually becomes a land dweller. On land, the frogs and toads occupy a variety of habitats. Many frogs stay close to the body of water where they hatched. The tree frogs and toads are often wanderers and may be found many kilometers away from any body of water.

Figure 24.9: The bullfrog is one of the largest amphibians in North America. It eats almost any animal that will fit into its large mouth, including other frogs. The bullfrog may live up to 15 years and spends practically all its life in water, even though it is an air-breather as an adult.

Figure 24.10: The tuatara is the lone survivor of a group of reptiles that lived with the dinosaurs. These animals have very slow body processes and may live a century or more.

Most reptiles are extinct. The **reptiles** are like the amphibians in one way: Most of them are extinct. There is also a major difference in the history of the two groups. The amphibians were never a major group of animals. The reptiles were. The reptiles were the dominant animals on the Earth for hundreds of millions of years. Now only four of the sixteen major groups of reptiles are represented. And of the four groups that are represented, one of them is about to disappear. Only a small population remains of the *tuatara*, which is the only surviving member of one major group of reptiles. The tuatara (Figure 24.10) lives on islands south of New Zealand, and is being carefully protected.

The reptiles are adapted for terrestrial life. The reptiles were the first major group of animals to "invade" the land and live on it successfully. What did this require? The main thing that was needed was an effective way of obtaining oxygen from the air and getting it to the body cells. The reptiles have lungs for obtaining the oxygen. They also have a good transport system that carries it to the body cells.

The reptile's method of reproduction is also adapted to terrestrial life. Reptiles reproduce by large yolk-filled eggs. Most reptiles deposit these eggs outside the body. Such eggs

reptile:
a member of one of the five major groups of vertebrates. Reptiles have lungs for breathing on land and are covered with scales or horny plates.

441

Organisms

Laboratory Activity

How Far Will a Cold Frog Jump?

PURPOSE Many species of reptiles and amphibians which roamed the Earth have become extinct. In this activity, you will investigate the effect of temperature on a frog. This can help you to understand one possible reason for the disappearance of many species of reptiles and amphibians.

MATERIALS
chalk
frog
container
sponge
watch with second hand
meter stick
mixing bowl
water
ice
wide-mouthed jug

PROCEDURE
1. Draw a 10-cm chalk circle on the floor. Remove the frog from its container. Place it in the circle.
2. Gently touch the frog on its back with a damp sponge to make it move. Your partner should begin to record the time the moment you place the frog in the circle. When 10 seconds have passed, mark the path of the frog on the floor. After 10 seconds, return the frog to its container.
3. Measure the distance the frog jumped. Measure the distance the frog jumps in two more trials. (Repeat Step 2.) Average the three distances.
4. Fill a bowl half full with ice and water. Place the frog in the ice water. Do not hold the frog. After 5 seconds, remove the frog from the ice water.
5. Immediately place the frog in the chalk circle. Gently touch the frog on its back with a damp sponge. Measure the distance the frog jumps in 10 seconds.
6. Repeat Steps 4 and 5 for two more trials. Average the three distances.
7. What is the average distance the frog jumped after being exposed to cold?

QUESTIONS
1. What effect does cold temperature have on a frog? What do you think happens to a frog in the winter?
2. How is the frog's reaction to cold a possible explanation for the disappearance of many species of reptiles and amphibians?
3. How might what you learned in this investigation apply to the survival of fish?

Figure 24.11: A female snapping turtle comes to the land to lay her eggs, then returns to her aquatic habitat. The young turtles will hatch and find their way to the water.

have a thick shell, which protects them. In this way, they are different from the eggs of fish and amphibians, which do not have this protection. The reptile egg also has plenty of food in the yolk for the developing embryo while it is inside the shell. Other reptiles retain their eggs inside the female's body. The embryo develops in the egg, and when its development is complete it is born alive.

Tortoises and turtles have shells. The *tortoises* (tôr′təs) and *turtles* make up one major group of reptiles. These animals have the upper and lower parts of their body covered with a hard shell.

What is a tortoise and what is a turtle? Unfortunately, both are common names and there are no strict rules for how they are used. In the United States, the reptiles that have a shell and live in or near water are usually called turtles. If the shelled reptile is strictly terrestrial, then it may be called a tortoise. Americans seem to be much more fond of the name "turtle" than "tortoise." (The name "tortoise" is more common in England.)

Turtles and tortoises have many interesting characteristics. One outstanding one is their long life span. Some large tortoises are known to live for 150 years!

Crocodiles and alligators are aquatic animals. It is interesting that many of the reptiles that have survived over the years are aquatic. Turtles, *crocodiles*, and *alligators* have been able to compete better in the aquatic environment. How-

Figure 24.12: An alligator basks in the sun. These animals are in danger of extinction, partly because hunters are illegally killing them and selling their hides.

ever, they are still linked to the land. They still return to the land to lay their eggs.

The crocodiles and alligators are effective predators. A look at their jaws and teeth is enough to convince anyone of that fact. They feed on a variety of animals. Even people have been known to be a part of their diet. Crocodiles live in Africa, Asia, and South America. Alligators live in the southern United States.

Lizards and snakes are the most successful reptiles. The largest group of reptiles is the group that includes the *lizards* and *snakes*. The members of this group are widely distributed everywhere but the polar areas.

The name "lizard" also includes animals called *geckos*, *skinks*, *iguanas* (ī **gwä′**nə), and *chameleons* (kə **mē′**lyən). Lizards feed on small animals—mostly insects—and on plant matter. Some of the lizards that look the most frightening are harmless vegetarians. (See Figure 24.13.)

Snakes are often thought of as legless lizards. But they differ from lizards in one important way. Snakes have a different type of jaw. This jaw allows a snake to open its mouth very wide.

Figure 24.13: This prehistoric-looking reptile is an iguana. It lives in Mexico and Central and South America. The iguana lives near water and can be 2 meters long. The iguana is a vegetarian.

This adaptation has a very great effect upon the snake's method of feeding. Snakes are not limited to eating small animals as are the lizards. Snakes can and do eat animals that are larger than themselves. The snake stretches its mouth over its prey. Then, after the prey is swallowed, the body of the snake stretches as the large mass of food moves through its digestive tract. (See Figure 24.15.)

Snakes kill their prey in three ways. Some snakes are called "constrictors." These wrap themselves around their prey and

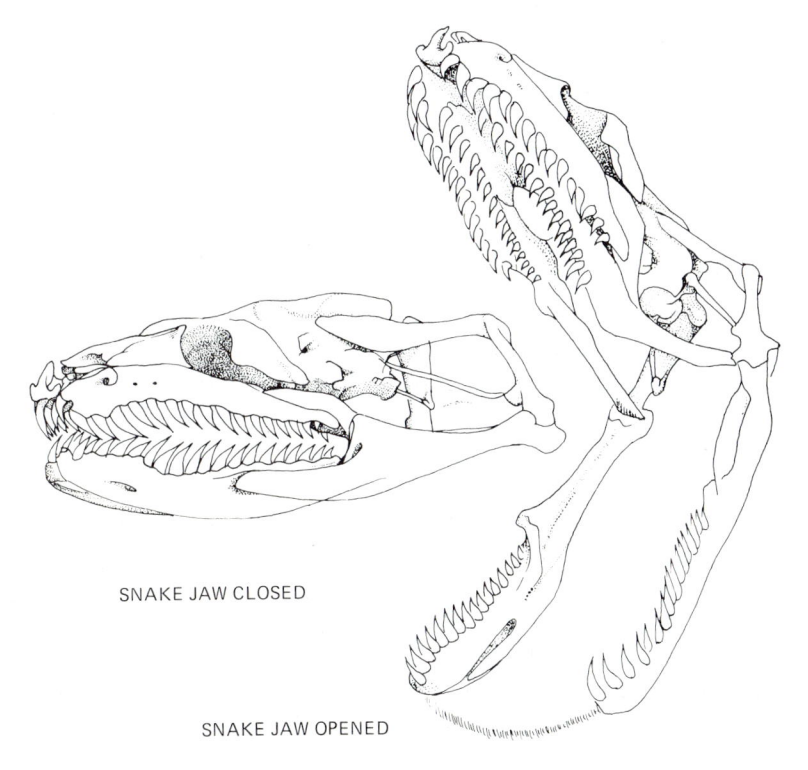

SNAKE JAW CLOSED

SNAKE JAW OPENED

Figure 24.14: The jaw of a snake differs from that of other vertebrates. The snake can "unhinge" its jaw to swallow animals larger around than itself. Backward-pointing teeth help the snake in swallowing its prey.

Figure 24.15: A 10-cm minnow is captured by a 45-cm northern water snake. The snake gradually swallows the fish, the process taking about 4 minutes. The water snake has no venom.

Vertebrates

Figure 24.16: This black-tailed rattlesnake in a desert habitat is ready to strike. The rattlesnake kills its prey with powerful venom ejected through two needlelike fangs.

squeeze, or constrict them to death. The *boa constrictor* is an example. Some snakes are venomous. These have a venom, or poison, that they inject into their prey. Some of the venoms only paralyze the prey. Others kill the prey. The rattlesnake is an example of a venomous snake. People have always had a fear of venomous snakes.

The snake's third method of killing its prey is the simplest. It is used by snakes like the *garter snake*. It simply catches its prey with its mouth. Then it swallows the prey alive.

LESSON REVIEW *(Think. There may be more than one answer.)*

1. Amphibians
    a. are mostly extinct.
    b. may live on land.
    c. may live in water.
    d. may live on land or in the water.

447

Organisms

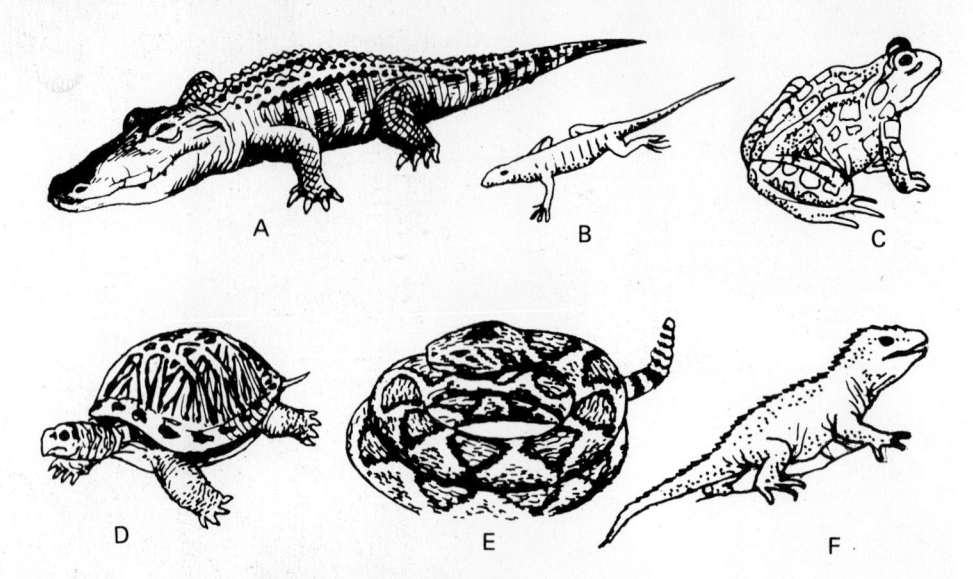

A   B   C

D   E   F

2. Of the animals shown,
   a. B and E are amphibians.
   b. A, D, and F are reptiles.
   c. A and D only live in aquatic habitats.
   d. F is a tuatara and is the last survivor of a major group of reptiles.

3. If an amphibian
   a. is shaped like a worm, it is the rarest type of amphibian.
   b. has gills, it could be a salamander.
   c. lived in the water, it later may live on land.
   d. can live on land, it can never again live in water.

4. The reptiles
   a. are mostly extinct.
   b. include frogs, lizards, and turtles.
   c. are adapted mainly for aquatic life.
   d. are all poisonous to humans.

5. Some reptiles may
   a. reproduce from eggs laid in the environment.
   b. reproduce from eggs held within the female's body.
   c. reproduce live reptiles that have not been in eggs.
   d. live twice as long as the human's average lifetime.

6. Snakes
   a. can eat animals that are larger than themselves.
   b. may kill their prey by catching it with their mouth and swallowing it alive.
   c. may kill their prey by injecting poisonous venom.
   d. may kill their prey by squeezing it to death.

KEY WORDS

amphibian (p. 439)
reptile (p. 441)

Figure 24.17: The ostrich (left) weighs up to 135 kg and stands over 2 meters tall. Although it cannot fly, it can run very fast. Ostrich feathers were once used for decoration. This female cardinal (right) is standing at her nest while her young open their mouths for food. Like all birds, the cardinal is warm-blooded. It can maintain a constant body temperature.

BIRDS AND MAMMALS

Birds have feathers, wings, and beaks. There are three easy ways to recognize birds. They all have feathers. They all have wings, or parts of wings. And they all have a beak. (See Figure 24.17.)

Birds are not very different from reptiles. The feathers of a bird are merely reptilelike scales that have become very specialized. The wings are specialized front legs. The beak is probably the most different kind of structure. Reptiles have jaws with teeth. Fossil birds also had jaws with teeth. The modern bird's beak, which is light in weight and has no teeth, is much better suited to a flying animal.

Many adaptations enable birds to fly. The birds are not the only animals that fly. Insects and bats also have that ability. But the birds have the body that is the most highly specialized for flying. For example, the bird does not need heavy teeth in its beak. Its food is mechanically digested in the *gizzard*, which is located under the wings. Another adaptation is the bird's bone structure. Many of the bones are thin and

449

Organisms

Figure 24.18: Bird beaks are specifically adapted for different feeding niches. These bird beaks are adapted for straining food from water, cracking seeds, tearing meat, stabbing insects, scooping fish from water, and chopping holes in wood. Match the beaks shown to the adaptations described.

hollow. This makes the bird lighter. A bird also has extra air spaces connected to the lungs. These are used in breathing. Then the breast bone, or keel, has deep grooves on both sides. These grooves provide the large, powerful flight muscles with much more area for attachment to the bone.

Birds have many other adaptations. In Figure 24.18, you can see a wide variety of beak shapes. Each of these different shapes is an adaptation that allows different birds to

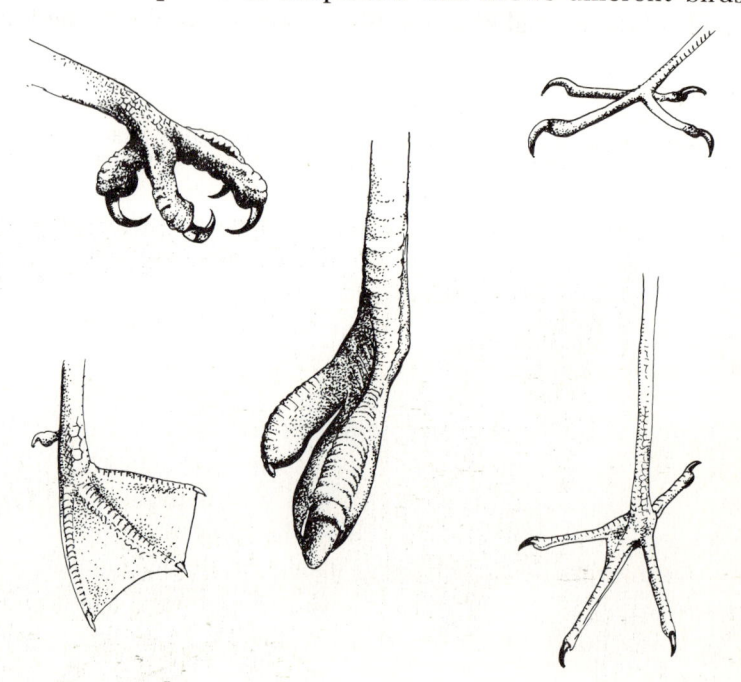

Figure 24.19: These bird feet are adapted for paddling in the water, grasping tree trunks and limbs, running, scratching, and grasping prey. Match the bird feet shown with the adaptations described.

eat special kinds of food. Figure 24.19 shows a variety of leg and foot adaptations. By looking at the beak and the feet of a bird, it is usually easy to describe its general niche and the type of ecosystem it is from.

Birds and mammals are warm-blooded animals.
Birds also differ from reptiles in one other way. Birds are warm-blooded, while reptiles are cold-blooded. What does that mean?

A **cold-blooded animal** is one that cannot regulate its own body temperature. Fish, amphibians, and reptiles are all cold-blooded. Their body temperature is very close to that of their surrounding environment. This is one reason why the terrestrial amphibians and reptiles never lived in cold polar areas. Their bodies would be frozen most of the time!

A **warm-blooded animal** is able to maintain a body temperature that is nearly the same all the time. The birds can do this. So can the mammals, which we will discuss next. This ability opened up new habitats for the birds and mammals. It also enabled them to be active during cold winter months in temperate zones. In much of the United States, for example, terrestrial amphibians and reptiles must hibernate through the winter.

Mammals are the newest animals. Mammals have two main characteristics that make them different from other animals. They all have hair, and all female mammals have milk-producing glands for suckling their young. The fossil record shows that the mammals are also the newest of the animals. In a parade of the animal kingdom, the mammals would be the last animals to appear.

cold-blooded animal: one that cannot regulate its body temperature which is therefore always close to that of its environment

warm-blooded animal: one that can regulate its body temperature to a nearly constant value, even though its environmental temperature may vary greatly

mammal: any member of one of the five major groups of vertebrates. A mammal has hair and a cerebrum that is large relative to the rest of its brain. The female mammal nurses its young.

Figure 24.20: A young bison, or buffalo, is taking milk from its mother. Like all mammals, this young bison is completely dependent upon its mother for food.

Organisms

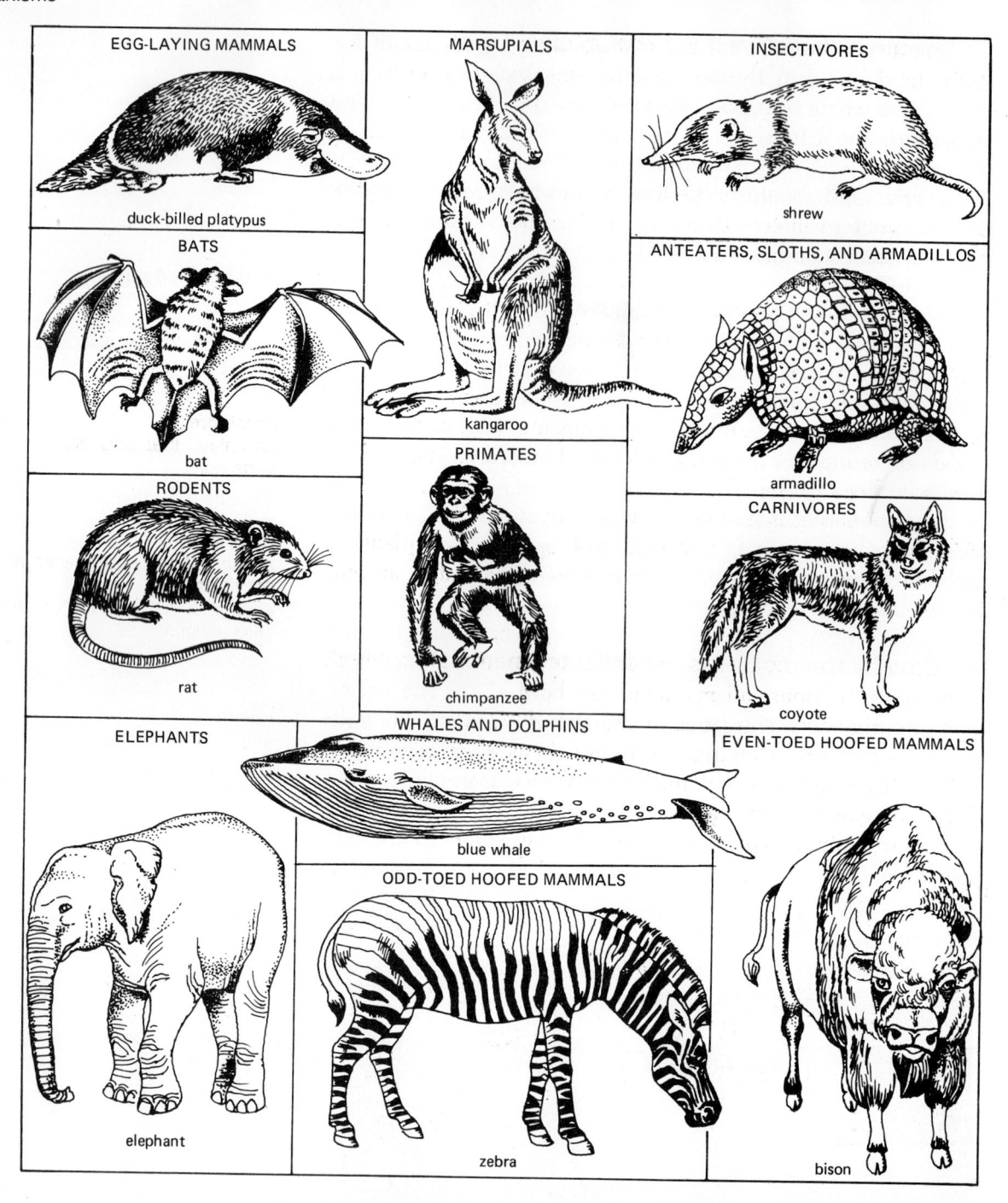

Figure 24.21: Almost all mammals belong to these twelve groups. Mammals are highly varied in size and habitat—from tiny shrews to elephants, from polar bears to tropical sloths. They are native to every continent but Antarctica and are found on land and in the sea. Sea mammals are often larger than land mammals. The blue whale, nearing extinction from whaling, is the largest animal that ever lived—larger than any dinosaur.

Vertebrates

Mammals have a variety of habitats. Mammals are found in the sea, in the air, and on the land. Sea mammals include the whale, dolphin, porpoise, and seal. The true flying mammal is the bat, although certain other mammals, such as the flying squirrel, can glide from trees. Even the land mammals use a variety of habitats. Some live in trees and some live in the ground. Some even cut down trees and build their own houses. Can you name two mammals that do that?

Mammals can be classified by what they eat. Some mammals eat plants. Some mammals eat animals. Some mammals eat both plants and animals. There are names that are often used to describe each type of mammal. A plant-eating mammal is called a **herbivore** (hûr′bə vôr). A grazing animal, such as a buffalo or zebra, is a good example of a herbivore. What niche does a herbivore occupy?

A meat-eating mammal, such as a coyote or lion, is called a **carnivore**. A mammal that eats both animals and plants is called an **omnivore**. Can you name a mammal that is an omnivore?

Care by the mother is essential for mammal survival. With two exceptions, all mammals are born alive. The exceptions are the egg-laying mammals, the *duck-billed platypus* and *spiny anteater*. Coming into the world as a newborn mammal creates a special problem. The newborn mammal is completely dependent upon a mother. (The human newborn is a possible exception.) This is one way that the mammals are very

herbivore:
an animal that eats plants or plant parts only

carnivore:
an animal that eats other animals
omnivore:
an animal that eats both plants and other animals

Figure 24.22: This is a female opossum with the young in her pouch exposed for the photographer. The opossum is the only marsupial in the United States.

453

different from all other vertebrates. Female fish, amphibians, and reptiles often leave their newborn young to get along by themselves. A female bird must tend its newborn chicks for a short time. But the father can help. This is not true with the mammals. A newborn mammal must have a female to care for it and, most important, to feed it. The mother's milk is needed for the survival of newborn mammals. In the marsupial group (the pouched mammals), the young are carried in the mother's pouch. This is the most extreme type of dependence. (See Figure 24.22.)

Mammals are the most intelligent animals. If you compared the brains of all five groups of vertebrates, you would learn an interesting fact. The cerebrum, which is the center for conscious thought, is the smallest in fish. It is next largest in the amphibians. It gets larger in the reptiles, and still larger in the birds. In the mammals, the cerebrum is the largest—when compared to the rest of the brain. What does this mean? It means that the mammals are the most intelligent animals on Earth.

Which mammals are the most intelligent? Humans, by far, are the most intelligent mammals. The *chimpanzees* and other apelike animals would rank second—maybe. Life scientists have had some surprises in the past few years. One of the measures of intelligence is the ability to learn new things. In recent years, marineland trainers in California and Florida have taught dolphins and porpoises to do amazing things. Just how intelligent are dolphins and porpoises? And how intelligent are the whales, which are their close relatives? No one knows the answers to these questions. But there is one curious mammal —the human—who is determined to find the answers.

LESSON REVIEW *(Think. There may be more than one answer.)*

1. In the illustration on page 455, which heads go with which feet?
    a. Head A goes with feet E.
    b. Head B goes with feet D.
    c. Head C goes with feet F.
    d. Head A goes with feet D.

2. Which of the following best describes the feeding adaptation of the birds pictured?
    a. The bird with beak A wades in water and spears fish with its sharp beak.

A   B   C

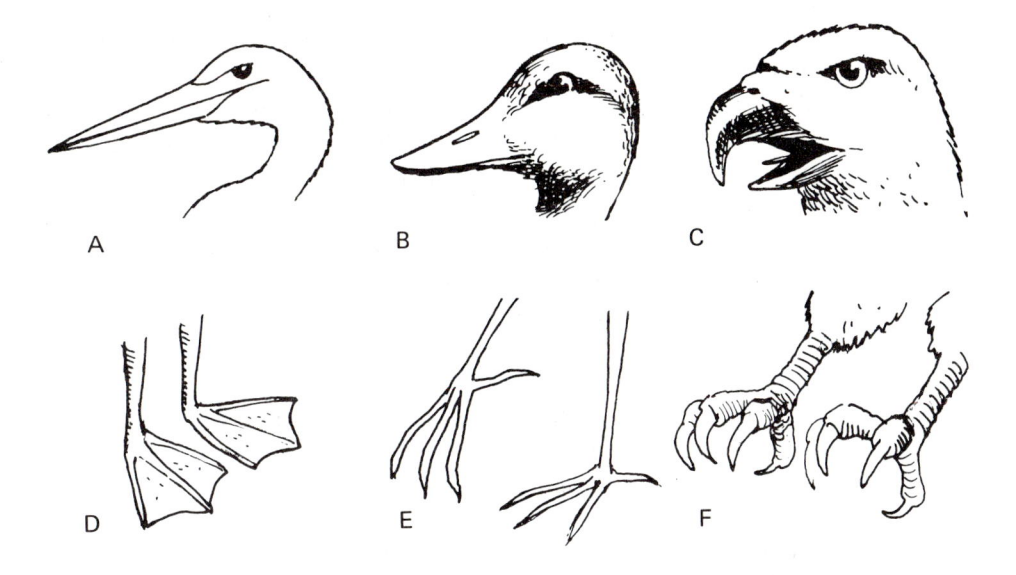

D   E   F

   b. The bird with feet D walks on sand and scoops insects out of holes.
   c. The bird with beak C eats large nuts such as acorns and walnuts.
   d. The bird with feet F grasps prey such as rodents or fish so that it can carry them away.
3. Birds
   a. have feathers.
   b. have wings, or parts of wings.
   c. have teeth in the lower part of their mouth.
   d. have thin, lightweight bones.
4. A bird
   a. that has long, thin legs adapted for wading could be an osprey.
   b. that has a beak adapted for cracking seeds could be a parakeet.
   c. that is cold-blooded could live in a cold, polar area.
   d. that is warm-blooded could be very unusual.
5. Mammals
   a. all have hair.
   b. all have milk-producing glands.
   c. all give birth to living young.
   d. are the most intelligent animals.
6. Mammals can
   a. live everywhere but underwater.
   b. run, jump, and even glide out of trees, but none can actually fly.
   c. often be completely independent shortly after birth.
   d. be classified into three groups based upon what they eat.

Organisms

7. Which of the following terms best matches each phrase?
   a. cold-blooded animal   d. herbivore
   b. warm-blooded animal   e. carnivore
   c. mammal   f. omnivore

   _____ an animal that eats plants and other animals

   _____ an animal with hair and relatively large cerebrum; the female nurses her young

   _____ a first-order consumer

   _____ an animal that can regulate its body temperature

   _____ an animal that eats other animals

   _____ an animal that cannot regulate its body temperature

KEY WORDS

cold-blooded animal (p. 451)   herbivore (p. 453)
warm-blooded animal (p. 451)   carnivore (p. 453)
mammal (p. 451)   omnivore (p. 453)

Applying What You Have Learned

1. Near or in what environment would you be most likely to find amphibians?
2. What other animal group has been as successful as the mammals in filling many different niches?
3. Why are mammals considered the most intelligent animals in the world?
4. How does being warm-blooded open up new environments to birds and mammals?
5. How are the skeletons of lampreys, hagfish, sharks, and rays different from the skeletons of other vertebrates?
6. Most of the reptiles and amphibians that once existed have become extinct. What may be one reason these vertebrates have disappeared?
7. What kind of feet and legs are you likely to find on a bird that feeds and wades in shallow water?
8. What major adaptive change allows most amphibians, reptiles, birds, and mammals to move around on land?
9. What is one reason amphibians have not successfully adapted to land?
10. Why do mammals have a better chance of developing to maturity than other vertebrates?

Glossary

ABO blood type: a way to classify blood based on the type of protein on the surface of red blood cells
absorption: the process by which digested food is taken into the bloodstream
adaptation: a trait that particularly helps an organism to survive in its environment
adaptive evolution: a change in the gene pool of a population that better suits it to survive in its environment
addiction: the strongest type of drug dependence. A person in this condition cannot function without the habit-forming drug.
adolescence: roughly the period between 12 and 18 years of age. It is noted for its rapid development and as the time of sexual maturation.
adrenal (ə drēn′əl) gland: one of two glands that produces adrenalin. Each adrenal gland is on top of one of the two kidneys in the body.
adrenalin (ə drēn′ə lin): a hormone produced when a person is frightened or angry. It causes changes in the body that ready the body for "fight or flight."
afterbirth: the placenta and the membranes attached to it that are expelled from the uterus after a baby is born
agar (ā′gĕr): a jellylike substance made from the cell walls of red algae
agriculture: most simply, farming, including the raising of livestock
alcoholic: a person with alcoholism. That is a disease that makes the alcoholic dependent on drinks made with grain alcohol. The only successful cure for alcoholism is self-control. The alcoholic *must not drink* beverages containing grain alcohol.
algae (ăl′jē): simple plantlike organisms. All algae carry on photosynthesis.
alveolus (ăl vē′ə ləs): one of many small air sacs in the lungs through which oxygen and carbon dioxide are exchanged
amphetamine (ăm fĕt′ə mēn′): one of a family of drugs that stimulates, or peps up, the user. Amphetamines can create addiction.
amphibian: a vertebrate that lives part of its life in water and part on land
annelid: any of a major group of worms with a body divided into ringlike segments
anther: a sac in a flowering plant that produces male reproductive cells. The anthers are at the ends of stamens.
antibiotic (ăn′tē bī ŏt′ĭk): a natural or laboratory-made chemical used to kill disease-causing bacteria
antibody: any of several proteins in blood plasma that destroy certain disease-causing bacteria and viruses
aorta: attached to the left ventricle, it is the largest artery in the body. It is the beginning of the systemic circulation outside the heart.

arachnid (ə răk′nĭd): a terrestrial arthropod that has four pairs of legs
artery: a blood vessel that carries blood *away from* the heart
arthropod: one of a major group of animals with jointed legs and feet, and an exoskeleton
asexual reproduction: a process by which some organisms reproduce their own kind. In asexual reproduction, cells divide and grow to form another organism.
atherosclerosis (ă thə rə sklə rō′sĭs): a disease of the arteries in which fats build up in the arterial walls
atria (ā′trē ə): (singular, "atrium"), the two upper chambers of the heart
attractant: in some female insects, a chemical with an odor that attracts males during the breeding season

bacteria: microscopic living things. Some cause disease.
barbiturate (bär bĭch′ər ĭt): one of a group of chemically related compounds used mainly as central nervous system depressants or to produce sleep
benthos (bĕn′thŏs): all together, the things that live on the bottom of a body of water
bilateral symmetry: a pattern in which parts are arranged around a line, so that each half of the pattern is a mirror image of the other
bile: a digestive fluid that comes into the small intestine from the gall bladder
biochemistry: the study of chemicals that make up living things
biological control: a pest-control method in which pests are killed by natural enemies
biological magnification: the buildup of a persistent chemical as it passes through a food chain
biosphere: the community of *all* living things on Earth
bivalve mollusk: a mollusk that has no head inside the two halves of its shell
brain: a large mass of nerve tissue enclosed in the skull. The brain interprets sensory input, directs body functions and activities, and is the center of thought and feeling.
bronchial (brŏng′kē əl) tubes: the tubes in the lungs that transport air to and from the alveoli

callus (kăl′əs): a thickened area of the epidermis
cancer: a tissue disease in which cells grow and divide out of control. Cancerous tissue loses its normal functions.
capillary: a blood vessel, one cell thick, through which blood passes from an artery to a vein
cardiac muscle: one of several types of muscle. Cardiac muscle is found only in the heart.
carnivore (kär′nə vôr): an animal that eats other animals
cartilage: a semihard material that may develop into bone. Some parts of the body

are formed mainly from cartilage, such as the nose and the ears.
cell: the basic structural unit of organisms
cell division: a process by which cells reproduce themselves by splitting apart
cell membrane: the outer part of a cell. It encloses and regulates whatever enters or leaves the cell. In a plant cell, the cell membrane is inside the cell wall.
cell wall: the stiff, outer covering of most plant cells
centipede: a wormlike arthropod with one pair of legs on each body segment
central nervous system depressant: a drug that affects the function of the brain and the spinal cord. The effect is for body reflexes and functions to slow down.
centriole: an organelle in an animal cell that helps divide the DNA into two equal parts when the cell divides
cephalopod (sĕf′ə lə pŏd′): the largest of the mollusks. The foot of a cephalopod is wrapped around its head and divided into tentacles.
cerebellum (sĕr ə bĕl′əm): a major region of the brain. It controls and coordinates muscular responses.
cerebrum (sĕr′ə brəm): the large upper layer of the brain. It is the brain region where all thought occurs and where input from the senses is interpreted.
cervix: the opening of the uterus
chiton (kī′tən): a mollusk with a shell made of eight overlapping plates
chlorophyll: the green substance in cells that carry on photosynthesis. In plants, chlorophyll is found in the chloroplasts.
chloroplast: a tiny, green body in the cytoplasm of a plant cell. Photosynthesis takes place in the chloroplast.
chromosome: one of the small bodies in a cell's nucleus that contains DNA tied to protein
cilia (sĭl′ē ə): hairlike structures in organisms. In animal breathing systems, they "sweep" dust and other particles out of inhaled air.
ciliate: a protozoan that moves by waving the cilia that surround its cell wall
clotting: the process by which proteins in blood plasma cause a network of fibers to form over a bleeding area
coelenterate (sĭ lĕn′tə rāt): an aquatic animal that has three distinct features—radial symmetry, stinging cells, and a large hollow cavity into which food enters and wastes leave
coitus (kō ī′təs): in sexual reproduction, the process by which sperm and semen are introduced by the male into the female reproductive system
cold-blooded animal: one that cannot regulate its body temperature, which is therefore always close to that of its environment
collar cell: a specialized cell in a sponge that moves water through its body, which helps feed the sponge
community: a group of populations that lives in a particular area

conifer (kŏn′ə fər): a plant (or tree) that produces its seeds in cones
connective tissue: tissue that holds internal body parts together
consumer: an animal that eats a producer or the product of a producer. Consumers make up one of the three basic niches in an ecosystem.
coral: a polyp-form coelenterate that produces a hard limestone case around its body
coronary heart attack: the death of heart tissue served by an artery that is clogged by a blood clot
cortisone: an adrenal hormone that helps to control the body's use of water and minerals
crustacean (krŭ stā′shən): usually, an aquatic arthropod with an exoskeleton
cyst (sĭst): the hard, shell-like covering protozoans form in time of drought
cytoplasm (sī′tə plaz′əm): the material outside the nucleus in a living cell

data: observed facts, including measurements. More generally, "data" has come to mean information of any kind.
decomposer: a living thing that digests dead producers and consumers. Decomposers make up one of the three basic niches in an ecosystem.
deficiency disease: any of several diseases caused by a lack of an essential nutrient
degree Celsius: a metric temperature equal to 9/5 of a degree Fahrenheit. To convert from degrees Celsius (°C) to degrees Fahrenheit (°F), use the formula °F = 9/5 °C + 32°. To convert from °F to °C, use the formula °C = 5/9 (°F − 32).
dermis: the inner of two layers that make up the skin
desert: a major terrestrial ecosystem marked by its very low moisture
development: the process by which the body grows and changes form
diabetes: a disease caused by a shortage of insulin in the body. That results in excess sugar in the blood, which, if uncontrolled, can lead to very serious problems.
diaphragm (dī′ə frăm′): a large muscle below the lungs that aids breathing by moving up and down
diatom (dī′ə tŏm′): one of a large number of golden-brown algae
dicot: a member of one of two groups of flowering plants. Most dicots have leaves with netted veins, and flower parts usually occur in fours or fives.
digestion: the mechanical and chemical breakdown of food into nutrients the body can use
digestive enzyme (ĕn′zīm): a chemical, usually a protein, that digests food
dinoflagellate (dī nō flă′jə lăt): a kind of single-celled marine phytoplankton
diversity: most simply, variety. In life science, diversity is typical of a habitat with many different kinds of populations.

DNA: an abbreviation for the complex chemical in all nuclei that carries "instructions" for heredity. All the body's proteins are "built" according to these instructions.
dominant gene: a gene that most strongly determines a trait that is the result of two or more genes
duodenum (doo′ə dē′nəm): the beginning portion of the small intestine

echinoderm (ĭ kī′nə dûrm′): any of a major group of marine animals with a limestone-like skeleton just beneath a bumpy or spiny skin. Echinoderms have tube feet.
ecology (ĭ kŏl′ə jē): the study of ecosystems
ecosphere: the biosphere interacting with the Earth's total physical environment
ecosystem (ĕk′ō sys′təm): a community together with its physical environment
egg cell: a female reproductive cell
embryo: the earliest stage in the development of organisms reproduced sexually
emphysema (ĕm′fĭ sē′mə): a lung disease in which the elastic tissue, which normally helps push out air, has lost its "squeezing" ability
endocrine (ĕn′də krĭn) **gland:** any one of several glands that produce hormones
endocrine system: one of ten organ systems in most animals. It is made up of glands that produce hormones, which are control "chemical messengers" of the body.
endoplasmic reticulum (rə tĭk′yoo ləm): the network of thin membranes across the cytoplasm of cells
environment: all things outside the bodies of living things that affect their lives
enzyme: any one of many proteins that affect chemical changes in the body
epidermis (ĕp′ĭ dûr′mĭs): the outer of two layers that make up the skin
esophagus (ē sŏf′ə gəs): the organ that moves food from the mouth to the stomach
estrogen (ĕs′trə jən): a hormone produced by the ovaries. It is largely responsible for the onset of sexual maturity in females.
eutrophication (yōō trəf′ĭ kā′shĭn): the natural process by which a body of water ages
evolution: the process of change in a population as its gene pool changes
exoskeleton: a kind of skeleton on the outside of an animal's body
extinction: the death of an entire population of plants or animals

Fallopian tube: either of two tubes that are part of the female reproductive system. Each leads into the uterus and serves as a path for an egg cell, which drops into it from an ovary.
feces (fē′sēz): partly decomposed waste materials that pass out of the body after digestion has taken place. Feces are semisolid, much of the water having been absorbed in the large intestine.
fern: a nonflowering vascular plant whose stem grows under the ground
fertilization: in sexual reproduction, the uniting of a male reproductive cell with a female reproductive cell
fertilized egg cell: the single cell formed when a male reproductive cell unites with a female reproductive cell
fetus (fē′təs): an unborn baby from about the third month of development in the uterus until birth. During the first 2 months of prebirth development, the unborn child is called an embryo.
filament: any threadlike structure
flagellum (flə jĕl′əm): a small, whiplike structure. In the collar cells, it helps to move water through the body of a sponge.
flatworm: a kind of worm with a flat body
flower: the reproductive system of all seed plants except conifers
fluke: a member of one of the three main groups of flatworms. All flukes are parasitic.
follicle (fŏl′ĭ kəl): the tiny sac in the skin from which a hair grows
food chain: a description of who eats whom in an ecosystem
food pyramid: a way to show the relationship among decomposers, producers, and consumers in an ecosystem. Decomposers are at the base of the pyramid. High-order consumers are at its top.
food web: a description of how all the populations in an ecosystem feed on each other
fossil: an impression, found in rock or other material, of the skeleton or other parts of a dead organism
freshwater habitat: an area in a body of water other than an ocean where a plant or an animal population has its home
fruit: the ripened ovary of a seed plant, in which the ovules have become seeds
fungi (fŭn′jī): simple, plantlike organisms that lack chloroplasts. The singular is fungus (fŭn′gəs).

gall bladder: a saclike organ within the liver. It produces bile.
ganglion: a tissue made up of groups of nerve cells
gastropod: a member of the largest group of mollusks. A large muscular foot is the common characteristic of all gastropods.
gastrotrich (găs′trō trĭk′): a microscopic, freshwater animal related to roundworms. Gastrotrichs have cilia on their bodies that give them a fuzzy appearance and help propel them through water.
gastrovascular cavity: a large hollow space in the body of a coelenterate with one opening through which food enters and wastes leave
gene pool: the total of all genes in a population
genes (jēnz): units in cell nuclei that control the ways in which an individual will develop. Genes contain instructions for building all body protein.
genetic disease: any of several diseases, caused by a faulty gene, that can be inherited

gill: a breathing organ by which fish and certain other aquatic animals take oxygen from water
gland: a tissue that produces one or more substances. These serve specific purposes in an animal.
Golgi (gōl′jē) complex: an organelle that prepares proteins for "export" from cells
grassland: a major terrestrial ecosystem of low moisture. Grasses and very low plants dominate this ecosystem, in which large populations of grazing animals used to live.
growth: the process by which something gets larger
growth hormone: a pituitary hormone that controls the growth and division of cells in the bones and muscles

habitat: the area where a plant or an animal population has its home
heart: a hollow muscle near the center of the chest. Its squeezing motions keep blood moving through the body.
heart rate: the number of times the heart contracts, or beats, in one minute; also called the pulse
heart valve: any of the valves in the heart that open and close to let blood flow, from an atrium into a ventricle, or out of a ventricle
hemoglobin: the oxygen-carrying substance in blood that gives it its red color
herb: a plant without a woody stem
herbivore (hûr′bə vôr): an animal that eats plants or plant parts only
heredity: the physical traits passed on to an individual from its parents. Heredity is controlled by the individual's genes, half of which come from its "father," the other half from its "mother."
hernia: a tear in one or more of the abdominal muscles
hibernate: to spend the winter in a sleep-like state
hormone: a chemical released within the body by a gland. Each hormone carries a specific chemical message. The message produces a specific bodily response.
host: the organism on or in which a parasite lives

immunity: protection against a disease by antibodies produced in the blood in reaction to the disease or by immunization
immunization: the process of causing the body to produce antibodies by injecting it with dead or weakened disease-causing viruses or bacteria. Sometimes immunization involves the injection of antibodies from another animal.
insulin: the hormone produced by the pancreas that causes excess blood sugar to be stored in the liver
integeument (in tĕg′yə mənt): the skin; an organ of the integumentary system
intertidal organism: an organism that lives near the ocean's shore, where the water comes and goes with the tides
involuntary control: control in the endocrine system and part of the nervous system that is automatic

joint: the point at which bones come together or are connected

kelp: a large, floating seaweed, the most common form of brown algae
kidney: one of two bean-shaped organs, on either side of the lower backbone, which removes many waste materials from the blood
kilo (kē′lō): short for kilogram, a metric weight equal to 2.2 pounds
kilometer: a metric distance equal to 0.6 mile
kingdom: one of four large groups in which all organisms are classified. The groups are: Monera, Protista, plant, and animal.

labor: rhythmic contractions of the uterus that force a baby out of its mother at the time of birth
large intestine: the last part of the digestive system, also called the colon. Water is absorbed here and bacterial decay of waste takes place.
larva: the second of four developmental stages in certain insects. It is wormlike in nature.
larynx: an organ between the throat and the trachea. The larynx contains the *vocal cords*, which are the basic source of speech production.
latent virus: an inactive virus in a cell. It may at some other time begin to reproduce and destroy the cell.
leech: a parasitic annelid that attaches itself to its animal host by suckers
legume: a plant of the pea family. All legumes have their seeds in pods.
levels of organization: a way to describe the increasing complexity in the way life is organized
life cycle: the series of changes in a living thing from birth to death. A key change occurs when the living thing can reproduce its own kind.
ligament: a band of connective tissue that holds bones together
lung: one of two expandable organs in the chest in which gas exchange takes place. Carbon dioxide leaves the blood here and oxygen enters it.
lymph: a liquid that contains blood plasma from the tissues as well as white blood cells and antibodies
lymph node: an enlarged area in the tubes of the lymphatic system where lymph is stored
lymphatic system: a network of tubes that carries blood plasma from tissues back to the blood
lysosome (lī′sō sōm): an organelle that digests food in an individual cell

mammal: any member of one of the five major groups of vertebrates. A mammal has hair and a cerebrum that is large relative to the rest of its brain. The female mammal nurses its young.
mantle: a tissue sheet over the top of a mollusk's body that usually produces a hard shell
marine (mə rēn′) habitat: an ocean area where a plant or an animal population has its home
marrow (măr′ō): soft tissue in certain bones. The marrow produces blood cells.
medulla (mə dŭl′ə): a major region of the brain, also called the brain stem. Located at the base of the brain and connected to the spinal cord, it controls many internal organs.
medusa (mə dōō′sə): a coelenterate with tentacles that hang from its body and the opening to its gastrovascular cavity on its "bottom" side
membrane: a thin, flexible layer of cells. It usually covers a cell's surface. It also separates different internal parts of an organism.
menstrual cycle: a series of events in a sexually mature female. It includes preparation of the uterus for a potential fertilized egg. That is followed by ovulation. Then, when fertilization does not take place, vaginal discharge during the menstrual period completes the cycle.
mesoglea (mē zō glē′ə): a layer of protein material between the two layers of cells in a coelenterate. In medusa-form coelenterates, the mesoglea acts like an internal skeleton.
microhabitat: a small habitat of one kind that is within a larger habitat of another kind
microorganism: an organism too small to be seen with the naked eye. Usually one-celled, it can be seen through a microscope.
millipede: a wormlike arthropod with two pairs of legs on each body segment
mitochondrion (mī tə kŏn′drē ən): an organelle in which most energy in a cell is produced
moist coniferous forest: a major terrestrial ecosystem marked by a moist climate with moderate temperatures. Large conifers dominate this forest ecosystem.
mold: a type of fungus
mollusk: a small animal that has a soft body. Some, but not all, mollusks have a shell, tentacles and eyes, and a muscular foot.
moneran (mŏn′ə rən): a simple organism of the kingdom Monera. There is no distinct nucleus in a moneran cell.
monocot: a member of one of two groups of flowering plants. Most monocots have leaves with parallel veins and flower parts that occur in threes or multiples of three.
monoculture: the practice of planting only one kind of crop in a field
moss: a small plant that grows in mats or clumps and is most abundant in moist habitats
mountain ecosystem: a major terrestrial

ecosystem made up of different ecosystems that change with altitude

mucus (myōō′kəs): a slippery liquid produced by some glands and membranes in the body. It serves to moisten and protect the membranes.

muscle tone: the condition that keeps a muscle ready for action. It is observed as a slight, but constant, contraction.

mutation: any change in a gene. It may lead to a change in an organism.

nekton (něk′tən): as a group, all animals larger than plankton that live free in a body of water

nematocyst (něm′ə tə sĭst′): the stinging cell of a coelenterate. The nematocyst shoots out a barbed thread coiled inside a capsule to sting prey.

nematode (něm′ə tōd): a member of the largest group of roundworms

neritic zone: the shallow-water region above the continental shelf in a marine ecosystem

nerve: a bundle of nerve cell branches enclosed in a sheath

nerve cell: a cell that carries electrical "messages" throughout the body. The messages produce specific body responses.

nerve cell body: the central part of a nerve cell. It contains the nucleus and most of the organelles.

nerve cell branch: the long part of a nerve cell that extends away from the nerve cell body

neuston (nōō′stən): all together, the things that live on the surface of a body of water

niche (nĭch): the total way of life for a living thing

nucleus (nōō′klē əs): usually the central portion of a cell. It controls the growth, reproduction, and heredity of the cell.

nutrient: most simply, food. Anything that is needed by a plant or an animal for growth.

nymph: the second of three developmental stages in certain insects. It is the wingless insect that hatches from an egg.

oceanic zone: the deep-water region in a marine ecosystem

oligochaete (ŏl′ə gō kēt′) **worm:** annelid that has few bristles, which are too small to be seen, but can be felt

omnivore (ŏm′nə vôr): an animal that eats both plants and other animals

opioid (ō′pē oid): opium, or any drug made from opium, which comes from a particular poppy plant

organ: a group of different tissues that form a specialized structure in a plant or an animal. The structure, or organ, does one or more specific jobs.

organ system: a group of organs that work together to do a specific job

organelle: any one of several parts of a cell that does a specific job

organism: an individual living thing of any kind

ovary: the female reproductive organ that produces egg cells

ovule (ō′vyōōl): a rounded body inside the ovary of a flowering plant that holds the female reproductive cell

oxygen debt: a condition in which skeletal muscle cells have less oxygen in them after muscular work has been done

pancreas: an organ near the stomach that produces digestive enzymes. It also produces chemicals that help control the body's blood-sugar level.

paranoid (păr′ə noid′): relating to a mental disorder in which a person feels persecuted or has imagined fears

parasite: an organism that lives at the expense of another organism. A parasite does not make its own food, but lives on or in living plants or animals.

parathyroid gland: any of four small glands that are found within the thyroid gland. The parathyroid glands produce a hormone that controls the body's use of calcium.

penis: the external organ of the male reproductive system, for the ejection of sperm and semen. It is also an organ of the male excretory system, for the passage of urine.

periphyton (pĕr′ĭ fī′tən): all together, the things that live clinging to shallow-water plants

persistent chemical: a chemical that persists, or lasts unchanged, for a long period of time

pest: a plant or an animal whose population is so large that it is a nuisance to people

pesticide: a chemical used to kill pests

phagocyte (făg′ə sīt): a type of white blood cell that attacks and digests some disease-causing organisms

photochemical smog: a polluted air condition caused by the reaction of sunlight with automobile exhaust gases and other gaseous products of industry

photosynthesis: the process by which certain individuals make food and produce oxygen

physical dependence: a condition in which a person's body loses its ability to function normally without a drug

physical environment: any part of an ecosystem that is not alive

phytoplankton (fī′tō plăngk tən) plantlike plankton

pituitary gland: the "master" endocrine gland located below the brain. Pituitary hormones control most of the other endocrine glands.

placebo (plə sē′bō): a fake drug or treatment used as a control in medical experiments. Often a placebo brings about a "cure" because the patient thinks it is real.

placenta: a large mass of tissue that attaches the embryo to the uterus. It is the place where materials are exchanged between them. It also produces hormones needed by the mother.

planarian (plə nâr′ē ən): a nonparasitic freshwater flatworm

plankton (plăngk′tən): very small plantlike or animallike organisms that live free in a body of water

plasma: the pale yellowish, liquid part of blood

platelet: one of the solid parts of the blood. The platelets help in the clotting process.

pollen: small particles that contain a plant's male reproductive cells

pollution: the products of human activity that harm—or destroy—the quality of the physical environment

polychaete (pŏl′ə kēt′) **worm:** an annelid that has many bristles and that usually lives in an intertidal marine habitat

polyp (pŏl′ĭp): a coelenterate with upright tentacles and the opening to its gastrovascular cavity on its "top" side

population: in life science, a group of living things of the same kind that are found in a particular place

population explosion: a rapid and great increase in the population of a certain living thing

pregnancy: the time during which the uterus houses a developing organism

producer: a living thing that makes food and releases oxygen. Producers make up one of the three basic niches in an ecosystem.

prostate: a male reproductive organ that produces semen. Strong contractions of the prostate eject sperm and semen through the penis.

protein (prō′tēn): one of a group of complex chemicals that are the building blocks of cells. They do most cellular work.

protist: a simple organism of the kingdom Protista. Protist cells have a nucleus.

protozoan: an animallike protist

psychedelic (sī′kĭ dĕl′ĭk) **drug:** any drug that causes a change in normal perception. The strongest reaction to such a drug is a hallucination.

puberty: the time of sexual maturity in humans, when reproduction first becomes possible

pulmonary circulation: the route the blood follows from the heart to the lungs and back to the heart

pupa (pyōō′pə): the third of four developmental stages in certain insects. It is the stage in which a larva changes into an adult.

radial symmetry: a pattern in which parts are arranged around a central point or line so that any vertical cut through the center divides the whole into two identical halves

red blood cell: one of the solid parts of the blood. The red blood cells carry oxygen.

red tide: a "bloom" of a poisonous, red dinoflagellate
regeneration: the process by which some organisms grow back a lost part of the body
reptile: a member of one of the five major groups of vertebrates. Reptiles have lungs for breathing on land and are covered with scales or horny plates.
ribosome (rī′bə sōm): an organelle that helps to make protein
rotifer (rō′tə fər): a microscopic freshwater roundworm. Rotifers have bands of cilia on their head ends that look like spinning wheels.
roundworm: any of a major group of worms that has a round body

saprophyte (săp′rə fīt′): an organism, such as a fungus, that gets nourishment from dead organisms and in the process helps decay them
scrotum: the saclike pouch that houses the male animal's testes
sebaceous (sĭ bā′shəs) **gland:** one of the tiny glands in the skin that secretes oil into a hair follicle
secondary sexual characteristic: a developmental change that appears at puberty
semen: a white liquid produced by the male reproductive system. It provides a liquid for sperm.
seminal vesicles (věs′ĭ kəlz): male reproductive organs that produce semen
sexual reproduction: a process by which organisms reproduce their own kind. In sexual reproduction, a male reproductive cell unites with a female reproductive cell.
shrub: a woody plant, usually shorter than a tree, with several stems growing out of the ground. Also called a bush.
sinus: one of the air spaces behind the nose. The sinuses warm air that is breathed in and add moisture to it if necessary.
skeletal muscle: one of several types of muscle. The skeletal muscles control all outside-the-body movements.
small intestine: the main organ of the digestive system. It is between the stomach and the large intestine. Most digestion and absorption occurs in it.
smog: a heavily polluted condition of the air. It usually is made up of smoke particles and poisonous sulfur and nitrogen gases.
smooth muscle: one of several types of muscles. Smooth muscles are found only in internal organs.
social insect: an insect that lives in organized groups. Within the groups, specific jobs are done by specific members.
soil erosion: the wearing away of soil by water or wind
sperm cell: a male reproductive cell
spinal cord: a thick, long body of nerve tissues that extends from the base of the brain, branching to form smaller nerves that serve various parts of the body
sponge: an aquatic creature that has the simplest structure of all the animals. A sponge has specialized cells, but no tissues, organs, or systems.
spore: a reproductive cell that develops into an organism
stamen: the stalk in a flower that has a sac (anther) at its end in which male reproductive cells are produced
stigma: the uppermost part of a style on which pollen first lands
stomach: a main organ of digestion. It is a saclike structure between the esophagus and small intestine.
stroke: the death of brain tissue served by an artery that is clogged by a blood clot. Paralysis, loss of speech, or death usually results.
style: the region between the ovary and the stigma in a flower
sweat gland: one of the tiny glands in the skin that secretes sweat (perspiration) through openings, or pores, in the skin
systemic circulation: the route the blood follows from the heart to all parts of the body—except the lungs—and back to the heart

taiga (tī gä′): the northernmost forest ecosystem of Earth, also called the northern coniferous forest
tapeworm: a member of one of the three main groups of flatworms. All tapeworms live as parasites in animal intestines.
temperate deciduous (dĭ sĭj′ōō əs) **forest:** a major terrestrial ecosystem marked by seasonal changes. Deciduous trees dominate this ecosystem.
temperature inversion: a condition in which a layer of warm air above the Earth's surface traps air and pollutants below it. The trapped, polluted air cannot rise.
tendon: a band of connective tissue that attaches muscles to bones
terrestrial (tə rěs′trē əl) **habitat:** an area on land where a plant or an animal population has its home
testes (těs′tēz): male reproductive organs, also called testicles; usually occur as a pair and produce sperm cells
testosterone (těs tŏs′tə rōn): a hormone produced by the testes. It is largely responsible for the onset of sexual maturity in males.
thermal pollution: the upset of a freshwater ecosystem when its temperature is raised by the addition of heated waste water to it
thyroid (thī′roid′) **gland:** an endocrine gland located in front of and to both sides of the trachea. It produces thyroxin.
thyroxin (thī rŏk′sĭn): a hormone produced by the thyroid gland. It contains iodine and stimulates the release of cellular energy.
timberline: the height above which trees do not grow on a mountain
tissue: a group of similar cells that work together to do a specific job
trachea (trā′kē ə): the windpipe through which air moves between the larynx and the lungs
trait: a physical characteristic, such as hair or eye color, that may be inherited
tranquilizer: a drug prescribed to calm the nerves or to relieve anxiety. Two such drugs—Librium and Valium—are the most heavily prescribed drugs in the world.
tree: an upright plant, usually with one strong, tall, woody main stem, or trunk
tropical rain forest: a major terrestrial ecosystem that is very warm and wet year round. It is the ecosystem with the greatest variety of living things.
tropical savanna (sə văn′ə): a major terrestrial ecosystem that is a warm, moist grassland. It has one rainy season a year, but is dry the rest.
tube feet: feetlike organs on echinoderms that have suction cups at their ends. Tube feet help echinoderms to move and to get food.
tundra: the coldest major terrestrial ecosystem
turbellarian (tûr bə lâ′rē ən): a member of one of the three main groups of flatworms. Turbellarians live in aquatic or moist terrestrial ecosystems.

umbilical cord: a flexible extension of tissue that joins the embryo to the placenta. It is filled with blood vessels. These transport blood from the embryo to the placenta and back.
urinary bladder: a saclike container connected to the kidneys that stores liquid wastes until they are passed out of the body as urine
urine: liquid waste materials removed from the blood by the kidneys
uterus (yōō′tər əs): a thick, hollow organ in females. The human embryo and fetus develop within the uterus. Also known as the womb (wōōm).

vagina (və jī′nə): an organ of the female reproductive system. It serves as a path to the uterus and Fallopian tubes for sperm cells ejected into it. It also serves as a birth canal through which a baby is born from its mother.
values: those things a person or society believes to be important
vascular plant: a plant that has a tubular transport system
vein: a blood vessel that carries blood *back to* the heart
ventricles: the two lower chambers of the heart

vertebrae (vûr′tə brē): the bones of the spine, or backbone
vertebrate: an animal with a spinal column that is usually made of bony units called vertebrae. In some, the vertebrae are made of cartilage.
villi (vĭl′lē): very small fingerlike projections on the wall of the small intestine. Digested food is absorbed through the villi into the bloodstream.
virus: a borderline form of life, much smaller than a cell. It acts like a living thing only when inside a living cell.

voluntary control: control in part of the nervous system that is not automatic. The control is brought about by conscious thought.

wandering cell: a specialized cell in a sponge that moves through the body of the animal, distributing food to other cells
warm-blooded animal: one that can regulate its body temperature to a nearly constant value, even though its environmental temperature may vary greatly

white blood cell: one of the solid parts of the blood. The white blood cells help fight "invaders" of the body, such as bacteria and viruses.
withdrawal: the step-by-step removal of a drug from an addicted person. It is usually accompanied by severe physical and mental upset.

zooplankton (zō′ə plăngk tən): very small animals and animallike plankton

Index

Note that italicized page references indicate an illustration and boldface page references indicate definitions.

abdomen, 133
ABO blood types, 236, 238–239
absorption, **146,** 148–149
accumulation of cell products, 244
"Adam's apple," 154
adaptation(s), **234,** 241, 407
  of desert plants and animals, 100
  of marine organisms, 86, 407
  of vertebrates, 85, 434, 441–443, 445, 449–451, *450*
adaptive evolution, **233**
addiction, **289,** 293–295
adolescence, 249–254, *250*
  growth and development during, *246,* 249–254
  puberty, **250**
adrenal glands, **170,** 171, 174–175
adrenalin, **170**–171, 174–175
afterbirth, **222**
agar, 337
agriculture, 23
  environmental effects of, 23, 30, 98, *310,* 314
  pest population problems, 23, 25–26, 30–31, 352
air, 40, 41, 46, 153–155
air pollution, **300,** 317–323
  automobile exhaust fumes and, 319, 320–322
  breathing system damage from, 321–322
  combustion and, 317, 319, 320
  fog and, *319*
  smog, 303, 319, 482
  sulfur–oxygen compounds and, 319
  temperature inversions and, 318–319
alcohol, 132, 349
  as a drug, **283,** 292, 295
  lethal with barbiturates, 295
alcoholics and alcoholism, **283, 283,** 290, 292, 295
algae, 67, 70, 331–337, 383, 389
  blue-green (monerans), 331–332, *334–335*
  diatoms (golden-brown), 70, *111,* 335–336, 337, 394
  in eutrophication, 310, *311,* 312
  green, *334–336,* 383
  kelps (red), 337
  protists, 334–337
alligators, 58, 443–**444**
alveolus, **154**–155, *155,* 163
ameba, 342
amphibians, **439**–440, 451, 482
amphetamines, **295**
anesthetics, **285,** 295
angina pectoris, 281
animals, 41, 48, 91, 96–98, 379–454
  arctic, 29–30, 46, *90*
  burrowing, 40, 90, 98

cells of, 108, 110, *115,* 120, *121,* 122
cold-blooded and warm-blooded, **451**
as consumers, 45
grazing, *45,* 70, *98,* 99, 453
organs and organ systems of, **131,** 135–136
population control of, natural, 29–30
tissues of, 128–129, *132*
annelids, **409,** *410*–412, 413
anteaters, 452
anther, 364
antibiotics, **27,** 267–268, 349
  and bacteria populations, 27, *268*
antibodies, **158,** 162, 267, 269–270, 274
ants, 420, *421*
aorta, 163
appendix and appendicitis, *149,* 268
aquatic ecosystems, 44, 45, 70, 90, 437
  populations in, 89–91, 382–383, 422
  producers in, 70, 91, 335–337
  (see also **water pollution**)
arachnids, **422,** *424*
arctic region, *90,* 95–96
  population explosions in, 29–30, *96*
arteries, **160,** 162, 163, 274–275
arthropods, **416,** 417–**422**
Ascaris, 262, 400
asexual reproduction, **209,** *209,* 210, 215
  of bread mold, *210,* 348
  of ferns, 354, *355*–356
  of hydra, 332
  of protozoans, 343
  of yeast, 212
atherosclerosis, **276,** 276–277
"athlete's foot," 260–261
atmosphere, 84, 90, 94, 320, 321
atomic radiation, 34, 35, 130, 315
atoms, 136, 229, **230**
attractants, 31, 34
atria, **159,** *159,* 162, 163
automobiles, pollution from, 319, 320, *322*

baboons, *432*
baby, human
  birth of, 217, 221–223
  needs and care of, 223
  prenatal development of, *245,* 265, 296
baby monkeys and "mothers," 222–223
bacteria, **5,** 133, 330–331, 337, 349
  antibiotics and, 27, 267–**268**
  antibodies and, 158, 267–269
  decomposers, **46,** 259–260, *307, 330,* 339, 342
  disease agents, 27, 259–260, 267, 268, 307, 331, 349
barbiturates, **285, 295,** 295
barnacles, 422
bass, 437
bats, *452,* 453
beefsteak, 128–*129*
bees, 365, 420, 422
beetles, 18, *31,* 66, 67, 419
benthos, 67–68

bilateral symmetry, **394**
bile, 133, 148, 172
biochemistry, **5**
biofeedback, *168*
biological control, 30–31
biological magnification, **53,** 53–54, 437
biology, 4–5
biosphere, 41–**42,** 137, 330, 336
birds, **449**–450, 451, 454
  in food chains, 50, 52, 53
  pesticide effects on, 9, 52, 53
birth, **217,** *220*–222
bison, 451
bivalve mollusks, **405**
black groupers, 20
bladder, urinary, *164,* 215
"bleeders," 158, 263
blood, 157–164
  antibodies in, **158,** 162, 267, 269–282
  in breathing process, 155
  circulation, 162–163
  clots in arteries, 275, *276*
  clotting process, **158,** 171, 263
  functions of, 155, 157–158
  plasma, **155,** 157, *158,* 160–162, 164
  platelets, 157, 158
  proteins, 158, 236, 263
  red cells of, *108,* 129, 132, **155,** 157–158, 236
  sugar level, 172
  transfusion, 236, 263
  waste removal from, 155, 163–164
  white cells in, 135, 157, 158, 162, 267
blood fluke disease, 396
blood flukes, 395–396
blood types, 236, 238–239
blood vessels, 136, 155, 160–161
  (see also **arteries; capillaries; veins**)
"blooms," phytoplankton, 84–85
blue-green algae, 331, 332, *334–335*
blue whale, 452
bluebonnet, *371*
boa constrictor, 447
body, knowing your own, 200–201
"body odor," 189
bones, 91, *121,* 122, 135, 171, *197,* 434 (see also **cartilage; skeletons**)
  backbones, 406, 433
  growth of, 244–245, 247
  marrow of, **196**
"Boy Named Sue, A," 25–26
brackish water, 76
brain, *176, 177,* **180,** *180*–184
  of cephalopods, 407
  growth of, 247
  protein deficiency disease damage to, 264–265
  stroke damage to, *274*–275
  of vertebrates, 454
bread, **209,** 348, 352
breathing system, 135, *144,* 153–156
  of amphibians, 439–440
  of fish, 437
  gas exchange in, 155, 162–163, *321*
  pollutant damage to, 321–322

brittle stars, 430
bronchial tubes, **154**–155, *154*
"bud" reproduction method, *386,* 388
buffalo, 45, 98, *451*
bugs, true, *421*
bullfrog, *440*
butterfly, *419, 421*

caddis fly larva, 67, *68*
calcium, 64, 122, 171, 244
callus, **188**
Canada lynx, 37
Canadian goose, *303*
cancer, **130,** *130,* 137, *208,* 283, *293,* 322
cannabis plant, *296*
"Cap-Chur" gun, *8*
capillaries, 155, *160*–161, 163
carbohydrates, 263, 264
carbon dioxide, 41, 46, 352
  in body processes, 155, 162, 164
  in photosynthesis, 44, 46, 331
carbon monoxide, 292, 320
cardiac muscle, *191*–192
cardinal, *449*
cardiopulmonary resuscitation, *170*
carnivores, *452,* 453
cartilage, **197,** 245, 437
cell biologist, *13*
cell division, 120, **206,** *206*–207, *228, 230,* 244
cell enlargement, 206–207, *244,* 247
cell membrane, **109,** *117,* 118, *123,* **206**
cell products, *107,* 112–114, *119, 120*–*121,* 122, 127, 136, 244
cell walls, *111, 123,* 336–337
  of animals, *111*
  of plants, *109,* 111, 118, 121
cells, 13, *107*–**108,** *109*–111, *112*–*115, 117*–120, *121, 123*
  animal, *108, 111,* *112*–*115,* 119, *121*–122
  blood (see **red blood cells; white blood cells**)
  brain, 247, 265
  chloroplasts in, *107,* **109,** *110,* 111, 117, *128*
  compared with viruses, 329–330
  cytoplasm in, **109,** 110–111, 117–119
  "daughter" and "parent," **206**
  fat, 113
  fibroblasts, *115*
  growth processes in, 244
  muscle, 113
  nerve, *110,* 175–177
  nucleus of, **109,** 117, 119, 120, 123
  organelles in, *117,* 120–123
  plant, *106, 110,* 121
  reproduction of, 205–208
  virus destruction of, *261,* 330
centipedes, *418,* 418–419
central nervous system, **183** (see also **nervous system**)
  depressants, **292,** 295
  stimulants, 295
centrioles, **120**
cephalopods, *406*–407
Ceratium, *83*
cerebellum, *180,* **181**

cerebrum, *180, 182,* **182**, 454
cervix, *221,* **222**, *222*
   cancer of, *208*
change processes
   adaptation, **234**
   evolution, 233–234
   growth and development, 243–254
   metamorphosis, 67, 419–422, 440
   mutation, **233**, *233*–234
chemical poisons (*see* pesticides)
chicken cholera microbes, 272–273
chicken wing, dissection of, 138–139
chickenpox, 261
chimpanzees, *452,* 454
chitons, *405,* 405
chlorophyll, 331, 334–335, 347
chloroplasts, *107,* **109**, *110, 111,* 128
chromosomes, *206,* **228**, 229
cigarettes, 130, 283–*284,* 288, 292–293, 322
cilia
   in human breathing system, **321**
   of paramecium, 339
   of rotifers, 401–402
ciliate, 342
circulatory system (*see* transport system)
cirrhosis of liver, 292
clams, 64, 405, 427–428
clamworm, *410*
cloning, 215
Clostridium tetani, *260*
clotting of blood, **158**, 171, 263
coal, 319, 323, 355
cockroaches, *34*
cocoon, 420
codeine, 294
codling moth, 26
coelenterates, 379, **384**, *384*–389
coitus, 216
cold, common, 261
cold-blooded animals, 451
collar cells, 382–*383*
colon, 149
combustion, *317,* 320
community, *38,* 39–42, 137
composite family, 371, 373, 374
cones, 359–360
conifers, *88,* 96–*97,* 359–*360,* *361*–363
connective tissue, **197**
consumers, **45**, 61, 81, 84, 90, 93, 389 (*see also* food chains)
   ecosystem niche of, *45–46,* 52
   interdependency of, 50–51
   orders of, *44–45, 49,* 70
   protists, 63, 70, *339*–343
continental shelf, 82, *83*
control systems, 169–178
   hormone control, 169–173
   nerve control, 174, *177*–178
   self-regulating, 169–178
   voluntary and involuntary, 177
controversial terms, 302
coral, **389**, *389*
coral reefs, 389
cork, *107*–*108,* 110
coronary artery, 163, 274–275
coronary heart attack, 274–275, 293
cortisone, **171**
cottony cushion scale, 31

crayfish, 67, *422*
crocodiles, 443–444
cross pollination, 365
crustaceans, **422**, *422*
currents
   freshwater, 63
   marine, 77–78, 79–83
Currier, Mary Jean, *2,* 4–11, 7
cyst, 342–343
cytoplasm, **109**, 111, 117, 118–*119,* 343

dandelion, 366, 371
data, 7, 12–14
DDT, 29, 31, 51, *52*–53
   magnified in food chains, 51–54, 55, 56, 303
death, 136–137
deciduous plants, 97
decomposers, **46**, 80, 82, *331,* 339, 349
   ecosystem niche of, *45–46,* 47–48, 50
   in eutrophication, 307–311
   as oxygen users, 63, 306–*311,* 312
deer, 45, 98
deficiency disease, 263, 264
degree Celsius, **12**
dermis, **188**
desert, 92, **100**, *100*
detergent pollution, 313
development, 244–248, *249*–254
   in adolescence, 249–254
   before birth, 215–217, 221, 244, 265, 296
   of boys and girls, 242, *249*–250, 251–253
   in childhood, 245–247, 249–250
   slow developers, 251, *254*
   stages of, 248–*249*, 250–251
diabetes, **172**, *172*
diaphragm, 156
diatoms, 70, **83**, *83, 111,* 335–*336,* 337, 394
dicots, 368–*369,* 371, 373
diet, 247, 263–264, 265, 274
digestion, 145–149
   chemical, 145–149
   hormones in, 171–172
   mechanical, 145, *146*
digestive enzymes, **146**, 147–148, 171–172, 386
digestive system, 133, 136, 145–*146,* 147–*149,* 170
dinoflagellates, 83–84
dinosaurs, 434
diphtheria, 274
disc flower, *374*
disease, 259–262, 265, 267–268, 270, 274–275, 276
   causes of (*see* disease agents)
   deficiency-based, **263**–264
   genetic, **263**
   immunity to, 269–270, 271–273, 274
   mind and, 280–281
   resisters, 134, 158, 162, 267–270, 274
   treatment and prevention, *258,* 260, 267–274, 280–282, 283
disease agents, 259–262, 263
   bacteria, 27, 158, 259–260, 267–268, 330–331, 349
   diet deficiencies, 263, *264*–265

   fungi, **260**–261, 268
   genes, faulty, 263
   pollution, **301**–302, *303,* 307, 310
   stress, 262, 276
   tobacco, 130, 292–293
   worm parasites, 258, *262,* 268, 395, *396*–397, *400*–401
   viruses, 130, **261**–262, 267–270, 329–330
diversity, 29–30
DNA, **119**, *119*
   double helix pattern in, *230*
   genes in, 228, *229*–230, 236
   in monerans, 331
   in protists, 334
dogtooth violet, 370
dolphins and porpoises, *452,* 454
dominant gene, **236**
dormant plant embryo, 366–367
double helix, *230*
dragonfly, 421
drug abuse, 288
drug dependence, 288–297
   addiction, **289**, 291–295
   breaking of, 283, 289–290
   and crime, 289, 294
   physical, **290**, 291, 294, 295
   withdrawal, 292–295
drug problem, 282–297
drugs, 278–297
   America as a "drug society," 285–287, 295–297
   "good" or "bad," 284
   placebos, **279**–281
duck-billed platypus, *452,* 453
duckweed, 65
duodenum, 148, 171, 172
dust bowls, 98

eagles, 45, 50
   bald, 45, *52*
   effects of DDT on, *52,* 53
earthworms, *410*–411, *412*–413
echinoderms, 417, **426**, *427*–430
ecology, 39
ecosphere, **42**, *42*
ecosystem, 39–**40**, 41–42, *44*–46, 48–50, 51–54
   biological magnification in, *52*–53, **53**
   community in, 39, 39–40
   consumers in, **45**, *45,* 51, 52
   decomposers in, **46**, *46*–47
   food chains in, **49**, *49*–52, 53–55
   food pyramid in, **50**–53
   interdependency in, 48–53
   niches in, **21**–23, 44–46
   persistent chemicals in, 51–55
   physical environment in, 40–42
   pollution and, 303
   producers in, 44–45
ecosystems, kinds of (*see* specific ecosystems)
eel, lamprey, *435*
egg cells, **210**, *215*–217, 227–228, 354, 360, 365
egg-laying mammals, *452,* 453
elastic tissue, 156
electric rays (fish), 437
electrocardiograph, 281
electron microscope, **109**, *117*–121
elephantiasis, *401*
elephants, *452*
elk, 98

embryo
   human, 191, **220**–*221*
   plant, 365, *366*–367
   reptile, 443
emphysema, **156**, *156,* 293, 322
endocrine glands, **171**–173
endocrine system, 135–136, **171**–173
   and nervous system, 175, 177
endoplasmic reticulum, **118**, *118*–119, *123*
energy, 44, 117, 133, 170, 171, 323
Engelmann spruce tree, *360*
environment, 3, 40–42
   change and survival in, *211,* 213, 234
   and genes in individual development, 236–237
   (*see also* pollution)
enzymes, **146**–148, 171–172, **229**–231, 386, 428
epidermis, **187**–188
esophagus, 146–147
estrogen, **216**–217, 250
estuary, 76
Euglena, *111*
eutrophication, **307**–311
evergreens (*see* conifers)
evolution, **233**
   adaptive, **233**
excretory system, 135, 148, 163–164
exercise, 276
   and muscle growth, *193*–*194,* 195, 245–246
exhaling, 155, 156
exoskeleton, 417, 422
experiment, 12, 13
extinction, 137, 441
eye, reflex action of, 178–179

Fallopian tube, **216**, 217
fat (body)
   in arteries, 274
   DDT in, 52
fats (in foods), 263, 276
feather stars, *430*
featherduster worm, *410*
feces, **149**
female, 210, 214, **216**
ferns, *346,* 355–356
fertilization, **210**, 217–218, 228, 244, 354, 365
fertilized egg cell, **210**, 217, 228, 354, 360, 365
fertilizers, 310, *313*
fetus, 191, **220**–221, 222
"fever sores," 261–262
fibers, 122
fibroblasts, *115*
"fight or flight," 170, 171
filaments, 331, 335, 348, 352
filaria worm, 347, 401
fish, 80, 84, 396, 397, 433, **434**–437, 454,
   adaptations of, *86,* 434
   breathing process, 437
   eutrophication and, 310
   in food chains, 45, *49,* 53, 314
   in freshwater ecosystems, 63, 64, 68–70
   pesticide effects on, *52*–53, *314*–315
   water pollution and, 304, 307, *314*–315

flagellum, 328
"flashbacks," 296
flatworms, 67, 393–397
    parasitic, 395–397
flower, 364
flowering plants, 362–371
    flower as reproductive system of, 126, 135, 364–367, 369–370
    fruit of, 365–366, 371
    kinds of, 368–371, 374
    seeds of, 366
flukes (flatworms), 395, 395–396
    as disease organisms, 396
fly, 421
fog, 318–319, 320
follicle, 189
food chain, 49, 49–55, 70, 83, 93, 314, 335, 359, 370
    biological magnification in, 51–53, 55, 437
    DDT in, 51–53, 55, 304
    and interdependency in ecosystem, 49–51, 52–53
    diatoms in, 70, 83, 335, 336–337
    "top dogs" in, 53, 303, 437
food manufacturing, (see photosynthesis)
food pyramid, 50, 50
food web, 49–50, 69
forest ecosystems, 44–45, 349
forest
    community in, 39–40, 41
    moist coniferous, 96–97
    northern coniferous, 96
    temperate deciduous, 97, 97
    tropical rain, 29, 99, 99–100
fossils, 433–434, 439, 449, 451
fresh water, 59–62, 63–64
freshwater ecosystems, 45, 59–60, 62–70, 90, 335, 380
    eutrophication in, 307, 310–313
    microhabitats in, 63, 67–68
    minerals in, 64
    oxygen supply and, 63, 68, 70
    population groups in, 65–70
    producers in, 44–45, 63, 70, 334–337
    temperature in, 60–62, 72, 90
    transparency and light in, 62
freshwater habitat, 20
frogs, 440, 442
fruit, 365–366, 371
fruit fly, 234
fungi, 260, 346–352
    characteristics of, 347–348
    decomposers, 46, 46, 48, 259, 349, 352
    disease agents, 259–260, 268
    parasites, 260, 348
    saprophytes, 348

gall bladder, 148–149
ganglion, 175, 177
garter snake, 447
gas exchange in breathing, 153–155, 162–163, 321
gastropods, 405–406
gastrotrichs, 402, 402
gastrovascular cavity, 386, 386
geckos, 444
gene pool, 232, 232–233
genes, 206, 209, 210–211, 227–228, 229
    and ABO blood type, 236, 238–239
    in DNA, 228, 230, 236
    dominant, 236
    growth "instructions" in, 247
    and heredity, 211, 213, 226, 227–228, 229–230, 232–234, 235, 236–239, 263
    mutations of, 233–234
    protein instructions in, 230, 231, 233, 263
    virus, 330
genetic disease, 263
giant sequoias, 96, 97, 362
giant water bugs, 69
gill, 437
glands, 129, 130, 135
    skin, 189
glacier lily, 370
Golgi complex, 119–120, 120, 123
"goose bumps," 189
grasses, 98, 361, 369–370
grasshopper, 420, 421
grassland, 20, 44, 50, 97–98, 98
great blue heron, 47, 49, 53
Great Salt Lake, 64
green algae, 334–336, 383
ground water, 315
growth, 173, 242–254, 244
    of bones, 244–246
    in boys and girls, 242, 248–249, 250
    causes of, 247
    cell activities in, 244
    of muscles, 245–246
    protein in diet and, 247
    rate of, 252–253
    stages of, 248, 249, 250–253
growth hormone, 173, 247
guard cells, 128
Gulf Stream, 77

habitat, 20, 20
    micro-, 63, 67
hagfish, 434–435, 437
hair, 188–189
hands, 246
hardwood, 362
Harlequin bug, 421
hashish, 296
heart, 133, 137, 158, 170, 268–269
    blood circulatory routes, 162–163, 164
    chambers of, 159 (see also atria; ventricles)
    muscle of, 191–192
    in transport system, 132, 136, 159–161, 162–163
heart disease, 268–269, 276, 277, 281, 283, 293
heart rate, 160
heart surgery, 268–269, 281
heart valves, 159, 159
HeLa cells, 137
hemoglobin, 155
hemophilia, 158, 263
hepatitis, 307
herbivores, 453
herbs, 361, 369
heredity, 211
    and blood type, 236–237, 238–239
    genes and, 211, 213, 227–239, 263
    genetic diseases, 263
    and personal appearance, 227, 228, 236
    prediction of inheritance, 236–237, 238–239
hernia, 251
heroin, 294
hibernate, 8
hip joints, 198
honeybees, 420
hoofed mammals, 452
hookworms, 262
hormone control, 169–173, 178
hormones, 135–136, 247, 250
    as control agents, 169–173, 178
    reproductive, 173, 215–216, 220
host, 396
hydra, 66, 67, 386, 387–388
hypnosis, 280

iguanas, 444, 445
immunity and immunization, 259, 269–270, 271–272, 273–276
influenza (flu) 261, 270
inhaling, 153–156
inheritance, 226–239
insect attractants, 31, 34
insectivores, 452
insects, 69, 70, 91, 96, 417, 419–421, 422–424
    in food chain, 48, 53, 314
    niches of, 419–422
    pesticides and, 26–29, 53
    pests, 27, 31
    pollination by, 365, 421
    using food to attract, 32–33
    (see also pest control; pesticides; pests)
insulin, 172
integument, 187
integumentary system, 135, 187–189
intelligence, 237, 265
    of mammals, 454
interdependency, 48–53
intertidal environment, 80–81, 85
intertidal organism, 85
intestines, 147–148, 149, 397
involuntary control, 177
iodine, 171
irradiation, insect, 34

Japanese beetles, 18
jellyfish, 386, 387
joints, 197–198
Joplin, Janis, 279, 283
jungle, 99–100
junipers, 363

kelps, 237, 237
kidneys, 133, 135, 164, 164, 171
kilogram, 8
kilometer, 11
kingdoms, 331
knees, 246

labor, 222, 222
lactic acid, 194–195
ladybird beetles, 31
Lake Erie pollution, 304, 306
lakes, 56, 63, 64, 69
    eutrophication in, 310, 311–313
    temperature layers, 62, 63
lampreys, 434–435, 437
large intestine, 148–149
largemouth bass, 437
larva, 67–68, 419, 420
larynx, 154, 154

latent virus, 261–262, 261
leaves, 128, 131–132, 368–369
leeches, 411, 411
legumes, 371
leukemia, 5
levels of organization, 136–137, 334
life cycle, 6
life scientists, work of, 2, 4–11, 12, 13, 14
ligament, 197
light, 62, 80, 82
light microscope, 107, 108–109, 121, 128
lily family, 370
liver, 132–133, 148, 170, 173, 292
liver flukes, 395
lizards, 444–445
"lockjaw" (tetanus), 260
LSD, 296
lung cancer, 293, 322
lungs, 135, 154, 154–156, 440, 441
    air pollution damage to, 321, 322
    in blood circulation, 162, 162–163
lymph, 161–162, 401
lymph node, 161, 401
lymphatic system, 161, 161–162
lysosomes, 120, 121

male, 210, 215–216
mammals, 432–433, 451–452, 453–454
    kinds of, 452, 453
    dependence on mother, 432, 451, 454
    pouched, 453
manta ray, 436
mantle, 404–406
marijuana, 296
marine ecosystems, 75–85
    currents in, 74–76, 77–83
    population "barriers" in, 76, 81–83
    populations in, 82–85
    producers in, 44–46, 76, 83, 335–336, 389
    tides in, 80–81
marine habitat, 20
marrow, 196
marsupials, 453, 454
mayfly nymph, 49, 69
measles, 258, 261, 269, 274
medulla, 180, 181
medusa, 386, 386–387
menstrual cycle, 217, 217–218, 250
mesoglea, 386–387, 389
metamorphosis
    of insects, 67, 419–421
    of toads and frogs, 440
metric system, 8, 11
microhabitats, 63, 67
microorganisms, 110–111
microscopes, 107–108, 109–121, 128
millipedes, 418, 418
mind, protective and destructive powers of, 280–281
minerals, 63, 64, 263
mitochondrion, 117–118, 123
moisture, 41
    in terrestrial ecosystems, 89–90, 93, 97–100
mold spores, 349, 350–351

molds, *210*, 260, 268, **348**, *348–349*
molecules, 136
  DNA, 228, *230*
  protein, 229, *230–231*
mollusks, 392, 393, **404**–407
monerans, **331**, 334
monkeys, 221, *222*
monocots, **368**, *368–371*, 373
monoculture, 30
morphine, 6, 294
mosquito, anopheles, *421*
mosses, 346, **353**–*354*
  reproduction of, *354–355*
moth, *421*
mountain ecosystems, *88*, **100**–*101*
mountain lions, *2*, *4*, *6–11*, *9*, *10*
mucus, 147, 321
multicellular organisms, 110
mumps, 261, 269, 274
muscle tissue, 129
muscle tone, **194**
muscles, 91, 135, 156, 171, 175, *186*, 191–195, 264–265
  of adolescent boys, 250–251
  growth of, 245–246
muscular system, 91, 135, **191**–**195**
mushrooms, 349, 352
mutation, **233**, *233–234*

nails (on fingers, on toes), *188–189*
nautilus, 406
Navajo medicine, *280*
nekton, **68**–69
nematocysts, *384*, *385*, *386*
nematodes, 399, *399*, *400*–401
neritic zone, *82*, *83–85*
nerve, *176*–**177**, **177**
nerve cell body, **176**
nerve cell branch, **176**
nerve cells, 110, 174, **175**, *175–177*
  cell body and branches, *175–177*
  "messages" of, 175
nerve control, **174**–**179**
nerve tissue, *176*
nervous system, 135, **174**–179, *189*
net, plankton, *69*
neuston, *65*
newts, **439**–*440*
niches, **21**–23, **44**–46
  of coelenterates, 389
  of fish, *434*
  in freshwater ecosystems, 65, 70
  of insects, *419–421*, 422
  of mosses, 354
  of sponges, 382–383
  vacant, 26–27
nicotine, 292
nitrates and nitrogen, 46–47
nitrogen dioxide, 321
nocturnal animals, 100
nucleus, **109**, 110, 119, 120, 121, *123*, *343*
  genes in, 228, *230*
  of moneran cells, 331, 334
  of protist cells, *334*
nutrients, **47**, 92
nymph, **420**, *420*

observation method, 13
oceanic zone, *83*, *83*
  (*see also* **marine ecosystems**)
octopus, 392, 406
oil spills, 315
oligochaetes, *410–411*

omnivores, **453**
opioids, 293–294
opossum, *453*
orchard mites, 27
orchid family, 371
organ systems, **135**, 136, **138**–**139**
  self-regulating, 169–173
organelles, **117**–**122**, 136, *329*
organisms, **107**, 110–*111*, **112**–**113**, *326*–*454* (*see also* **cells**)
organs, 131–134, *137*
Oscillatoria, *70*, *331*
ostracods, *422*
ostrich, *449*
ovaries
  of flowers, *364*, 365, 370
  human, **216**, *216*–218
ovulation, 217, 218
ovules, *364*, 365
oxygen, 83–84, 90, 101, 145, 260, 336, 359
  in breathing processes, 68, 153–155, *162–163*, 437, 440, 441
  decomposers as users of, 63, *306*, *307*, 310, *311*
  in ecosystems, 63, 68, 70
  photosynthesis and, 41, 44–45, 70, 83–84, 336, 359
  thermal pollution, 63, *313*, *314*
  in blood, 155, 157, 158, 162–163
oxygen debt, **195**
oysters, 405
ozone, *320*, 321

pancreas, 148, *172–173*
paramecium, *111*, *339*, *342–343*
paranoid behavior, 295, 296
parasites, 262, *262–263*, 348, 352, 395–397, *400*–401
parathyroid gland, 171, *172*
pea family, 371
pelicans, 52, *53*
penicillin, 268, 285, 349
penis, *216*
perception and drugs, 296
peregrine falcons, 53
periphyton, *66–67*
persistent chemicals, **51**, 53, *55*, *66*
personality, 237, 295, 296
pest control, **28**–**31**, **34**–**35**
  biological method, *30–31*
  with chemical poisons, 25–27
pesticides, **25**, 27, 35, 365
  biological magnification of, **53**, *53*, *55*
  broad-spectrum, 34
  danger, in food chain, 51–53
  DDT, 29–31, 51, 52, *53*, 303
  effects of, on birds, 9, 52–53, *56*
  and fish kills, 314–315
  and populations, 9, 25–30, *31*
  vacant niche problems, 26, 27
pests, **23**, 25–31, 34–35, *330*, 352, *422*
  population explosions of, **29**–30
petals, 365, *368*, *369*
phagocyte, **267**, *267*
phosphates, 310, *313*
photochemical smog, 319–321
photosynthesis, **44**, 46, 63, 70, 109, 131, *132*, 319, 331, 348, *359*, 389
  chloroplasts and, 109, *110*
  diatoms and, 70, 83, 336
physical dependence, **290**
physical environment, 40, 42

in freshwater ecosystems, 59–64
in marine ecosystems, 75–81
in terrestrial ecosystems, 89–93
phytoplankton, **70**, *70*, 83–85, *87*
pigeons, 21
pigments, 334–335
pine trees, *88*, 360, *361*
pinworms, *400*
pituitary gland, 173, 247
placebos, **279**–**281**
placenta, **220**, *222*
plague, 302, *303*
planarians, 394, *394*
plankton, **68**–69, 70, *80*
plant embryo, 365, *366–367*
plants, 65, 66, 96–97, 100, **368**–**369**
  cells of, 107, 109, *111*–114, 121, *128*
  in ecosystems, 41, 44–46, 49
  support systems of, 90–*91*, 355
  tissues of, *128*, 131, *132*
plasma, **155**, 157–158, 160–162, 164
platelets, **157**, **158**, *159*
pneumonia, 330
"pointing the bone" ceremony, 281
polar bear, *90*
pollen, 359, 365
pollination, 365
pollution, **300**–**323**
  ecosystem survival and, 51–53, *84*, 302–304
  as health hazard, 302–*303*, 307, 319, 321–322
  scenic (visual), 301–302, *304*, *305*
  (*see also* **air pollution**; **water pollution**)
polychaetes, *409–410*
polyps, **386**, *386–387*, 389
ponds, 38, *39*, 60, 62, *311*
  food chains in, *48*, *49*
  organisms in, 65–70, 110–111, 334–336, 339, 401
  pollution of, *313*
population explosions, **29**–30, *84*–85
populations, **11**–13, **19**–35, **48**–55, *136*, *137*
  change and survival of, *211*, *213*, 232–234, 303
  control of, in nature, 29–30, *31*
  freshwater, 65–70
  marine, 82–85
  pesticides and, 25–31, *32*–33
Portuguese man-of-war, *385*
prairie, 98
predators, **70**, *419*, **444**–**446**
pregnancy, **220**–*221*, *222*
prenatal development, 220, *221–222*, 245, 265, *296*
primates, 452
producers, **44**, *44–47*, *50–51*, *70*, *76*, 91–93
  protists, 83–84, 334, *334*–*337*
pronghorn antelope, *45*
prostate, *215*, *216*
protein deficiency disease, **263**–265
proteins, **47**, 119, *119*–121, **229**–230, 236, 263, 387 (*see also* **antibodies**)
  in diet, *247*, 263, 264, *265*
  enzymes, **229**–231
  genetic "instructions" for

building, 229, *230*, 231, *233*, *263*
protists, **328**, **334**–**337**, 339, 342–343, 348
  consumers, *70*, 339, 340, 342, 343
  nucleus of, *333*
  producers, *83*, 334–337
protozoans, 67, **339**, 342–343, 394
psychedelic drugs, 296
puberty, **250**
puffballs, 352
pulmonary circulation, **162**, *162–163*
pulse, 160
pupa, 67, **419**, *420*
pyramid of numbers, *50–51*

radial symmetry, **384**–**385**, 426
radiation, 34, *35*, 130, 315
radioactive wastes, 315
rain forest, 29, *99*, *99–*100
rattlesnakes, *447*
ray flower, *374*
rays, 435, *436–437*
reactions, controlled, **178**–**179**
recycling, 46–47, *306–307*, 400
Red Baron, 279, 283
red blood cells, *129*, 132, **155**, 157, **158**, *159*, 236
red tide, **84**, *84*
redwood trees, *96*, *97*, 361, 362
reflex actions, **178**–**179**
regeneration, 394, **428**–**430**
rejection
  of disease agents, 134, 158, 162, 267, 269–270, 274
  of transplants, 133–134
reproduction, 34, 48, 53, **205**–**223** (*see also* **asexual reproduction; sexual reproduction**)
  cell activities in, **205**, *206–207*
  of ferns, *355–356*
  human, 215–218, 220–223
  of molds, 348
  of mosses, 354
  of reptiles, *441*, *443*
  of yeast, 212
reproductive system, 136
  of earthworms, *413*
  of flowering plants, *126*, 135, *364–367*, *369–370*, 372–373
  human female, *216–217*
  human male, 215–216
reptiles, **441**–**449**, 451, 454
  adaptations of, *441*, *443*
  reproduction of, *441*, *443*, *444*
ribosomes, *118*, **119**, 121, *123*, *230*
rickets, 265
ringworm, *260–261*
RNA, 330
rodents, 452
rose family, 371
rotifers, 67, **401**–**402**
roundworms, 262, 399, *399–400*, *401–402*
  parasitic, 262, *400–401*
rusts and smuts, 352

salamanders, **439**–440
salinity, 76
saliva, 146, *175*
salivary glands, *130*, 146, *175*
Salmonella typhosa, *310*
salt, 63, 76, 171

sand dollars, *429*
saprophytes, 348
savanna, 98, *99*
scallops, 405
screwworm, *35*
scrotum, **215**, *215*
sea anemone, 386
sea cucumbers, 430
sea lilies, 430
"sea monsters," 406
sea urchins, *428*–430
seaweeds, 337
sebaceous gland, **189**
sebum, 189
secondary sexual characteristics, **250**
sedatives, 295
"seed cones," *360*
seeds and seed plants, *91*–93, 358–374
  dormant plant embryo, *366–367*
  flowering, 364–374
segmented worms, 409, *410–412*, 413
self-destruction, 280–284, 292–295
self-regulating systems, 169–178
semen, 215
seminal vesicles, **215**, *215*
sepals, *364*, **365**, *368*
sexual characteristics, 216–217, 250
sexual reproduction, 209, **210**–*211*, 213, 234
  and individual differences, 210–*211*
  human, 215–222
  of mosses, 354–355
  as safety mechanism, 211–213
sharks, *435*–437
shells, 64, *404–406*, 443
shrubs, 361, *363*, 369
single-celled organisms, **111**, 330–332
sinuses, 153–*154*
skates, 436
skeletal muscle, **192**–*193*, 194
skeletal system, 135, 196–*197*, 198
skeletons, 91, 389, 426
  cartilage, *436*, 437
  exoskeletons, **417**, 422
  growth process in, *187–188*, 189, 245, 246
skin, 132–135
skin diseases, 260–261, 262, 396
skinks, 444
slow developers, 251, 254
slugs, 393, *406*
small intestine, 147–148, *149*
smallpox, 270, 271
smog, **303**, 319, *320*–321
smoking, 130, 156, 292–293, 322
smooth muscle, **191**
smuts, 352
snails, 64, 67, 393, *404*, 406
  hosts for parasites, 262, 395, 396
snakes, 98, 444, *445–447*
social insects, **420**
softwood, 362
soil, 40, 41, 91–93, 100
  nutrients in, 46–47, 91–93
soil erosion, *91*, 93, 99, *314*, 355
specialization, 207, 220
sperm cells, **210**–217, 227, 228, 354
spiders, *416*, 422, 424
spinal cord, 110, 135, 176, 178, **182**

spiny anteater, 453
sponges, 66, *378*, 379, **380**, *380*–382
spores, **209**, *210*, *348*, 350–351, 355–356
squids, 393, *406*, *407*
squirrels, 48
stamens, *364*, **365**, *368*, 370, *372–374*
starfish, 426, *427–428*, 429
stigma, *364*, **365**, *374*
sting rays, 437
stomach, 136, **147**, *147*, 171
"stomach foot," *404–406*
stress, 262, 276
"stroke," 274–275
style, **365**, *372*, 373
sugar, 170, 172, 173, 349
sulfur–oxygen compounds, 319
sunfish, *49*, *52*
sunflower, 371
sunlight, 44, 63, 77, 82, 307, 310
support systems, 90–91
surgery, *159*, *268*, 281, 285, 295
  heart, 268–269, 281
  transplants, 133–134, 137, 268–269
sweat gland, **189**
"swimmer's itch," 262–263
systemic circulation, 163

tadpoles, 69, 440
taiga, 96
Talman, William, *293*
tapeworms, 397–398
temperate grassland, 97–98
temperature, 41, 372, 437
  body, 157
  in freshwater ecosystems, *60*–62, 90
  in marine environment, 76
  in terrestrial ecosystems, 90, 93, 97, 99, 101
temperature inversion, *318*–319
tendon, 197
tentacles, 385, 389
  of mollusks, 406
  of polyp and medusa, 386, 387
termites, 420
terrestrial ecosystems, 44, 45, 46, 88–93, 95–101, 349, 354
terrestrial habitat, **20**
testes, **215**, *215*
testosterone, **216**, 250
tetanus ("lockjaw"), *260*, 274
thermal pollution, 313–314
thyroid gland, **171**, *171*, 172, 173, 247
thyroxin, 171, 173, 247
tides, *80–81*
tightrope walker, *181*
tiger shark, 436
timberline, **101**, *101*
timothy, 370
tissues, **127**, *129–130*, 131–133, 136
  elastic, 156
  nerve, 175, *176*
  transplanted, 133
toads, 440
toadstools, *352*
tobacco, 130, 283–284, 292–293, 322
tobacco mosaic viruses, *330*
"top dogs," *53*, 303
tortoises, 443

trachea, **154**, *154*
traits, inheritance of, *226–237*
tranquilizer, 288
transplants, *133*–134, 268–269
transport system, 136
  blood and blood vessels, 157–164
  circulation routes, 157–164
  heart, 132, 158–160, *162*, 163
  lymphatic system and, **161**, *161*–162
trees, *40*, *42*, 44, 48, 97–99, 110, 237
  conifers, *88*, **96**–97, 359–361, 362–363
  deciduous, 97
  dicot and monocot, 368–369
  sequoias, redwoods, 96–97, 361–362
trichina worm, 262, 401
tropical rain forest, 29, *99*, *99*–100
trumpeter swans, *204*
tuatara, *441*
tube feet, **427**, *427*
tube worm, 410
tundra, 90, **95**–96
turbellarians, 394
turtles, 69, 70, *443*
typhoid fever, 307

ulcers, 280–281
umbilical cord, **220**
upwelling, 77, *80*, 81
urea, 189
urinary bladder and urine, **164**, *215*
uterus, 91, **191**, *216*, 217, *220*, 221, 222

vaccines, 258, 269 (see also immunization)
vagina, *216*, 217
values, 304
valves, heart, **159**, *159*, 162–163, 268–269
vascular plants, **355**
veins, 159, **160**–163
ventricles, **159**, 162, 163
vertebrae, **182**
vertebrates, 396, 433–454
villi, **148**, *148*
viruses, *133*, *329*–330
  disease agents, 130, 158, *261*–262, 267–270, 330
  in immunization 269–*270*, 273
vitamins, 263, *264*
voluntary control, **177**
vorticella, *328*

warm-blooded animals, 96, **451**
wandering cells, *382*
wastes, 63, 303–304, 306, 310, 311
  as nutrients in eutrophication, *312*
  radioactive, 315
wastes (body), 307, 310
  breathing system and, 154, 162, 163, *321*
  excretory system and, 135, 148–149, 163–164
  transport system and, 145, 157–158, 163–164
water hyacinth, *65*, 66
water lilies, 66–67
water pollution, 306, *307*, 308–310, *311*, 313–315

decomposers and oxygen consumption, 63, 306, 307–310, *311*, 313
disease organisms and, 302, 310–*311*
eutrophication, 308–310, *311*, 313
and fish, 304, *307*, 309–*311*, 313–315
in Lake Erie, 304, 306
oil spills, 315
oxygen content and, 64, 306–307, *311*–314
by pesticides, *314*, 315
and radioactive waste disposal, 315
soil erosion and, *314*
thermal, 313–*314*
water snakes, 69, 446
water striders, 66
weeds, 30, 69
whales, *452*–454
wheat, 98, 99, 364
"wheel animals" (*see* rotifers)
whirligig beetles, 66
white blood cells, 134, **158**, *159*, 162, 267
wild flowers, 370, 371
windpipe, 154
witch doctors, *280*
withdrawal, *292*–296
worm parasites, 260, 262, 393–402
  disease agents, 259, 262, 268, 396–397, 400–*401*
  tapeworms, 397–398
wormlike amphibians, **439**
worms, 393–402, 407–413
  flatworms, 67, *393*–397
  roundworms, 262, 399, *399*–400, *401*–402
  segmented (annelids), 409, *410*–*412*, 413

X-ray, 130

yeasts, 349, 352
  asexual reproduction of, *212*

zones, marine, *82–83*
zooplankton, 70

Illustration Credits

Unit I: xii–1: Edward Currier. **Chapter 1:** 2, 7, 8, 9, 10, 11: Edward Currier; 12: (left) Andy McGowan, St. Luke's Hospital, New York, (right) Yoram Kohana/Peter Arnold Photo Archives; 13: (left) Dr. J. F. Gennaro, Jr., N.Y.U. Laboratory, Cellular Biology. **Unit II:** 16–17: Robert Clemenz. **Chapter 2:** 18, 20: Allan Roberts; 21: L. T. Cerone; 27: Grant Heilman; 29: Swedish Information Service; 30: W. H. Hodge/Peter Arnold Photo Archives; 31: William L. Smallwood; 34: Martin Jacobson, USDA, ARS; 35: USDA. **Chapter 3:** 38, 40: Allan Roberts; 41: John Sumner/Monkmeyer; 42: NASA; 44, 45: Allan Roberts; 46: Dr. Edmund Tylutki, Department of Biological Sciences, University of Idaho; 52: Allan Roberts. **Chapter 4:** 58: Grant Heilman; 62: R. E. Joseph; 63: Allan Roberts; 65: (left) William L. Smallwood, (right) C. J. Anderson, American Museum of Natural History; 66: B. B. Jones, *The Life of the Pond*, McGraw-Hill, 1967; 68: Edward S. Ross, *The Life of Rivers and Streams*, McGraw-Hill, 1967; 69: (top) Woods Hole Oceanographic Institution, (bottom) Dr. E. R. Degginger. **Chapter 5:** 74: Stan Wayman/Photo Researchers; 76: NASA; 79: Scripps Institution of Oceanography; 80: Karl W. Kenyon, *The Life of the Ocean*, McGraw-Hill, 1967; 83: D. P. Wilson, *The Life of the Ocean*, McGraw-Hill, 1967; 84: Carleton Ray/Photo Researchers; 85: Ruth Kirk, *The Life of the Seashore*, McGraw-Hill, 1967; 86: American Museum of Natural History. **Chapter 6:** 88: Josef Muench; 90: Frederick Baldwin, National Audubon Society; 91: Charles Belinky; 94–95: Eric G. Hieber Associates, Inc., adapted by R. E. Joseph; 96: (left) Dr. John Marr, *The Life of the Forest*, McGraw-Hill, 1967, (right) Leland J. Prater, U.S. Forest Service; 97: State of Vermont; 98: (left) Durwood L. Allen, *The Life of Prairies and Plains*, McGraw-Hill, 1967, (right) A. Boker, American Museum of Natural History; 99: W. H. Hodge/Peter Arnold Photo Archives; 100: Myron D. Sutton, U.S. Forest Service, *The Life of the Desert*, McGraw-Hill, 1969; 101: J. C. Pemberton. **Unit III:** 104–105: Robert J. Capece/McGraw-Hill. **Chapter 7:** 106: B. Panessa, N.Y.U.; 108–109: (top left) National Library of Medicine, Bethesda, Md., (others) Dr. J. F. Gennaro, Jr., N.Y.U. Laboratory, Cellular Biology; 110: Runk/Schoenberger/Grant Heilman; 115: American Red Cross; 118: Dr. George E. Palade; 120: M. Yoder, N.Y.U.; 121: (top) Dr. George E. Palade, (bottom) Carolina Biological Supply Company. **Chapter 8:** 126: Allan Roberts; 128: (left) Dr. J. F. Gennaro, Jr., N.Y.U. Laboratory, Cellular Biology, (right) Allan Roberts; 129: Dr. J. F. Gennaro, Jr., N.Y.U. Laboratory, Cellular Biology; 130: Stanton H. Hall, D.D.S.; 137: (bottom) Dr. Samuel Silverstein, Rockefeller University. **Unit IV:** 142–143: Jerry Cooke/*Sports Illustrated*. **Chapter 9:** 144: Minoru Aoki/Monkemeyer; 146: Dr. J. F. Gennaro, Jr., N.Y.U. Laboratory, Cellular Biology; 147: Dr. Richard Marshak; 148: David E. Birk, N.Y.U. Laboratory, Cellular Biology; 156: American Lung Association; 158: (right) American Red Cross National Headquarters; 159: (top) L. Grillone, N.Y.U. Laboratory, Cellular Biology. **Chapter 10:** 168: Andrew J. Cannistraci; 170: American Heart Association; 172: Visiting Nurse Service of New York; 176: Carolina Biological Supply Company; 177: (right) ITT; 181: UPI. **Chapter 11:** 186: John Hanlon/*Sports Illustrated*; 191–192: Dr. J. F. Gennaro, Jr., N.Y.U. Laboratory, Cellular Biology; 194: Larry Gibbons; 198: Joe Baker/*Medical World News*; 199: American Association of Health, Physical Education & Recreation, Girls & Women in Sports. **Chapter 12:** 204: Allan Roberts; 206: Carolina Biological Supply Company; 208: American Cancer Society; 210: Allan Roberts; 221–222: *Birth Atlas*, Maternity Center Association, New York; 223: University of Wisconsin Primate Laboratory. **Chapter 13:** 226: Mimi Forsyth/Monkmeyer; 229: (top) Carolina Biological Supply Company, (bottom) J. H. Tjio and T. T. Puck, *Proc. Natl. Acad. Sci.*, 44, 1229, 1958; 230: (left) Pfizer; 234: Carolina Biological Supply Company; 236: Martin H. Rotker/Taurus; 237: Allan Roberts. **Chapter 14:** 242: Peter Vadnai; 245: All Linda Ferrer Rogers/Woodfin Camp, except (top right) Paul S. Conklin/Monkmeyer; 246: Charles Robertson, M.D.; 247: National Dairy Council; 249: Dunn/D.P.I.; 250: (right photos) Peter Vadnai; 254: (left) Allan W. Richards, (right) UPI. **Unit V:** 256–257: John V. A. Neale/Photo Researchers. **Chapter 15:** 258: Center for Disease Control; 260: Lester V. Bergman and Assoc.; 261: St. Luke's Hospital, New York; 262: Carolina Biological Supply Company; 265: UNICEF; 267: Pfizer; 268: (top) Pfizer, (bottom) Andy McGowan, St. Luke's Hospital, New York; 269: John Senzer/*Medical World News*; 270: Dupont; 271: Parke, Davis & Co.; 274: Lester V. Bergman & Assoc. **Chapter 16:** 278: Peter Vadnai; 280: American Museum of Natural History; 283: Sybil Shelton/Monkmeyer; 285: (top left) *Cleveland Plain Dealer*, (top right) Hugh Rogers/Monkmeyer, (bottom) Peter Vadnai; 287: Peter Vadnai; 289: John Collier/Black Star; 293: American Cancer Society; 294: Joe Toto/Young & Rubicam International, Inc., The Mayor's Narcotics Control Council, NY; 296: Gene Anthony/Black Star. **Chapter 17:** 300: Research-Cottrell, Inc.; 303: (top) New York Public Library, (bottom) Ashley National Forest, Utah; 304–305: George Hall/Woodfin Camp; 310: Robley C. Williams & Ralph W. G. Wycoff, *Proc. Soc. Exptl. Biol. Med.* 59, 265–270, 1945; 311: Allan Roberts, 312: Environmental Protection Agency; 313, 314: Allan Roberts; 317: Will Faller, Boy Scouts of America; 318: Bernard P. Wolff/Photo Researchers; 320: (top) Elliott Erwitt/Magnum, (bottom) Martin M. Rotker/Taurus; 322: (top) Ford Motor Company. **Unit VI:** 326–327: Robert Clemenz. **Chapter 18:** 328: Carolina Biological Supply Company; 330: (top) Virus Lab, University of California at Berkeley; 331: Russ Kinne/Photo Researchers; 332: Robert Clemenz, *The Life of the Pond*, McGraw-Hill, 1967; 334, 335: Allan Roberts; 336: (top) Dr. J. F. Gennaro, Jr., N.Y.U. Laboratory, Cellular Biology, (bottom) Russ Kinne/Photo Researchers; 337: Jack White/National Audubon Society; 341: Robert J. Capece/McGraw-Hill; 342: Carolina Biological Supply Company; 343: (top) Carolina Biological Supply Company, (bottom) Eric V. Grave. **Chapter 19:** 346: Grant Heilman; 349, 352: Allan Roberts; 354: Bego Mintegui; 355: C. G. Maxwell/Photo Researchers; 356: William Douglas. **Chapter 20:** 358: Charles E. Rotkin for PFI; 360: (top) William L. Smallwood, (bottom) Allan Roberts; 361: Leland Prater, U.S. Forest Service; 362: Grant Heilman; 366: Allan Roberts; 369: Grant Heilman; 370: Allan Roberts; 371: (top) Helen Cruickshank/National Audubon Society, (bottom) Allan Roberts; 374: (top) Allan Roberts. **Chapter 21:** 378: Douglas Faulkner; 380: Carolina Biological Supply Company; 381: (top) Carolina Biological Supply Company, (bottom) Douglas Faulkner, *The Life of the Ocean*, McGraw-Hill, 1967; 382: Allan Roberts; 385: D. P. Wilson; 387: (left) L. P. Madin/Woods Hole Oceanographic Institution; 389: Allan Roberts. **Chapter 22:** 392: Douglas Faulkner; 394, 395: Carolina Biological Supply Company; 396: H. Zaiman, M.D.; 397: Allan Roberts; 399: H. Zaiman, M.D.; 400: (top) Martin M. Rotker/Taurus; 401: Pan American Sanitary Bureau Photo; 402: (top) R. H. Head/Photo Researchers, (bottom) Dr. Royal B. Brunson/University of Montana; 404: Allan Roberts; 405: (left) Dr. Royal B. Brunson/University of Montana, (right) L. T. Grigg, FPG, *The Life of the Ocean*, McGraw-Hill, 1967; 406: Allan Roberts; 407: Jesus Uriarte Camara; 410: (top left) Carolina Biological Supply Company, (top right) Allan Roberts, (bottom) Martin M. Rotker/Taurus; 411: Allan Roberts. **Chapter 23:** 416: Grant Heilman; 418: Allan Roberts; 422: (left) Henry C. Johnson, *The Life of the Pond*, McGraw-Hill, 1967, (right) William L. Smallwood; 424: Allan Roberts; 427: (top and bottom left) Allan Roberts, (bottom right) Russ Kinne/Photo Researchers; 428: (bottom) Ruth Kirk, *The Life of the Seashore*, McGraw-Hill, 1967; 429; Paul E. Taylor/Photo Researchers; 430: D. P. Wilson. **Chapter 24:** 432: Harry Redl/Black Star; 435: (top) Karl H. Maslowski, National Audubon Society, (bottom) Douglas Faulkner; 436: Douglas Baglin/National Audubon Society; 437: C. C. Lockwood/Photo Researchers; 440, 441, 443, 444, 445: Allan Roberts; 446: Allan Roberts, *The Life of the Pond*, McGraw-Hill, 1967; 447: D. J. Lyons/Biologicals Unlimited; 449: (left) Michael T. Smith/National Audubon Society, (right) Allan Roberts; 451, 453: Allan Roberts.